Objective Biology for NEET

Volume 1

Board of Editors

Universities Press

OBJECTIVE BIOLOGY FOR NEET, VOLUME 1

UNIVERSITIES PRESS (INDIA) PRIVATE LIMITED

Registered Office
3-6-747/1/A & 3-6-754/1, Himayatnagar, Hyderabad 500 029, Telangana, India
info@universitiespress.com; www.universitiespress.com

Distributed by
Orient Blackswan Private Limited

Registered Office
3-6-752 Himayatnagar, Hyderabad 500 029 Telangana, India

Other Offices
Bengaluru / Chennai / Guwahati / Hyderabad / Kolkata
Mumbai / New Delhi / Noida / Patna / Visakhapatnam

ISBN 978-93-89211-70-2

Typeset in Times Ten LT Std 10 points *by*
Cameo Corporate Services Limited, Coimbatore

501790

Printed in India by
The Print Park, Chennai 600 117

Published by
Universities Press (India) Private Limited
3-6-747/1/A & 3-6-754/1, Himayatnagar, Hyderabad 500 029, Telangana, India

Preface

Objective Biology for NEET, Volumes 1 and 2 are primarily aimed at guiding students who aspire to gain admission to Medical or Dental colleges by boosting their score to a competitive level, thereby enabling them to achieve success in the NEET exam.

The book provides exhaustive coverage of the National Council of Educational Research and Training (NCERT) syllabus using a topic-wise approach. Volume 1 deals with the Class 11 syllabus, while Volume 2 covers the topics studied in Class 12.

More than 5450 exercise questions, comprising multiple choice, true/false, fill in the blank, match the following and diagram-oriented questions, have been presented. About 570 NCERT Exemplar questions have been included for a quick review of concepts. The practice test papers, comprising more than 1750 questions, can be used for self-evaluation and thorough revision. Solutions to all questions have been provided to resolve doubts and improve clarity of understanding in students.

This book has been framed by incorporating the current trends followed in NEET. We hope that students will find the study material useful.

Board of Editors

Contents

1 The Living World

EXERCISES

What is Living?, Diversity in the Living World

1. All living organisms have a feature, without exception, which is not shown by non-living objects. This feature is:
(a) Reproduction (b) Intrinsic growth
(c) Metabolism (d) All except (a)

2. Non-living objects show:
(a) Irreversible intrinsic growth
(b) Extrinsic growth
(c) Self-regulation
(d) All except (b)

3. Read the following statements with respect to growth and select the correct option:
A. It is a fundamental characteristic
B. In plants, growth by cell division occurs continuously during their lifespan
C. In the majority of higher plants and animals, growth and reproduction are mutually exclusive events
(a) Only (B) is correct
(b) (A) and (B) are correct
(c) (A), (B) and (C) are correct
(d) (A) and (C) are correct

4. Synthesis of proteins from amino acids and breakdown of sugars to liberate energy are:
(a) Anabolic processes
(b) Catabolic processes
(c) Biochemical reactions
(d) Present only in multicellular organisms

5. Select the incorrectly matched pair:
(a) *Hydra* Budding
(b) *Planaria* True regeneration
(c) *Amoeba* Regeneration
(d) Mosses Protonema

6. Growth is extrinsic in all, except:
(a) Snow mountains (b) Sand dunes
(c) Crystals (d) Prokaryotes

7. Reproduction is synonymous with growth in:
(i) Bacteria
(ii) *Hydra*
(iii) *Planaria*
(iv) *Amoeba*
(v) Unicellular algae
(a) (i) (iv) and (v) (b) (i) (iii) and (iv)
(c) (i), (ii) and (iii) (d) All of the above

8. Which set of organisms multiplies by fragmentation?
(a) Earthworm, *Amoeba* and Fungi
(b) Filamentous algae, Fungi and Moss protonema
(c) *Hydra, Planaria* and Bacteria
(d) *Amoeba, Hydra*, Yeast

9. Multicellular organism in which new individuals are produced by budding is:
(a) *Hydra* (b) Yeast
(c) *Planaria* (d) Both (a) and (b)

10. Which of the following is regarded as the defining property of all living organisms?
(a) Self-consciousness
(b) Sensitivity and awareness
(c) Reproduction
(d) Both (a) and (b)

11. Which of the following processes are mutually inclusive events in unicellular organisms?
(a) Growth and reproduction
(b) Consciousness and self-consciousness
(c) Growth and metabolism
(d) Reproduction and respiration

12. State true (T) or false (F) for the given statements with respect to metabolism and select the correct option:
A. The sum total of all the chemical reactions in our body is called metabolism
B. No non-living objects exhibit metabolism
C. Metabolic reactions can be demonstrated outside the body in cell-free systems
D. Isolated in vitro metabolic reactions are living things but surely non-living reactions

	A	B	C	D
(a)	T	T	T	T
(b)	T	T	F	F
(c)	T	T	T	F
(d)	T	T	F	T

13. The most obvious and technically complicated feature of living organisms is:
(a) Consciousness (b) Reproduction
(c) Growth (d) Self-consciousness

14. Consciousness is the property of:
(a) All animals but not plants
(b) All complex animals only
(c) All complex plants and animals only
(d) All living organisms

15. Human beings are the only organisms that are aware of themselves and this property is referred to as:
(a) Self-regulation
(b) Self-replication
(c) Self-consciousness
(d) Self-organisation

16. Read the following statements carefully:
A. Reproduction cannot be considered as the defining property of living organisms
B. There are many living organisms like worker bees, mules and infertile human couples who do not reproduce at all
(a) Only (A) is correct
(b) Only (B) is correct
(c) Both (A) and (B) are correct
(d) Both (A) and (B) are incorrect

17. Consider the following features:
(i) Intrinsic growth (ii) Extrinsic growth
(iii) Metabolism (iv) Cellular organisation
(v) Consciousness (vi) Self-consciousness
How many of the above features are defining ones of all living beings?
(a) (iii) and (v) only (b) (i), (iii), (iv) (v)
(c) (i), (iii), (v), (vi) (d) (iii) only

18. The properties of tissues:
(a) Are present in the constituent cells
(b) Are due to similar cells in them
(c) Are due to their similar origin
(d) Arise as a result of interactions among the constituent cells

19. A. Properties of cellular organelles are present in the molecular constituents of the organelles
B. Interaction among the molecular components of the organelles results in the properties of cell organelles
C. Biology is the story of life on earth
D. Biology is the story of the evolution of living organisms on earth

	A	B	C	D
(a)	F	T	T	T
(b)	T	F	T	T
(c)	T	T	F	T
(d)	F	T	T	F

20. All living organisms present, past and future are:
(a) Not linked to any extent
(b) Linked to one another in the apoplast of the cell
(c) Linked to one another by the sharing of common genetic material, but to varying degrees
(d) Linked to one another by the sharing of common genetic material to 100% extent

21. The number of species that are known and described range between:
(a) 1.7 and 1.8 billion
(b) 1.7 and 1.8 million
(c) 17 and 18 million
(d) 7 and 8 million

22. Mark the mismatched pair:
(a) Father of taxonomy – Linnaeus
(b) Systematics – Derived from Latin word
(c) Phylogeny – Evolutionary history
(d) Taxonomic hierarchy – Ascending arrangement of taxonomic categories

23. Binomial nomenclature of scientific naming of organisms was developed by:
(a) Charles Darwin (b) Carolus Linnaeus
(c) John Ray (d) Ernst Mayr

24. Scientific rules for naming of animals are set by:
(a) ICZN (b) ICNZ
(c) ICBN (d) ICNB

25. ICBN stands for:
(a) International Congress of Biological Names
(b) International Code of Biological Nomenclature
(c) International Code of Botanical Nomenclature
(d) Indian Committee of Botanical Nomenclature

26. For both the words in a biological name, ___A___ is separately underlined or ___B___ in italics to indicate their ___C___ origin:

	A	B	C
(a)	Printed	Handwritten	Latin
(b)	Handwritten	Printed	Latin
(c)	Handwritten	Printed	Italic
(d)	Printed	Handwritten	Greek

27. Which of the following represents the correct binomial epithet?
(a) *Mangifera indica*
(b) *Mangifera indica* Linn
(c) *Mangifera Indica* Linn
(d) *mangifera Indica* Linn

28. The binomial name is printed in italics. It:
(a) Makes them conspicuous
(b) Indicates their Latin origin
(c) Indicates their standardisation
(d) Represents their taxonomic value

29. Select the incorrect statement with respect to binomial nomenclature:
(a) The first word in the biological name represents the genus
(b) Name of the author is written in abbreviated form
(c) Binomial epithet is collectively formed by two words
(d) Scientific names are latinised or derived from Latin, irrespective of their origin

30. In binomial nomenclature:
(i) Generic name, specific epithet and author citation collectively form the binomial epithet
(ii) Scientific names are underlined separately when handwritten
(iii) Specific epithet is followed by generic name
(a) (i) and (ii) are correct
(b) Only (ii) is correct
(c) (ii) and (iii) are correct
(d) (i), (ii) and (iii) are correct

31. Arrangement of organisms into convenient categories on the basis of similarities and differences in certain easily observable but fundamental characters is:
(a) Taxonomy (b) Classification
(c) Nomenclature (d) Identification

32. Taxonomy is the branch dealing with all, except:
(a) Identification (b) Characterisation
(c) Distribution (d) Nomenclature

33. First act in taxonomy is:
(a) Identification (b) Classification
(c) Naming (d) Description

34. If the biological name of mango is written as *Mangifera indica* Linn, the abbreviation after the specific epithet indicates that:
(a) This is a variety of India
(b) This is a subspecies of mango
(c) This species was first described by Linnaeus
(d) Name of this species is derived from Latin

35. Linnaeus gave the principle of binomial nomenclature in:
(a) *Species Plantarum*
(b) *Genera Plantarum*
(c) *Philosophica Botanica*
(d) *Hortus Uplandicus*

36. Systematics is different from taxonomy. It deals with:
(a) Identification (b) Phylogeny
(c) Nomenclature (d) Classification

37. A taxon is the unit of:
(a) Order (b) Species
(c) Genus (d) Taxonomy

38. *Species Plantarum* and *Systema Naturae* were written by:
(a) Hooker (b) Linnaeus
(c) Aristotle (d) Engler

39. Read the following statements and select the correct option:
A. The word 'systematic' is derived from the Latin word 'systema', which means systematic arrangement of organisms
B. Linnaeus used '*Systema Naturae*' as the title of his publication
(a) Only (A) is correct
(b) Only (B) is correct
(c) Both (A) and (B) are correct
(d) Both (A) and (B) are incorrect

40. The name of plants is standardised by:
(a) ICBN (b) ICZN
(c) ICNB (d) ICNCP

41. Modern taxonomy includes the study of:
(a) Internal structure and external structure
(b) Cell structure
(c) Developmental process and ecological information
(d) All of the above

42. Taxonomic studies of various species of plants, animals and other organisms are useful in:
(a) Knowing our bioresources
(b) Agriculture
(c) Forestry
(d) All of the above

43. In classical taxonomy, species are delimited on the basis of:
(a) Physiology, cytology
(b) Biochemistry, genetics
(c) Morphology, external features
(d) Ecology, evolution

44. Modern taxonomy studies are similar to classical ones in dealing with:
(a) Cell structure and function
(b) Internal and external features
(c) Developmental and ecological processes
(d) Morphological features

45. The science that deals with diversity of organisms and all their comparative and evolutionary relationship is:
(a) Under the field of Science rather than Taxonomy
(b) Referred to as Systematics
(c) Based on Phylogeny
(d) All of the above

46. Read the following statements:
A. Scientific name *Rattus rattus* represents tautonyms
B. Tautonyms are not valid in plant nomenclature
(a) Only (A) is correct
(b) Only (B) is correct
(c) Both (A) and (B) are correct
(d) Both (A) and (B) are incorrect

Taxonomic Categories

47. What is a taxonomic category?
(a) A rank in the taxonomic hierarchy
(b) Fundamental unit of classification
(c) A group of unrelated species
(d) All of the above

48. How many obligate taxonomic categories are there in the taxonomic hierarchy?
(a) 9 (b) 5
(c) 7 (d) 6

49. The higher the category in a taxonomic hierarchy:
A. The greater the number of organisms in it
B. The fewer the number of common characters
(a) Only (A) is correct
(b) Only (B) is correct
(c) Both (A) and (B) are correct
(d) Both (A) and (B) are incorrect

50. All of the following taxonomic categories are the same for plants as well as animals, except:
(a) Division (b) Family
(c) Class (d) Order

51. The lowest taxonomic category in the taxonomic hierarchy is:
(a) Kingdom (b) Order
(c) Variety (d) Species

52. Select the odd one out with respect to broad categories of the taxonomic hierarchy:
(a) Kingdom, Phylum, Division
(b) Order, Family, Genus
(c) Tribe, Variety, Subspecies
(d) Family, Genus, Species

53. The taxon Sapindales represents which one of the following categories in the taxonomic hierarchy?
(a) Order (b) Family
(c) Division (d) Class

54. Which of the following pairs does not have taxa of the same hierarchy?
(a) Chordata and Arthopoda
(b) Polymoniales and Sapindales
(c) Primata and Mammalia
(d) Solanaceae and Hominidae

55. Mark the incorrect match with respect to wheat:
(a) Family – Poaceae
(b) Class – Monocotyledonae
(c) Order – Polymoniales
(d) Division – Angiospermae

56. Fishes, amphibians and reptiles constitute the next higher category, called:
(a) Class (b) Phylum
(c) Division (d) Kingdom

57. How many of the following taxa are related to kingdom Animalia?

Primata, Chordata, Muscidae, *Petunia*, Poales, *Felis*, Carnivora, *Musca*

(a) Five (b) Seven
(c) Four (d) Six

58. The taxon that has the highest number of common characters is:
(a) Diptira (b) Muscidae
(c) Arthopoda (d) Animalia

59. Select the odd one out with respect to polytypic genus:
(a) *Solanum* (b) *Brassica*
(c) *Homo* (d) *Panthera*

60. Taxon is:
(a) Any type of taxonomic grouping like Species, Family, Phylum, based on similarity of traits
(b) A rank in hierarchical classification
(c) A group of interbreeding populations
(d) A group of organisms found in different geographical regions

61. "Taxa can indicate categories at different levels". This statement is justified by which of the following groups of organisms?
(a) Wheat, Maize, Rice
(b) Lion, Tiger, Leopard
(c) Plant, Monocot, Rice
(d) Mango, Wheat, Maize

62. Mayer's biological concept of species is mainly based on:
(a) Reproduction isolation
(b) Morphological traits
(c) Geographical distribution
(d) Phylogenetic relationship

63. Plant families Convulvulaceae and Solanaceae are placed under one order ___A___ on the basis of ___B___:

	A	B
(a)	Sapindales	Vegetative characters
(b)	Polymoniales	Floral characters
(c)	Polymoniales	Vegetative characters
(d)	Sapindales	Floral characters

64. *Petunia, Datura* and brinjal are placed in family Solanaceae on the basis of:
(a) Habit and habitat
(b) Vegetative characters only
(c) Reproductive characters only
(d) Vegetative and reproductive characters

65. Members of which of the following groups are not concerned with the same taxon?
(a) Brinjal, Mango
(b) Hominidae, Muscidae
(c) Insects, Housefly
(d) Diptera, Poales

66. In which of the following will there be great difficulty in determining the relationship with other taxa at the same level?
(a) Phylum (b) Order
(c) Family (d) Class

67. Select the correct option to complete the given table:

Name	Family	Order	Class
Humans	Hominidae	A	Mammalia
Housefly	Muscidae	Diptera	B
Mango	C	Sapindales	Dicotyledonae
D	Poaceae	Poales	Monocotyledonae

	A	B	C	D
(a)	Carnivora	Insecta	Convulvulaceae	Brinjal
(b)	Primata	Insecta	Anacardiaceae	Wheat
(c)	Primata	Arthopoda	Anacardiaceae	Wheat
(d)	Chordata	Arthopoda	Solanaceae	Brinjal

68. Wheat, Monocot and Plants represent:
(a) Same taxa at same level
(b) Different taxa at same level
(c) Different categories of the same taxa
(d) Taxa at different levels

69. Fishes, amphibians, reptiles and mammals are placed in the same phylum Chordata due to common features like:
A. Presence of notochord
B. Presence of dorsal, hollow neural system
C. Presence of dorsal, solid neural system
The correct statements are:
(a) Only (A) (b) Only (C)
(c) (A) and (B) (d) (A) and (C)

70. The animal order ___A___ includes families ___B___ and Felidae:

	A	B
(a)	Primata	Hominidae
(b)	Primata	Canidae
(c)	Chordata	Muscidae
(d)	Carnivora	Canidae

71. From Species towards Kingdom, the number of common characteristics ___A___ so that the problem of classification becomes ___B___ complex:

	A	B
(a)	Increases	More
(b)	Decreases	More
(c)	Increases	Less
(d)	Decreases	Less

72. Categories which include related orders are:
(a) Anacardiaceae, Poaceae
(b) Poales, Sapindales
(c) Insecta, Mammalia
(d) Primata, Diptera

73. Wheat and mango have maximum similarity at the level of:
(a) Division (b) Kingdom
(c) Class (d) Order

74. Arrange the following taxa with respect to the descending order of their taxonomic categories and select the correct option:
A. Chordata B. Amphibia
C. Solanaceae D. Sapindales
(a) (A), (B), (D), (C)
(b) (C), (D), (B), (A)
(c) (B), (D), (A), (C)
(d) (C), (B), (A), (D)

75. Which one is the incorrectly matched pair?
(a) *Panthera*: Lion, leopard
(b) Canidae: Cat, dog
(c) Primata: Gorilla, gibbon
(d) Chordata: Mammalia, Reptilia

76. Which of the following is not recognised by botanists?
(a) Phylum (b) Subspecies
(c) Division (d) Both (a) and (b)

Taxonomic Aids

77. Taxonomic aids are storehouses of information which are used in study and reference. Identify A and B:
A. Storehouse of collected plant specimens that are dried, pressed and preserved on sheets
B. Objective is public exhibition to understand wildlife, recreation and education

	A	B
(a)	Herbarium	Museum
(b)	Herbarium	Zoological Park
(c)	Botanical Garden	Zoological Park
(d)	Museum	Botanical Garden

78. Which of the following taxonomic aids has a collection of preserved animal specimens for study and reference?
(a) Zoological Park (b) Museum
(c) Key (d) Animal Herbarium

79. Select the correct option to fill in the blanks (A), (B), (C) and (D):
(i) ____A____ serves as a quick source of reference in taxonomic studies.
(ii) ____B____ is the place used for storage, preservation and exhibition of plants and animals.
(iii) ____C____ is the place where plant species are grown for identification purposes.
(iv) ____D____ are useful in providing information for the identification of names of species found in an area.

	A	B	C	D
(a)	Herbarium	Museum	Botanical garden	Manuals
(b)	Herbarium	Museum	Botanical garden	Flora
(c)	Botanical garden	Herbarium	Flora	Monograph
(d)	Botanical garden	Herbarium	Herbarium	Manuals

80. Match the following columns and select the correct option:

Column I	Column II
A. Botanical garden	(i) Preserved plant and animal specimens
B. Zoological park	(ii) Artificial analytical device
C. Key	(iii) Collection of living plants for study and reference
D. Museum	(iv) Wild animals are kept in a protected environment

(a) A – (iv), B – (iii), C – (i), D – (ii)
(b) A – (i), B – (iv), C – (ii), D – (iii)
(c) A – (iii), B – (iv), C – (i), D – (ii)
(d) A – (iii), B – (iv), C – (ii), D – (i)

81. Mark the incorrect option with respect to taxonomic key:
(a) Based on contrasting characters
(b) It represents the choice between two opposite options
(c) Used for the identification of plants only
(d) Each statement is called a lead

82. In taxonomic keys, the pairs of contrasting characters are known as:
(a) Doublet (b) Couplet
(c) Lead (d) Index

83. The famous botanical garden of Kew is located in:
(a) Germany (b) India
(c) France (d) England

84. Indian botanical garden is located in:
(a) Howrah (b) Lucknow
(c) Mumbai (d) Mysore

85. Taxonomic literature dealing with all available information about a single taxon is:
(a) Catalogue (b) Flora
(c) Manual (d) Monograph

86. Mark the correctly matched pair:
(a) Zoological park – In-situ conservation
(b) Museum – Non-preservative solution
(c) Herbarium – Dried plant specimens
(d) Manual – Actual account of habitat and distribution of plants

87. Ex-situ conservation strategies of plants are associated with:
A. Herbaria B. Zoological park
C. Botanical garden D. Flora
(a) (A) and (C) (b) (C) only
(c) (B) and (C) (d) (C) and (D)

88. In zoological parks:
A. Animals and plants are kept for exhibition
B. We can study the food habits and behaviour of animals
C. Preserved wild animals are kept in protected areas
D. For protection, all animals are provided a totally artificial habitat, which is not similar to their natural habitat
The correct statements are:
(a) Only (B)
(b) (A), (B) and (C)
(c) (B) and (C)
(d) (A), (B) and (D)

89. How many of the given features are associated with taxonomic key and herbarium?

Analytical, Preservation of specimen, Lead, Couplet, Formaline, Use of 0.1% $HgCl_2$, Quick reference system, Holotype

	Taxonomic Key	Herbarium
(a)	3	5
(b)	4	5
(c)	4	3
(d)	5	4

90. Taxonomic key is prepared for:
(a) Class only
(b) Genus only
(c) Species only
(d) For all taxonomic categories

91. Which of the following pairs contain a recorded description of organisms without any specimens?
(a) Monograph, Museum
(b) Flora, Herbarium
(c) Manual, Herbarium
(d) Flora, Catalogue

92. Read the following statements with respect to taxonomic aids and state whether true (T) or false (F):
A. Museum has a collection of preserved plant and animal specimens
B. Specimens are preserved in a container or jar in preservative solution
C. Plants and animals are always preserved in a liquid solution of FAA
D. Museum does not usually have a collection of animal skeletons

	A	B	C	D
(a)	T	T	T	F
(b)	T	T	F	F
(c)	F	F	T	T
(d)	T	F	F	T

93. Ex-situ conservation strategies of organisms are related to:
(a) Botanical garden (b) Zoological park
(c) Museum (d) All except (c)

94. Read the following statements and identity the taxonomic aid:
(i) Storehouse of collected plant specimens
(ii) Used as quick reference in taxonomic studies
(a) Herbarium (b) Botanical garden
(c) Museum (d) Flora

95. Information for the identification of names of species found in an area is mentioned in:
(a) Manuals (b) Flora
(c) Monographs (d) Catalogue

96. Herbarium sheets are arranged according to a universally accepted system of classification. In India, herbarium sheets are arranged according to ________ system of classification:
(a) Linnaeus (b) Eichler
(c) Bentham and Hooker (d) Hutchinson

97. National Botanical Research Institute is situated at:
(a) Dehradun (b) Bhopal
(c) Lucknow (d) Delhi

98. One can learn about the food habits and behaviour of animals, along with accurate identification, from:
(a) Zoological park (b) Animal herbarium
(c) Botanical garden (d) Both (A) and (B)

99. Which of the following information is not provided by a herbarium sheet?
(a) Date on which the specimen was collected
(b) Size of the herbarium sheet
(c) Local name of the specimen
(d) Family of the specimen

100. Objectives of zoological park are all, except:
(a) Public exhibition to understand wildlife
(b) Breeding of rare fauna
(c) In-situ conservation
(d) Recreation

101. Botanical gardens:
A. Have plant species labelled with their botanical name and family
B. Have endemic and exotic plant species
C. Are a means of ex-situ conservation strategies
(a) Only (A) (b) (A) and (B) only
(c) (B) and (C) only (d) (A), (B) and (C)

NCERT EXEMPLAR QUESTIONS

1. As we go from Species to Kingdom in a taxonomic hierarchy, the number of common characteristics:
(a) Will decrease
(b) Will increase
(c) Remain the same
(d) May increase or decrease

2. Which of the following 'suffixes' used for units of classification in plants indicates a taxonomic category of 'family'?
(a) –ales (b) –onae
(c) –aceae (d) –ae

3. The term 'systematics' refers to:
(a) Identification and study of organ systems
(b) Identification and preservation of plants and animals
(c) Diversity of types of organisms and their relationship
(d) Study of habitats of organisms and their classification

4. Genus represents:
(a) An individual plant or animal
(b) A collection of plants or animals
(c) Group of closely related species of plants and animals
(d) None of the above

5. The taxonomic unit 'phylum' in the classification of animals is equivalent to which hierarchical level in the classification of plants?
(a) Class (b) Order
(c) Division (d) Family

6. Botanical gardens and zoological parks have:
(a) Collection of endemic living species only
(b) Collection of exotic living species only
(c) Collection of endemic and exotic living species
(d) Collection of only local plants and animals

7. Taxonomic key is one of the taxonomic tools used in the identification and classification of plants and animals. It is used in the preparation of:
(a) Monographs (b) Flora
(c) Both (a) and (b) (d) None of the above

8. All living organisms are linked to one another because:
(a) They have common genetic material of the same type
(b) They share common genetic material but to varying degrees
(c) All have common cellular organisation
(d) All of the above

9. Which of the following is a defining characteristic of living organisms?
(a) Growth
(b) Ability to make sounds
(c) Reproduction
(d) Response to external stimuli

10. Match the following and choose the correct option:

Column I	Column II
A. Family	(i) *Tuberosum*
B. Kingdom	(ii) Polymoniales
C. Order	(iii) *Solanum*
D. Species	(iv) Plantae
E. Genus	(v) Solanaceae

(a) A – (v), B – (iv), C – (ii), D – (i), E – (iii)
(b) A – (iv), B – (iii), C – (v), D – (ii), E – (i)
(c) A – (iv), B – (iii), C – (v), D – (i), E – (ii)
(d) A – (iv), B – (iii), C – (ii), D – (v), E – (i)

PRACTICE TEST PAPER

1. The feature of living organisms which is present only in human beings is:
(a) Metabolism (b) Reproduction
(c) Consciousness (d) Self-consciousness

2. Systematics:
A. Is derived from a Latin word
B. Is considered as taxonomy along with evolutionary history
C. Deals with identification, nomenclature and classification
(a) Only (A) and (C) are correct
(b) Only (A) and (B) are correct
(c) Only (C) is correct
(d) (A), (B) and (C) are correct

3. Which of the following features can be said to be true defining features of living beings without any exception?
(a) They can digest their food
(b) All of them can reproduce
(c) They can regenerate
(d) They can respond to external stimuli

4. Who wrote *Systema Naturae* and provided a basis for the classification of plants and animals?
(a) Charles Darwin (b) Carolus Linnaeus
(c) Robert Hooke (d) Leeuwenhoek

5. Which of the following are defining features of living organisms?
A. Intrinisic growth
B. Reproduction
C. Metabolism
D. Cellular organisation
E. Consiousness
(a) (A), (C) and (D) only
(b) (B), (C) and (D) only
(c) (A), (C), (D) and (E)
(d) (A), (B), (C), (D) and (E)

6. A scientist connected with the study of external and internal structures, structure of the cell, developmental processes and ecological information of organisms along with classification is known as:
(a) Modern taxonomist
(b) Classical taxonomist
(c) Conventional taxonomist
(d) Old taxonomist

7. Match the entities in Column I with their mode of reproduction in Column II:

Column I	Column II
A. Fungi	(i) True regeneration
B. Yeast	(ii) Fragmentation
C. *Planaria*	(iii) Sporulation
D. Moss protonema	(iv) Budding

(a) A – (iii), B – (i), C – (ii), D – (iv)
(b) A – (iii), B – (iv), C – (i), D – (ii)
(c) A – (ii), B – (iv), C – (i), D – (iii)
(d) A – (ii), B – (iii), C – (i), D – (iv)

8. A taxon is:
(a) A group of related families
(b) A group of related species
(c) A type of living organism
(d) A taxonomic group of real organisms of any ranking

9. Which taxonomic aid gives comprehensive information about a taxon?
(a) Taxonomic key (b) Herbarium
(c) Flora (d) Monograph

10. The plant *aestivum* is a species of the genus:
(a) *Mangifera* (b) *Triticum*
(c) *Solanum* (d) *Pisum*

11. Botanical gardens provide:
(a) Natural habitat for wildlife
(b) Beautiful area for recreation
(c) Ex-situ conservation of germplasm
(d) In-situ conservation of rare plants

12. Binomial nomenclature was given by:
(a) RH Whittaker (b) Linnaeus
(c) Aristotle (d) Theophrastus

13. Museums contain:
(a) Stuffed animals (b) Animal skeletons
(c) Live animals (d) All except (c)

14. Read the following statements and select the correct option:
A. Living organisms are considered as a closed system
B. Energy of living organisms cannot be lost or gained from the external environment
(a) Only (A) is correct
(b) Only (B) is correct

(c) Both (A) and (B) are incorrect
(d) Both (A) and (B) are correct

15. A taxon which includes related species is:
(a) Genus (b) Family
(c) Order (d) Class

16. The 'Birds' taxonomically represent:
(a) Family (b) Order
(c) Class (d) Phylum

17. Two organisms of the same division but different orders will be kept under the same:
(a) Genus (b) Class
(c) Species (d) Family

18. According to binomial nomenclature, a biological name consists of:
(a) Generic name and specific epithet
(b) Species and subspecies
(c) Generic name, specific epithet and author citation
(d) Generic name, specific epithet and subspecies

19. In *Mangifera indica L.*, the L represents:
(a) Latin language (b) Linnaeus
(c) Lecto type (d) Variety name

20. Read the statement carefully: "Wheat is a monocot and monocots are plants". In this statement, wheat, monocot and plants represent:
(a) Taxa at different levels
(b) Same taxa of different categories
(c) Same category of different taxa
(d) Taxa of genus, family and kingdom, respectively

21. How many species have been identified and described till now?
(a) 17 million (b) 1.7–1.8 million
(c) 0.2 million (d) 5 million

22. Select the incorrect statement for binomial nomenclature system
(a) Biological names are generally taken from the Latin language
(b) The name of the author is written in roman letters after the name of the specific epithet
(c) The biological name contains two components
(d) Tautonyms are not valid in animal nomenclature

23. Which of the following taxonomic aids is used for the identification of plants only?
(a) Museum (b) Key
(c) Flora (d) Zoological park

24. Which of the following pairs of taxa do not have a similar category in hierarchy?
(a) Solanaceae, Anacardiaceae
(b) Primata, Diptera
(c) Hominidae, Muscidae
(d) Mammalia, Arthopoda

25. Of the following taxa, which one represents the highest rank in taxonomic hierarchy?
(a) Dicotyledonae (b) Angiospermae
(c) Sapindales (d) Anacardiaceae

26. Which of the following represents a group of related orders?
(a) Muscidae (b) Primata
(c) Insecta (d) Arthopoda

27. A place with a collection of living plant specimens is:
(a) Botanical garden (b) Musium
(c) Herbarium (d) Key

28. Read the following features.
(i) Taxonomic aid is used for identification of plants only
(ii) Each plant is labelled by its botanical name and its family

Identify the taxonomic aid on the basis of the above information:
(a) Key (b) Botanical garden
(c) Museum (d) Herbarium

29. Which of the following organisms is not classified under the order Polymoniales?
(a) Potato (b) Tomato
(c) Mango (d) *Datura*

30. For which of the following organisms are growth and reproduction synonymous?
(a) Worker bee (b) *Hydra*
(c) Mule (d) Bacteria

31. In ascending order of taxonomic categories:
A. Common character increases
B. Number of organisms increases
C. Complexity of classification decreases
(a) Only (A) and (B) (b) Only (B)
(c) Only (A) and (C) (d) (A), (B) and (C)

32. Select the incorrectly matched pair with respect to the housefly:
(a) Kingdom – Animalia
(b) Order – Primata
(c) Class – Insecta
(d) Genus – *Musca*

33. Choose the incorrect statement:
(a) Taxonomic categories are distinct biological entities.
(b) Keys are analytical in nature.
(c) Herbarium sheets are arranged according to Bentham and Hooker's system of classification.
(d) Properties of tissues are present in the constituent cells.

34. Taxonomy includes all, except:
(a) Identification (b) Nomenclature
(c) Phylogeny (d) Classification

35. In keys, a pair containing a set of contrasting characters is called a:
(a) Lead (b) Couplet
(c) Monograph (d) Taxon

36. Potato and mango belong to the same:
(a) Family (b) Order
(c) Genus (d) Class

37. Read the following statements and select the correct option:
A. Human beings are organisms who have self-consciousness
B. Self-consciousness is a defining property
(a) Only (A) is correct
(b) Only (B) is correct
(c) Both (A) and (B) are correct
(d) Both (A) and (B) are incorrect

38. Which of the following categories includes all the others?
(a) Division (b) Family
(c) Order (d) Class

39. Identify the following statements as true (T) or false (F) and choose the option accordingly:
A. Museums and herbariums are means of ex-situ conservation
B. Living organisms are self-replicating, evolving and self-regulating interactive systems
C. Fungi, filamentous algae and protonema of mosses multiply by sporulation
D. Isolated metabolic reactions occurring in vitro are living things but surely living reactions

	A	B	C	D
(a)	T	T	F	T
(b)	F	T	F	F
(c)	T	F	T	T
(d)	F	F	T	F

40. As we go from Kingdom to Species in a taxonomic hierarchy, the number of common characters:
(a) Will increase
(b) Will decrease
(c) Remain the same
(d) May increase or decrease

41. Interbreeding is possible between two members of the same:
(a) Class (b) Family
(c) Species (d) Genus

42. All are polytypic genera, except:
(a) *Solanum* (b) *Panthera*
(c) *Brassica* (d) *Homo*

43. Plant families like Convolvulceae and Solanaceae are placed in order ____**A**____ on the basis of ____**B**____ characters:

	A	B
(a)	Polymoniales	Floral
(b)	Polymoniales	Vegetative
(c)	Sapindales	Vegetative
(d)	Poales	Floral

44. The equivalent taxonomic categories of Poales and Monocotyledonae (with respect to animals) are:
(a) Primata and Chordata
(b) Carnivora and Muscidae
(c) Diptera and Insecta
(d) Carnivora and Animalia

45. Select the incorrect match with respect to taxonomic categories of wheat:
(a) Kingdom – Angiospermae
(b) Class – Monocotyledonae
(c) Order – Poales
(d) Genus – *Triticum*

ANSWER KEYS

Exercises

1. (d)	**2.** (b)	**3.** (c)	**4.** (c)	**5.** (c)	**6.** (d)	**7.** (a)	**8.** (b)	**9.** (a)	**10.** (b)
11. (a)	**12.** (c)	**13.** (a)	**14.** (d)	**15.** (c)	**16.** (c)	**17.** (b)	**18.** (d)	**19.** (a)	**20.** (c)
21. (b)	**22.** (d)	**23.** (b)	**24.** (a)	**25.** (c)	**26.** (b)	**27.** (b)	**28.** (b)	**29.** (c)	**30.** (a)
31. (b)	**32.** (c)	**33.** (a)	**34.** (c)	**35.** (a)	**36.** (b)	**37.** (d)	**38.** (b)	**39.** (c)	**40.** (a)
41. (d)	**42.** (d)	**43.** (c)	**44.** (d)	**45.** (d)	**46.** (c)	**47.** (a)	**48.** (c)	**49.** (c)	**50.** (a)
51. (d)	**52.** (c)	**53.** (a)	**54.** (c)	**55.** (c)	**56.** (b)	**57.** (d)	**58.** (b)	**59.** (c)	**60.** (c)
61. (c)	**62.** (a)	**63.** (b)	**64.** (d)	**65.** (c)	**66.** (a)	**67.** (b)	**68.** (d)	**69.** (c)	**70.** (d)
71. (b)	**72.** (c)	**73.** (a)	**74.** (a)	**75.** (b)	**76.** (d)	**77.** (b)	**78.** (b)	**79.** (a)	**80.** (d)
81. (c)	**82.** (b)	**83.** (d)	**84.** (a)	**85.** (d)	**86.** (c)	**87.** (b)	**88.** (a)	**89.** (a)	**90.** (d)
91. (d)	**92.** (b)	**93.** (d)	**94.** (a)	**95.** (a)	**96.** (c)	**97.** (c)	**98.** (a)	**99.** (b)	**100.** (c)
101. (d)									

NCERT Exemplar Questions

1. (a)	**2.** (c)	**3.** (c)	**4.** (c)	**5.** (c)	**6.** (c)	**7.** (c)	**8.** (b)	**9.** (d)	**10.** (a)

Practice Test Paper

1. (d)	**2.** (d)	**3.** (d)	**4.** (b)	**5.** (a)	**6.** (a)	**7.** (b)	**8.** (d)	**9.** (d)	**10.** (b)
11. (c)	**12.** (b)	**13.** (d)	**14.** (c)	**15.** (a)	**16.** (c)	**17.** (b)	**18.** (c)	**19.** (b)	**20.** (a)
21. (b)	**22.** (d)	**23.** (c)	**24.** (d)	**25.** (b)	**26.** (c)	**27.** (a)	**28.** (b)	**29.** (c)	**30.** (d)
31. (b)	**32.** (b)	**33.** (d)	**34.** (c)	**35.** (b)	**36.** (d)	**37.** (a)	**38.** (a)	**39.** (b)	**40.** (a)
41. (c)	**42.** (d)	**43.** (a)	**44.** (c)	**45.** (a)					

SOLUTIONS

What is Living?, Diversity in the Living World

1. Reproduction is not a defining feature of living organisms. There are organisms which do not reproduce, like mules, worker bees and infertile human couples.
2. Non-living objects can also grow if we take increase in mass as a criterion of growth, but this growth is extrinsic.
3. (A), (B) and (C) are correct.
4. The sum total of biochemical reactions (anabolic and catabolic) occurring in our body is called metabolism.
5. *Amoeba* reproduce by binary fission.
6. Living organisms show intrinsic growth.
7. In unicellular organisms (I, IV, V), growth and reproduction are synonymous.
8. In fragmentation, the body of the organism breaks up into two or more parts, each of which grows into a new individual. Fragmentation is quite common in filamentous algae, fungi and protonema of mosses.
9. Unicellular organism (Yeast) and multicellular organism (*Hydra*) reproduce by bud formation.
10. Sensitivity and awareness is regarded as the defining property of living organisms. Human beings are the only organisms that have self-consciousness, but it is not a defining feature.
11. In unicellular organisms, growth and reproduction are mutually inclusive events. But, in the case of the majority of higher plants and animals, growth and reproduction are mutually exclusive events.
12. D is false. Isolated metabolic reactions, in-vitro, are not living things, but a living reaction.
13. Consciousness
14. Consciousness is the ability to sense the surroundings and respond to stimuli. It is shown by all organisms, from prokaryotes to the most complex eukaryotes.
15. Self-consciousness
16. Both (A) and (B) are correct.
17. Intrinsic growth, metabolism, cellular organisation and consciousness are the defining features of all living forms.
18. Properties of tissues are not present in the constituent cells but arise as a result of interactions among the constituent cells.
19. A is false. Properties of cellular organelles are not present in constituent cells but arise as a result of interactions among the constituent cells.
20. All living organisms are linked to one another by the sharing of common genetic material, but to varying degrees.
21. 1.7–1.8 million
22. Descending arrangement of taxonomic categories is called taxonomic hierarchy.
23. Binomial nomenclature was developed and popularised by Carolus Linnaeus.
24. Scientific rules for nomenclature of animals are standardised by ICZN (International Code of Zoological Nomenclature).
25. International Code of Botanical Nomenclature
26. A – Handwritten, B – Printed, C – Latin
27. *Mangifera indica* Linn
28. The scientific names are printed in italics to indicate their Latin origin.
29. In a biological name, all the three words (generic name + species epithet + author citation) collectively form the binomial epithet.
30. Only (i) and (ii) are correct. Each organism is given only one name consisting of two components: generic name and specific epithet. Generic name is followed by specific epithet.

31. Classification is the grouping of organisms into convenient categories on the basis of observable characters.

32. Basic elements of taxonomy are (i) Characterisation and Identification, (ii) Nomenclature, (iii) Classification

33. Identification is the first act in taxonomic studies.

34. In *Magnifera indica* Linn, Linn indicates that this species was first described by Linnaeus.

35. Linnaeus proposed the scientific name of plants in his book *Species Plantarum.*

36. Systematics = Taxonomy + Phylogeny. Phylogeny is the evolutionary history of an organism.

37. A taxon is the unit of taxonomy; taxon represents any level of grouping of organisms based on easily observable common characters.

38. Linnaeus

39. Both (A) and (B) are correct.

40. ICBN

41. External and internal structure along with the structure of the cell, development process and ecological information of organisms are essential and form the basis of modern taxonomic studies.

42. Option (d) is correct.

43. Classical taxonomy is based on observable morphological characters.

44. Modern taxonomy is similar to classical taxonomy in dealing with morphological features.

45. Systematics deals with diversity of all organisms and the evolutionary relationship (phylogeny) among them.

46. Both (A) and (B) are correct. When generic name and specific epithet is the same, it is called tautonym. For example, *Rattus rattus.* Tautonyms are not valid in plant nomenclature, but valid in animal nomenclature.

Taxonomic Categories

47. Taxonomic category is a rank in the taxonomic hierarchy.

48. There are seven obligate taxonomic categories: Kingdom, Division/Phylum, Class, Order, Family, Genus and Species.

49. Both the statements are correct.

50. Division for plants and Phylum for animals

51. Species is the lowest taxonomic category. It is the basic and smallest unit of classification.

52. Tribe, Varieties and Subspecies are subcategories.

53. Sapindales is the order of mango.

54. Primata (Order) and Mammalia (Class)

55. Order – Poales

56. Fishes, amphibians and reptiles are placed together in phylum Chordata.

57. Six taxons are related to kingdom Animalia (Primata, Chordata, Muscidae, *Felis,* Carnivora, *Musca*).

58. Muscidae (Family). The lower the category, the higher will be the number of common characters, when we travel from Kingdom to Species in a taxonomic hierarchy.

59. A genus with more than one species is called a polytypic genus. For example, *Solanum, Brassica, Panthera,* etc. *Homo* is a monotypic genus.

60. Taxon is a group of living organisms assigned to a particular category. It is a rank in hierarchical classification.

61. Plant, Monocot and Rice are taxa of Kingdom, Class and Species level, respectively.

62. Mayer's proposed biological concept of species defines species as a group of naturally interbreeding populations that are reproductively isolated from other groups.

63. A – Polymoniales, B – Floral Characters

64. Plant families are characterised on the basis of both vegetative and reproductive characters.

65. Insects, Housefly

66. The higher the category, the greater the difficulty in determining the relationship with other taxa at the same level.

67. A – Primata, B – Insecta, C – Anacardiaceae, D – Wheat

68. Wheat, Monocot and Plant represent taxa at different levels (Species, Class and Kingdom, respectively).

69. Phylum Chordata includes features like presence of notochord and presence of dorsal hollow neural system.

70. A – Carnivora, B – Canidae

71. A – Decreases, B – More

72. Class is a group of related orders.

73. Wheat and mango belong to division Angiospermae.

74. A(Phylum) – B(Class) – D(Order) – C (Family)

75. Canidae (Dogs), Felidae (Cats)

76. Phylum and Subspecies are not recognised by botanists; they use Division and Variety for the same.

Taxonomic Aids

77. A – Herbarium, B – Zoological park

78. Museums contain collections of preserved plant and animal specimens for study and reference. Specimens are preserved in jars or containers in preservative solutions like formalin.

79. A – Herbarium, B – Museum, C – Botanical garden, D – Manuals

80. Option (d) is correct.

81. Keys are used for identification of plants and animals.

82. Keys are based on the set of contrasting characters generally in a pair, known as couplet.

83. Royal Botanical Gardens, Kew (England)

84. Howrah

85. Monograph contains information about any one taxon.

86. Herbarium is a storehouse of collected plant specimens which are dried, pressed and preserved on sheets.

87. Botanical gardens are a means of ex-situ conservation of plants.

88. Only (B) is correct. Zoological parks are used for ex-situ conservation of animals and they are kept in a protected environment which is made similar to their natural habitats as much as possible.

89. Taxonomic Key – 3 (Analytical lead, Couplet), Herbarium – 5 (Preservation of specimen, Formalin, Use of 0.1 % $HgCl_2$, Quick reference system, Holotype)

90. Separate taxonomic keys are required for each taxonomic category.

91. Flora, Catalogue

92. C and D are false. Skeletons of some larger animals are also preserved in various museums.

93. Botanical Gardens and Zoological Parks

94. Herbarium

95. Manuals are useful in providing information for the identification of species found in an area.

96. Bentham and Hooker system of classification

97. National Botanical Research Institute (NBRI) is located in Lucknow.

98. Zoological park

99. Size of herbarium sheet is not mentioned on the herbarium sheet along with other information.

100. Ex-situ conservation

101. (a), (b) and (c) are correct.

Biological Classification

EXERCISES

Introduction to Classification

1. Read the following statements and select the correct option:
   A. Aristotle was the earliest to attempt a more scientific basis for classification
   B. Aristotle classified plants on the basis of their morphological characters and categorised them into herbs, shrubs and trees
   C. Aristotle divided animals into two groups: those which are motile and those which are not
   (a) Only (A) is correct
   (b) (A) and (B) are correct
   (c) (A) and (C) are correct
   (d) (A), (B) and (C) are correct

2. The Two Kingdom classification system was proposed by:
   (a) Aristotle (b) Haeckel
   (c) Linnaeus (d) Whittaker

3. The Two Kingdom system of classification did not distinguish between:
   (a) Unicellular and multicellular organisms
   (b) Photosynthetic and non-photosynthetic organisms
   (c) Prokaryotic and eukaryotic organisms
   (d) All of these

4. The Five Kingdom system of classification was proposed by:
   (a) RH Whittaker (b) Haeckel
   (c) Linnaeus (d) Copeland

5. Which of the following is not a criterion of the Five Kingdom classification:
   (a) Cell structure
   (b) Body organisation
   (c) Mode of nutrition
   (d) Cell wall composition

6. The number of criteria used in classifying organisms in the Five Kingdom classification is:
   (a) 5 (b) 4
   (c) 3 (d) 6

7. In the Five Kingdom classification, multicellular and eukaryotic organisms with heterotrophic mode of nutrition are included in:
   (a) Plantae (b) Fungi
   (c) Animalia (d) All except (a)

8. According to the classification system of RH Whittaker, how many kingdoms comprise eukaryotes?
   (a) One (b) Four
   (c) Three (d) Two

9. According to __A__ classification, *Chlamydomonas* and *Chlorella* are kept along with __B__ under kingdom __C__.

	A	B	C
(a)	RH Whittaker	*Paramecium*	Protista
(b)	Linnaeus	*Paramecium*	Monera
(c)	Linnaeus	Angiosperms	Algae
(d)	RH Whittaker	Yeast	Fungi

10. Consider the following table and select the appropriate answer:

Characters	Monera	Protista	Fungi	Plantae	Animalia
Cell type	A	Eukaryotic	Eukaryotic	Eukaryotic	Eukaryotic
Nuclear membrane	Absent	Present	B	Present	Present
Body organisation	Cellular	C	D	Tissue/ Organ	Tissue system

	A	B	C	D
(a)	Prokaryotic	Present	Multicellular	Loose tissue
(b)	Prokaryotic	Present	Cellular	Loose tissue
(c)	Prokaryotic	Absent	Multicellular	Tissue/Organ
(d)	Prokaryotic	Present	Loose tissue	Organ system

11. Select the incorrect match with respect to Whittaker's system of classification:
(a) Monera: Autotrophic or heterotrophic organisms with non-cellulosic cell wall
(b) Protista: Unicellular eukaryotic organisms with tissue level of body organisation
(c) Plantae: Autotrophic organisms with cellulosic cell wall and tissue level body organisation
(d) Fungi: Multicellular eukaryotic organisms with loose tissue body organisation

Kingdom Monera

12. Bacteria, archaebacteria and *Mycoplasma* belong to:
(a) Protista (b) Eukaryotes
(c) Monera (d) Plantae

13. Fill in the blanks and choose the correct option:
(i) ___A___ are the sole members of kingdom Monera
(ii) The most common shape of bacteria is ___B___

	A	B
(a)	Bacteria	Bacillus
(b)	Cyanobacteria	Cocci
(c)	Protozoans	Bacillus
(d)	Archaebacteria	Vibrio

14. Identify the diagrams labelled as A, B, C, D, respectively:

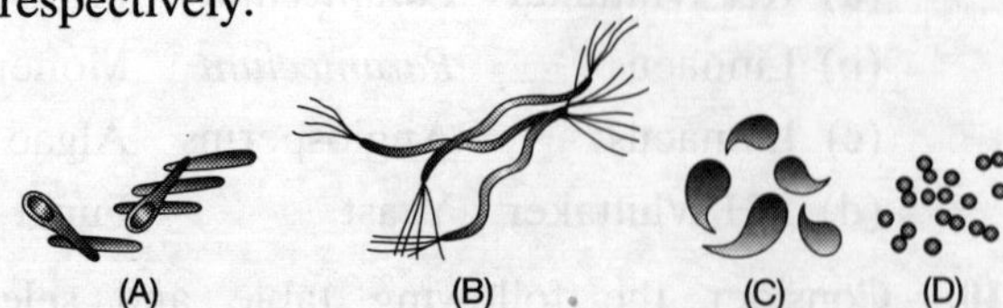

	A	B	C	D
(a)	Bacilli	Spirilla	Vibrio	Cocci
(b)	Bacilli	Spirilla	Cocci	Vibrio
(c)	Vibrio	Bacilli	Spirilla	Cocci
(d)	Vibrio	Cocci	Spirilla	Bacilli

15. The cell wall of monerans is usually made up of:
(a) Protein
(b) Cellulose
(c) Amino acid + Polysaccharides
(d) Cellulose, hemicellulose and pectin

16. The most extensive metabolic diversity is found in:
(a) Monera (b) Protista
(c) Fungi (d) Plantae

17. Read the following statements carefully and select the correct option for bacteria:
A. They are the sole members of kingdom Monera
B. They are the most abundant microorganisms
C. They cannot live in snow and deep oceans
D. No bacteria live in or on other organisms as parasites
(a) (A) and (B) (b) (B) and (C)
(c) (C) and (D) (d) (A) and (D)

18. Bacteria is a group of prokaryotic organisms that is characterised by:
(a) Peptidoglycan cell wall
(b) Absence of membrane-bound cell organelle
(c) Presence of 70S ribosomes
(d) All of the above

19. In prokaryotes, respiratory enzymes are found associated with:
(a) Genophore (b) Mesosomes
(c) Mitochondria (d) Ribosomes

20. The most common mode of reproduction in bacteria is:
(a) Binary fission
(b) Reproduction
(c) Budding
(d) Gene recombination

21. The *Spirillum* bacterium is:
(a) Comma shaped and flagellated
(b) Spiral shaped with flagella at one end
(c) Spiral shaped with flagella at both ends
(d) Spiral shaped with no flagella

22. State true (T) or false (F) for the given statements and select the correct option:
A. Hundreds of bacteria are present in a handful of soil
B. Bacteria are grouped under three categories based on their shape
C. Bacteria are very simple in structure and very complex in behaviour
D. Most bacteria synthesize their own food from inorganic substances

	A	B	C	D
(a)	T	F	T	T
(b)	T	F	T	F

(c) F T T T
(d) F F T T

23. Majority of bacteria are nutritionally ______.
(a) Photoautotrophs
(b) Chemoautotrophs
(c) Heterotrophs
(d) Symbiotics

24. Cyanobacteria are:
A. Mostly multicellular, colonial, filamentous
B. Also referred to as blue-green algae
C. Fresh water, marine or terrestrial algae
(a) Only (A) is correct
(b) Both (A) and (B) are correct
(c) (B) and (C) are correct
(d) (A), (B) and (C) are correct

25. ______ often form algal bloom in polluted water bodies:
(a) Cyanobacteria (b) PPLO
(c) Thermoacidophiles (d) Halophiles

26. Identify the figure and select the incorrect statement:

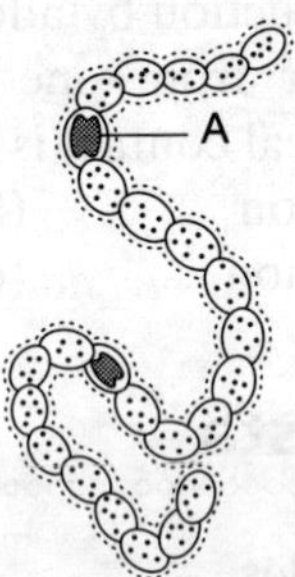

(a) Filamentous cyanobacteria as anoxygenic phototroph
(b) It has chlorophyll a, similar to higher plants
(c) It is surrounded by mucilaginous sheath
(d) The labelled structure is specialised for nitrogen fixation

27. What is true for archaebacteria?
(a) They are all heterotrophs
(b) They have membrane-bound organelles
(c) They differ from eubacteria in the sequence of 16S rRNA genes
(d) They have a peptidoglycan cell wall

28. Select the incorrect statement with respect to cyanobacteria:
(a) Absence of flagellum throughout the life cycle
(b) Presence of chlorophyll a, similar to higher plants
(c) Surrounded by mucilaginous sheath
(d) Cannot perform nitrogen fixation in free-living condition

29. Heterocyst present in *Nostoc* is specialised for:
(a) Reproduction
(b) Nitrogen fixation
(c) Storage of food material
(d) Carbon reduction

30. Cholera causing bacteria are grouped under which of the following categories:

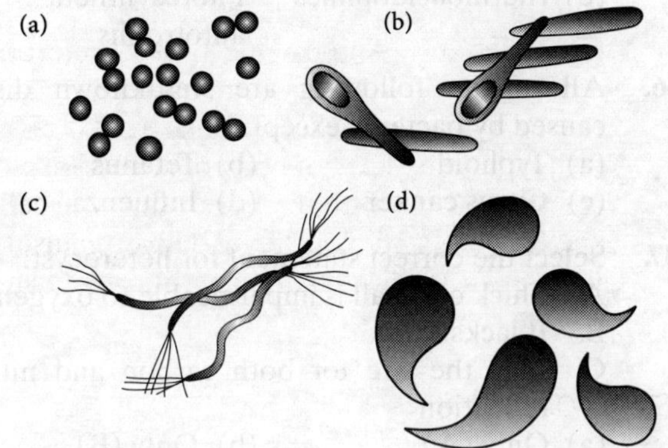

31. The monerans that live in the most harsh habitats show/are:
(a) Cyanobacteria
(b) Presence of different cell wall structure
(c) Branched chain lipid in their cell membrane
(d) All except (a)

32. Which of the following plays an important role in biogas production?
(a) *Mycoplasma*
(b) Methanogens
(c) Halophiles
(d) Thermoacidophiles

33. *Mycoplasma* is:
A. Facultative anaerobe
B. Sensitive to penicillin
(a) Only (A) is correct
(b) Only (B) is correct
(c) Both (A) and (B) are correct
(d) Both (A) and (B) are incorrect

34. The largest bacteria that was discovered in the intestine of a brown surgeonfish is:
(a) *E.coli*
(b) *Dialister pneumosintes*
(c) *Epulopiscium fishelsoni*
(d) *Spirillum*

35. Read the following statements:
(A): Primitive organisms present in the gut of ruminant animals are responsible for the production of biogas
(B): They oxidise various inorganic substances such as nitrate, nitrite and ammonia and release energy for their food production

	Statement A	Statement B
(a)	Methanogens	Heterotrophs
(b)	Methanogens	Chemosynthetic autotrophs
(c)	Thermoacidophiles	Chemosynthetic autotrophs
(d)	Thermoacidophiles	Photosynthetic autotrophs

36. All of the following are well-known diseases caused by bacteria, except:
(a) Typhoid (b) Tetanus
(c) Citrus canker (d) Influenza

37. Select the correct statement for heterocyst:
A. Thick cell wall is impermeable to oxygen
B. It lacks PSII
C. It is the site for both carbon and nitrogen reduction
(a) Only (A) (b) Only (B)
(c) Both (A) and (B) (d) (A), (B) and (C)

38. *Mycoplasma* is resistant to the action of which antibiotic?
(a) Tetracyclin (b) Streptomycin
(c) Penicillin (d) Erythromycin

39. Which feature of archaebacteria enables them to survive in extreme environmental conditions?
(a) Habitat (b) Cell wall structure
(c) Mode of nutrition (d) Cell type

40. An ancient group of bacteria of marshy habitats and living in exclusively anaerobic conditions are:
(a) Filamentous bacteria
(b) Halophiles
(c) Methanogens
(d) Thermoacidophiles

41. Which of the following is an essential feature of sexual reproduction in bacteria?
(a) Formation and fission of gametes
(b) Meiosis and syngamy
(c) A primitive type of DNA transfer from one bacterium to another
(d) All except (c)

42. The organisms called halophiles are most abundant in:
(a) Marshy areas (b) Salty areas
(c) Hot springs (d) Polluted streams

43. Fill in the blanks and choose the correct option:
(i) __A__ bacteria are used in the production of antibiotics
(ii) __B__ bacteria play a major role in recycling nutrients like nitrogen, iron, sulphur and phosphorous

	A	B
(a)	Heterotrophic	Chemosynthetic autotrophic
(b)	Autotrophic	Chemosynthetic autotrophic
(c)	Chemoautotrophic	Photoautotrophic
(d)	Chemoautotrophic	Heterotrophic

44. Read the following statements:
A. These are the smallest known living cells
B. They cannot survive without oxygen
C. They completely lack cell wall
D. They are pathogenic in animals only
Choose the option which includes incorrect statements regarding *Mycoplasma:*
(a) (A) and (C) (b) (A) and (D)
(c) (B) and (C) (d) (C) and (D)

45. Select the incorrectly matched pair:
(a) Energy parasites – *Chlamydia*
(b) Filamentous bacteria – Actinomycetes
(c) Joker of plant kingdom – *Mycoplasma*
(d) Complementary chromatic adaptation – Archaebacteria

46. Sexual reproduction by adopting primitive type of DNA transfer from one bacterium to another without physical contact is called:
(a) Conjugation (b) Transformation
(c) Transduction (d) Transfaction

Kingdom Protista

47. Protista includes:
(a) Unicellular eukaryotes
(b) Unicellular prokaryotes
(c) Virus, lichen, mycorrhiza
(d) Cyanobacteria and *Mycoplasma*

48. Which of the following kingdoms has no well-defined boundaries?
(a) Monera (b) Protista
(c) Fungi (d) Metazoa

49. Which of the following kingdoms forms a link with others dealing with plants, animals and fungi?
(a) Monera (b) Protista
(c) Plantae (d) Fungi

50. The walls are embedded with silica, and thus, are indestructible in:
(a) Diatoms (b) Slime moulds
(c) Dinoflagellates (d) Euglenoids

51. Members of Protista are primarily:
(a) Aquatic (b) Parasitic
(c) Terrestrial (d) Saprotrophic

52. Nutritionally, protists are:
(a) Photoautotrophs (b) Heterotrophs
(c) Myxotrophs (d) All of the above

53. Which of the following are placed under Protista?
(a) Chrysophytes and dinoflagellates
(b) Euglenoids
(c) Slime moulds and protozoans
(d) All of the above

54. Diatoms and dinoflagellates have similar:
(a) Cell wall structure
(b) Type of photosynthetic segments
(c) Type of food store
(d) All except (b)

55. The given organism is:

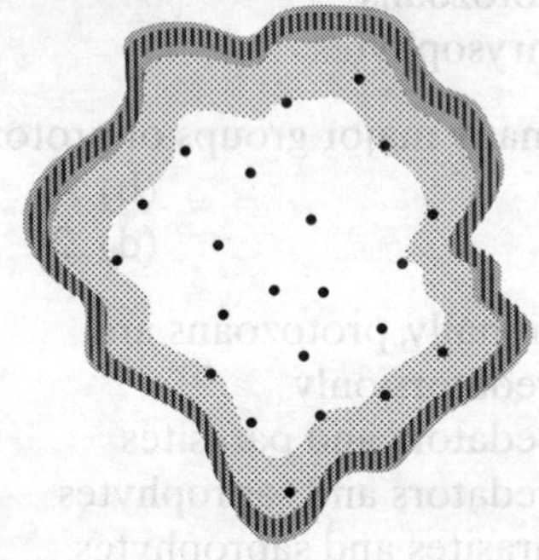

(a) Protozoan protist
(b) Laboratory weed
(c) Bread mould
(d) Consumer-decomposer protist

56. Which of the following shows holophytic mode of nutrition?
(a) Dinoflagellates
(b) Sporozoans
(c) Consumer-decomposer protists
(d) Protozoans

57. Select the mismatched pair:
(a) Euglenoids – Mixotrophic nutrition
(b) *Gonyaulax* – Red tide
(c) Sporozoans – Endoparasites
(d) *Trypanosoma* – Ciliated protozoans

58. Locomotory structure in protists is:
(a) Flagella (b) Cilia
(c) Pseudopodia (d) All of the above

59. The chief producers in oceans are:
(a) Dinoflagellates (b) Diatoms
(c) Euglenoids (d) Green algae

60. Which of the following forms fruiting bodies like fungi, lacks chlorophyll and has saprobic or phagotrophic mode of nutrition, and a wall-less vegetative body?
(a) Slime moulds (b) Euglenoids
(c) Protozoan protist (d) Desmides

61. Photosynthetic protists include:
(a) Euglenoids and slime moulds
(b) Desmides and ciliates
(c) Diatoms, dinoflagellates and euglenoids
(d) Dinoflagellates, desmides and zooflagellates

62. Read the following statements carefully:
A: Dinoflagellates have two flagella: one lies longitudinally and the other lies transversely in a furrow between the cell plates
B: Dinoflagellates do not harm marine life
(a) Only (A) is correct (b) Both are correct
(c) Only (B) is correct (d) Both are incorrect

63. Red tide in warm coastal waters develops due to super dominance of:
(a) Diatoms and desmides
(b) Dinoflagellates
(c) Euglenoid forms
(d) Few species of *Chlamydomonas*

64. In which of the following does the cell wall have a stiff cellulose plate on the outer surface?
(a) Desmides (b) Diatoms
(c) Dinoflagellates (d) Euglenoids

65. Diatomaceous earth is:
(a) Indestructible frustule of diatoms
(b) Gritty in nature
(c) Used in polishing, filtration of oil and syrups
(d) All of the above

66. The mode of nutrition present in *Euglena* is referred to as:
(a) Osmotrophy (b) Mixotrophy
(c) Phototrophy (d) Chemotrophy

67. Choose the incorrect match:
(a) *Gonyaulax* – Red tide
(b) Euglenoids – Heterotrophs when deprived of sunlight
(c) Desmides – Pyrrophytes
(d) *Paramecium* – Sleeper animalcule

68. Which of the following release toxins that may even kill other marine animals like fishes?
(a) *Gonyaulax* (b) *Paramecium*
(c) Euglenoids (d) Sporozoans

69. Identify the figure and select incorrect statements:

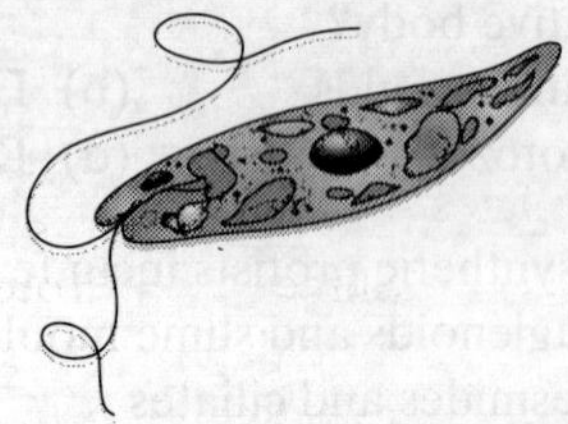

(a) Found in fresh stagnant water
(b) Pressure of proteinacious layer called pellicle instead of cell wall
(c) Photosynthetic pigments are identical to red algae
(d) Show flagellar locomotion and metaboly

70. Which of the following statements regarding slime moulds are correct?
A. They are saprophytic protists
B. Under unfavourable conditions, they form an aggregation called *Plasmodium*
C. Their spores possess true walls
D. Their vegetative phase resembles plants
(a) (A) and (B) (b) (A) and (C)
(c) (B) and (C) (d) (C) and (D)

71. What may be a photosynthetic protist to one biologist may be a plant to another. Which of the given features of slime moulds show linkage with plants?
(a) Pressure of holophytic nutrition
(b) Slime moulds have cellulosic cell wall around spores
(c) Formation of fruiting bodies
(d) Both (a) and (b)

72. A. It is a ciliated protozoan
B. It is an actively moving organism
C. It has a cavity (gullet) that opens to the outside of the surface

The above statements are assigned to:

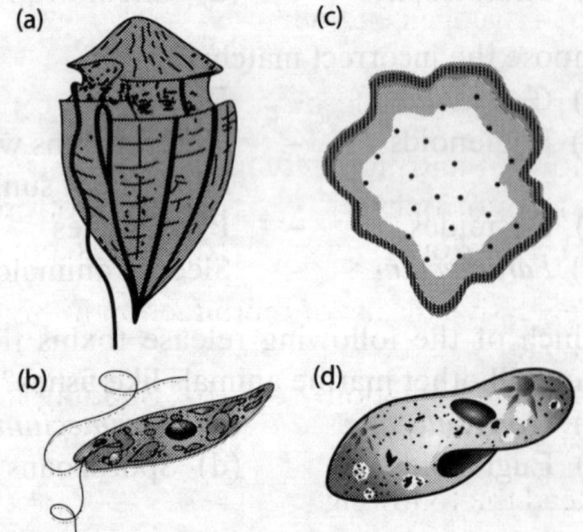

73. The spores are dispersed by ______ in slime moulds:
(a) Air current
(b) Water current
(c) Explosive mechanism
(d) All except (a)

74. The slimy mass of protoplasm with nucleus forms the body of the slime mould and is called:
(a) *Plasmodium* (b) Pyxamoeba
(c) Sporocytes (d) Myxoamoeba

75. Which of the following is considered to be a primitive relative of animals?
(a) Dinoflagellates
(b) Consumer-decomposer protists
(c) Protozoans
(d) Chrysophytes

76. How many major groups of protozoans are there?
(a) 3 (b) 4
(c) 5 (d) 2

77. Nutritionally, protozoans are:
(a) Predators only
(b) Predators and parasites
(c) Predators and saprophytes
(d) Parasites and saprophytes

78. Which of the following are protozoans?
(a) Diatoms, Flagellates and Ciliates
(b) Desmides, Euglenoids and Zooflagellates
(c) Desmides, Amoeboids and Sporozoans
(d) Amoeboids, Flagellates, Ciliates and Sporozoans

79. Select the incorrect statement with respect to amoeboid protozoans:
(a) They live in freshwater, seawater or moist soil
(b) Freshwater forms have silica shells on their surfaces
(c) They have pseudopodia for locomotion and capture prey
(d) Few members such as entamoeba are parasites

80. Match the following columns and select the correct option:

Column I	Column II
A. Amoeboid protozoan	(i) Parasite causes sleeping sickness
B. Flagellated protozoan	(ii) Marine forms with silica shells on their surface
C. Ciliated protozoan	(iii) Exclusively endoparasites
D. Sporozoan	(iv) Possess definite region of ingestion and egestion

(a) A – (iv), B – (iii), C – (i), D – (ii)
(b) A – (ii), B – (iii), C – (iv), D – (i)
(c) A – (ii), B – (iv), C – (i), D – (iii)
(d) A – (iii), B – (iv), C – (i), D – (iv)

81. Flagellated protozoans are:
(a) Free living
(b) Parasites
(c) Either free living or parasites
(d) Causative agents of malaria

82. *Plasmodium* (malarial parasite) is a:
(a) Sporozoan
(b) Ciliated protozoan
(c) Amoeboid protozoan
(d) Flagellated

83. Nuclear dimorphism is shown by:

(a) (c)

(b) (d)

84. Choose the odd one out with respect to the features of protists:
(a) Members of Protista are primarily aquatic
(b) Cell structure is eukaryotic
(c) Sexual reproduction involves gene recombination
(d) Reproduce asexually by binary fission

85. Protists with silica shells are:
(a) Chrysophytes
(b) Marine dinoflagellates
(c) Marine amoeboid protozoans
(d) Both (a) and (c)

86. Fill in the blanks and choose the correct option:
(i) Microscopic and holophytic organisms that float passively in water currents are ____**A**____
(ii) In ____**B**____, true cell wall is present only in the reproductive stage
(iii) Nutritionally diverse unicellular eukaryotic organism is ____**C**____
(iv) Marine predator protozoans having silica shells on their surface are ____**D**____

	A	B	C	D
(a)	Diatoms	Slime moulds	Euglenoids	Sarcodines
(b)	Dinoflagellates	Euglenoids	Euglenoids	Ciliates
(c)	Diatoms	Slime moulds	Protozoans	Sporozoans
(d)	Dinoflagellates	Euglenoids	Diatoms	Amoeboid protozoans

87. Slime moulds undergo aggregation:
(a) Under favourable conditions
(b) To form *Plasmodium*
(c) Both (a) and (b)
(d) To form fruiting bodies

Kingdom Fungi

88. Fungi show:
A. Great diversity of morphology and habitat
B. Abundance in dark, warm and humid places
C. Cell wall made up of chitin and polysaccharide
The correct option is:
(a) Only (A) (b) Only (A) and (B)
(c) Only (C) (d) (A), (B) and (C)

89. Which of the following is the correct sequence of steps in the sexual cycle of fungi?
(a) Plasmogamy → Karyogamy → Meiosis
(b) Karyogamy → Plasmogamy → Meiosis
(c) Meiosis → Karyogamy → Plasmogamy
(d) Plasmogamy → Meiosis → Karyogamy

90. Select the odd one out with respect to asexual spores:
(a) Conidia (b) Sporangiospores
(c) Zoospores (d) Oospores

91. Reserve food material in fungi is:
(a) Starch
(b) Glycogen and oil
(c) Chitin
(d) Fungal cellulose

92. Mode of nutrition in fungi is:
(a) Autotrophs (b) Heterotrophs
(c) Holozoic (d) All except (b)

93. Vegetative means of reproduction in fungi are all, except:
(a) Fragmentation (b) Budding
(c) Sporulation (d) Fission

94. Read the following:
(i) Multicellular, eukaryotic organisms
(ii) Cosmopolitan

(iii) Haploid body
(iv) Heterotrophic mode of nutrition
(v) Show diversity of morphology and nutrition
(vi) Presence of membrane-bound organelles
(vii) Presence of unicisternal Golgi bodies
(viii) Cytoplasmic 70S ribosomes

How many statements are correct with respect to fungi?
(a) Six (b) Five
(c) Seven (d) Four

95. Which of the following is not the basis of classification of fungi into various classes?
(a) Morphology of mycelium
(b) Type of fruiting bodies
(c) Spore formation
(d) Mode of nutrition

96. Read the following statements carefully:
A. Found in aquatic habitats and on decaying woods
B. Has aseptate and coencyte mycelium
C. Produces zoospores for asexual reproduction

Identify the class of fungi on the basis of the above features:
(a) Ascomycetes (b) Zygomycetes
(c) Phycomycetes (d) Deuteromycetes

97. Ascospores and basidiospores are similar in:
(a) Being produced endogenously
(b) Being meiospores
(c) Having flagella
(d) Both (a) and (b)

98. Match Column I and Column II and select the correct option:

Column I	Column II
A. *Ustilago*	(i) Laboratory weed
B. *Aspergillus*	(ii) Absence of sexual stage
C. Toadstools	(iii) Smut fungi
D. *Alternaria*	(iv) Poisonous mushroom

(a) A – (iii), B – (i), C – (iv), D – (ii)
(b) A – (ii), B – (i), C – (iv), D – (iii)
(c) A – (iii), B – (i), C – (ii), D – (iv)
(d) A – (iii), B – (iv), C – (ii), D – (i)

99. Dolipore septum is found in:
(a) Rust fungi (b) Smut fungi
(c) *Mucor* (d) *Agaricus*

100. An extremely important research tool in genetics and biochemistry is:
(a) *Penicillium* (b) *Rhizopus*
(c) *Neurospora* (d) *Aspergillus*

101. Fungi with coencytic mycelium:
(a) Are found in aquatic habitat
(b) Reproduce asexually by motile or non-motile spore
(c) Belong to the class Phycomycetes
(d) All of the above

102. In the sexual cycle of fungi, which of the following is the first step?
(a) Meiosis (b) Plasmogamy
(c) Karyogamy (d) Dikaryotic

103. Yeast reproduces vegetatively by:
(a) Ascospore (b) Conidia
(c) Budding (d) All except (b)

104. Choose the incorrect match:
(a) Phycomycetes – Cellulosic cell wall
(b) Ascomycetes – Coprophilous
(c) Basidiomycetes – Clamp connection
(d) Deuteromycetes – Prolonged dikaryotic phase

105. The most common method of reproduction in the members of class Ascomycetes is:
(a) Conidia (b) Sporangiospore
(c) Zoospores (d) Plasmospores

106. All of the given fungi are useful for human beings, except:
(a) *Saccharomyces* (b) *Neurospora*
(c) *Agaricus* (d) Toadstool

107. Identify the figure and select the incorrect option:

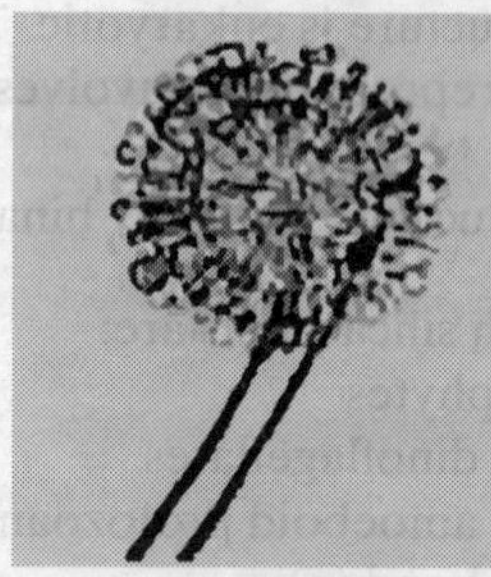

(a) It belongs to the class Ascomycetes
(b) Possesses branched septate mycelium
(c) Asexually reproduces by conidia formation
(d) Most extensively used in genetic and biochemical research

108. Delicacies like morels and truffles belong to the class:
(a) Sac fungi (b) Club fungi
(c) Algal fungi (d) Conjugation fungi

109. Which of the following features are common to Ascomycetes and Deuteromycetes:
A. Cell wall composition
B. Pressure of branched septate mycelium
C. Pressure of dolipore septa and clamp connection
D. Pressure of intervening dikaryotic stage (n + n nuclei per cell)
E. Absence of asexual stage
F. Presence of exogenously produced sexual spores
(a) (A), (B), (D) and (F)
(b) (B), (C), (D) and (E)
(c) (A) and (B)
(d) (A), (B), (D) and (E)

110. All of the fungi are members of class Basidiomycetes, except:
(a) *Ustilago* (b) *Puccinia*
(c) Toadstool (d) *Claviceps*

111. Identify the correct set of features for the organism shown in the figure:

(a) *Agaricus*, short (n + n) stage, non-motile basidiospore
(b) *Agaricus*, prolonged (n + n) stage, endogenously produced sexual spores
(c) Mushroom, longer (n + n) stage, absence of asexual spores
(d) Mushroom, absence of dikaryotic, exogenously produced meiospores

112. In which class of fungi are most of the members decomposers of litter?
(a) Phycomycetes (b) Ascomycetes
(c) Deuteromycetes (d) Basidiophycetes

113. Select the incorrect match from the given pairs:
(a) Penicillin – Green mould
(b) *Neurospora* – *Drosophila* of plant kingdom
(c) *Puccinia* – White rust on mustard
(d) Morels – Edible sac fungi

114. Once the perfect stages of members of class Deuteromycetes were discovered, they were often moved to:
(a) Oomycetes and Zygomycetes
(b) Oomycetes and Ascomycetes
(c) Phycomycetes and Basidiomycetes
(d) Ascomycetes and Basidiomycetes

115. Members of class Deuteromycetes reproduce asexually by ___A___ produced exogenously on special mycelium called ___B___:

	A	B
(a)	Sporangia	Sporangiospore
(b)	Zoospore	Zoosporangium
(c)	Conidia	Conidiophore
(d)	Ascospore	Asci

116. Absence of motile structure as well as sex organs are characteristic of which of the following classes of fungi:
(a) Ascomycetes (b) Basidiomycetes
(c) Deuteromycetes (d) Phycomycetes

117. Select the incorrect match:
(a) Dikaryotic mycelium is the dominant phase of life cycle – *Agaricus*
(b) Source of antibiotic – *Penicillium*
(c) Causal agent of disease which is characterised by white blisters on mustard leaf – *Albugo*
(d) Sexual reproduction by planogametic copulation – *Alternaria*

118. Ascospores and basidiospores are similar in:
(a) Being sexual spores
(b) Being meiospores
(c) Being produced exogenously
(d) Both (a) and (b)

119. Read the following statements:
(i) Pressure of branched and septate mycelium
(ii) Produce conidia for asexual reproduction
(iii) Sexual reproduction occurs by somatogamy
Identify the class of fungi on the basis of the above features:
(a) The algal fungi
(b) The sac fungi
(c) The club fungi
(d) The conjugation fungi

120. Club fungi, which grow in living plant bodies as parasites, are:
(a) *Agaricus, Ustilago*
(b) Bracket fungi, toadstool
(c) *Puccinia, Ustilago*
(d) Puffballs, *Agaricus*

121.

Fungi	Spores	Development
Phycomycetes	A	Zoospores
Ascomycetes	Sexual	B
Basidiomycetes	C	Basidiospores
Deuteromycetes	Conidia	D

	A	B	C	D
(a)	Asexual	Endogenously	Sexual	Exogenously
(b)	Sexual	Endogenously	Asexual	Exogenously
(c)	Sexual	Exogenously	Asexual	Endogenously
(d)	Asexual	Exogenously	Sexual	Endogenously

122. Hyphae are continuous tubes filled with multinucleate cytoplasm:
(a) *Rhizopus*, Yeast
(b) *Rhizopus, Mucor*
(c) *Puccinia, Aspergillus*
(d) *Ustilago, Agaricus*

Kingdoms Plantae and Animalia

123. Insectivorous plants are:
(a) Parasitic
(b) Autotrophic
(c) Partially heterotrophic
(d) Chemoautotrophic

124. Which of the following is an example of an insectivorous plant?
(a) Venus flytrap (b) Bladderwort
(c) *Nepenthes* (d) All of the above

125. *Cuscata* is:
(a) An insectivorous plant
(b) A prokaryotic organism
(c) A partially stem parasite
(d) A totally stem parasite

126. Select the incorrect statement with respect to plants:
(a) Kingdom Plantae includes eukaryotic, heterotrophic, chlorophyll containing organisms
(b) Plant cells have distinct nucleus, prominent chloroplast and chitinous cell wall
(c) Includes algae, bryophyte, pteredophyte, gymnosperm and angiosperm
(d) Plants show alternation of generation

127. Kingdom Animalia includes:
(a) Heterotrophs
(b) Eukaryotes
(c) Multicellular organisms that lack cell wall
(d) All of the above

128. Organisms of kingdom Animalia:
(i) All forms are capable of locomotion
(ii) They digest their food in an internal cavity
(iii) Show sexual reproduction by copulation of male and female followed by embryological development
(iv) Reserve food material is similar to fungi
(v) Higher and lower forms show elaborate sensory and neuromotor mechanism

How many of the statements given above are correct?
(a) Four (b) Three
(c) Two (d) Five

Viruses, Viroids, Prions and Lichen

129. The term virus was coined by:
(a) Pasteur (b) Beijerinck
(c) Ivanowsky (d) Stanley

130. The term virus means:
(a) Non-cellular (b) A parasite
(c) A killer (d) Venom or poison

131. Viruses are non-cellular organisms, but replicate themselves once they infect the host cell. To which of the following kingdoms do viruses belong?
(a) Monera (b) Protista
(c) Plantae (d) None of the above

132. Contagium virium fluidum was proposed by:
(a) D Ivanowsky (b) MW Beijerinck
(c) Stanley (d) Robert Hook

133. Viruses are:
A. Nucleoprotein particles
B. Connecting link between living and non-living
C. Smaller than bacteria and can pass through bacteria-proof filter

The correct statements are:
(a) (A) only (b) (A) and (B) only
(c) (A) and (C) only (d) (A), (B) and (C)

134. ____A____ showed that viruses could be crystallised and they consist largely of ____B____:

	A	B
(a)	Beijerinck	Protein
(b)	Stanley	Nucleic acid
(c)	Stanley	Protein
(d)	Iwanowsky	Nucleic acid

135. The most common type of genetic material present in bacteriophages is:
(a) dS RNA (b) SS RNA
(c) dS DNA (d) SS DNA

136. Identify the figure and select the incorrect option:

(a) Rod shaped
(b) SS RNA as genetic material
(c) Causes mosaic disease in tobacco plant
(d) Infectious protein coat is called capsid which protects the genetic material

137. Viruses did not find a place in classification since:
(a) They are acellular
(b) They are obligate parasites
(c) They replicate within the host cell
(d) Complex life cycle

138. Viruses are non-cellular organisms that are characterised by:
(a) Having no nucleic acid
(b) Lacking protoplast
(c) Ability to live independently of a living cell
(d) All except (a)

139. Potato spindle tuber disease is caused by:
(a) Virus (b) Viroid
(c) Prion (d) *Mycoplasma*

140. Read the following statements with respect to viruses and state true (T) or false (F):
A. Viruses are obligate parasites
B. They have an inert crystalline structure outside the host cell
C. A virus is nucleoprotein and the genetic material is infectious
D. The protein coat of the virus is made up of small subunits called peplomers

	A	B	C	D
(a)	T	T	F	F
(b)	F	T	F	T
(c)	T	T	F	T
(d)	T	T	T	F

141. The causal agent of which of the following diseases is a retrovirus?
(a) Influenza (b) Herpes
(c) Smallpox (d) AIDS

142. Which of the following is the infectious component of the virus?
(a) Nucleic acid (b) Protein
(c) Envelope (d) Capsid

143. In general, plant viruses have ______ as genetic material:
(a) SS DNA (b) dS DNA
(c) dS RNA (d) SS RNA

144. Choose the incorrect option with respect to lichen:

	Phycobiont	Mycobiont
(a)	Photoautotrophs	Heterotrophs
(b)	Green algae or BGA	Ascomycetes/ Basidiomycetes
(c)	Carbon dioxide fixer	Nitrogen fixer
(d)	Food manufactured	Body structure and covering

145. All are viral diseases, except:
(a) Tobacco mosaic, potato leaf roll
(b) Influenza, mumps
(c) Citrus canker, crown gall
(d) AIDS, smallpox

146. In plants, mosaic formation, leaf rolling and curling, yellowing and vein clearing, dwarfing and stunted growth are symptoms of:
(a) Viral disease (b) Fungal disease
(c) Bacterial disease (d) Mycloplasmal disease

147. In the year 1971, TO Diener discovered a new infectious agent:
(i) That was smaller than virus
(ii) It was found to be a free RNA
(iii) It lacked the protein coat

The above statements can be assigned to:
(a) Viruses (b) Viroids
(c) Virusoids (d) Prions

148. Read the following statements with respect to prions and select the correct option:
A. The prion particles do not contain any nucleic acids
B. These are infectious, abnormally folded protein particles
C. Prions are known to cause several neurodegenerative diseases in mammals
D. No prion disease of plants is known so far
(a) Only (B) is correct
(b) All are correct except (D)
(c) (A) and (C) are incorrect
(d) (A), (B), (C) and (D) are correct

149. The protein coat of viruses is called ___A___, which is made up of small units called ___B___:

	A	B
(a)	Capsid	Capsomere
(b)	Capsid	Peplomere
(c)	Envelope	Capsomere
(d)	Glycocalyx	Peplomere

150. Given below is a diagram of a bacteriophage. In which of the options are all the four parts correct?

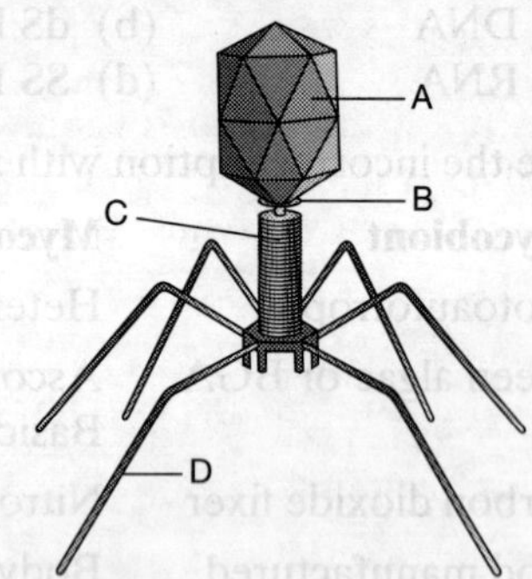

	A	B	C	D
(a)	Head	Collar	Sheath	Tail fibres
(b)	Head	Tail	Sheath	Tail fibres
(c)	Head	Sheath	Collar	Tail fibres
(d)	Collar	Sheath	Head	Tail fibres

151. Lichens are the symbiotic association between two components, which are, respectively:
(a) Acellular and cellular organisms
(b) Living and non-living organisms
(c) Algae and fungi
(d) Chemosynthetic and parasite

152. Which of the following statements is not correct about viroids?
(a) They were discovered by TO Diener
(b) They are of high molecular weight
(c) Smaller than virus
(d) Infectious SS RNA molecules

153. Which of the following infectious agents lack nucleic acid?
(a) Virus (b) Viroids
(c) Virusoids (d) Prion

154. Which of the following infectious agents replicate within the host cell but do not cause any infection?
(a) Prions (b) Virusoids
(c) Viroids (d) Both (a) and (c)

155. Identify the figure and select the incorrect option:

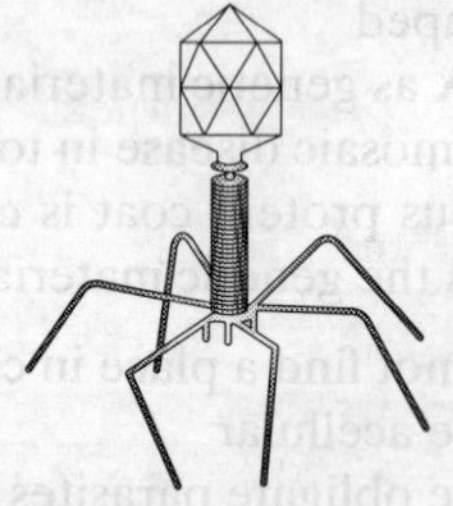

(a) Shows tadpole-like structure
(b) Has polyhydral head and helical tail
(c) Generally has dS DNA as genetic material
(d) Has tail pin and tail fibres for motility

156. Prions are:
(a) Infectious RNA particles
(b) Similar in size to viruses
(c) Causal organisms of mad cow disease
(d) All except (a)

157. Mutually beneficial association of fungi and orchid roots is known as ______:
(a) Mycorrhizae (b) Lichens
(c) Helotism (d) Protocooperation

158. A decrease in lichen population of an area is indicative of:
(a) Air pollution (b) Water pollution
(c) Soil pollution (d) Radioactive pollution

NCERT EXEMPLAR QUESTIONS

1. All eukaryotic unicellular organisms belong to:
(a) Monera (b) Protista
(c) Fungi (d) Bacteria

2. The Five Kingdom classification was proposed by:
(a) RH Whittaker (b) C Linnaeus
(c) A Roxberg (d) Virchow

3. Organisms living in salty areas are called as:
(a) Methanogens (b) Halophiles
(c) Heliophytes (d) Thermoacidophiles

4. Naked cytoplasm, multinucleated and saprophytic are the characteristics of:
(a) Monera (b) Protista
(c) Fungi (d) Slime moulds

5. An association between roots of higher plants and fungi is called:
(a) Lichen (b) Fern
(c) Mycorrhiza (d) BGA

6. A dikaryon is formed when:
(a) Meiosis is arrested
(b) The two haploid cells do not fuse immediately
(c) Cytoplasm does not fuse
(d) None of the above

7. Contagium vivum fluidum was proposed by

(a) DJ Ivanowsky (b) MW Beijerinek
(c) Stanley (d) Robert Hook

8. Associations between mycobiont and phycobiont are found in:
(a) Mycorrhiza (b) Root
(c) Lichens (d) BGA

9. Difference between virus and viroid is:
(a) Absence of protein coat in viroid but present in virus
(b) Presence of low molecular weight RNA in virus but absent in viroid
(c) Both (a) and (b)
(d) None of the above

10. With respect to fungal sexual cycle, choose the correct sequence of events:
(a) Karyogamy, Plasmogamy and Meiosis
(b) Meiosis, Plasmogamy and Karyogamy
(c) Plasmogamy, Karyogamy and Meiosis
(d) Meiosis, Karyogamy and Plasmogamy

11. Viruses are non-cellular organisms but replicate themselves once they infect the host cell. To which of the following kingdoms do viruses belong?
(a) Monera (b) Protista
(c) Fungi (d) None of the above

12. Members of Phycomycetes are found in:
(i) Aquatic habitats
(ii) Decaying wood
(iii) Moist and damp places
(iv) Plants as obligate parasites
Choose from the following options:
(a) None of the above (b) i and iv
(c) ii and iii (d) All of the above

PRACTICE TEST PAPER

1. According to the Five Kingdom system, all single-celled eukaryotes are placed in kingdom________
(a) Mycota (b) Protista
(c) Monera (d) Plantae

2. Which kingdom represents the presence of all the ecological types of organisms: producer, consumer and decomposer?
(a) Monera (b) Fungi
(c) Protista (d) Both (a) and (c)

3. Consider the following statements with respect to the classification system proposed by RH Whittaker:
A. All prokaryotic organisms were grouped together under kingdom Monera
B. *Paramecium* and *Amoeba* lack cell wall and were placed in kingdom Animalia
C. *Chlorella* and *Chlamydomonas* both have cell wall and were placed in kingdom Protista
D. Multicellular eukaryotic organisms are placed in three kingdoms
(a) Only (A)
(b) (A), (C) and (D)
(c) Only (A) and (B)
(d) (A), (B), (C) and (D)

4. Select the correct statement with respect to *Nostoc*:
A. It is a filamentous BGA
B. Performs oxygenic photosynthesis
C. Has the ability to fix nitrogen in a specialised cell called heterocyst
(a) Only (A) and (C) (b) Only (B) and (C)
(c) Only (C) (d) (A), (B) and (C)

5. In the light of the recent classification of living organisms into three domains of life (Bacteria, Archaea and Eukaryea), which of the following statements is true about Archaea?
(a) Archaea completely differ from prokaryotes
(b) Archaea resemble Eukaryea in all respects
(c) Archaea have some novel features that are absent in other prokaryotes and eukaryotes
(d) Archaea completely differ from both prokaryotes and eukaryotes

6. Which of the following organisms completely lack cell wall, are the smallest living cells known and can survive without oxygen?
(a) *Mycoplasma* (b) Euglenoids
(c) Slime moulds (d) Actinomycetes

7. Consider the given statements and select the correct option:
A. Cocci are rounded or spherical bacteria
B. Have simple structure and are complex in behaviour
C. Mainly reproduce by binary fission in unfavourable conditions
D. Show most extensive metabolic diversity

(a) Only (A) and (C)
(b) Only (B), (C) and (D)
(c) Only (A), (B) and (D)
(d) (A), (B), (C) and (D)

8. Mesokaryon condition is found in the members of:
(a) Diatoms (b) Dinoflagellates
(c) Euglenoids (d) Ciliates

9. Identify the particle agent that shows the following characteristics:
A. Abnormally folded protein
B. Causes neurological diseases
(a) Prion (b) Virion
(c) Virusoid (d) Viroid

10. In which group of organisms does the cell wall form two thin overlapping shells which fit together?
(a) Slime moulds (b) Chrysophytes
(c) Euglenoids (d) Dinoflagellates

11. How many of the given features of Protista are similar to prokaryotes:
Cell membrane composition, Cell type, Body organisation, Asexual reproduction, Cell wall composition, Cyclosis, Zygote formation
(a) Four (b) Five
(c) Three (d) Two

12. Archaebacteria of marshy habitats show/are:
(a) Pseudopeptidoglycan in cell wall
(b) Obligate anaerobes
(c) Also present in gut of ruminant mammals
(d) All of the above

13. Select the odd one out with respect to dinoflagellates:
(a) Mostly marine water forms
(b) Stiff cellulose plates
(c) Two heterokont laterally inserted flagella
(d) Cause red tides

14. Fill in the blanks and select the correct option:

Fungi	Spores	Development	Fruiting body
Ascomycetes	Asexual	A	Present
Basidiomycetes	Sexual	B	Present
Deuteromycetes	C	Exogenously	D

	A	B	C	D
(a)	Exogenously	Endogenously	Asexual	Absent
(b)	Exogenously	Exogenously	Asexual	Absent
(c)	Endogenously	Exogenously	Sexual	Present
(d)	Endogenously	Endogenously	Sexual	Present

15. Aquatic fungi are placed in class:
(a) Phycomycetes (b) Ascomycetes
(c) Basidiomycetes (d) Deuteromycetes

16. Nuclear dimorphism is a characteristic feature of:
(a) Euglenoids (b) Ciliates
(c) Zooflagellates (d) Sarcodians

17. The protein coat of the virus is made up of subunits called:
(a) Shell (b) Envelope
(c) Capsomere (d) Capsid

18. Which feature of archaebacteria enables them to survive extreme environmental conditions?
(a) Mode of nutrition (b) Nucleoid
(c) Ribosomal proteins (d) Cell wall

19. In plants, the symptoms caused by virus can be:
A. Mosaic formation
B. Leaf rolling and leaf curling
C. Yellowing and vein clearing
(a) Only (A)
(b) Only (B) and (C)
(c) Only (A) and (C)
(d) (A), (B) and (C)

20. Where there are two haploid nuclei per cell in some fungi before the formation of diploid, the stage is called:
(a) Diplotene (b) Diplophase
(c) Dikaryophase (d) Bivalent

21. Diatoms do not decay easily because:
(a) They have siliceous walls
(b) Their body is impervious to water
(c) They are abundant in saline water
(d) They are chitinous

22. Select the incorrectly matched pair:
(a) Methanogens – Archaebacteria
(b) Desmides – Chrysophyta
(c) *Trypanosoma* – Ciliated protozoan
(d) Club fungi – Basidiomycetes

23. Which organism is responsible for making the sea appear red by rapid multiplication?
(a) *Euglena* (b) *Noctiluca*
(c) *Gonyaulax* (d) *Physarum*

24. Which disease is not caused by fungus?
(a) White rust (b) Red rust
(c) Late blight (d) Ergot

25. Consider the following statements and select the correct ones for viroids:
A. Discovered by T Diener
B. They are naked S DNA and devoid of protein coat
C. Similar to the size of viruses

D. Causes potato spindle tuber disease
(a) (A) and (B) (b) (A), (B) and (D)
(c) (A) and (D) (d) (B) and (C)

26. Which of the following sets of fungi belong to the same class?
(a) *Alternaria, Colletotrichum, Aspergillus*
(b) Yeast, Neurospore, *Claviceps*
(c) Morels, truffles, mushroom
(d) *Puccinia, Ustilago, Phytophthora*

27. A parasitic fungi which causes black rust of wheat does not show:
(a) Chitinous cell wall
(b) Branched septate mycelium
(c) Asexual reproduction
(d) Prolonged dikaryophase

28. Select the incorrect statement with respect to the given figure:

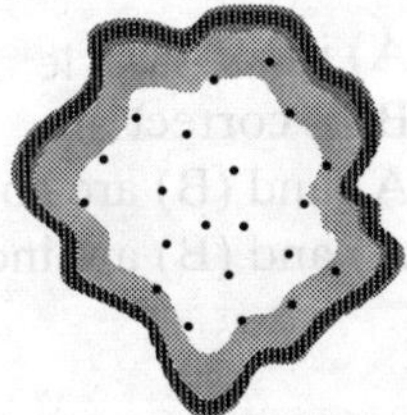

(a) Spores are produced under unfavourable conditions
(b) Spore dispersal occurs by air current
(c) Possesses characteristics of both plants and animals
(d) Possesses true cellulosic cell wall in vegetative phase

29. Select the correct match:

(a)	Plant virus	– Generally as dS DNA
(b)	"Contagium vivum fluidum"	– Infectious non-living fluid
(c)	Viruses, viroids and lichens	– Acellular organisms
(d)	*Cuscuta*	– Partially heterotrophic plant

30. Which of the following statements about heterocyst cells is incorrect?
(a) Site of nitrogen reduction
(b) Presence of thick cell wall
(c) Absence of carbon reduction
(d) Presence of PSII

31. All prokaryotic cells lack:
(a) Cell wall
(b) Well-defined nucleus
(c) Membrane-bound cell organelles
(d) Both (b) and (c)

32. Which one exhibits the character of a protozoan during one phase of its life cycle and the character of a fungi in another phase of its life cycle?
(a) Water molecules (b) Slime moulds
(c) Diatoms (d) Sporozoans

33. Protista differ from Monera in having:
(a) Cell wall
(b) Flagella
(c) Nuclear membrane
(d) Autotrophic nutrition

34. State true (T) or false (F) for the following statements and select the correct option:
A. Fungi prefer to grow in dark, warm and humid places
B. Members of class Phycomycetes could be found in aquatic habitats
C. All prokaryotes are heterotrophic and live as predators or parasites
D. Diatoms are microscopic and float actively in water current

	A	B	C	D
(a)	F	F	T	T
(b)	T	F	F	T
(c)	T	T	F	F
(d)	T	T	T	F

35. Spores of slime moulds:
(a) Are always diploid
(b) Have cellulosic cell wall
(c) Are formed under favourable conditions
(d) All of the above

36. Edible sac fungi are:
(a) Morels (b) *Agaricus*
(c) Toadstool (d) All except (c)

37. All the given features are common to members of Ascomycetes and Deuteronmycetes, except:
(a) Branched and septate mycelium
(b) Reproduce by conidia formation
(c) Heterotrophic mode of nutrition
(d) Endogenously produced sexual spores

38. Which of the following does not have coenocytic mycelium?
(a) *Rhizopus* (b) Dung mould
(c) *Albugo* (d) Blue-green moulds

39. Identify the mismatched pair:

(a) Atmospheric nitrogen fixation – *Nostoc*
(b) Production of antibiotics – Heterotrophic bacteria
(c) Making of curd from milk – Autotrophic bacteria
(d) Source of LSD or ergot – *Claviceps*

40. Gritty soil of commercial use is formed by cell wall deposits of:
(a) Diatoms (b) Dinoflagellates
(c) Pyrrophytes (d) Euglenoids

41. The given statements describe a group of organisms:
A. Absence of locomotary structure
B. Endoparasites
C. Infectious stage is spore-like structure
Which of the following groups is referred to here?
(a) Slime moulds (b) Ciliated protozoans
(c) Sporozoans (d) Flagellated protozoans

42. Find the incorrect match:
(a) *Puccinia graminis tritici* – Black rust of wheat
(b) *Ustilago nuda tritici* – Smut of wheat
(c) *Colletotrichum* – Red rot of sugarcane
(d) *Alternaria solani* – Late blight of potato

43. The longest dikaryotic phase is present in members of class:
(a) Ascomycetes (b) Basidiophycetes
(c) Deuteromycetes (d) Phycomycetes

44. Lichens are very sensitive to:
(a) Sewage (b) N_2
(c) CO (d) SO_2

45. Read the following statements and select the correct option:
A. *Euglena* is considered as the connecting link between plants and animals
B. In the absence of sunlight, *Euglena* starts photosynthesis, while in the presence of sunlight, it behaves like a heterotroph
(a) Only (A) is correct
(b) Only (B) is correct
(c) Both (A) and (B) are correct
(d) Both (A) and (B) are incorrect

ANSWER KEYS

Exercises

1. (b)	**2.** (c)	**3.** (d)	**4.** (a)	**5.** (d)	**6.** (a)	**7.** (d)	**8.** (b)	**9.** (a)	**10.** (b)
11. (b)	**12.** (c)	**13.** (a)	**14.** (a)	**15.** (c)	**16.** (a)	**17.** (a)	**18.** (d)	**19.** (b)	**20.** (a)
21. (c)	**22.** (b)	**23.** (c)	**24.** (c)	**25.** (a)	**26.** (a)	**27.** (c)	**28.** (d)	**29.** (b)	**30.** (d)
31. (d)	**32.** (b)	**33.** (a)	**34.** (c)	**35.** (b)	**36.** (d)	**37.** (c)	**38.** (c)	**39.** (b)	**40.** (c)
41. (c)	**42.** (b)	**43.** (a)	**44.** (a)	**45.** (d)	**46.** (b)	**47.** (a)	**48.** (b)	**49.** (b)	**50.** (a)
51. (a)	**52.** (d)	**53.** (d)	**54.** (b)	**55.** (d)	**56.** (a)	**57.** (d)	**58.** (d)	**59.** (b)	**60.** (a)
61. (c)	**62.** (a)	**63.** (b)	**64.** (c)	**65.** (d)	**66.** (b)	**67.** (c)	**68.** (a)	**69.** (c)	**70.** (b)
71. (b)	**72.** (d)	**73.** (a)	**74.** (a)	**75.** (c)	**76.** (b)	**77.** (b)	**78.** (d)	**79.** (b)	**80.** (c)
81. (c)	**82.** (a)	**83.** (d)	**84.** (c)	**85.** (d)	**86.** (a)	**87.** (c)	**88.** (d)	**89.** (a)	**90.** (d)
91. (b)	**92.** (b)	**93.** (c)	**94.** (a)	**95.** (d)	**96.** (c)	**97.** (b)	**98.** (a)	**99.** (d)	**100.** (c)
101. (d)	**102.** (b)	**103.** (c)	**104.** (d)	**105.** (a)	**106.** (d)	**107.** (d)	**108.** (a)	**109.** (c)	**110.** (d)
111. (c)	**112.** (c)	**113.** (c)	**114.** (d)	**115.** (c)	**116.** (b)	**117.** (d)	**118.** (d)	**119.** (b)	**120.** (c)
121. (a)	**122.** (b)	**123.** (c)	**124.** (d)	**125.** (d)	**126.** (b)	**127.** (d)	**128.** (b)	**129.** (a)	**130.** (d)
131. (d)	**132.** (b)	**133.** (d)	**134.** (c)	**135.** (c)	**136.** (d)	**137.** (a)	**138.** (b)	**139.** (b)	**140.** (d)
141. (d)	**142.** (a)	**143.** (d)	**144.** (c)	**145.** (c)	**146.** (a)	**147.** (b)	**148.** (d)	**149.** (a)	**150.** (a)
151. (c)	**152.** (b)	**153.** (d)	**154.** (b)	**155.** (d)	**156.** (d)	**157.** (a)	**158.** (a)		

NCERT Exemplar Questions

1. (b)	**2.** (a)	**3.** (b)	**4.** (d)	**5.** (c)	**6.** (b)	**7.** (b)	**8.** (c)	**9.** (a)	**10.** (c)
11. (d)	**12.** (d)								

Practice Test Paper

1. (b)	**2.** (c)	**3.** (b)	**4.** (d)	**5.** (c)	**6.** (a)	**7.** (c)	**8.** (b)	**9.** (a)	**10.** (b)
11. (c)	**12.** (d)	**13.** (c)	**14.** (b)	**15.** (a)	**16.** (b)	**17.** (c)	**18.** (d)	**19.** (d)	**20.** (c)
21. (a)	**22.** (c)	**23.** (c)	**24.** (b)	**25.** (c)	**26.** (b)	**27.** (c)	**28.** (d)	**29.** (c)	**30.** (d)
31. (d)	**32.** (b)	**33.** (c)	**34.** (c)	**35.** (b)	**36.** (a)	**37.** (d)	**38.** (d)	**39.** (c)	**40.** (a)
41. (c)	**42.** (d)	**43.** (b)	**44.** (d)	**45.** (c)					

SOLUTIONS

Introduction to Classification

1. (A) and (B) are correct. Aristotle divided animals into two groups: those that had red blood and those that did not.

2. Linnaeus proposed the Two Kingdom system of classification.

3. Option (d) is correct.

4. RH Whittaker

5. The main criterion for classification used by Whittaker included cell structure, body organisation, mode of nutrition, reproduction and phylogenetic relationship.

6. Option (a) is correct.

7. In the Five Kingdom system of classification, multicellular organisms with heterotrophic mode of nutrition are placed in Fungi and Animalia.

8. Eukaryotes belong to four kingdoms (Protista, Fungi, Plantae and Animalia)

9. A – RH Whittaker, B – *Paramecium*, C – Protista

10. A – Prokaryotes, B – Present, C – Cellular, D – Loose tissue

11. Protista is a kingdom of unicellular eukaryotic organisms. Tissue-level organisation is not present in the members of kingdom Protista.

Kingdom Monera

12. Monera is a kingdom of unicellular, colonial or multicellular (such as filamentous BGA) prokaryotic organisms and include bacteria, archaebacteria and *Mycoplasma*.

13. A – Bacteria, B – Bacillus

14. Option (a) is correct.

15. The cell wall of monerans is made up of peptidoglycan or murein. It is a polymer consisting of sugars and amino acids.

16. Monera as a group shows the most extensive metabolic diversity.

17. (A) and (B) are correct. Bacteria occur almost everywhere, they also live in extreme habitats such as hot springs, deserts, snowy areas and deep oceans. Many of them live in or on the body of other organisms as parasites.

18. Option (d) is correct.

19. In prokaryotes, respiratory enzymes are present in mesosomes (folding of plasma membrane).

20. Bacteria mainly reproduce by binary fission.

21. *Spirillum* bacterium is a spiral-shaped bacteria with flagella at both ends.

22. (B) and (D) are false. Bacteria are grouped under four categories based on their shape (Cocci, Bacilli, Vibrio and Spirilla). Most bacteria are heterotrophic and the majority are important decomposers.

23. Heterotrophs

24. (B) and (C) are correct.

25. Cyanobacteria often form bloom in polluted water bodies.

26. Cyanobacteria show oxygenic photosynthesis. It is believed that the origin of oxygen on earth was due to cyanobacteria.

27. In archaebacteria, the sequence of 16S rRNA genes is different from eubacteria.

28. Cyanobacteria can perform nitrogen fixation in free-living conditions.

29. Heterocyst are specialised for nitrogen fixation.

30. Cholera is caused by *Vibrio cholerae*.

31. Archaebacteria live in some of the most harsh habitats such as extreme salty areas, hot springs and marshy areas. They are characterised by the presence of different cell wall structure.

32. Methanogens are present in the gut of several ruminant animals and are responsible for the production of methane (bio gas) from the dung of these animals.

33. Only (A) is correct. *Mycoplasma* can survive without oxygen, so they are facultative anaerobes. Due to the absence of peptidoglycan cell wall, they are resistant to penicillin, but sensitive to other antibiotics like streptomycin and chloramphenicol.

34. *Epulopiscium fishelsoni*

35. A – Methagens, B – Chemosynthetic autotrophs

36. Influenza, commonly known as the flu, is an infectious disease caused by an influenza virus.

37. Only (A) and (B) are correct. Heterocyst is a site of nitrogen reduction (N_2 fixation). Heterocyst lacks PS II, so photolysis of water cannot occur. PS I is present and forms ATP for nitrogen fixation.

38. Penicillin

39. Cell wall structure

40. Methanogens

41. Bacteria show a sort of sexual reproduction by adopting a primitive type of DNA transfer from one bacterium to the other.

42. Halophiles are most abundant in salty areas.

43. A – Heterotrophic, B – Chemosynthetic autotrophs

44. Only (A) and (C) are correct. Many *Mycoplasma* are pathogenic in plants and animals.

45. Complementary chromatic adaptation is shown by cyanobacteria. They can change their colour according to different wavelengths of light falling on them; this feature is due to the presence of phycobilin.

46. Transformation

Kingdom Protista

47. All single-celled eukaryotes are placed under kingdom Protista.

48. The boundaries of kingdom Protista are not well defined as what may be a photosynthetic protist to one biologist may be a plant for another.

49. Kingdom Protista

50. Silica cell wall of dead diatoms are nearly indestructible.

51. Protistans are mostly aquatic organisms.

52. Option (d) is correct.

53. Option (d) is correct.

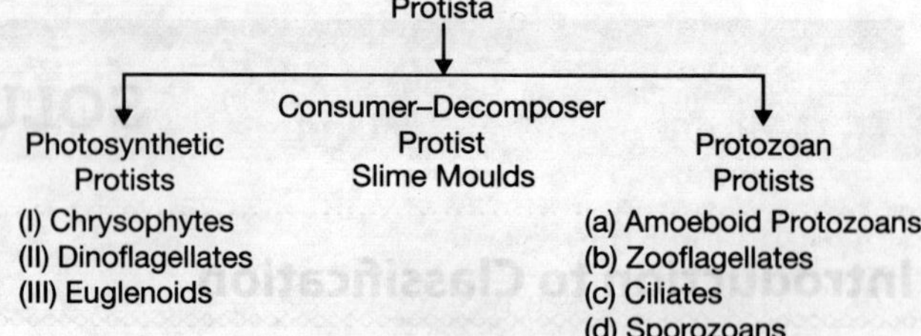

54. Presence of chlorophyll a, chlorophyll c

55. Slime mould (consumer–decomposer protist)

56. Dinoflagellates show holophytic mode of nutrition.

57. *Trypanosoma* is a flagellated protozoan.

58. In kingdom Protista, movement is brought about by flagella, cilia and pseudopodia.

59. Diatoms are chief producers in the ocean.

60. Slime moulds

61. Diatoms, Dinoflagellates and Euglenoids

62. Only (A) is correct. Some dinoflagellates release toxins and may even kill other marine animals such as fishes.

63. Red dinoflagellates like *Gonyaulax* undergo rapid multiplication so that they make the sea appear red, called red tides.

64. In dinoflagellates, the cell wall contains stiff cellulose plates and pectin.

65. Diatomaceous earth is an indestructible frustule of diatoms. Being gritty, this is used in polishing and the filtration of oil and syrups.

66. Mixotrophy

67. Desmides – include under Chrysophyta along with diatoms

68. *Gonyaulax*

69. The pigments in euglenoids are identical to those present in higher plants.

70. A and C are correct. Vegetative phase of slime moulds resemble animals because they lack cell wall under unfavourable conditions. *Plasmodium*, under differentiation, forms fruiting bodies bearing spores at their tips.

71. Slime moulds have cellulosic cell wall around spores.

72. *Paramecium*
73. Air current
74. Under suitable conditions, slime moulds form aggregations to form *Plasmodium,* which is without cell wall and represents the vegetative phase.
75. Protozoans
76. 4 (Amoeboid protozoan, Ciliates, Zooflagellates, Sporozoans)
77. All protozoans are heterotrophs and live as predators or parasites.
78. Option (d) is correct.
79. Marine forms have silica shells on their surface.
80. Option (c) is correct.
81. Flagellated protozoans are either free living or parasites.
82. *Plasmodium* (malarial parasite) is a sporozoan.
83. *Paramecium*
84. In protists, sexual reproduction involves formation and fusion of gametes.
85. Chrysophytes (Diatoms) and marine amoeboid protozoans
86. A – Diatoms, B – Slime moulds, C – Euglenoids, D – Sarcodines (amoeboid protozoans)
87. Slime moulds undergo aggregation under favourable conditions to form *Plasmodium.*

Kingdom Fungi

88. (A), (B) and (C) are correct.
89. Sexual reproduction involves three stages in sequence: Plasmogamy, Karyogamy and Meiosis.
90. Oospores are sexual spores of class Oomycetes.
91. Glycogen and oil
92. Fungi are heterotrophs, either saprotrophs or parasites.
93. In fungi, vegetative reproduction occurs by three methods – fragmentation, budding and fission.
94. Six statements (i, ii, iii, iv, vi, vii) are correct. Fungi show great diversity of morphology and habitat, and cytoplasmic ribosomes are of the 80S type.
95. Mode of nutrition is not a criterion of fungal classification.
96. Phycomycetes
97. Ascospores and basidiospores are thin-walled, non-motile, sexual spores and meiospores (N). Ascospores are produced endogenously while basidiospores are exogenous in origin.
98. *Ustilago* – Smut fungi, *Aspergillus* – Laboratory weed, Toadstools – Poisonous mushroom, *Alternaria* – Absence of sexual stage
99. Hyphae of members of class Basidiomycetes are septate, with a complex pore called dolipore.
100. *Neurospora* (pink bread mould) is used extensively in biochemical and genetic work.
101. Option (d) is correct.
102. Plasmogamy is the first step in the sexual cycle in Fungi.
103. Yeast vegetatively reproduce by budding.
104. In members of class Deuteromycetes, sexual reproduction is either absent or not discovered.
105. In class Ascomycetes, asexual spores are conidia, produced exogenously on the special mycelium called conidiophores.
106. Toadstool (poisonous mushroom)
107. The fungi shown in the figure is *Aspergillus.*
108. Sac fungi (Ascomycetes)
109. (A) and (B) are correct.
110. *Claviceps* belongs to class Ascomycetes.
111. Mushroom, presence of prolonged dikaryotic stage (n+n) and absence of asexual spores
112. In class Deuteromycetes, a large number of members are decomposers of litter and help in mineral cycling.
113. *Puccinia* causes black rust of wheat.
114. Ascomycetes and Basidiomycetes
115. A – Conidia, B – Conidiophore
116. Class – Basidiomycetes
117. In *Alternaria* (class Deuteromycetes), sexual reproduction is absent.
118. Both (a) and (b) are correct.
119. Sac fungi (class Ascomycetes)
120. *Puccinia* and *Ustilago* are parasitic species of class Basidiomycetes.
121. A – Asexual, B – Endogenously, C – Sexual, D – Exogenously
122. *Rhizopus* and *Mucor* species have coenocytic mycelium (aseptate and multinucleate).

Kingdoms Plantae and Animalia

123. Insectivorous plants are partially heterotrophs.

124. Venus flytrap, bladderwort and *Napenthes* are insectivorous plants. Their nitrogen need is fulfilled by sucking the juice of insects.

125. *Cuscuta* is a totally stem parasite.

126. Plants cells are characterised by the presence of distinct nucleus, prominent chloroplast and cellulosic cell wall.

127. Option (d) is correct.

128. Three statements (ii, iii and iv) are correct. Most animal forms are capable of locomotion and only higher forms show elaborate sensory and neuromotor mechanism.

Viruses, Viroids, Prions and Lichen

129. Pasteur coined the term virus.

130. Virus means venom or poisonous fluid.

131. Viruses are non-cellular and so did not find a place in the classification.

132. MW Beijerinek demonstrated that the extract of the infected plants of tobacco could cause infection in healthy plants, and called it contagium vivum fluidum (infectious living fluid).

133. (A), (B) and (C) are correct.

134. A – Stanley, B – Protein

135. Bacteriophages usually have double-stranded DNA as genetic material.

136. In viruses, genetic material is infectious.

137. Viruses are acellular or non-cellular.

138. Lacking protoplast

139. Potato spindle tuber disease is caused by viroids. Viroids are infectious ssRNA molecules.

140. Only (D) is false. The protein coat of the virus is made up of small subunits called capsomeres.

141. AIDS

142. In virus, nucleic acid (DNA or RNA, but not both) is the infectious component.

143. In general, plant viruses have ssRNA as genetic material.

144. In lichens, the fungal partner (mycobiont) does not perform the function of nitrogen reduction.

145. Citrus canker and crown gall are caused by bacteria.

146. Viral disease

147. Viroids

148. (A), (B), (C) and (D) are correct.

149. A – Capsid, B – Capsomere

150. A – Head, B – Collar, C – Sheath, D – Tail fibres

151. Lichens are a symbiotic association between algae and fungi. The algal component is called phycobiont and the fungal component is called mycobiont.

152. The RNA of the viroid has low molecular weight.

153. Prions are infectious protein particles without nucleic acids.

154. Virusoids cannot replicate on their own and require the help of a helper virus; therefore, they are called satellite viruses.

155. Bacteriophages have tail pins and tail fibres for attachment on the host surface.

156. Prions are infectious proteins without nucleic acid. The most notable disease caused by prions is bovine spongiform encephalopathy (BSE), commonly called mad cow disease.

157. Mycorrhiza is symbiotic association between fungus and the roots of higher plants (gymnosperms and angiosperms).

158. Lichens are very good pollution indicators (air pollution, especially SO_2 pollution), they do not grow in polluted areas.

3 Plant Kingdom

EXERCISES

Classification System

1. Artificial system of classification given by Linnaeus considers:
   (a) Androecium characters
   (b) Ultrastructure and anatomy
   (c) Cell wall structure
   (d) Evolutionary characters

2. Artificial classification is based upon:
   (a) Habit and habitat
   (b) Vegetative characters
   (c) One or few morphological characters
   (d) All of the above

3. Natural system of classification was based on all, except:
   (a) External and internal features
   (b) Anatomy and ultrastructure
   (c) Phylogeny
   (d) Embryology and phytochemistry

4. Who among the following gave the classification for flowering plants, which is based on internal features like ultrastructure, anatomy and embryology:
   (a) Linnaeus
   (b) George Bentham and Joseph Dalton Hooker
   (c) Ernst Mayr
   (d) Eichler

5. In which of the following systems of classification is equal weightage given to both vegetative and sexual characteristics?
   (a) Artificial system (b) Phylogenetic system
   (c) Natural system (d) Both (a) and (c)

6. All of the following are included under cytotaxonomy, except:
   (a) Chromosome number
   (b) Nucleotide sequence in DNA
   (c) Chromosome structure
   (d) Chromosome behaviour during cell division

7. Select the incorrect statement with respect to numerical taxonomy:
   (a) All possible characters are compared by computers
   (b) Numbers and codes are assigned to all the characters
   (c) Only one character can be considered at a time
   (d) Organisation and analysis of date forms the core of this taxonomy

8. Natural system of classification is based on all, except:
   (a) Ultrastructure (b) Phytochemistry
   (c) Embryology (d) Phylogeny

9. Examine the following matches:
   (i) Chemotaxonomy – Chemical nature of proteins
   (ii) Cytotaxonomy – Chromosomal behaviour
   (iii) Karyotaxonomy – Chromosome structure
   (iv) Numerical taxonomy – Uses of number and code
   (v) Phenetics – Adansonian taxonomy
   How many of these are correctly matched?
   (a) Two (b) Three
   (c) Five (d) Four

10. Phylogenetic classification systems:
   (a) Are based on a few morphological features
   (b) Were used by Bentham and Hooker
   (c) Separated closely related species
   (d) Assume that organisms belonging to the same taxa have a common ancestor

11. The system of classification that evaluates the resemblances and differences or primitiveness and advancement through statistical methods based on the character obtained from all disciplines of biology followed by computer analysis is:
(a) Classical taxonomy
(b) Numerical taxonomy
(c) Karyotaxonomy
(d) Chemotaxonomy

12. Read the following statements carefully and select the correct option:
A: Phylogenetic classification is based on possible evolution of different traits
B: The artificial system of classification gives information about both natural and phylogenetic relationships
(a) Only A is correct
(b) Only B is correct
(c) Both (A) and (B) are correct
(d) Both (A) and (B) are incorrect

13. A system of classification in which a large number of traits are considered is:
(a) Natural system (b) Artificial system
(c) Phylogenetic system (d) Cladistic system

14. Natural system of classification was proposed by:
(a) Linneaus
(b) Bentham and Hooker
(c) Engler and Prantl
(d) Aristotle

Algae (Chlorophyceae, Pheophyceae, Rhodophyceae)

15. Select the incorrect statement with respect to algae:
(a) Chlorophyll-containing, simple, thalloid; largely aquatic organisms
(b) They are mostly aquatic
(c) The form and size is highly variable
(d) Algae form associations with fungi called mycorrhizae

16. Select the mismatched pair:
(a) Unicellular – *Chlamydomonas*
(b) Colonial – *Volvox*
(c) Filamentous – *Ulothrix*
(d) Kelps – *Porphyra*

17. Asexual reproduction occurs by production of different types of spores in algae. The most common types of spores produced by algae are:
(a) Zoospores (b) Aplanospores
(c) Hypnospores (d) Akinetes

18. Oogamy is seen in:
(a) *Volvox* (b) *Fucus*
(c) *Ulothrix* (d) Both (a) and (b)

19. Sex organs of algae are generally:
(a) Multicellular and jacketed
(b) Multicellular and non-jacketed
(c) Unicellular and non-jacketed
(d) Unicellular and jacketed

20. In green algae, the chloroplast can be:
(a) Discoid or plate-like
(b) Cup-shaped
(c) Spiral or ribbon-shaped
(d) All of the above

21. Pyrenoids in green algae:
(a) Are inclusion bodies
(b) Are located in the chloroplast
(c) Contain protein besides starch
(d) All except (a)

22. Floridean starch is the reserve food material of:
(a) *Laminaria* and *Fucus*
(b) *Porphyra* and *Polysiphonia*
(c) *Chlorella* and *Spirullina*
(d) *Dictyota* and *Porphyra*

23. Which of the following is mismatched?
(a) Kelps: Red algae form massive plant bodies
(b) *Udorina*: Anisogamous sexual reproduction
(c) *Eudorina*: Reproduce by non-motile spores
(d) *Ectocarpus*: Simple branched filamentous brown algae

24. Select true (T) or false (F) for the given statements with respect to the characters of brown algae and select the correct option:
A. They show great variation in size and form
B. Food is stored as complex carbohydrates, which is very similar to amyloprotein and glycogen
C. Zoospores are pyriform with two laterally inserted heterokont flagella
D. Sexual reproduction is of oogamous type and accompanied by complex post-fertilisation development

	A	B	C	D
(a)	F	T	F	T
(b)	F	T	F	F
(c)	T	F	T	F
(d)	T	F	T	T

25. Which of the following unicellular green algae is/are used as food supplements by space travellers?
(a) *Chlorella* (b) *Spirulina*
(c) *Porphyra* (d) Both (a) and (b)

26. Algin and carrageenan hydrocolloids are obtained, respectively, from:
(a) Red algae and brown algae
(b) Brown algae and red algae
(c) Brown algae and green algae
(d) Green algae and red algae

27. A research student collected certain algae and found that its cell contained chlorophyll a and chlorophyll c, as well as fucoxanthim. The algae belongs to class ________:
(a) Rhodophyceae (b) Cyanophyceae
(c) Phaeophyceae (d) Chlorophyceae

28. Which of the following is not a unicellular algae?
(a) *Spirulina* (b) *Ulothrix*
(c) *Chlorella* (d) Both (a) and (b)

29. Identify the algae on the basis of the given information and select the correct option:

Reserve food material is very similar to amylopectin and glycogen.

Cell wall is made up of cellulose, pectin and polysulphated esters.
(a) *Eudorina* (b) *Laminaria*
(c) *Ectocarpus* (d) *Polysiphonia*

30. The given figure is an example of brown algae. A, B and C refer to:

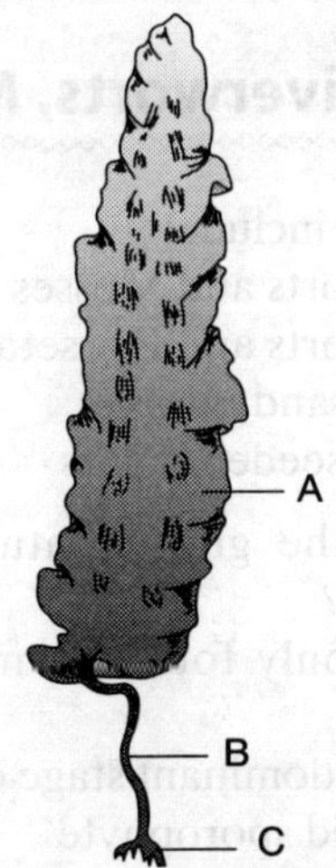

	A	B	C
(a)	Frond	Stipe	Holdfast
(b)	Stipe	Holdfast	Frond
(c)	Frond	Holdfast	Stipe
(d)	Stipe	Frond	Holdfast

31. Reproduction by pyreform zoospore with laterally inserted unequal flagella is a feature related to an algae class that is:
(a) Phaeophyceae (b) Rhodophyceae
(c) Chlorophyceae (d) Bacillariophyceae

32. Identify the mismatched pair with respect to algae:

	Algae	Pigment	Number of flagella
(a)	*Chlamydomonas*	Chl a, b	2, equal, apical
(b)	*Laminaria*	Chl a, c	2, unequal, lateral
(c)	*Porphyra*	Chl a, d	Non-flagellate
(d)	*Fucus*	Chl a, c	Non-flagellate

33. Characteristic colour of class Rhodophyceae is due to the predominance of:
(a) Chlorophyll a (b) Fucoxanthin
(c) R-phycoerythrin (d) Chlorophyll d

34. Identify the given figure and select the correct statement:

(a) It has major pigments such as chlorophyll a and d
(b) Stores food material in the form of laminarin and mannitol
(c) Rich in proteins and used as food supplements even by space travellers
(d) It is found primarily in freshwater habitats

35. Consider the following statements:
A. Vegetative cells have a cellulosic wall usually covered on the outside by a gelatinous coating of algae

B. Zoospores are characterised by the presence of 2–8 isokont anteriorly attached flagella

C. Reserve food material is Floridean starch, which is very similar to amylopectin and glycogen

These statements are correct for which class of algae?

	A	B	C
(a)	Chlorophyceae	Phaeophyceae	Rhodophyceae
(b)	Phaeophyceae	Chlorophyceae	Rhodophyceae
(c)	Rhodophyceae	Chlorophyceae	Phaeophyceae
(d)	Phaeophyceae	Rhodophyceae	Chlorophyceae

36. Which among the following has a cell wall made up of polysulphated esters?
(a) Rhodophyceae (b) Chlorophyceae
(c) Phaeophyceae (d) Both (a) and (c)

37. Presence of phycocolloid-like agar, photosynthetic pigment like chlorophyll a, chlorophyll d and phycobilins and Floridean starch as reserve food are the characteristic features of:
(a) Chlorophyaceae (b) Phaeophyceae
(c) Phycophyceae (d) Rhodophyceae

38. Non-flagellated isogamy is seen in:
(a) *Spirogyra* (b) *Chlamydomonas*
(c) *Eudorina* (d) *Volvox*

39. The algae shown in the figures belong to class:

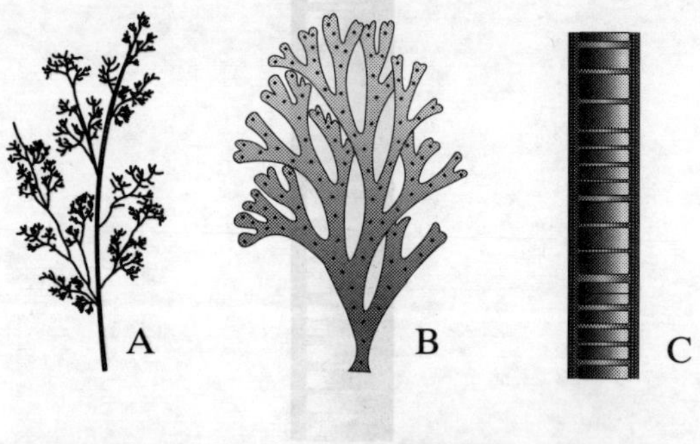

	A	B	C
(a)	Rhodophyceae	Chlorophyceae	Phaeophyceae
(b)	Chlorophyceae	Rhodophyceae	Phaeophyceae
(c)	Rhodophyceae	Phaeophyceae	Chlorophyceae
(d)	Rhodophyceae	Phaeophyceae	Cyanophyceae

40. At least half the total CO_2 fixation on earth is carried out through photosynthesis by ________:
(a) Marine algae (b) Angiosperms
(c) Bryophytes (d) Both (a) and (c)

41. Species of *Porphyra*, *Laminaria* and *Sargassum* are among the ________ species of algae:
(a) 90 (b) 70
(c) 25 (d) 10

42. Two species of algae are said to be closely related when:
(a) They reproduce asexually
(b) They are found in the same habitat
(c) They are photosynthetic
(d) They have the same type of photosynthetic pigments

43. Identify the algae given below and choose the odd one out with respect to the algae class:

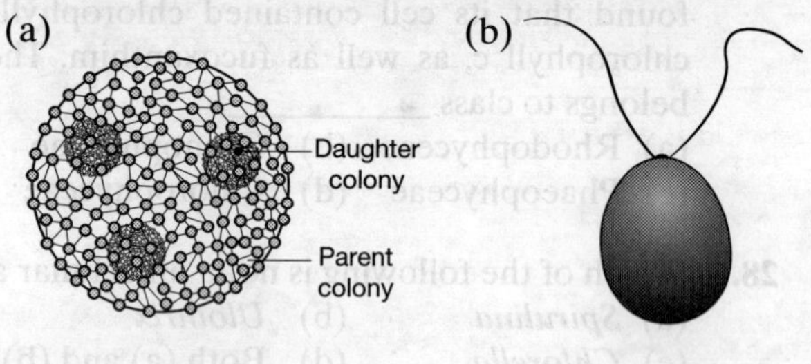

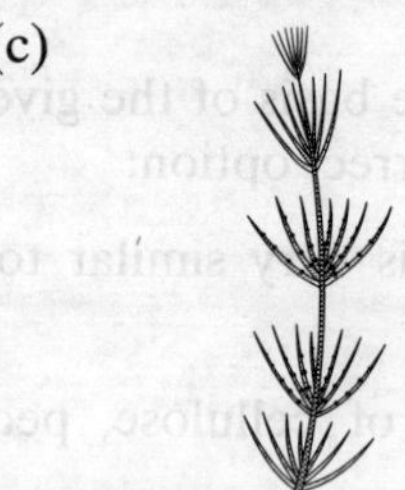

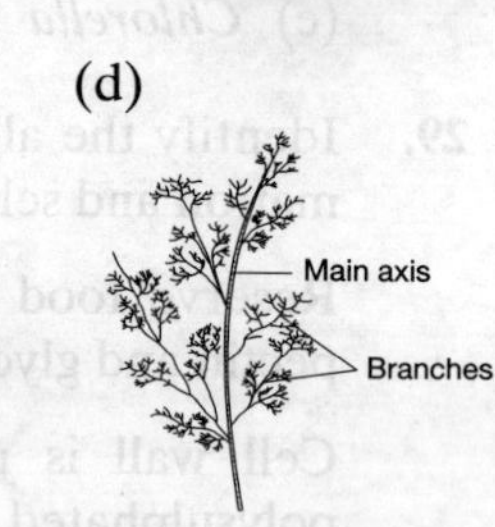

44. Internal fertilisation is observed in all of these, except:
(a) *Spirogyra* (b) *Fucus*
(c) *Chlamydomonas* (d) *Volvox*

Bryophyta (Liverworts, Mosses)

45. Bryophytes include:
(a) Liverworts and Mosses
(b) Liverworts and Horsetail
(c) Mosses and Ferns
(d) Naked seeded plants

46. Which of the given features is not related to bryophytes?
A. Commonly found in moist, shaded areas in hills
B. The predominant stage of the life cycle of moss is diploid sporophyte
C. Plant body is more differentiated than that of algae
D. They were the first land vascular plants

(a) (A) and (B) (b) (B) and (C)
(c) (B) and (D) (d) (C) and (D)

47. Bryophytes are called "amphibians of the plant kingdom" because:
(a) They need water for seed dispersal
(b) The plants live in the soil but are dependent on water for sexual reproduction
(c) Vascular tissue xylem and phloem are absent
(d) They are found in water only

48. Bryophytes differ from algae in:
(a) Being aquatic
(b) Having a more differentiated plant body
(c) Having primitive vascular tissue
(d) Both (b) and (c)

49. The unique feature of bryophytes compared to other plant groups is that:
(a) They are found in cool, damp and shady places
(b) Their sex organs are multicellular and jacketed
(c) Their sporophyte is partially dependent on the gametophyte
(d) They require water for sexual reproduction

50. Complete the following sequences by selecting the correct option for *Funaria*:

Sporophyte —**A**→ Spore —Germination→ **B**

	A	**B**
(a)	Mitosis	Primary protonema
(b)	Reductional division	Leafy stage
(c)	Meiosis	Primary protonema
(d)	Meiosis	Thalloid gametophyte

51. *Funaria* consists of upright slender axes bearing ______ arranged leaves:
(a) Spirally (b) Oppositely
(c) Alternately (d) Superimposed

52. Protonema and leafy stage are the predominant stages of the life cycle of:
(a) *Marchantia* (b) *Riccia*
(c) *Funaria* (d) *Polysiphonia*

53. Which of the following bryophytes has the simplest sporophyte with no foot and seta?
(a) *Marchantia* (b) *Funaria*
(c) *Anthoceros* (d) *Riccia*

54. The predominant stage in the life cycle of bryophyte is:
(a) Gametophyte (b) Sporophyte
(c) Primary protonema (d) Prothallus

55. In *Marchantia*, gemmae are:
(a) Non-green, multicellular vegetative structures
(b) Green, multicellular sexual bud
(c) Green, multicellular asexual bud
(d) Non-green, multicellular sexual bud

56. Embryonic development in bryophytes takes place in the ______:
(a) Capsule (b) Sporongium
(c) Archegonium (d) Oogonium

57. ______ is a liverwort:
(a) *Funaria* (b) *Marchantia*
(c) *Polytrichum* (d) *Sphagnum*

58. Adult gametophyte in *Funaria* has:
(a) Upright slender axes bearing spirally arranged leaves
(b) Multicellular and branched rhizoids
(c) Multicellular jacketed sex organs
(d) All of the above

59. The juvenile gametophyte of *Funaria*:
(a) Develops directly from the spore
(b) Is called primary protonema
(c) Is the creeping, green, branched, filamentous stage
(d) All of the above

60. Examine the figure given below and select the correct labels A–E:

A
B
C
D
E

	A	B	C	D	E
(a)	Capsule	Seta	Leaves	Main axis	Multicellular rhizoids
(b)	Capsule	Seta	Leaves	Main axis	Unicellular rhizoids
(c)	Strobilus	Seta	Sporophyll	Stem	Roots
(d)	Capsule	Foot	Leaflets	Stem	Adventitious roots

61. Which of the following cannot be a use of bryophytes?
(a) Provide food for herbaceous mammals, birds and other animals
(b) Used in the preparation of ice cream and jellies
(c) Decomposition of rocks to make substrate suitable for the growth of higher plants
(d) *Sphagnum* used as packing material for trans-shipment of living material

62. Which bryophyte is related to fuel production?
(a) *Sphagnum* (b) *Polytrichum*
(c) *Anthoceros* (d) *Pogonatum*

63. Which of the following features is not true for bryophytes?
(a) Sporic meiosis
(b) Independent sporophyte
(c) Dependent sporophyte
(d) Zygotic meiosis

64. Read the following statements carefully and state whether true (T) or false (F):
(a) *Marchantia* is a monoecious bryophyte
(b) Leafy stage in moss develops from secondary protonema as a lateral bud
(c) The sporophyte in moss is more elaborate than that of liverworts
(d) Liverworts have an elaborate mechanism of spore dispersal

	A	B	C	D
(a)	T	T	T	F
(b)	F	F	F	T
(c)	F	T	T	F
(d)	T	T	F	F

65. Which of the following plant species is used as packing material for the trans-shipment of living material?

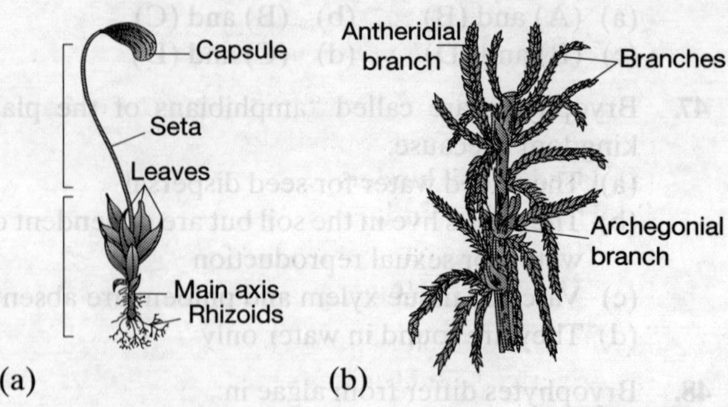

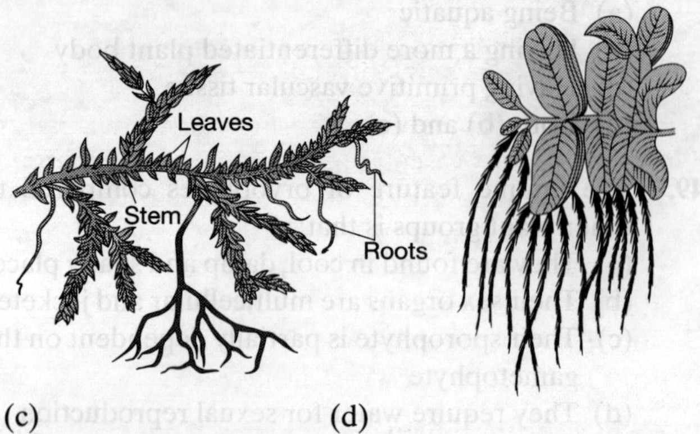

66. Which of the following statements is/are incorrect?
A. Mosses are attached to the soil through multicellular and unbranched rhizoids
B. In mosses, spores are formed after meiosis
C. Mosses are ecologically more important but economically less important
(a) Only (A) and (B) (b) Only (B)
(c) Only (A) (d) (A), (B) and (C)

67. Asexual reproduction in *Marchantia* can occur by:
(a) Spores
(b) Gemmae
(c) Primary protonema
(d) Secondary protonema

68. An elaborate mechanism of spore dispersal is present in:
(a) *Marchantia* (b) *Riccia*
(c) *Chara* (d) *Funaria*

69. Ploidy level in rhizoids, spore, seta and spore mother cells in moss are, respectively:
(a) N, N, 2N, 2N (b) N, 2N, 2N, N
(c) N, N, N, 2N (d) 2N, N, 2N, 2N

Pteridophyta

70. Pteridophytes are:
(a) First land plants
(b) Vascular cryptogams

(c) Non-vascular amphibians
(d) First embryophytes

71. Which of the following is not an economic importance of pteridophytes?
(a) Used as soil binders
(b) Used as air binders
(c) Grown as ornamentals
(d) Source of protein

72. Pteridophytes include:
(a) Liverwort and Mosses
(b) Mosses and Ferns
(c) Ferns and Horsetail
(d) Hornwort and Horsetail

73. A. Pteridophytes were the first terrestrial plants to possess vascular tissue xylem and phloem
B. Leaves in pteridophytes are small in *Selaginella* or large as in fern
(a) Only (A) is correct
(b) Only (B) is correct
(c) Both (A) and (B) are correct
(d) Both (A) and (B) are incorrect

74. Which of the following features is not related to pteridophytes?
A. First terrestrial plant to possess vascular tissue
B. Found in cool, damp and shady places
C. Majority of pteridophytes are heterosporous
D. Gametophytic generation is highly reduced and dependent on photosynthetic gametophyte
(a) (A) and (C) (b) (B) and (D)
(c) (C) and (D) (d) (B) and (C)

75. In fern, spores are formed in:
(a) Sporangium (b) Archegomium
(c) Capsule (d) Oogonium

76. Sporangia bearing leaves are called:
(a) Sporophyll (b) Lax
(c) Cone (d) Strobilus

77. Complete the following sequences by selecting the correct option for *Dryopteris*:

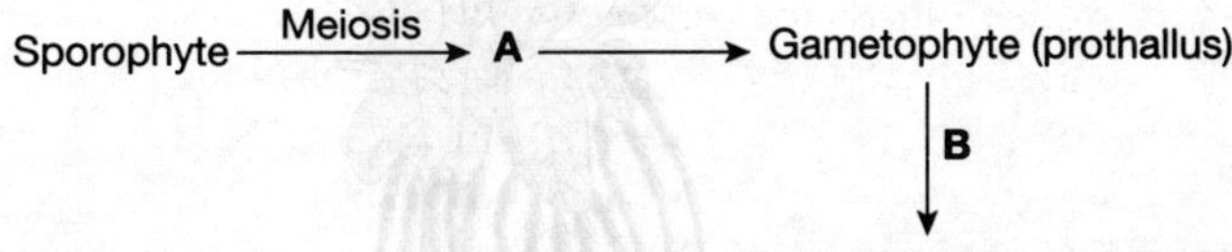

	A	B
(a)	Spore	Mitosis
(b)	Spore	Meiosis
(c)	Protonema	Mitosis
(d)	Sorus	Mitosis

78. Identify the given figure and select the incorrect statement:

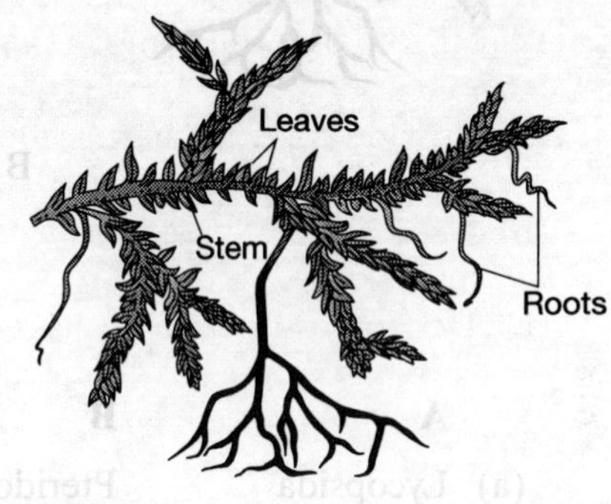

(a) It is heterosporous
(b) It shows seed habit
(c) It is an aquatic fern
(d) It has strobili or cones

79. A water fern which is used as biofertiliser in rice fields is:
(a) *Salvinia* (b) *Azolla*
(c) *Marsilia* (d) *Adiantum*

80. Select the incorrectly matched pair:
(a) *Azolla* – Biofertiliser
(b) *Adiantum* – Walking fern
(c) *Equisetum* – Horsetail
(d) *Dryopteris* – Heterosporous

81. Strobili or cones are found in:
(a) *Selaginella, Equisetum*
(b) *Azolla, Selaginella*
(c) *Equisetum, Sphagnum*
(d) *Selaginella, Salvinia*

82. Prothallus is:
(a) Inconspicuous, multicellular, photosynthetic and thalloid gametophyte
(b) Dominant phase of pteridophyte
(c) Green and multicellular but is partially dependent on sporophyte
(d) Creeping, green filament stage like algae

83. Identify (A) and (B) with respect to pteridophytes and their respective class:

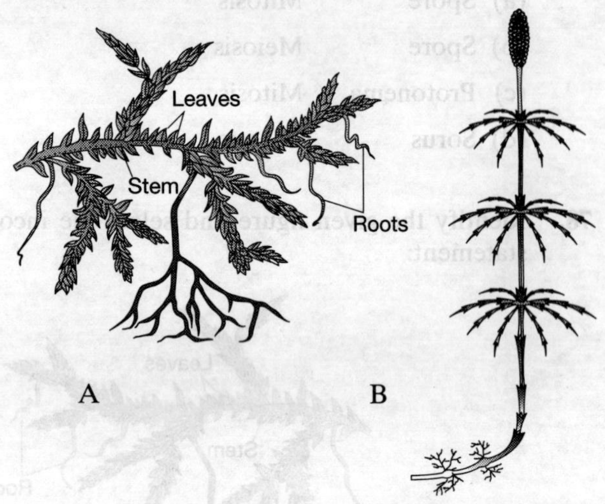

	A	B
(a)	Lycopsida	Pteridopsida
(b)	Lycopsida	Sphenopsida
(c)	Sphenopsida	Pteridopsida
(d)	Sphenopsida	Lycopsida

84. ____________ is considered an important step in evolution:
(a) Free-living gametophyte
(b) Sporic meiosis
(c) Seed habit
(d) Homospory

85. Pteridophyte showing tendency towards seed habit is:
(a) *Pteris* (b) *Selaginella*
(c) *Salvinia* (d) All except (a)

86. Identify the incorrectly matched pair:

	Class	Example	Feature
(a)	Psilopsida	*Psilotum*	Heterosporous
(b)	Lucopsida	*Selaginella*	Seed habit
(c)	Sphenopsida	*Equisetum*	Strobilus
(d)	Pteridopsida	*Dryopteris*	Macrophylls

87. What kind of division will occur in zygote and spore mother cell, respectively, in pteridophyte?
(a) Meiosis and Mitosis
(b) Mitosis and Meiosis
(c) Meiosis and Reductional
(d) Reductional and Equational

88. The distribution of pteridophyte is restricted to a narrow geographical region because:
(a) The gametophyte requires cool, damp, shady places to grow
(b) Water is essential for fertilisation
(c) Motile female gametes
(d) Both (a) and (b)

89. The prothallus of pteridophytes is/has:
A. Small but multicellular
B. Conspicuous
C. Free-living and photosynthetic
D. Thalloid
E. Differentiated plant body
F. Vascular tissue
(a) (A), (C), (D)
(b) (A), (B), (C), (E) and (F)
(c) (A), (B), (C) and (F)
(d) (B), (C), (E) and (F)

90. Identify the figure and select the incorrect option:

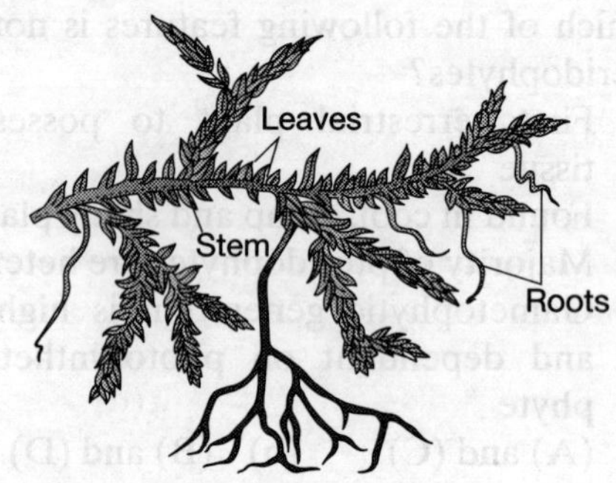

(a) Shows cespitose habit
(b) Produces two kinds of spores
(c) Has larger leaves
(d) Forms distinct strobili or cones

91. Identify the pteridophyte in the figure given below:

(a) *Dryopteris* (b) *Pteris*
(c) *Selaginella* (d) *Salvinia*

92. Seed habit is considered an important step in evolution in:
(a) *Adiantum* (b) *Pteris*
(c) *Dryopteris* (d) *Salvinia*

93. Prerequisite for origin of seed habit includes all, except:
(a) Heterospory
(b) Development of zygote into embryo takes place in the female gametophyte
(c) Retention of female gametophyte on the parental plant for variable period
(d) Formation of embryo in life cycle

94. Heterospory is the production of:
(a) Sexual and asexual spore
(b) Micro and megaspore
(c) Haploid and diploid spore
(d) Diploid and tetraploid spore

Gymnosperms

95. In which of the following plant groups would you place a plant which produces seeds but lacks flowers and fruit?
(a) Bryophytes (b) Pteredophytes
(c) Gymnosperm (d) Angiosperm

96. Gymnosperms are called naked seeded plants because they lack:
(a) Cotyledon (b) Endosperm
(c) Ovary wall (d) Seed coat

97. Which is the tallest gymnosperm plant?
(a) Redwood tree (*Sequoia*) (b) *Pinus*
(c) *Ginkgo* (d) *Cycas*

98. Select true (T) or false (F) for the given statements with respect to gymnosperm and select the correct option:
A. The main plant body is a sporophyte, which is differentiated into true root, stem and leaves
B. Gymnosperms include medium-sized trees or tall trees or shrubs
C. They generally possess a tap root system.
D. They are heterosporous and produce haploid microspores and diploid megaspores

	A	B	C	D
(a)	T	F	F	T
(b)	T	T	T	F
(c)	F	F	T	T
(d)	T	F	T	F

99. ______ represent(s) the reproductive structure among gymnosperms:
(a) Cones (b) Gammae
(c) Prothallus (d) Capsule

100. *Cycas* does not have:
(a) Pollen grains (b) Endosperm
(c) Female cones (d) Male cones

101. Ploidy level of endosperm in gymnosperm is:
(a) 4n (b) 3n
(c) 2n (d) n

102. The endosperm in gymnosperm represents:
(a) A male gametophyte
(b) A female gametophyte
(c) Seed
(d) Embryo

103. The vascular bundles in gymnosperms have:
(a) Xylem with vessels and phloem with companion cells
(b) Xylem without vessels and phloem with companion cells
(c) Xylem without vessels and phloem without companion cells
(d) Xylem with vessels and phloem without companion cells

104. Match column (I) and column (II) and select the correct option:

Column I	Column II
A. Coralloid root	(i) *Ginkgo*
B. Ephedrine	(ii) Medicinal value
C. Sago	(iii) Nitrogen fixation
D. Pagoda tree	(iv) *Cycas revoluta*

(a) A – (iii), B – (ii), C – (iv), D – (i)
(b) A – (iv), B – (ii), C – (i), D – (iii)
(c) A – (ii), B – (iii), C – (iv), D – (i)
(d) A – (iii), B – (iv), C – (ii), D – (i)

105. Mycorrihizal root having symbiotic fungi is found in:
(a) *Cycas* (b) *Pinus*
(c) *Ginkgo* (d) *Cedrus*

106. Where does the pollen tube discharge its male gametes in gymnosperm?
(a) On the megasporophyll
(b) In the ovary
(c) Near the mouth of archegonia
(d) In the embryo sac

107. Male and female cones are borne on the same tree in:
(a) *Cycas* (b) *Pinus*
(c) *Ginkgo* (d) All of these

108. Identify the given figure and select the incorrect statement:

Dwarf shoot
Long shoot
Seeds

(a) Ovules are not enclosed by any ovary wall
(b) Multicellular female gametophyte is retained within the megasporangium
(c) Possess endosperm as post-fertilisation tissue
(d) Represents a living fossil

109. Select the incorrect match:
(a) Tallest gymnosperm – *Sequoia*
(b) Unbranched stem – *Pinus*
(c) Pinnate leaves – *Cycas*
(d) Living fossil – *Ginkgo*

110. Gymnosperm are ___A___; they produce ___B___ microspores and megaspores:

	A	B
(a)	Homosporous	Haploid
(b)	Homosporous	Diploid
(c)	Heterosporous	Haploid
(d)	Heterosporous	Diploid

111. The adaptation(s) in conifers to reduce water loss is/are the presence of:
(a) Needle-like leaves
(b) Thick cuticle
(c) Sunken stomata
(d) All of the above

112. Pollination in gymnosperms is:
(a) Direct and Entomophilous only
(b) Indirect and Anemophilous mainly
(c) Direct and Anemophilous only
(d) Indirect and Entomophilous mainly

113. Chilgoza seeds are edible and are obtained from a species of:
(a) *Cycas* (b) *Pinus*
(c) *Ephedra* (d) *Taxus*

114. Select the correct option to fill in the blanks in the following statements:
(i) In gymnosperm, the male and female ___A___ do not have a/an ___B___ free-living existence.
(ii) The stems are unbranched in ___C___ and branched in ___D___.

	A	B	C	D
(a)	Gametophyte	Independent	*Cycas*	*Pinus*
(b)	Sporophyte	Independent	*Pinus*	*Cycas*
(c)	Sporophyte	Dependent	*Cycas*	*Pinus*
(d)	Gametophyte	Independent	*Cedrus*	*Pinus*

115. *Pinus* and *Dryopteris* resemble each other in having:
(a) Free-living existence of gametophyte
(b) Siphonogamous internal fertilisation
(c) Flagellated gametes
(d) Archegonia

116. Choose the odd one out with respect to post-fertilisation structure in gymnosperms:
(a) Seed (b) Perisperm
(c) Endosperm (d) Embryo

117. In *Pinus*, the ovules are borne on ___A___, which may be clustered to form ___B___:

	A	B
(a)	Microsporophyll	Male strobili
(b)	Megasporophyll	Female lax
(c)	Megasporophyll	Male cone
(d)	Microsporophyll	Male lax

118. In gymnosperms, the microspore develops into a male gametophytic generation, which is:
A. Highly reduced
B. Known as pollen grain
C. Carrier of male gametes
(a) Only (A) is correct
(b) Only (B) and (C) are correct
(c) Only (A) and (C) are correct
(d) (A), (B) and (C) are correct

119. Endosperm of gymnosperm is:
A. Haploid
B. Developed from the megaspore mother cell directly
C. Female gametophyte
D. Triploid and post-fertilisation tissue
The correct statements are:
(a) Only (D) (b) (A) and (C)
(c) (A), (B) and (C) (d) (B) and (D)

120. Winged pollen grain is a characteristic feature of:
(a) *Cycas* (b) *Pinus*
(c) *Gnetum* (d) *Ephedra*

121. Number of cotyledons in *Cycas* and *Pinus* seeds are:
(a) 1, 2 (b) 2, 1
(c) 2, Many (d) Many, 2

Angiosperms

122. Floral features are chiefly used in angiosperm identification because:
(a) Flowers are nice to work with
(b) Flowers have various colours
(c) Reproductive parts are more stable and conservative than vegetative parts
(d) All of the above

123. Microscopic angiosperm is:
(a) *Wolffia* (b) *Sequoia*
(c) *Zamia* (d) *Eucalyptus*

124. Male and female sex organs in angiosperms are located in:
(a) Sporophyll (b) Cone
(c) Flower (d) Strobilus

125. Angiosperms are important as they provide:
(a) Food (b) Fuel
(c) Medicine (d) All of the above

126. Angiosperms have dominant flora primarily because of their:
(a) Power of adaptability in diverse habitats
(b) Property of producing large number of seeds
(c) Nature of self-pollination
(d) Domestication by humans

127. Select the incorrect statement with respect to plants producing flowers and fruits during their lifespan:
(a) Large group of plant kingdom occurs in a wide range of habitats
(b) Show great variation in size, from small microscopic plants to tall trees
(c) Pollination is direct and occurs by wind or other agencies
(d) They show double fertilisation event during sexual reproduction

128. A feature which is not common to gymnosperm and angiosperm:
(a) Sporophytic plant body which is differentiated into root, stem and leaves
(b) Absence of independent free-living existence of male and female gametophytes
(c) Development of zygote into embryo takes place within female gametophyte
(d) Post-fertilisation development of endosperm

129. Which of the following structures of the embryo sac is three-celled?
(a) Antipodals (b) Egg apparatus
(c) Polar nuclei (d) All except (c)

130. Sporophyte generation in angiosperms is not represented by:
(a) Style, Stigma and Nucellus
(b) Embryo, Suspensor and Cotyledons
(c) Microspore, Embryo sac and Male gamete
(d) Anther, Megaspore mother cell and Integument

131. Select the incorrectly matched pair with respect to angiosperms:
(a) Embryo sac – Female gametophyte
(b) Antipodals – Degenerate after fertilisation
(c) Endosperm – Triploid nutritive tissue
(d) Secondary nucleus – Haploid

132. Which among the following features does not belong to angiosperms?
(a) Each anther consist of four microsporangia
(b) Ovules are born inside the ovary
(c) Fertilisation occurs twice in the female gametophyte
(d) Heterosporous and meroblastic embryo development

133. Angiosperms and gymnosperms:
(i) Differ in their dominant phase
(ii) Exhibit similar life cycle pattern
(iii) Exhibit sporic meiosis
(iv) Are similar in their mode of nutrition
(v) Show monosporic development of female gametophyte
(vi) Are archegoniate spermatophytes

Select the correct option:
(a) Two statements are incorrect
(b) Three statements are incorrect
(c) Only one statement is incorrect
(d) All statements are correct

134. Egg apparatus is:
(a) Two-celled and diploid
(b) Positioned at micropyle end of embryo sac
(c) Three-celled, one egg cell and two antipodals
(d) Both (b) and (c)

135. The embryo sac of flowering plants is also known as:
(a) Megasporangium
(b) Megasporaphyll
(c) Female gametophyte
(d) Endosperm

136. Identify the incorrect statement about angiosperms:
(a) Double fertilisation is a characteristic feature of angiosperms
(b) Endosperm is triploid and nutritive tissue
(c) The ovary develops into fruit after fertilisation
(d) Triple fusion involves fusion of two male gametes with polar nuclei

137. Read the following statements with respect to angiosperms and state whether true (T) or false (F):
A. Embryo sac formation is preceded by meiosis
B. Pollen grains germinate on the ovule and the resulting pollen grows through the tissues of stigma and style
C. Synergids, antipodals and PEN degenerate after fertilisation
D. After fertilisation, the ovule develops into a seed and the ovaries develop into fruit

	A	B	C	D
(a)	T	F	F	T
(b)	T	T	T	T
(c)	F	T	T	T
(d)	F	T	F	T

138. Which of the following plant groups show multicellular dominant diploid phase in life cycle with few-celled triploid phage?
(a) Bryophytes (b) Pteridophytes
(c) Gymnosperms (d) Angiosperms

139. Dicotyledons differ from monocotyledons in:
(a) Presence of two cotyledons
(b) Absence of parallel venation
(c) Presence of tetra or pentamerous flowers
(d) All of the above

140. An angiosperm differs from a gymnosperm by possessing:
(a) Seeds (b) Tracheids in xylem
(c) Flower and fruit (d) All except (b)

141. Which of the following is not true about angiosperms?
(a) Archegoniate spermatophyte
(b) Siphonogamous fertilisation
(c) Meroblastic embryo development
(d) Embryophyte

Plant Life Cycles and Alternation of Generations

142. The haploid plant body produces gametes by:
(a) Meiosis (b) Mitosis
(c) Amitosis (d) Reductional division

143. In diplontic life cycle, sporophyte is:
(a) Diploid (b) Dominant
(c) Photosynthetic (d) All of the above

144. Select the mismatched pair with respect to life cycle pattern:
(a) *Funaria* – Haplodiplontic
(b) Lily – Diplontic
(c) *Chlamydomonas* – Haplontic
(d) *Dryopteris* – Diplontic

145. The gametophyte phase is not free living and represented by single to few cell stage in:
(a) Diplontic life cycle of seed bearing plants
(b) Haplontic life cycle of algae
(c) Haplodiptontic life cycle of fern
(d) Haplodiplontic life cycle of mosses

146. Which kind of life cycle pattern is exhibited by seed bearing plants?
(a) Haplontic (b) Diplohaplontic
(c) Diplontic (d) Haplodiplontic

147. Algae, bryophytes and pteridophytes resemble each other in which of the following features?
(a) Presence of zygotic meiosis
(b) Water is essential for sexual reproduction
(c) Exhibit haplodiplontic life cycle pattern
(d) Show zygotic meiosis

148. Select the statement which is true for all spermatophytes:
(a) Sporophyte generation is represented only by one-celled zygote
(b) Sporophyte is partially or completely dependent upon photosynthetic gametophyte
(c) Male and female gametophytes do not have an independent free-living existence
(d) Sporic meiosis and haplodiplontic life cycle pattern

149. Consider the following statements about gametophytic generation:
A. Generation that produces the gametes
B. Generation that produce the spores

C. Generation that possesses vascular tissue
D. Haploid generation
Select the incorrect statements:
(a) Only (A) (b) (B) and (C)
(c) (A) and (D) (d) (A), (C) and (D)

150. The life cycle of bryophyte and pteridophyte is:
(a) Haplontic
(b) Diplontic
(c) Monomorphic haplodiplontic
(d) Dimorphic diplohaplontic

151. Which of the following features belongs to the concerned plant group?
A. Plant with protonema stage
B. Presence of seed habit
C. Ovules remain exposed both before and after fertilisation

	A	B	C
(a)	Bryophyte	Pteridophyte	Gymnosperm
(b)	Bryophyte	Gymnosperm	Angiosperm
(c)	Algae	Bryophyte	Angiosperm
(d)	Bryophyte	Gymnosperm	Pteridophyte

152. Fill in the blanks in the given table by selecting the correct option:

Feature	Algae	Bryophyte	Gymnosperm
Vascular tissue	Absent	A	Present
Meiosis	B	Sporic	C
Life cycle	Haplontic	D	Diplontic

	A	B	C	D
(a)	Absent	Zygotic	Sporic	Haplontic
(b)	Present	Gametic	Zygotic	Diplontic
(c)	Absent	Zygotic	Sporic	Haplodiplontic
(d)	Present	Sporic	Gametic	Haplodiplontic

153. Observe the given figures and select the correct statements:

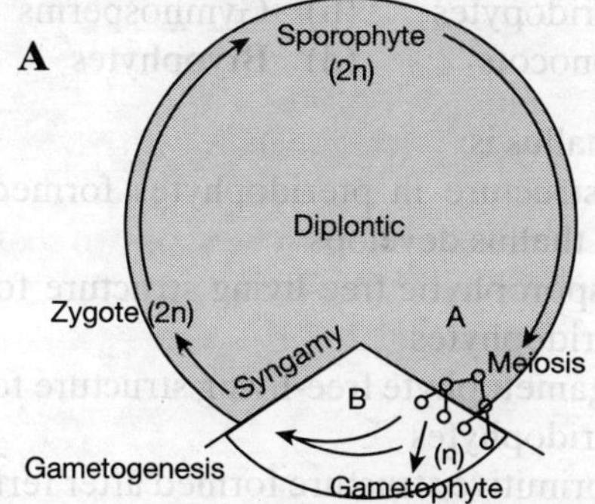

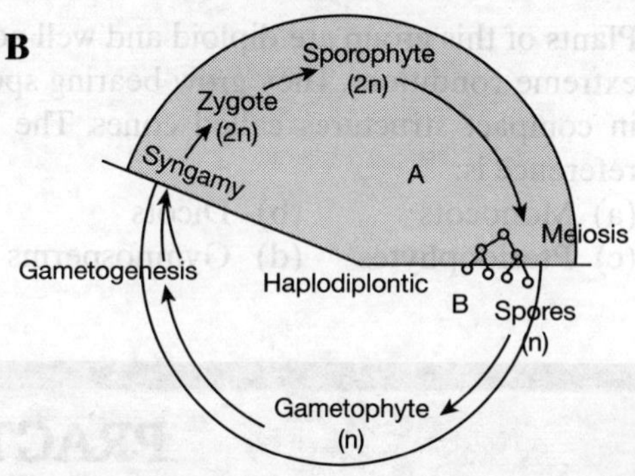

(a) A: Sporic meiosis; represented by bryophyte
(b) A: Gametic meiosis; represented by flowering plants
(c) B: Sporic meiosis, represented by gymnosperms and angiosperms
(d) B: Sporic meiosis, represented by bryophytes and pteridophytes

154. How many of the following organisms show haplodiplontic life cycle?
Volvox, Fucus, Kelp, *Ectocarpus, Polysiphonia, Funaria, Marchantia, Selaginella, Dryopteris, Cycas, Pinus*
(a) Seven (b) Eight
(c) Six (d) Five

NCERT EXEMPLAR QUESTIONS

1. Cyanobacteria are classified under:
(a) Protista (b) Plantae
(c) Monera (d) Algae

2. Fusion of two motile gametes which are dissimilar in size is termed as:
(a) Oogamy (b) Isogamy
(c) Anisogamy (d) Zoogamy

3. Holdfast, stipe and frond constitute the plant body in the case of:
(a) Rhodophyceae (b) Chlorophyceae
(c) Phaeophyceae (d) All of these

4. A plant shows thallus level of organisation. It shows rhizoids and is haploid. It needs water to complete its life cycle because the male gametes are motile. It may belong to:

(a) Pteridopytes (b) Gymnosperms
(c) Monocots (d) Bryophytes

5. A prothallus is:
(a) A structure in pteridophytes formed before the thallus develops
(b) A sporophytic free-living structure formed in pteridophytes
(c) A gametophyte free-living structure formed in pteridophytes
(d) A primitive structure formed after fertilisation in pteridophytes

6. Plants of this group are diploid and well adapted to extreme conditions. They grow bearing sporophylls in compact structures called cones. The group in reference is:
(a) Monocots (b) Dicots
(c) Pteridophytes (d) Gymnosperms

7. The embryo sac of an angiosperm is made up of:
(a) 8 cells (b) 7 cells and 8 nuclei
(c) 8 nuclei (d) 7 cells and 7 nuclei

8. If the diploid number of chromosomes of a flowering plant is 36, what would be the chromosome number in its endosperm?
(a) 36 (b) 18
(c) 54 (d) 72

9. Protonema is:
(a) Haploid and is found in mosses
(b) Diploid and is found in liverworts
(c) Diploid and is found in pteridophytes
(d) Haploid and is found in pteridophytes

10. The giant redwood tree (*Sequoia sempervirens*) is a/an:
(a) Angiosperm (b) Free fern
(c) Pteridophyte (d) Gymnosperm

PRACTICE TEST PAPER

1. Classification of flowering plants given by Bentham and Hooker is a/an:
(a) Natural system
(b) Phylogenetic system
(c) Artificial system
(d) Phenetic system

2. Cytological information like chromosome number, structure and behaviour is included in:
(a) Chemotaxonomy
(b) Biochemical taxonomy
(c) Cytotaxonomy
(d) Classical taxonomy

3. Select the odd one out with respect to unicellular algae:
(a) *Chlorella* (b) *Chylamydomonas*
(c) *Spirulina* (d) *Ectocarpus*

4. Select the correct features for algae:
A. Have simple, thalloid, autotrophic plant bodies
B. They occur in aquatic habitat, both in fresh and marine water
C. Kelps are massive plant bodies seen primarily in freshwater habitat
D. *Chlorella* is a multicellular, green algae rich in proteins and is used as a food supplement by space travellers
(a) (A), (B) and (D)
(b) (A) and (B) only
(c) (C) and (D) only
(d) (A), (B), (C) and (D)

5. At least half of the total CO_2 fixation on earth is carried out by:
(a) Algae (b) Gymnosperms
(c) Angiosperms (d) Mosses

6. Great diversity in body size and shape of chloroplast is seen in members of __________ and __________, respectively:
(a) Brown algae, Green algae
(b) Red algae, Green algae
(c) Green algae, Red algae
(d) Red algae, Brown algae

7. A moss differs from a fern in having:
(a) Flagellated antherozoids
(b) Archigonia
(c) Haploid spores
(d) Dependent sporophyte

8. In which of the following groups would you place a plant which produces spores and embryos but lacks seeds and vascular tissues?
(a) Bryophytes (b) Pteridophytes
(c) Algae (d) Both (a) and (c)

9. The term prothallus refers to:
(a) The stage before thallus formation
(b) Reduced sporophyte
(c) A moss plant which is not differentiated into root, stem and leaves
(d) Reduced gametophyte of fern

10. In which of the following is fertilisation possible without water?
(a) *Ulothrix* (b) *Funaria*
(c) *Cycas* (d) *Dryopteris*

11. Which of the following classification systems gave equal weightage to vegetative and sexual characteristics?
(a) Natural classification system
(b) Artificial classification system
(c) Phylogenetic classification system
(d) Cladistic classification system

12. In which of the following algae is sexual reproduction oogamous and accompanied by complex post-fertilisation development?
(a) *Fucus* (b) *Chara*
(c) *Ectocarpus* (d) *Polysiphonia*

13. Gametophyte of bryophyte is:
(a) Green, thalloid and vascular
(b) Dependent on sporophyte
(c) Independent and multicellular
(d) Dominant phase and produces spores

14. Isogamous sexual reproduction by involvement of non-motile gametes is found in:
(a) *Spirogyra* (b) *Eudorina*
(c) *Porphyra* (d) *Chlamydomonas*

15. Which of the given features is not related to pteridophytes?
A. First terrestrial plants to possess vascular tissues
B. Main plant body is gametophyte
C. Sporangia are subtended by leaf-like appendages called sporophyll
D. Majority of pteredophytes produce two kinds of spores, microspores and megaspores
(a) (A) and (B) only
(b) (A) and (C) only
(c) (A), (B) and (C)
(d) (A), (B), (C) and (D)

16. Match the following columns and select the correct option:

Column I	Column II
A. *Ectocarpus*	(i) Unbranched filaments algae
B. *Ulothrix*	(ii) Coenobium
C. *Volvox*	(iii) Branched filamentous algae
D. *Chlamydomonas*	(iv) Microscopic unicellular form

(a) A – (iii), B – (i), C – (ii), D – (iv)
(b) A – (i), B – (iii), C – (ii), D – (iv)
(c) A – (iv), B – (i), C – (iii), D – (ii)
(d) A – (iv), B – (iii), C – (i), D – (ii)

17. Which of these in *Cycas* seeds represents parental sporophytic generation?
(a) Perisperm and endosperm
(b) Endosperm and embryo
(c) Perisperm and seed coat
(d) Female gametophyte and seed coat

18. *Dryopteris, Pteris* and *Adiantum* belong to which of the following classes of pteridophytes?
(a) Psilopsida (b) Pteridopsida
(c) Sphenopsida (d) Bryopsida

19. Which of the following is considered an important step in the development of seed habit?
(a) Free-living sporophyte
(b) Haplodiplontic life cycle
(c) Sporic meiosis
(d) Formation of two kinds of spores

20. Which of the following is haploid in the life cycle of *Funaria*?
(a) Embryo (b) Capsule
(c) Protonema (d) Foot

21. Which of the following post-fertilisation events is/are not associated with gymnosperm?
A. Zygote – Embryo
B. PEN – Endosperm
C. Ovules – Seeds
D. Ovary – Fruit
E. Nucellus – Perisperm
(a) Only (D) (b) Only (B)
(c) (B) and (D) (d) (B), (D) and (E)

22. Select the incorrect statement with respect to Rhodophyceae:
(a) Majority of red algae are marine, with greater concentration found in warmer areas
(b) Reserve food material is very similar to amylopectin and glycogen
(c) Cell wall is made up of cellulose, pectin and polysulphate esters
(d) Sexual reproduction is oogamous and occurs by fusion of motile male gamete with large static non motile female gamete

23. Motile spores in brown algae have:
(a) 2–8 isokont flagella
(b) 2 heterokont laterally inserted flagella
(c) 2 isokont laterally inserted flagella
(d) 2 heterokont apical flagella

24. An elaborate mechanism of spore dispersal is present in:
(a) *Marchantia* (b) *Sphagnum*
(c) *Funaria* (d) All except (a)

25. Some plant groups exhibit intermediate condition with respect to life cycle pattern. Which characteristic will not be exhibited by such kinds of plants?
(a) Haplodiplontic life cycle pattern
(b) They can differ in the dominant phase
(c) Gametophyte is dependent on sporophyte
(d) Sporic meiosis

26. How many statements are correct for bryophytes?
(a) Less differentiated than algae
(b) They are in general of great economic importance
(c) Zygotes do not undergo meiotic division immediately
(d) Sex organs are multicellular and without jacket layer
(a) One (b) Two
(c) Three (d) Four

27. Which of the following is incorrectly matched?
(a) Amylopectin-like reserve food – *Porphyra*
(b) Simple branched filamentous brown algae – *Ectocarpus*
(c) Zoospores with laterally inserted heterokont flagella – *Dictyota*
(d) Pyrenoids within chloroplasts – *Polysiphonia*

28. Source of hydrocolloid is:
A. Red algae B. Brown algae C. Green algae
(a) Only (A) (b) Only (B)
(c) (A) and (B) (d) (B) and (C)

29. Pollination in gymnosperm is:
(a) Hydrophilous
(b) Anemophilous
(c) Entomophilous
(d) Siphonogamous

30. Read the following statements.
A. Sporophyte is the dominant, photosynthesis-independent phase of a plant
B. Highly reduced female gametophyte as embryo sac is present in the ovule
C. The pollen grains are carried in air currents and come in contact with the opening of the ovules

How many statements are correct for gymnosperms and angiosperms, respectively?
(a) One and two (b) Zero and three
(c) Two and two (d) Three and two

31. Plant body is differentiated into holdfast, stipe and frond in:
(a) *Laminaria* (b) *Porphyra*
(c) *Spirogyra* (d) *Ulothrix*

32. Evolutionary pteridophytes were the first terrestrial plants to possess:
(a) Sporangia
(b) Vascular tissues
(c) Embryo
(d) Multicellular sex organs

33. Select the incorrect statement for gymnosperms:
(a) Naked seeded plant
(b) *Sequoia* is one of the tallest tree species
(c) Smallest group of the plant kingdom
(d) Show haplodiplontic life cycle pattern

34. Life cycle of seed plants is:
(a) Diplontic (b) Haplodiplontic
(c) Haplodiplontic (d) Diplohaplontic

35. Read the following statements and select the correct option:
A. Male and female gamelophytes in gymnosperms do not show an independent free-living existence
B. The female gametophyte remains within the megasporangia retained on sporophytes
(a) Only (A) is correct
(b) Only (B) is correct
(c) Both (A) and (B) are correct
(d) Both (A) and (B) are incorrect

36. A heterosporous pteridophyte is:
(a) *Dryopteris* (b) *Selaginella*
(c) *Pteris* (d) All except (b)

37. Winged seeds belong to:
(a) *Cycas* (b) *Pinus*
(c) *Wolffia* (d) *Ranunculus*

38. Which of the following algae has an abundance of the pigment fucoxanthin?
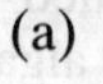
(a)

(b)
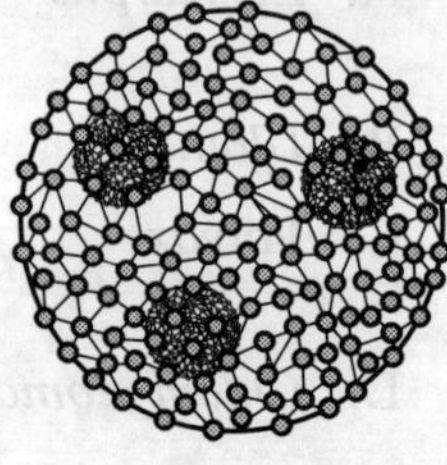

(c) C (d)

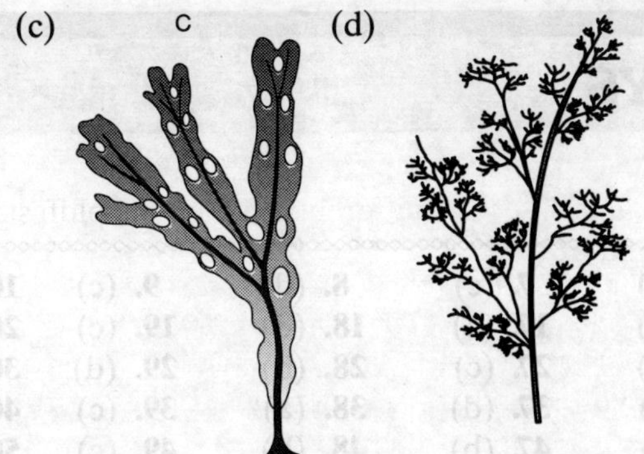

39. Identify the following statements as true (T) or false (F) and choose the option accordingly:
A. The sporophyte in mosses is more elaborate than that in liverworts
B. Pteridophytic gametophyte has protonema and a leafy stage
C. In gymnosperms, the female gametophyte is free living
D. Origin of seed habit can be traced in pteridophytes

	A	B	C	D
(a)	T	T	F	T
(b)	T	F	F	T
(c)	F	F	T	T
(d)	F	T	T	F

40. Which marine algae produces large amounts of hydrocolloids?
(a) Brown and green
(b) Red and brown
(c) Only red
(d) Red and blue green

41. In gymnosperms, reduced male gametophyte is called:
(a) Pollen grain (b) Endosperm
(c) Cone (d) Microsporophyll

42. Which of the following belongs to tracheophyte?
(a) Only spermatophyte
(b) Bryophyte and pteridophyte only
(c) Pteridophyte, gymnosperms and angiosperms
(d) Angiosperms only

43. Select the correct feature for *Pinus*:
(i) Monoecious plant
(ii) Pinnate leaves
(iii) Sunken stomata
(iv) Heterosporous
(a) (i), (ii) and (iv)
(b) (i), (iii) and (iv)
(c) (ii), (iii) and (iv)
(d) (i), (ii), (iii), (iv)

44. *Sphagnum* is commonly used as packing material for trans-shipment of living material due to:
(a) Capacity to hold water
(b) Easy availability
(c) Nature as it can grow anywhere
(d) All of the above

45. Vessels are present in which one of the following groups of plants?
(a) Bryophyta (b) Pteridophyta
(c) Gymnosperms (d) Angiosperms

ANSWER KEYS

Exercises

1. (a) **2.** (d) **3.** (c) **4.** (b) **5.** (a) **6.** (b) **7.** (c) **8.** (d) **9.** (c) **10.** (d)
11. (b) **12.** (a) **13.** (a) **14.** (c) **15.** (d) **16.** (d) **17.** (a) **18.** (d) **19.** (c) **20.** (d)
21. (d) **22.** (b) **23.** (a) **24.** (c) **25.** (a) **26.** (b) **27.** (c) **28.** (b) **29.** (d) **30.** (a)
31. (a) **32.** (d) **33.** (c) **34.** (d) **35.** (b) **36.** (a) **37.** (d) **38.** (a) **39.** (c) **40.** (a)
41. (b) **42.** (d) **43.** (d) **44.** (c) **45.** (a) **46.** (c) **47.** (b) **48.** (b) **49.** (c) **50.** (c)
51. (a) **52.** (c) **53.** (d) **54.** (a) **55.** (a) **56.** (c) **57.** (b) **58.** (d) **59.** (d) **60.** (a)
61. (b) **62.** (a) **63.** (b) **64.** (c) **65.** (b) **66.** (c) **67.** (b) **68.** (d) **69.** (a) **70.** (b)
71. (d) **72.** (c) **73.** (c) **74.** (c) **75.** (a) **76.** (a) **77.** (a) **78.** (c) **79.** (b) **80.** (d)
81. (a) **82.** (a) **83.** (b) **84.** (c) **85.** (d) **86.** (a) **87.** (b) **88.** (d) **89.** (a) **90.** (c)
91. (d) **92.** (d) **93.** (d) **94.** (b) **95.** (c) **96.** (c) **97.** (a) **98.** (b) **99.** (a) **100.** (c)
101. (d) **102.** (b) **103.** (c) **104.** (a) **105.** (b) **106.** (c) **107.** (b) **108.** (c) **109.** (b) **110.** (c)
111. (d) **112.** (c) **113.** (b) **114.** (a) **115.** (d) **116.** (c) **117.** (b) **118.** (d) **119.** (b) **120.** (b)
121. (c) **122.** (c) **123.** (a) **124.** (c) **125.** (d) **126.** (a) **127.** (c) **128.** (d) **129.** (d) **130.** (c)
131. (d) **132.** (c) **133.** (a) **134.** (b) **135.** (c) **136.** (d) **137.** (a) **138.** (d) **139.** (d) **140.** (c)
141. (a) **142.** (b) **143.** (d) **144.** (d) **145.** (a) **146.** (c) **147.** (b) **148.** (c) **149.** (b) **150.** (d)
151. (a) **152.** (c) **153.** (d) **154.** (a)

NCERT Exemplar Questions

1. (c) **2.** (c) **3.** (c) **4.** (d) **5.** (c) **6.** (d) **7.** (b) **8.** (c) **9.** (a) **10.** (d)

Practice Test Paper

1. (a) **2.** (c) **3.** (d) **4.** (b) **5.** (a) **6.** (a) **7.** (d) **8.** (a) **9.** (d) **10.** (c)
11. (b) **12.** (d) **13.** (c) **14.** (a) **15.** (b) **16.** (a) **17.** (c) **18.** (b) **19.** (d) **20.** (c)
21. (c) **22.** (d) **23.** (b) **24.** (d) **25.** (c) **26.** (a) **27.** (d) **28.** (c) **29.** (b) **30.** (c)
31. (a) **32.** (b) **33.** (d) **34.** (a) **35.** (c) **36.** (b) **37.** (b) **38.** (c) **39.** (b) **40.** (b)
41. (a) **42.** (c) **43.** (b) **44.** (a) **45.** (d)

SOLUTIONS

Classification System

1. Carolus Linnaeus proposed the artificial system of classification based exclusively on nature and the number of stamens and carpels. It was known as the sexual system of classification.

2. The artificial system was based on gross superficial morphological characters such as habit, colour, number of stamen/carpel, shape of leaves, etc. They were based mainly on vegetative characters or on androecium structure.

3. Natural system of classification was based on natural affinities among the organisms and considered not only the external features but also internal features like ultrastructure, anatomy, embryology and phytochemistry.

4. George Bentham and Joseph Dalton Hooker proposed the natural system of classification of flowering plants.

5. Artificial system gave equal weightage to vegetative and sexual characters.

6. Cytotaxonomy is based on cytological information like number, structure and behaviour of chromosomes at the time of cell division. Nucleotide sequence in DNA is the criteria of chemotaxonomy or biochemical taxonomy.

7. At the same time, more than 100 characters can be considered.

8. Phylogeny

9. All five matched pairs are correct.

10. Phylogenetic classification system is mainly based on genetic as well as evolutionary affinities. Organisms belonging to the same taxa are believed to have a common ancestor and can be represented in the form of a family tree called cladogram.

11. Numerical taxonomy or phenetics or taximetrics or Adansonean taxonomy

12. Only (A) is correct. Artificial system of classification is mainly based on one or two or few morphological characters.

13. Natural system

14. Phylogenetic system of classification was proposed by Engler and Prantl.

Algae (Chlorophyceae, Pheophyceae, Rhodophyceae)

15. Algae form associations with fungi called lichen.

16. Kelps are large brown algae reaching heights of more than 100 metres; *Porphyra* is a red algae.

17. Zoospores are most the common type of spores produced by algae; they are flagellated spores produced inside the zoosporangium, so they are endogenous in origin.

18. Fusion between one large non-motile (static) female gamete and a small motile male gamete is called oogamous reproduction. It is seen in *Volvox* and *Fucus*.

19. In algae, sex organs are unicellular and non-jacketed. Exceptionally multicellular and jacketed sex organs are present in *Chara*.

20. In green algae, the chloroplast may be discoid, plate-like, reticulate, cup-shaped, spiral or ribbon-shaped.

21. Most members of green algae have one or more storage bodies in the chloroplasts. Pyrenoids contain protein (central) besides starch (peripheral).

22. Floridean starch is the reserve material of red algae *(Prophyra, Polysiphonia)*.

23. Kelps are brown algae which form massive plant bodies.

24. (B) and (D) are incorrect for brown algae. Red algae are characterised by the presence of complex carbohydrate as storage material, which is very similar to amylopectin and glycogen. In red algae, sexual reproduction is of oogamous type and also shows post-fertilisation development.

25. *Chlorella* rich in proteins is used as food supplement by space travellers.

26. Brown algae and red algae

27. Members of class Pheophyceae are characterised by the presence of pigments like chl-a, chl-c and fucoxanthin.

28. *Ulothrix* is an unbranched filamentous green algae found in running freshwater bodies.

29. *Polysiphonia* (Red algae)

30. A – Frond, B – Stipe, C – Holdfast

31. Phaeophyceae

32. *Fucus* is a brown algae and contains chl-a and chl-c as photosynthetic pigments and flagellated zoospores. The zoospores are pyriform with two laterally inserted heterokont flagella.

33. The members of class Rhodophyceae are red because of the predominance of red pigment R-phycoerythrin.

34. The given figure represents the green algae *Ulothrix* (unbranched filamentous green algae) which is found in freshwater bodies.

35. A – Phaeophyceae, B – Chlorophyceae, C – Rhodophyceae

36. In class Rhodophyceae, the cell wall is made up of cellulose and polysulphate esters (sulphated hydrocolloids).

37. Rhodophyceae

38. Non-flagellated isogamy is seen in *Spirogyra*.

39. A – *Polysiphonia* (class Rhodophyceae), B – *Dictyota* (class Phaeophyceae), C – *Ulothrix* (class Chlorophyceae)

40. Marine algae

41. 70 species of marine algae are used as food.

42. Algae are classified mainly on the basis of photosynthetic pigments.

43. *Laminaria* (brown algae). *Chara, Chlamydomonas* and *Ulothrix* are green algae.

44. *Chlamydomonas* (unicellular, motile, green algae)

Bryophyta (Liverworts, Mosses)

45. Bryophyta include liverworts and mosses.

46. (B) and (D) are not related to bryophytes. The predominant stage of the life cycle of moss is haploid gametophyte. They were the first land plants but lack vascular tissues (xylem and phloem).

47. Bryophytes are called the amphibians of the plant kingdom as they live on soil but are dependent on water for sexual reproduction.

48. The plant body of bryophytes is more differentiated than that of algae.

49. Sporophyte of bryophytes is not free living but attached to a photosynthetic gametophyte (N) and derives nourishment from it.

50. In *Funaria* (moss), the sporophyte consists of three parts: foot, seta and capsule. In the capsule, the spore mother cells undergo meiotic division to form haploid spores (N). On germination, the spores give rise to primary protonema, which is a creeping, green, branched, frequently filamentous stage.

51. Spirally arranged leaves

52. *Funaria* (moss). In liverworts, the protonema stage is absent.

53. *Riccia*

54. The predominant phase in the life cycle of a bryophyte is gametophyte (N). It produces gametes, hence called gametophyte.

55. Gemmae are green, multicellular asexual buds which develop in small receptacles called gemma cups located on the thalli.

56. Archigonium. The fertilised egg develops into an embryo inside the archegonium.

57. *Marchantia* is a liverwort.

58. Adult gametophyte (leafy stage) consists of upright slender axes bearing spirally arranged leaves. They are attached to the soil by multicellular and branched rhizoids. This stage bears the sex organs which are multicellular and jacketed.

59. The juvenile gametophyte of *Funaria* is protonema.

60. A – Capsule, B – Seta, C – Leaves, D – Main axis, E – Multicellular rhizoids

61. Not used in the preparation of ice cream and jellies

62. Species of *Sphagnum* (a moss) provide peat that has long been used as fuel

63. In bryophytes, the sporophyte is not free living; it is dependent on photosynthetic gametophyte.

64. A and D are false. *Marchantia* is a dioecious bryophyte (male and female thallus are separate). Mosses have an elaborate mechanism of spore dispersal.

65. *Sphagnum* is used as packing material for trans-shipment of living material because of their capacity to hold water.

66. Only A is incorrect. In mosses, rhizoids are multicellular and branched.

67. Gemmae formation

68. *Funaria* (Moss)

69. Rhizoids (N), Spore (N), Seta (2N), Spore mother cell (2N)

Pteridophyta

70. Pteridophytes are vascular cryptogams because they are vascular plants that reproduce using spores rather than seeds.

71. Source of protein

72. Ferns and Horsetail

73. Both (A) and (B) are correct.

74. (C) and (D) are incorrect. Majority of pteridophytes are homosporous; only a few pteridophytes like *Selaginella* and *Salvinia* are heterosporous. In pteridophytes, both gametophyte and sporophyte are free-living independent stages.

75. In ferns, spores are produced in sporangia.

76. Sporangia bearing leaves are called sporophyll.

77.

Sporophyte —Meiosis→ Spore (N) → Prothallus (N) —Mitosis↓ Gamete

78. *Selaginella* occurs mostly in tropical forests.

79. *Azolla* is an aquatic fern which contains as endosymbiont the nitrogen fixing cyanobacterium *Anabaena azollae.*

80. *Dryopteris* is a homosporous pteridophyte.

81. In some pteridophytes like *Selaginella* and *Equisetum,* sporophyll form distinct compact structures called strobili or cones.

82. In ferns, the spores germinate to give rise to inconspicuous, small but multicellular, free-living photosynthetic thalloid gametophytes called prothallus.

83. A – *Selaginella* (Lycopsida), B – *Equisetum* (Sphenopsida)

84. Seed habit or heterospory is considered an important step in evolution.

85. *Selaginella* and *Salvinia* show seed habit. They produce two kinds of spores: microspores and megaspores.

86. *Psilotum* is a homosporous pteridophyte.

87. Mitosis, Meiosis

88. Gametophytes require cool damp shady places to grow. Because of this specific restricted requirement and need for water for fertilisation, the spread of living pteridophytes is limited and restricted to narrow geographical regions.

89. (A), (C) and (D) are correct.

90. In *Selaginella*, leaves are small (microphylls).

91. *Salvinia*

92. Heterosporous pteridophyte *Salvinia*

93. Embryo formation is not included in seed habit.

94. Heterospory is the production of two kinds of spores: microspores and megaspores; both are haploid spores.

Gymnosperms

95. Gymnosperms produce seeds but lack flowers and fruits; they are naked seeded plants because ovules are not enclosed by any ovary wall.

96. Ovary wall

97. The giant redwood tree *Sequoia* is the tallest tree among gymnosperms.

98. Only D is false. All gymnosperms are heterosporous and produce haploid micro-spores and megaspores.

99. Cones or strobili or lax represent the reproductive structure among gymnosperms. Sporangia are born on sporophylls, which are arranged spirally along the axis to form cone or strobili.

100. In *Cycas*, female cones are absent. Megasporophyll are attached directly at the apex of the plant body.

101. In gymnosperms, the endosperm represents the female gametophyte (N).

102. Female gametophyte

103. In gymnosperms, xylem lacks vessels and phloem lacks companion cells, but exceptionally, vessels are present in the members of order Gnetales, for example, *Gnetum, Ephedra*.

104. A. Coralloid root – Nitrogen fixation, B. Ephedrine – Medicinal value, C. Sago – *Cycas revoluta*, D. Pagoda tree – *Ginkgo*

105. In *Pinus*, the roots have fungal association in the form of mycorrhizae.

106. Pollen tubes discharge their male gametes near the mouth of archegonia.

107. *Pinus* is a monoecious plant having male and female cones on the same tree.

108. In *Ginkgo* (gymnosperms), the endosperm is a pre-fertilisation tissue and represents the female gametophyte.

109. Stem is unbranched in *Cycas*, while *Pinus* has a branched stem.

110. A – Heterosporous, B – Haploid

111. Conifer leaves show some xerophytic characters. These are (i) sunken stomata, (ii) thick cuticle and (iii) needle-like leaves.

112. Pollination in gymnosperms is direct as the stigma is absent and, therefore, pollen grains directly enter the micropyle of ovules. Pollination is most commonly anemophilous (by wind).

113. Option (b) is correct.

114. A – Gametophyte, B – Independent, C – *Cycas*, D – *Pinus*

115. *Pinus* (gymnosperms) and *Dryopteris* (pteridophyte) resemble each other in having archenogia. *Pinus* lack flagellated gametes and the free-living existence of the gametophytic stage. *Pinus* shows siphonogamous fertilisation (occurs through pollen tube) but it is absent in *Dryopteris*.

116. Endosperm is a pre-fertilisation tissue in gymnosperms.

117. A – Megasporophyll, B – Female lax

118. (a), (b) and (c) are correct.

119. Only (A) and (C) are correct. One cell of the nucellus of the ovule differentiates into megaspore mother cell, which undergoes meiosis to form four megaspores. Three of them degenerate and one functional megaspore divides many times to form the female gametophyte or endosperm(N). So it shows monosporic development.

120. In *Pinus*, the pollen grains are winged, which assist in anemophily or air pollination.

121. Number of cotyledons in the seeds of *Cycas* – 2, *Pinus* – Many.

Angiosperms

122. Floral characters are more conservative as compared to vegetative characters and not affected by the environment.

123. *Wolffia*

124. Flowers

125. Option (d) is correct.

126. The angiosperms are an exceptionally large group of plants occurring in a wide range of habitats because of their power of adaptibility in diverse habitats.

127. Pollination is indirect and occurs through various agencies like wind, water or insect.

128. In angiosperms, the endosperm develops after fertilisation and is triploid in nature.

129. A typical embryo sac is a 7-celled 8-nucleated structure. The arrangement of nuclei is 3+2+3. Three cells (egg cell and two synergids) are present at the micropylar end, 3 antipodal cells are present at the chalazal end, and 2 polar nuclei present at the centre.

130. Microspores, embryo sac and male gametes are haploid and represent gametophyte generation.

131. Before fertilisation, two polar nuclei fuse to form secondary nucelus.

132. Double fertilisation is a unique feature of flowering plants. It involves syngamy and triple fusion.

133. Only one statement (vi) is incorrect. Angiosperms lack archegonia.

134. Egg apparatus is present at the micropylar end of the embryo sac.

135. The embryo sac represents the female gametophyte.

136. Triple fusion involves fusion of male gamete and two polar nuclei.

137. (B) and (C) are false.

Pollen grains germinate on the stigma and the resulting pollen tubes grow through the tissue of the stigma and style and reach the ovules. After fertilisation, PEN develops into endosperms which provide nutrition to the developing embryo, while synergids and antipodals degenerate after fertilisation.

138 Angiosperms have triploid endosperm.

139. Dicotyledons are characterised by the presence of two cotyledons in the seed, reticulate venation in leaves and tetramerous and pentamerous flowers.

140. Gymnosperms and angiosperms are spermatophytes (seed bearing plants) but gymnosperms lack flowers and fruits.

141. An angiosperm's female sex organs lack archegonia.

Plant Life Cycles and Alternation of Generations

142. Gametes are haploid. So a haploid plant body produces gametes by mitotic division.

143. In a diplontic life cycle pattern, the diploid sporophyte is the dominant, photosynthetic and independent phase of the plant.

144. *Dryopteris* (pteridophyte) shows diplohaplontic life cycle pattern.

145. In seed bearing plants (gymnosperms and angiosperms), the gametophytic phase is represented by single- to few-celled haploid gametophytes.

146. Diplontic life cycle pattern

147. Algae, bryophytes and pteridophytes require water for sexual reproduction. Algae show zygotic meiosis while bryophytes and pteridophytes exhibit sporic meiosis.

148. In gymnosperms and angiosperms, male and female gametophytes are highly reduced and confined to a limited number of cells and do not have an independent free-living existence.

149. (B) and (C) are incorrect. Gametophyte generation produces gametes and vascular tissues are not found in gametophytic structures.

150. Dimorphic diplohaplontic life cycle pattern is represented by bryophytes and pteridophytes.

151. A – Bryophyte, B – Pteridophyte, C – Gymnosperms

152. A – Absent, B – Zygotic, C – Sporic, D – Haplodiplontic

153. Bryophytes and pteridophytes show sporic meiosis.

154. Seven plants (Kelp, *Ectocarpus*, *Polysiphonia*, *Funaria*, *Marchantia*, *Selaginella* and *Dryopteris*) show haplodiplontic life cycle pattern.

Animal Kingdom

EXERCISES

General Characters

1. Which of the following characters is exhibited by all members of kingdom Animalia?
(a) Coelom
(b) Cellular level of organisation
(c) Body segmentation
(d) Multicellularity

2. There is no diploblastic animal with:
(a) Asymmetry
(b) Bilateral symmetry
(c) Radial symmetry
(d) Acoelomate condition

3. Following is a diagrammatic sectional view of a body cavity. Identify the type of body cavity along with its correct examples:

(a) Acoelom, *Planaria*
(b) Pseudocoelom, *Ascaris*
(c) True coelom, *Planaria*
(d) Pseudocoelom, *Spongilla*

4. The phylum showing radial symmetry is/are:
(a) Annelida
(b) Platyhelminthes
(c) Arthropoda
(d) Coelenterata and Echinodermata

5. The symmetry found in an animal which can be divided into similar halves by one plane only into right and left mirror images is:
(a) Biradial (b) Radial
(c) Bilateral (d) Spherical

6. Segmentation is absent in:
(a) Annelids (b) Arthropods
(c) Molluscs (d) Chordates

7. Notochord is a rod-like structure formed on the dorsal side during embryonic development in chordates and is derived from:
(a) Ectoderm (b) Mesoderm
(c) Endoderm (d) Both (a) and (b)

8. In __________, the body cavity is not lined by mesoderm; instead, mesoderm is present as scattered pouches in between the ectoderm and endoderm:
(a) Platyhelminthes (b) Aschelminthes
(c) Annelida (d) Arthropoda

9. Identify the odd one out with respect to radial symmetry:
(a) Coelenterata (b) Ctenophores
(c) Echinodermata (d) Annelids

10. Tissue level of organisation is found in:
(a) Porifera (b) Coelenterata
(c) Platyhelminthes (d) Annelida

11. Identify the triploblastic acoelomate phylum:
(a) Annelida (b) Platyhelminthes
(c) Aschelminthes (d) Mollusca

12. Open circulatory system is not found in:
(a) Arthropoda (b) Annelids
(c) Molluscs (d) Hemichordates

13. Cell aggregate body plan is found in:
(a) Flatworms (b) Ctenophores
(c) Sponges (d) Nematodes

14. Haemocoelom is found in:
(a) Earthworm (b) Cockroach
(c) Tapeworm (d) Sea anemone

15. Neurons are absent in:
(a) Sponges (b) Cnidarians
(c) Ctenophores (d) Flatworms

16. Blind sac body plan is found in:
(a) Annelids and molluscs
(b) Coelenterates and Platyhelminthes
(c) Coelenterates and annelids
(d) Ctenophores and annelids

17. The digestive system is termed complete, if:
(a) Both mouth and anus are present
(b) Digestion occurs in digestive tract
(c) One opening serves as mouth and anus
(d) All digestive glands are present

18. Metamerism is a characteristic of:
(a) Coelenterata (b) Ctenophora
(c) Platyhelminthes (d) Annelida

19. Radial symmetry is often exhibited by animals having:
(a) One opening of alimentary canal
(b) Benthos and sedentary nature
(c) Aquatic mode of living
(d) Metamerism

20. Select the incorrect statement with respect to open circulatory system:
(a) Found in arthropods and non-cephalopod molluscs
(b) Blood moves very fast and at high pressure
(c) Blood flow to organs is not precisely regulated
(d) Capillaries are absent

21. The levels of organisation, body symmetry and coelom are similar in:
(a) Ctenophores and Platyhelminthes
(b) Annelids and arthropods
(c) Platyhelminthes and Aschelminthes
(d) Aschelminthes and annelids

22. Blood is conveyed to the organs through capillaries in:
(a) Arthropods (b) All molluscs
(c) Flatworms (d) Annelids

23. Which organism is not correctly matched with its body plan and type of coelom?
(a) Tube within tube, haemocoelom - honey bee
(b) Blind sac, acoelomate - flatworm
(c) Cell aggregate, coelomate - *Hydra*
(d) Tube within tube, pseudocoelom - *Ascaris*

24. Select the incorrect statement:
(a) All invertebrates are cold-blooded
(b) Non-chordates have a single hollow nerve cord
(c) All members of Animalia are multicellular
(d) Sponges, cnidarians and ctenophores are diploblastic while flatworms are triploblastic

25. When any plane passing through the central axis of the body divides the organism into two identical halves, it is called:
(a) Bilateral symmetry
(b) Radial symmetry
(c) Asymmetrical
(d) None of the above

26. Identify the organism and its corresponding symmetry with respect to the diagram given below:

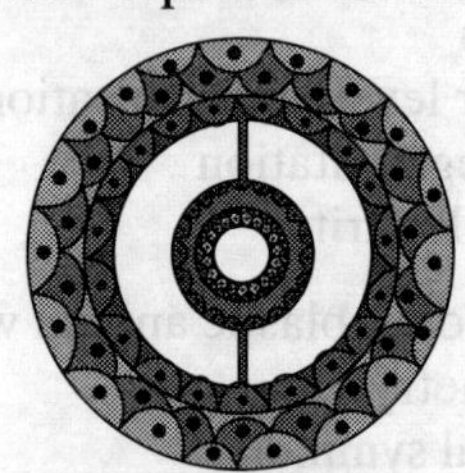

(a) *Taenia solium*, acoelom
(b) *Ascaris*, pseudocoelom
(c) *Pheretima*, true coelom
(d) *Periplaneta*, pseudocoelom

27. Identify the symmetry depicted in figures A and B, respectively:

(A)

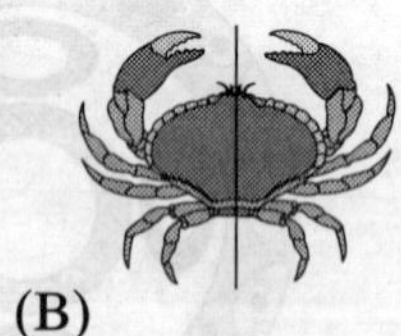

(B)

(a) Radial, bilateral
(b) Bilateral, radial
(c) Radial, asymmetry
(d) Bilateral, asymmetry

28. Animals with false body cavity (pseudocoelom) have/are:
(a) Metamerically segmented
(b) Radially symmetrical
(c) Incomplete digestive tract
(d) Tube within tube plan

29. Select the second largest phylum of the animal kingdom:
(a) Arthropoda (b) Mollusca
(c) Annelida (d) Echinodermata

30. Which of the following statements is correct about the diagram shown below?

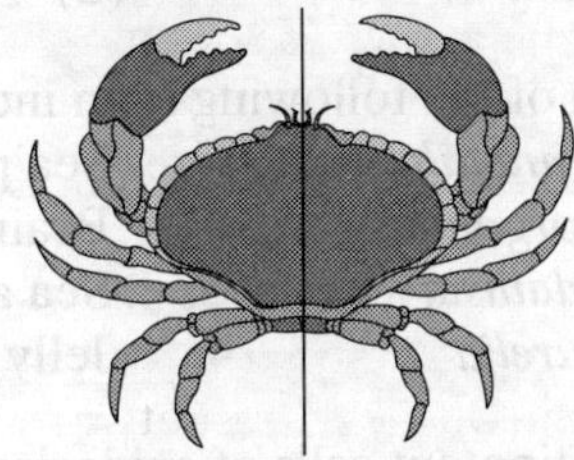

(a) Sea anemone represents this type of symmetry
(b) There are no right and left sides observed in the given animal
(c) The body of the animal can be divided into equal halves by any longitudinal plane passing through the centre
(d) The body can be divided into equal halves only by a single longitudinal plane passing through the centre

Porifera

31. All of the following are found in sponges, except:
(a) Choanocytes
(b) Spicules
(c) Aquatic mode of living
(d) Digestive tract and extracellular digestion

32. Which of the following is incorrect with respect to members of Porifera?
(a) Generally marine
(b) Mostly asymmetrical
(c) Cellular level of organisation
(d) Only sexual reproduction is seen

33. The pathway of water through the water canal system does not help in:
(a) Food gathering (b) Respiratory exchange
(c) Removal of waste (d) Locomotion

34. In sponges, the fertilisation is ___**A**___ and development is ___**B**___. Select the option to fill in the blanks A and B correctly:
(a) Internal, direct (b) External, indirect
(c) External, direct (d) Internal, indirect

35. Choanocytes:
(a) Are called collar cells
(b) Line the spongocoel
(c) Line the canals
(d) All of the above are correct

36. Sponges have:
(a) Exoskeleton made of spicules and spongin fibres
(b) Separate sexes
(c) Sexual and asexual reproduction
(d) More than one option is correct

37. Identify the freshwater sponge:
(a) *Sycon* (b) *Spongilla*
(c) *Euspongia* (d) *Aurelia*

38. The water current in the canals of sponges is maintained by:
(a) Pinacocytes (b) Choanocytes
(c) Archaeocytes (d) Myocytes

39. Venus flower basket is the name of:
(a) *Euspongia* (b) *Leucosolenia*
(c) *Adamsia* (d) *Euplectella*

40. The common bath sponge is:
(a) *Sycon* (b) *Euspongia*
(c) *Spongilla* (d) *Hyalonema*

41. ________ is called boring sponge and is detrimental to the pearl industry:
(a) *Cliona* (b) *Chalina*
(c) *Euplectella* (d) *Poterion*

42. The totipotent cells of sponges are:
(a) Thesocytes (b) Trophocytes
(c) Myocytes (d) Archaeocytes

43. Classification of sponges is based on:
(a) Reproduction (b) Symmetry
(c) Spicules (d) Branching

44. In sponges, the digestion of food takes place in:
(a) Choanocytes and amoebocytes
(b) Archaeocytes
(c) Pinacocytes
(d) Porocytes

45. Identify the diagram given below:

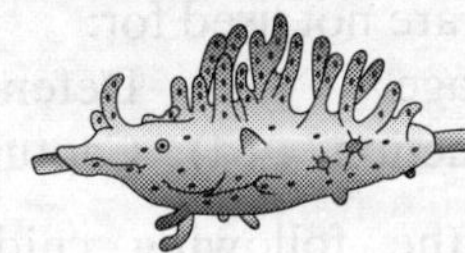

(a) *Spongilla* (b) *Euspongia*
(c) *Leucosolenia* (d) *Scypha*

Coelenterata

46. The diagrams below indicate the outline of the medusa and polyp forms.

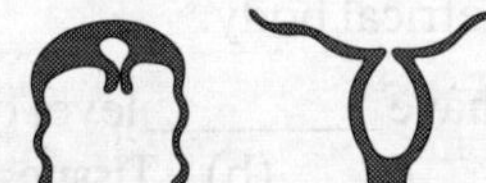

Which of the following cnidarians exist in both forms and exhibit metagenesis?
(a) *Hydra* (b) *Adamsia*
(c) *Obelia* (d) *Aurelia*

47. Which of the following cnidarians are corals and have a skeleton composed of calcium carbonate?
(a) *Aurelia* and *Obelia*
(b) *Meandrina* and *Gorgonia*
(c) *Hydra* and *Adamsia*
(d) *Hydra* and *Physalia*

48. Identify the incorrect option with respect to cnidarians:
(a) Mostly fresh water
(b) Radially symmetrical
(c) Sessile or free living
(d) Diploblastic

49. Cnidoblasts are present on/in:
(a) Tentacles
(b) Body
(c) Both (a) and (b)
(d) Gastrovascular cavity

50. The gastrovascular cavity of cnidarians is used for:
(a) Digestion and respiration
(b) Digestion and circulation
(c) Circulation and excretion
(d) Excretion and respiration

51. The medusa form of cnidarians is:
(a) Sessile (b) Cylindrical
(c) Umbrella-shaped (d) Both (a) and (b)

52. Which of the following is incorrect with respect to metagenesis in cnidarians?
(a) Polyps produce medusae asexually
(b) Medusae form polyps sexually
(c) Both are haploid
(d) Also called alternation of generations

53. Cnidoblasts are not used for:
(a) Anchorage (b) Defence
(c) Reproduction (d) Capture of prey

54. Which of the following cnidarians is called Portuguese Man-of-War?
(a) *Adamsia* (b) *Physalia*
(c) *Meandrina* (d) *Gorgonia*

55. Cnidarians have:
(a) Cell aggregate body plan
(b) Blind sac body plan
(c) Tube within tube plan
(d) Asymmetrical body

56. Cnidarians have _________ level of organisation:
(a) Cellular (b) Tissue
(c) Organ (d) Organ system

57. Digestion in cnidarians is:
(a) Intracellular (b) Extracellular
(c) Both (d) None

58. Which of the following is an incorrect match?
(a) *Pennatula* – Sea pen
(b) *Gorgonia* – Brain coral
(c) *Adamsia* – Sea anemone
(d) *Aurelia* – Jelly fish

59. The totipotent cells of cnidarians are called:
(a) Archaeocytes (b) Cnidoblasts
(c) Colloblasts (d) Interstitial cells

60. __________ larva is present in the life cycle of *Obelia*:
(a) Amphiblastula (b) Planula
(c) Nauplius (d) Cydippid

61. Polyp phase is absent in:
(a) *Aurelia* (b) *Obelia*
(c) *Hydra* (d) *Adamsia*

62. Identify the diagram given below:

(a) *Adamsia* (b) *Aurelia*
(c) *Physalia* (d) *Gorgonia*

Ctenophora

63. Ctenophores are:
(a) Fresh water (b) Marine
(c) Terrestrial (d) Both (a) and (b)

64. Ctenophores are also called:
(a) Sea walnuts (b) Comb jellies
(c) Cnidarians (d) Both (a) and (b)

65. In ctenophores, reproduction takes place:
(a) Only by asexual method
(b) Only by sexual method
(c) By both asexual and sexual method
(d) Only by fragmentation

66. Ctenophores do not have:
(a) Radial symmetry
(b) Organ level of organisation
(c) Diploblastic nature
(d) Indirect development

67. The body of ctenophores bears:
(a) Six external rows of ciliated comb plates
(b) Eight external rows of ciliated comb plates
(c) Eight external rows of flagellated comb plates
(d) Four external rows of flagellated comb plates

68. Digestion in ctenophores is:
(a) Intracellular only
(b) Extracellular only
(c) Both intracellular and extracellular
(d) None of the above

69. The property of ________ is well marked in ctenophores:
(a) Metamerism (b) Metagenesis
(c) Bioluminescence (d) Apolysis

70. Which of the following is incorrect with respect to ctenophores?
(a) Sexes are separate
(b) Fertilisation is external
(c) Indirect development
(d) Both (b) and (c)

71. Which of the following is not found in ctenophores?
(a) Cnidoblast (b) Colloblast
(c) Tentacle (d) Aquatic mode of life

72. Which of the following is a correct set with respect to examples of ctenophores?
(a) *Pleurobrachia and Ctenoplana*
(b) *Physalia* and *Fasciola*
(c) *Ascaris* and *Wuchereria*
(d) *Taenia* and *Adamsia*

73. The comb plates of ctenophores help in:
(a) Locomotion (b) Respiration
(c) Excretion (d) Reproduction

74. Select the incorrect statement:
(a) In ctenophores, the aboral surface has statocysts
(b) Ctenophores have colloblasts for adhesion
(c) Ctenophores have freshwater habitat and muscular foot for locomotion
(d) Members of Ctenophora and sponges are hermaphrodites

75. Identify the diagram given below:

(a) *Aurelia* (b) *Pleurobrachia*
(c) *Hydra* (d) *Fasciola*

Platyhelminthes

76. Flatworms are:
(a) Diploblastic and acoelomate
(b) Triploblastic and acoelomate
(c) Triploblastic and coelomate
(d) Diploblastic and coelomate

77. Platyhelminthes are mostly:
(a) Ectoparasites (b) Endoparasites
(c) Free living (d) Marine

78. ________ absorb nutrients from the host directly through the body surface:
(a) Tapeworm (b) Liver fluke
(c) *Planaria* (d) *Adamsia*

79. In flatworms, specialised cells called ________ help in osmoregulation and excretion:
(a) Renette cells
(b) Flame cells
(c) Chloragogen cells
(d) Archaeocytes

80. Which of the following is incorrect with respect to flatworms?
(a) Fertilisation is internal
(b) Sexes are not separate
(c) Development is direct
(d) Dorsoventrally flattened body

81. In tapeworms, respiration takes place:
(a) Through mouth
(b) Through suckers
(c) Anaerobically
(d) Through body surface

82. Liver fluke has:
(a) No mouth, no anus
(b) Single mouth, two anuses
(c) Double mouth, one anus
(d) Single opening acting as mouth and anus

83. Incomplete alimentary canal is present in:
(a) *Ascaris* (b) *Fasciola*
(c) *Wuchereria* (d) *Periplaneta*

84. ________ system is absent in tapeworm:
(a) Digestive (b) Excretory
(c) Nervous (d) Reproductive

85. Dog tapeworm is:
(a) *Taenia saginata*
(b) *Taenia solium*
(c) *Echinococcus granulosus*
(d) *Fasciola hepatia*

86. In which of the following flatworms is transverse binary fission observed?
(a) Liver fluke (b) *Planaria*
(c) Tapeworm (d) Bloodworm

87. Which of the following is incorrect with respect to the diagram given below?

(a) It is *Fasciola hepatica*
(b) Its primary host is sheep and secondary host is snail
(c) It is a hermaphrodite
(d) It has tissue level of organisation

88. Mark the incorrect statement with respect to the diagram and its structures A, B and C, illustrated below:

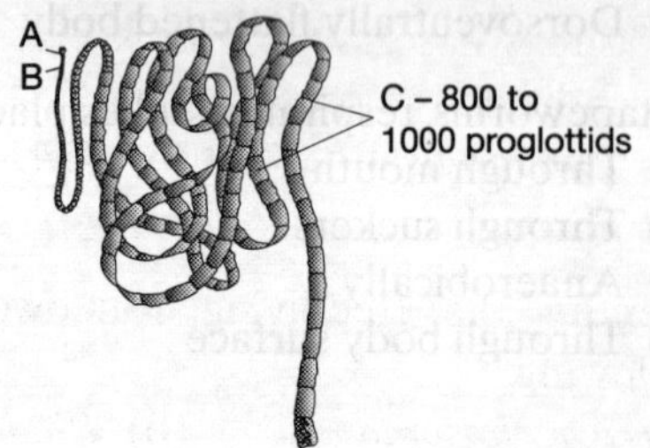

(a) It is *Taenia solium*
(b) Pigs are the primary host and humans are the intermediate host
(c) It absorbs nutrients from the host directly through the body surface
(d) The structure marked A is the scolex, which has hooks and suckers for attachment, B is the neck and C are body sections called proglottids

89. Which of the following is not the correct comparison between *Taenia* and *Fasciola*?
(a) Both are hermaphrodites
(b) Fertilisation is internal
(c) They have specialised cells called flame cells for excretion and osmoregulation
(d) *Taenia* shows true metameric segmentation, but it is absent in *Fasciola*

90. In which of these phylums is there a solid mesoderm?
(a) Cnidaria (b) Platyhelminthes
(c) Aschelminthes (d) Annelida

91. Which of the following shows higher power of regeneration?
(a) *Planaria* (b) Earthworm
(c) *Ascaris* (d) Cockroach

92. Choose the option that correctly matches the organism with its characters:
(a) *Obelia* – Triploblastic with metagenesis during life cycle
(b) *Ctenoplana* – Freshwater habitat with muscular foot for locomotion
(c) *Spongilla* – Marine and commonly called bath sponge
(d) Liver fluke – Digenetic parasite with suckers for attachment

Aschelminthes

93. Which of the following is found in lymph nodes?
(a) *Taenia* (b) *Ascaris*
(c) *Wuchereria* (d) *Trichinella spiralis*

94. Which of the following is incorrect with respect to roundworms?
(a) They are bilaterally symmetrical
(b) They are pseudocoelomates
(c) They are triploblastic
(d) All members are parasitic

95. The body of Aschelminthes is ___**A**___ in cross-section, and hence is called ___**B**___. Fill in the blanks A and B with the correct option:
(a) Circular, roundworm
(b) Dorsoventrally flattened, flatworm
(c) Circular, comb jelly
(d) Flattened, sponge

96. Which of the following is the correct reason for identification of sex of the animals A and B given in the diagram below?

(A) (B)

(a) A is male *Ascaris*, the posterior end is straight; and B is female *Ascaris* with curved posterior end
(b) A is female *Ascaris* with curved posterior end and B is male *Ascaris*
(c) A is female *Ascaris* with straight posterior end and B is male *ascaris*
(d) A is male *Ascaris* with curved posterior end and B is female *Ascaris* with straight posterior end

97. Which one of the following organisms has a mesoderm in the form of scattered pouches between ectoderm and endoderm during embryonic development?
(a) *Taenia* (b) Earthworm
(c) Sea anemone (d) *Enterobius*

98. The organism that shows bilateral symmetry, has pseudocoelom and muscular pharynx is:
(a) *Taenia* (b) *Ctenoplana*
(c) *Pheretima* (d) *Ascaris*

99. Identify the correct set with respect to examples of Aschelminthes:
(a) *Nereis, Pheretima, Hirudinaria*
(b) *Bombyx, Laccifer, Apis*
(c) *Pila, Loligo, Pinctada*
(d) *Ascaris, Ancylostoma, Wuchereria*

100. Infection with pinworm is called:
(a) Ascarias (b) Enterobiasis
(c) Oxyuriasis (d) Both (b) and (c)

101. Which of the following is a monogenetic parasite?
(a) *Taenia solium* (b) *Ascaris*
(c) *Fasciola* (d) *Pheretima*

102. All of the following are parasites of the small intestine in humans, except:
(a) *Ascaris* (b) *Taenia solium*
(c) *Ancylostoma* (d) *Enterobius*

103. Which of the following is a viviparous parasite?
(a) *Ascaris* (b) *Trichinella*
(c) *Enterobius* (d) *Ancylostoma*

104. The food of *Ascaris* in the intestine of the human host is:
(a) Lymph (b) Blood
(c) Chyme (d) Epithelial cell

105. Elephantiasis is transmitted by:
(a) Sandfly (b) Tabanus fly
(c) Housefly (d) *Culex* mosquito

106. Beneath the cuticle in *Ascaris* lies a thin epidermis having scattered nuclei and no separate cells, which is:
(a) Syncytial (b) Multicellular
(c) Uninucleate (d) Aseptate

107. Which of the following is incorrect with respect to Aschelminthes?
(a) Sexes are separate
(b) Often males are longer than females
(c) Fertilisation is internal
(d) Development may be direct or indirect

108. Which one of the following enters the body through the skin of feet in humans if he/she walks barefoot in contaminated soil?
(a) *Ascaris* (b) *Ancylostoma*
(c) *Enterobius* (d) *Trichuris trichura*

109. Which of the following is not a characteristic feature of the phylum Aschelminthes?
(a) They are dioecious
(b) Incomplete alimentary canal
(c) Pseudocoelomate animals
(d) Absence of segmentation

110. Which of the following is a monogenetic nematode parasite, whose adult worms live in the caecum, appendix and colon of humans?
(a) *Enterobius* (b) *Trichinella*
(c) *Ascaris* (d) *Loa loa*

111. Identify the free-living roundworm from the following:
(a) *Ancylostoma* (b) *Rhabditis*
(c) *Dracunculus* (d) *Taenia*

Annelida

112. Identify the incorrect option with respect to the characteristics of annelids:
(a) They may be aquatic
(b) They may be terrestrial
(c) They are always free-living and never parasitic
(d) They are triploblastic and coelomate

113. In annelids, locomotion takes place by:
(a) Longitudinal and circular muscles
(b) Jointed legs
(c) Wings
(d) Muscular foot

114. Aquatic annelids like _______ possess lateral appendages called parapodiae, that help in swimming:
(a) *Nereis* (b) *Pheretima*
(c) *Lumbricus* (d) *Hirudinaria*

115. Which of the following is incorrect with respect to annelids?
(a) They are metamerically segmented
(b) They exhibit organ system level of body organisation
(c) A closed circulatory system is present
(d) All annelids are monoecious

116. In annelids, ______ help in osmoregulation and excretion.
Fill in the blank with the correct option:
(a) Flame cells
(b) Green glands
(c) Malpighian tubules
(d) Nephridia

117. In annelids, respiration takes place through:
(a) Trachea (b) Book lungs
(c) Moist skin (d) Lungs

118. True coelom appeared for the first time in:
(a) Nematodes (b) Annelids
(c) Arthropods (d) Molluscs

119. Identify the character which is not common among *Nereis*, earthworm and leech:
(a) Segmented body
(b) Longitudinal and circular muscles in body wall
(c) Parapodiae and setae
(d) Triploblastic coelomate body

120. Identify the incorrect statement:
(a) *Nereis*, an aquatic form, is dioecious
(b) Earthworms and leeches are monoecious
(c) *Hirudinaria* is a blood-sucking leech
(d) Both asexual and sexual methods of reproduction are seen in annelids

121. Botryoidal tissue is present in the body cavity of:
(a) *Ascaris* (b) *Hirudinaria*
(c) *Nereis* (d) *Pheretima*

122. Which of the following is a matching set of phylum and its three examples?
(a) Ctenophora – *Gorgonia, Ctenoplana, Hormiphora*
(b) Platyhelminthes – *Taenia, Fasciola, Wuchereria*
(c) Aschelminthes – *Ascaris, Apis, Laccifer*
(d) Annelida – *Nereis, Pheretima, Hirudinaria*

123. Consider the diagrams (A) and (B) given below and choose the incorrect option:

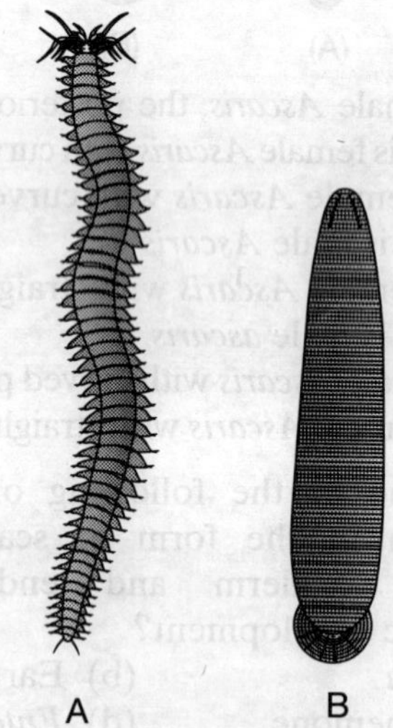

(a) A is *Nereis*, B is *Hirudinaria*
(b) A is a marine free-living annelid, while B is an ectoparasite
(c) A has parapodiae and setae while B does not
(d) A is monoecious while B is dioecious

Arthropoda

124. Over two-thirds of all named species on earth are:
(a) Annelids (b) Arthropods
(c) Molluscus (d) Echinoderms

125. Choose the incorrect match:
(a) Gregarious pest – *Locusta*
(b) Living fossil – *Limulus*

(c) Vectors of disease – *Laccifer*
(d) Economically important insects – *Apis*

126. Arthropoda are best described as:
(a) Radially symmetrical, having wings
(b) Triploblastic, having only abdominal appendages
(c) Having exoskeleton and jointed appendages
(d) Acoelomate animals having unjointed legs

127. Which of the following is a common character of earthworm and cockroach?
(a) Nephridia
(b) Closed circulatory system
(c) Malpighian tubule
(d) Solid ventral nerve cord

128. Mark the odd one out with respect to respiratory organs of arthropods:
(a) Book lungs (b) Gills
(c) Tracheal system (d) Antennal glands

129. Which of the following animals shows metameric segmentation and has an open circulatory system?
(a) Annelids and arthropods
(b) Only arthropods
(c) Only annelids
(d) Annelids and chordates

130. The body of arthropods is covered with thick, tough and non-living exoskeleton made of:
(a) Cellulose (b) Chitin
(c) Silica (d) Keratin

131. The excretory organs of crustaceans are:
(a) Nephridia
(b) Antennal glands
(c) Malpighian tubules
(d) Coxal glands

132. The vector for kala azar and oriental sore is:
(a) Tsetse fly (b) Sandfly
(c) Kissing fly (d) Fruit fly

133. Which of the following is incorrect with respect to arthropods?
(a) They are mostly dioecious
(b) Fertilisation is mostly external
(c) They are mostly oviparous
(d) Development may be direct or indirect

134. The study of insects is called:
(a) Ornithology (b) Entomology
(c) Ichthyology (d) Acarology

135. Economically important insects are:
(a) *Apis, Lepisma, Bombyx*
(b) *Apis, Bombyx, Laccifer*
(c) *Anopheles, Culex, Apis*
(d) *Culex, Limulus, Lepisma*

136. *Aedes* is the vector of:
(a) Dengue fever (b) Yellow fever
(c) Chikungunya (d) All of the above

137. Which of the following is also called hexapoda?
(a) Insecta (b) Crustacea
(c) Arachnida (d) Mollusca

138. What are the major respiratory and excretory organs of prawns?
(a) Gills and green glands
(b) Book lungs and coxal glands
(c) Gills and kidneys
(d) Trachea and malpighian tubules

139. Which of the following sets of animals belong to a single taxonomic group?
(a) Cuttlefish, jelly fish, silver fish
(b) Sea fan, sea cucumber, tsetse fly
(c) Housefly, sandfly, tsetse fly
(d) Glow worm, earthworm, sand worm

140. Which of the following is correct with respect to the organism, its circulatory system and respiratory system?

	Organism	Circulatory system	Respiratory organ
(a)	*Anopheles*	Open	Tracheal system
(b)	*Pheretima*	Open	Cutaneous respiration
(c)	*Locusta*	Closed	Tracheal system
(d)	*Nereis*	Open	Gills

141. One of the representatives of phylum Arthropoda is:
(a) Puffer fish (b) Cuttlefish
(c) Silver fish (d) Flying fish

142. Two common characters found in centipede, cockroach and crab are:
(a) Book lungs and antennae
(b) Green glands and tracheae
(c) Jointed legs and chitinous exoskeleton
(d) Gills and malpighian tubules

143. Which one of the following is not correctly matched?
(a) *Culex* – Filariasis
(b) *Aedes* – Yellow fever
(c) *Anopheles* – Leishmaniasis
(d) *Glossina* – Sleeping sickness

144. Which one of the following has an open circulatory system?
(a) *Pheretima* (b) *Periplaneta*
(c) *Nereis* (d) *Octopus*

145. Consider the figures of animals A, B, C and D. Select the correct answers with respect to the common characteristics of two of these animals:

A.  B. 

C. 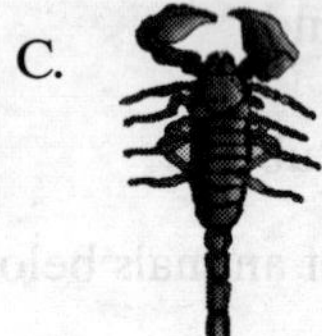 D. 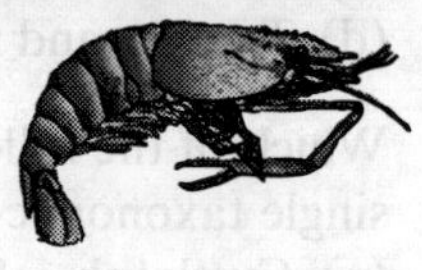

(a) A and C have book lungs for respiration
(b) C and D use malpighian tubules for excretion
(c) A and B have jointed appendages, tracheal respiration and excretion through malpighian tubules
(d) A and B have cutaneous respiration

Mollusca

146. Molluscs are:
(a) The largest phylum of Animalia
(b) Unsegmented
(c) Metamerically segmented
(d) Radially symmetrical

147. Identify the option whose members have haemocoelom:
(a) Arthropoda and Annelida
(b) Annelida and Mollusca
(c) Annelida and Aschelminthes
(d) Arthropoda and Mollusca

148. The body of molluscs is divided into:
(a) Head, thorax and abdomen
(b) Head, muscular foot and visceral hump
(c) Cephalothorax and abdomen
(d) Proboscis, collar and trunk

149. In molluscs, mouth contains a file-like rasping organ for feeding called:
(a) Chelicerae (b) Radula
(c) Osphradia (d) Ommatophore

150. Identify the incorrect statement with respect to Mollusca:
(a) They are usually dioecious
(b) They are oviparous
(c) They have indirect development
(d) Their digestive system is incomplete

151. Match the options given in the columns:

Column I	Column II
A. *Pila*	(i) Cuttlefish
B. *Pinctada*	(ii) Squid
C. *Sepia*	(iii) Apple snail
D. *Loligo*	(iv) Pearl oyster

(a) A – (ii), B – (i), C – (iii), D – (iv)
(b) A – (iv), B – (ii), C – (iii), D – (i)
(c) A – (iii), B – (iv), C – (i), D – (ii)
(d) A – (iii), B – (i), C – (iv), D – (ii)

152. Which of the following is not a characteristic of phylum Mollusca?
(a) They are bilaterally symmetrical and triploblastic
(b) Respiratory organs are feather-like gills and present in muscular foot
(c) Body is covered with calcareous shell
(d) They are terrestrial or aquatic (marine or freshwater)

153. Sea hare is:
(a) *Sepia* (b) *Turbinella*
(c) *Sepia* (d) *Aplysia*

154. Which of the following is mismatched?
(a) *Octopus* – Devil fish
(b) *Dentalium* – Tusk shell
(c) *Chaetopleura* – Freshwater mussel
(d) *Limulus* – King crab

155. Visceral mass undergoes torsion in:
(a) *Pila* (b) *Octopus*
(c) *Sepia* (d) *Unio*

156. Osphradium of *Pila* is meant for:
(a) Grinding food
(b) Excretion
(c) Sensing chemical nature of water
(d) Gaseous transport

157. Identify the odd one out with respect to cephalopod molluscs:
(a) *Octopus* (b) *Sepia*
(c) *Dentalium* (d) *Loligo*

158. Consider the diagrams given below and choose the correct option, identifying the labels (A) and (B), respectively:

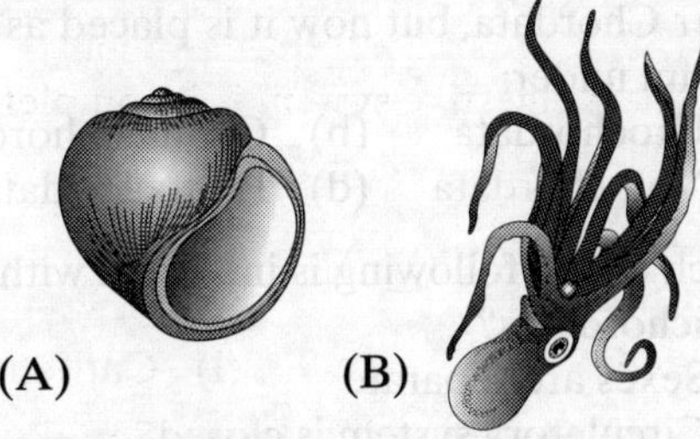

(a) *Pila, Sepia*
(b) *Pila, Octopus*
(c) *Dentalium, Octopus*
(d) *Aplysia, Sepia*

Echinodermata

159. In which of the following phylums are the larvae bilaterally symmetrical while adults are radially symmetrical?
(a) Mollusca (b) Echindermata
(c) Hemichordata (d) Annelida

160. Aristotle's lantern, which helps in mastication, is a characteristic feature of:
(a) *Asterias* (b) Sea urchin
(c) Sea cucumber (d) Brittle star

161. Echinoderms are:
(a) Exclusively marine
(b) Both marine and freshwater animals
(c) Terrestrial
(d) Metamerically segmented

162. Which of the following is not a function of the water-vascular system in echinoderms?
(a) Locomotion
(b) Respiration
(c) Capture and transport of food
(d) Reproduction

163. The vital system absent in echinoderms is:
(a) Reproductive system
(b) Excretory system
(c) Skeletal system
(d) Digestive system

164. Which of the following is incorrect with respect to echinoderms?
(a) Sexes are separate
(b) Both asexual and sexual reproduction is seen
(c) Fertilisation is usually external
(d) Development is indirect with free-swimming larva

165. Identify the incorrect match:
(a) *Asterias* – Starfish
(b) *Echinus* – Sea urchin
(c) *Antedon* – Brittle star
(d) *Cucumaria* – Sea cucumber

166. Identify the incorrect option with respect to echinoderms:
(a) They have exoskeleton of calcareous ossicles
(b) They are called spiny bodied animals
(c) They are triploblastic and coelomate
(d) Digestive system is complete

167. The organs of locomotion is echinoderms are:
(a) Parapodia (b) Tube feet
(c) Pseudopodia (d) Tentacles

168. Echinoderms have:
(a) Mesodermal endoskeleton
(b) Enterocoelom
(c) Deuterostomous development
(d) All of the above

169. Echinoderms are:
(a) Acoelomate, spiny and marine
(b) Coelomate, spiny and marine
(c) Coelomate, spiny and fresh water
(d) Jointed appendages, spiny and fresh water

170. Arms are absent in:
(a) Sea urchin (b) Starfish
(c) Brittle star (d) All of the above

171. Echinoderms lack head, heart and excretory organs, and yet, from an evolutionary point of view, they have been placed at the top of the invertebrate phyla because of:
(a) Power of regeneration
(b) Exclusively marine habitat
(c) Presence of enterocoelom
(d) Radial symmetry

172. The water-vascular system is found in:
(a) Only echinoderms
(b) Only molluscs
(c) Both echinoderms and Porifera
(d) None of the above

173. In which of the following echinoderms is the stalk found for attachment with the substratum:
(a) Starfish (b) Sea lily
(c) Sea cucumber (d) Brittle star

174. Echinodermata is a subgroup of Bilateria animals. What common feature does it share with *Hydra*?
(a) Acoelomate nature
(b) Direct development
(c) Radial symmetry
(d) Organ system level of organisation

175. The diagram below illustrates different animals. Which of these animals have an endoskeleton of calcareous ossicles and a spiny bony?

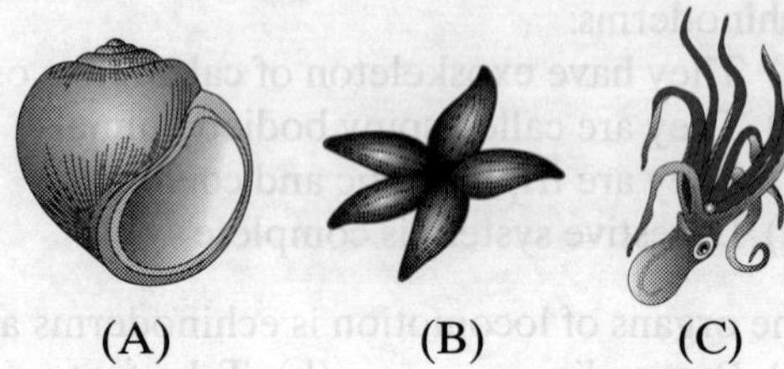

(a) A and B (b) B only
(c) C only (d) B and C

176. Which of the following is not a correct identification or characteristic feature of the organism depicted in the diagram?

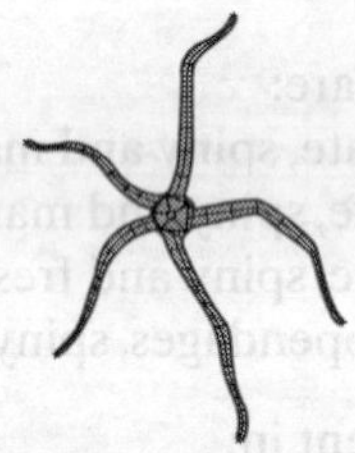

(a) The organism is *Ophiura* (brittle star)
(b) They are exclusively marine
(c) They are coelomate and segmented animals
(d) They are radially symmetrical but their larvae have bilateral symmetry

Hemichordata

177. Hemichordates are:
(a) Fresh water (b) Marine
(c) Terrestrial (d) Both (a) and (b)

178. The excretory organs of hemichordates are:
(a) Green glands
(b) Flame cells
(c) Proboscis glands
(d) Malpighian tubules

179. Examples of hemichordates are:
(a) *Balanoglossus* and *Asterias*
(b) *Saccoglossus* and *Echinus*
(c) *Ophiura* and *Chaetopleura*
(d) *Balanoglossus* and *Saccoglossus*

180. Hemichordata was earlier considered a subphylum under Chordata, but now it is placed as a separate phylum under:
(a) Urochordata (b) Cephalochordata
(c) Non-chordata (d) Protochordata

181. Which of the following is incorrect with respect to hemichordates?
(a) Sexes are separate
(b) Circulatory system is closed
(c) Development is indirect
(d) Fertilisation is external

182. The body of hemichordates is cylindrical and is composed of:
(a) Head, thorax and abdomen
(b) Anterior proboscis, collar and trunk
(c) Cephalothorax and abdomen
(d) Head and trunk

183. In hemichordates, respiration takes place through:
(a) Gills (b) Book lungs
(c) Book gills (d) Lungs

184. Hemichordates are:
(a) Radially symmetrical, coelomate and marine
(b) Bilaterally symmetrical, coelomate and marine
(c) Bilaterally symmetrical, acoelomate and marine
(d) Bilaterally symmetrical, coelomate and fresh water

185. The notochord-like structure of hemichordates is:
(a) Glomerulus (b) Stomochord
(c) Pallium (d) Mantle

186. Identify the organism in the diagram below, and its corresponding character:

(a) *Balanoglossus* – Body composed of anterior proboscis, a collar and long trunk
(b) *Saccoglossus* – Has a closed circulatory system
(c) *Asterias* – Worm-like, marine animals
(d) *Ophiura* – Spiny-bodied animal

Chordata

Protochordata and Agnatha

187. Which of the following is incorrect with respect to chordates?
(a) Heart is dorsal
(b) Pharynx is perforated by gills
(c) A post-anal tail is present
(d) Central nervous system is dorsal, hollow and single

188. Which of the following is not a subphylum of Chordata?
(a) Urochordata (b) Cephalochordata
(c) Vertebrata (d) Hemichordata

189. Which of the following is a member of Cephalochordata?
(a) *Ascidia* (b) *Salpa*
(c) *Branchiostoma* (d) *Doliolum*

190. Urochordata is also called:
(a) Tunicata (b) Vertebrata
(c) Agnatha (d) Cephalochordate

191. Identify the incorrect statement:
(a) All vertebrates are chordates
(b) All chordates are not vertebrates
(c) In cephalochordates, notochord is replaced by a cartilaginous or bony vertebral column
(d) Members of Chordata are characterised by the presence of a notochord

192. Subphyla Urochordata and Cephalochordata are often referred to as:
(a) Agnatha (b) Protochordata
(c) Vetebrata (d) Hemichordata

193. Vertebrates have:
(a) Ventral muscular heart with two, three or four chambers
(b) Kidneys for excretion and osmoregulation
(c) Paired appendages which may be fins or limbs
(d) All of the above

194. In urochordates, notochord is:
(a) Absent
(b) Present in adult life
(c) Present in only in larval tail
(d) Extends from head to tail region

195. All protochordates are:
(a) Freshwater forms
(b) Marine
(c) Both marine and fresh water
(d) Terrestrial

196. In cephalochordates, notochord is:
(a) Present only in larval tail
(b) Extends from head to tail region
(c) Persistent throughout life
(d) Both (b) and (c)

197. Cyclostomes are marine, but migrate to fresh water:
(a) After spawning
(b) For spawning
(c) After metamorphosis
(d) None of the above

198. Identify the correct set of jawless vertebrates:
(a) *Petromyzon* and *Clarias*
(b) *Myxine* and *Pristis*
(c) *Petromyzon* and *Myxine*
(d) *Ascidia* and *Amphioxus*

199. Retrogressive metamorphosis is shown by:
(a) *Herdmania* (b) *Myxine*
(c) *Petromyzon* (d) *Branchiostoma*

200. In _______, excretion is carried out by flame cells:
(a) Hemichordates
(b) Cyclostomes
(c) Cephalochordates
(d) Tunicates

201. Which of the following is a member of Tunicate?
(a) Sea fan (b) Sea squirt
(c) Sea lemon (d) Sea star

202. The chordate with open circulatory system is:
(a) Cephalochordates (b) Pisces
(c) Urochordate (d) Cyclostomes

203. Cranium and vertebral column are cartilaginous in:
(a) Cephalochordates (b) Urochordates
(c) Cyclostomes (d) Hemichordates

204. Select the incorrect statement with respect to Agnatha:
(a) They include lamprey and hag fish
(b) Circulation is of closed type
(c) All living members are endoparasites on some fish
(d) Jaw is absent

205. Body is devoid of scales and paired fins in:
(a) *Pristis* (b) Sea horse
(c) *Petromyzon* (d) *Exocoetus*

206. Acraniata includes:
(a) Urochordata
(b) Protochordata
(c) Cephalochordata
(d) All of the above

207. Notochord is absent in:
(a) *Myxine* (b) *Lancelet*
(c) Larval tunicate (d) Adult tunicate

208. Which of the following is incorrect with respect to characteristics of urochordates?
(a) Open circulatory system
(b) Cranium is absent
(c) Development is direct
(d) Neural gland as excretory organ

209. Identify the organism in the figure given below:

(a) *Petromyzon* (b) *Ascidia*
(c) *Balanoglossus* (d) *Myxine*

210. Select the incorrect statement for the organism in the diagram given below:

(a) They are marine but migrate for spawning into rivers
(b) They are ectoparasites on other fish
(c) They have elongated body with 6–15 pairs of gill slits
(d) After spawning, the adult migrates to the ocean where it lives

Pisces

211. *Carcharodon* is commonly called:
(a) Saw fish
(b) Great white shark
(c) Sting ray
(d) Rohu

212. Which of the following is a fish?
(a) Sea cow (b) Sea lion
(c) Sea horse (d) Sea hare

213. In Chondrichthyes, teeth are modified ______ scales, which are _____ directed:
(a) Placoid, backwardly
(b) Placoid, forwardly
(c) Ctenoid, forwardly
(d) Ctenoid, backwardly

214. Which of the following is incorrect with respect to members of Chondrichthyes?
(a) They are marine animals
(b) Mouth is located ventrally
(c) Notochord is persistent throughout life
(d) Swim bladder is present

215. An aquatic organism with prehensile tail is:
(a) *Chameleon* (b) *Hippocampus*
(c) *Salamandra* (d) *Rana*

216. The member of Chondrichthyes that possesses poison sting is:
(a) *Torpedo* (b) *Trygon*
(c) *Clarias* (d) *Pristis*

217. Select the incorrect statement with respect to members of Chondrichthyes:
(a) Sexes are separate
(b) In males, pectoral fins bear claspers
(c) They have internal fertilisation
(d) Many of them are viviparous

218. Four pairs of gills, covered by operculum, is a character of:
(a) *Pristis* (b) *Trygon*
(c) *Labeo* (d) *Scoliodon*

219. Which of the following is an edible marine fish?
(a) *Exocoetus* (b) *Labeo*
(c) *Catla* (d) *Clarias*

220. A fish that feeds on mosquito larvae and is used for biological control of malaria is:
(a) Bombay duck (b) *Pterophyllum*
(c) *Gambusia* (d) *Hilsa*

221. *Torpedo* is commonly called:
(a) Globe fish (b) Saw fish
(c) Magur (d) Electric ray

222. Isinglass is a gelatinous product obtained from the air bladder of:
(a) Dog fish (b) Saw fish
(c) *Catla* (d) Sting ray

223. The branch of biology dealing with the study of Pisces is:
(a) Ornithology (b) Herpetology
(c) Saurology (d) Ichthyology

224. Mark the odd one out with respect to characters of Osteichthyes:
(a) Mouth is mostly terminal
(b) Skin is tough, containing minute placoid scales
(c) They are mostly oviparous
(d) Development is direct

225. Which of the following is an aquarium fish?
(a) *Clarias* (Magur)
(b) *Catla* (Katla)
(c) *Pristis* (Saw fish)
(d) *Betta* (Fighting fish)

226. _______ is a kind of untanned leather with a rough granulated surface and obtained from the skin of shark:
(a) Isinglass (b) Shagreen
(c) Mermaid's purse (d) Pecten

227. Which of the following is not applicable for the organism whose diagram is given below?

(a) It is a freshwater edible fish
(b) Air bladder is present to regulate buoyancy
(c) It has internal fertilisation
(d) They are cold-blooded animals

228. Identify the animal in the diagram below:

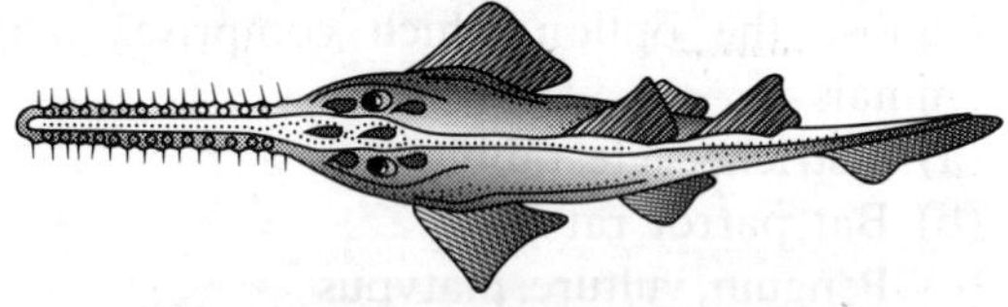

(a) *Carcharodon* (b) *Hippocampus*
(c) *Catla* (d) *Pristis*

Amphibia

229. Which of the following is a limbless amphibian?
(a) *Bufo* (b) *Rana*
(c) *Hyla* (d) *Ichthyophis*

230. The tree frog is:
(a) *Rana* (b) *Hyla*
(c) *Bufo* (d) *Salamandra*

231. Which of the following characters is not related to amphibians?
(a) Body is divisible into head and trunk
(b) A tympanum represents the ear
(c) Heart is three-chambered with two ventricles and one auricle
(d) Skin is moist without scales

232. Which of the following is not applicable to the animal in the diagram given below?

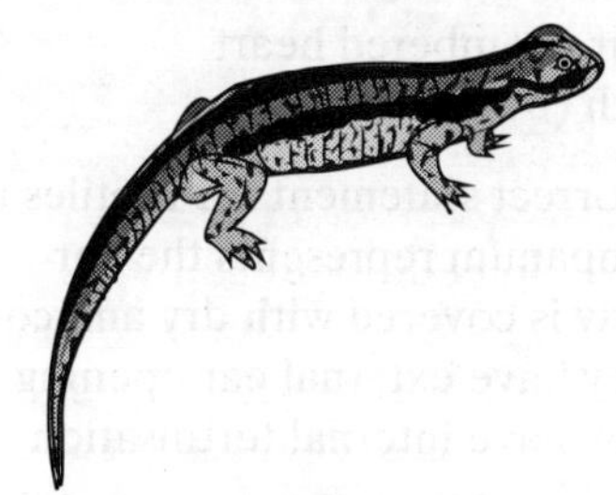

(a) The animal is *Salamandra*
(b) They are poikilothermous
(c) Fertilisation is external
(d) Development is direct

233. Frog is:
(a) Ammonotelic (b) Ureotelic
(c) Uricotelic (d) Aminotelic

234. Nucleated RBCs are present in:
(a) Rat (b) Rabbit
(c) Frog (d) Squirrel

235. Which of the following is not present in amphibians?
(a) Scales (b) Eyelids
(c) Kidneys (d) Tympanum

236. Which of the following is not true for all amphibians?
(a) All have tail
(b) Limbs
(c) Three-chambered heart
(d) Both (a) and (b)

237. Tadpole larva of frog respires through:
(a) Gills (b) Skin
(c) Lungs (d) All of the above

238. Amphibians are characterised by all of the following, except:
(a) Thin and moist skin
(b) Amnion
(c) Hepatic portal system
(d) Cloaca

Reptilia

239. The class name Reptilia refers to:
(a) Dentition
(b) Creeping or crawling mode of locomotion
(c) Habitat
(d) Dry and cornified skin

240. Which of the following is a common character of humans and crocodile?
(a) Enucleated RBCs
(b) Thecodont dentition

(c) Four-chambered heart
(d) Both (b) and (c)

241. The incorrect statement for reptiles is:
(a) Tympanum represents the ear
(b) Body is covered with dry and cornified skin
(c) They have external ear opening
(d) They have internal fertilisation

242. Find the incorrect match:
(a) *Testudo* – Tortoise
(b) *Naja* – Krait
(c) *Vipera* – Viper
(d) *Calotes* – Garden lizard

243. Which of the following shed their scales as skin casts?
(a) Crocodile and turtle
(b) Snakes and lizard
(c) Frog and crocodile
(d) Birds and crocodile

244. Reptiles, birds and mammals can be grouped under
(a) Amniotes
(b) Anamniotes
(c) Protochordates
(d) Homeothermous

245. Fangs of snakes are modified:
(a) Mandibular teeth (b) Scales
(c) Limbs (d) Maxillary teeth

246. Reptiles are successful as terrestrial animals due to all of the options, except:
(a) Body covered with scales
(b) Amnion
(c) Shelled eggs
(d) External fertilisation

247. Poisonous land snake is:
(a) *Python* (b) *Vipera*
(c) *Calotes* (d) *Chameleon*

248. Which of the following is the correct statement?
(a) Most lizards are viviparous
(b) Poison glands of snakes are modified parotid glands
(c) All reptiles possess three-chambered heart
(d) Viper is a neurotoxic snake

249. Jacobson's organ is a vomeronasal organ in the roof of the buccal cavity of snakes and lizards; these organs have:
(a) Gustatory receptors
(b) Olfactory receptors
(c) Thermoreceptors
(d) Propioreceptors

250. The largest living lizard in the world is:
(a) *Draco*
(b) *Hemidactylus*
(c) *Varanus komodensis*
(d) *Testudo*

251. Which of the following is/are neurotoxic snakes?
(a) *Naja* (cobra)
(b) *Bangarus* (krait)
(c) *Hydrophis* (sea snake)
(d) All of the above

252. Flying lizard is:
(a) *Draco* (b) *Hemidactylus*
(c) *Chameleon* (d) *Heloderma*

Aves

253. Which of the following is the correct statement with respect to aves?
(a) Air sacs help in excretion
(b) Skin of birds has many glands
(c) Preen gland is present at the base of tail
(d) Heart is three-chambered

254. Identify the incorrect statement regarding aves:
(a) Feathers are present
(b) Most of them can fly
(c) Hind limbs are modified into wings
(d) Long bones are hollow with air cavities

255. Choose the option which comprises oviparous animals:
(a) Ostrich, eagle, rabbit
(b) Bat, parrot, rat
(c) Penguin, vulture, platypus
(d) Whale, sparrow, dolphin

256. The digestive tract of birds has additional chambers called:
(a) Crop and hepatic caecae
(b) Crop and gizzard
(c) Hepatic caecae and rectal papillae
(d) Proventriculus and malphighian tubules

257. Which of the following is not a character of aves?
(a) In females, left ovary is functional
(b) Scales are present on hind limbs
(c) They are warm blooded
(d) Beak possesses homodont teeth

258. Find the incorrect match:
(a) *Neophron* – Eagle
(b) *Columba* – Pigeon
(c) *Struthio* – Ostrich
(d) *Pavo* – Peacock

259. In which of the following are the forelimbs modified into flippers?
(a) *Aptenodytes* (b) *Psittacula*
(c) *Corvus* (d) *Neophron*

260. The largest bird is:
(a) *Columba* (b) Emu
(c) Ostrich (d) Cassowary

261. Poisonous bird is:
(a) *Gallus*
(b) *Tyto*
(c) Hooded pitohui
(d) Swift

262. Feathers used for flying are:
(a) Down feathers (b) Coverts
(c) Contour (d) Quills

263. Smallest flightless bird is:
(a) *Struthio* (b) *Apteryx*
(c) *Dromaius* (d) Cassowary

264. Sound box of birds is called:
(a) Larynx (b) Syrinx
(c) Pygostyle (d) Urostyle

265. In birds, the sternum is modified into:
(a) Synsacrum (b) Pygostyle
(c) Keel (d) Furcula

266. Two clavicles of birds are fused to form a wishbone, also called:
(a) Furcula (b) Urostyle
(c) Pecten (d) Syrinx

Mammalia

267. An egg laying mammal is:
(a) *Rhinoceros*
(b) Kangaroo
(c) *Ornithorhynchus*
(d) *Macropus*

268. Corpus callosum is found in the brain of:
(a) Frog (b) Salamander
(c) Rabbit (d) Cobra

269. The most unique feature of mammals is the presence of:
(a) Limbs
(b) Four-chambered heart
(c) Mammary gland
(d) Homeothermy

270. All mammals have:
(a) Hair (b) Scrotum
(c) Pinna (d) Diaphragm

271. Which of the following is not a characteristic feature of mammals?
(a) Seven cervical vertebrae
(b) Ten pairs of cranial nerves
(c) Heterodont dentition
(d) Left aortic arch in circulatory system

272. Tusks of elephants are modified:
(a) Incisors (b) Canines
(c) Premolar (d) Molar

273. Dicondylic skull is found in:
(a) Aves and mammals
(b) Amphibians and reptiles
(c) Reptiles and mammals
(d) Amphibians and mammals

274. Scrotum is absent in:
(a) Platypus and elephant
(b) Sea cow and whale
(c) Whale and platypus
(d) All of the above

275. Identify the incorrect match:
(a) *Pteropus* – Monkey
(b) *Macropus* – Kangaroo
(c) *Equus* – Horse
(d) *Felis* – Cat

276. Which of the following is an oviparous mammal?
(a) *Ornithorhynchus* (b) *Rattus*
(c) *Balaenoptera* (d) *Delphinus*

277. Which of the following is a common character of aves and mammals?
(a) Three-chambered heart
(b) Enucleated RBCs
(c) Pneumatic bones
(d) Homeothermy

278. Which of the following is/are characters of *Ornithorhynchus?*
(a) Oviparous
(b) Testis is abdominal
(c) Pinna is absent
(d) All of the above

279. Mammary glands are modified:
(a) Sebaceous glands
(b) Sweat glands
(c) Tarsal glands
(d) Lacrimal glands

280. Choose the correct set of aquatic mammals:
(a) *Felis* and *Camelus*
(b) *Panthera tigris* and *Rattus*
(c) *Delphinus* and *Balaenoptera*
(d) *Pteropus* and *Macaca*

281. The presence of two uteri and vaginae are characteristic of:
(a) *Ornithorhynchus* (b) *Macropus*
(c) *Rattus* (d) *Canis*

NCERT EXEMPLAR QUESTIONS

1. In some animal groups, the body is found divided into compartments with serial repitition of at least some organs. This characteristic feature is called:
(a) Metastasis (b) Metamerism
(c) Metagenesis (d) Metamorphosis

2. Given below are types of cells present in some animals. Which of the following cells can differentiate to perform different functions?
(a) Choanocytes (b) Interstitial cells
(c) Gastrodermal cells (d) Nematocytes

3. Which one of the following sets of animals share a four-chambered heart?
(a) Amphibian, Reptiles, Birds
(b) Crocodiles, Birds, Mammals
(c) Crocodiles, Lizards, Turtles
(d) Lizards, Mammals, Birds

4. Which of the following pairs of animals has non-glandular skin
(a) Snake and Frog (b) *Chameleon* and Turtle
(c) Frog and Pigeon (d) Crocodile and Tiger

5. Birds and mammals share one of the following characteristics as a common feature:
(a) Pigmented skin (b) Pneumatic bones
(c) Viviparity (d) Warm blooded

6. Which one of the following sets of animals belongs to a single taxonomic group?
(a) Cuttlefish, Jellyfish, Silverfish, Dogfish, Starfish
(b) Bat, Pigeon, Butterfly
(c) Monkey, Chimpanzee, Humans
(d) Silkworm, Tapeworm, Earthworm

7. Which one of the following statements is incorrect?
(a) Mesoglea is present in between ectoderm and endoderm in *Obelia*
(b) Exhibits radial symmetry, *Asterias*
(c) *Fasciola* is a pseudocoelomate animal
(d) *Taenia* is a triploblastic animal

8. Which one of the following statements is incorrect?
(a) In cockroaches and prawns, excretion of waste material occurs through malpighian tubules
(b) In ctenophores, locomotion is mediated by comb plates
(c) In *Fasciola*, flame cells help in excretion
(d) Earthworms are hermaphrodites and yet cross fertilisation take place among them

9. Which one of the following is oviparous?
(a) Platypus (b) Flying fox (Bat)
(c) Elephant (d) Whale

10. Which one of the following is a non-poisonous snake?
(a) Cobra (b) Viper
(c) Python (d) Krait

11. Match the following list of animals with their level of organisation:

Column I	Column II
A. Organ level	(i) *Pheretima*
B. Cellular aggregate level	(ii) *Fasciola*
C. Tissue level	(iii) *Spongilla*
D. Organ system level	(iv) *Obelia*

(a) A – (iv), B – (i), C – (ii), D – (iii)
(b) A – (iv), B – (i), C – (iii), D – (ii)
(c) A – (ii), B – (iii), C – (iv), D – (i)
(d) A – (i), B – (iv), C – (iii), D – (ii)

12. Body cavity is the cavity present between body wall and gut wall. In some animals, the body cavity is not lined by mesoderm. Such animals are called:
(a) Acoelomates
(b) Pseudocoelomates
(c) Coelomates
(d) Haemocoelomates

13. Match the columns and choose the correct option:

Column I	Column II
A. Porifera	(i) Canal system
B. Aschelminthes	(ii) Water-vascular system
C. Annelida	(iii) Muscular pharynx
D. Arthropoda	(iv) Jointed appendages
E. Echinodermata	(v) Metameres

(a) A – (ii), B – (iii), C – (v), D – (iv), E – (i)
(b) A – (ii), B – (v), C – (iii), D – (iv), E – (i)
(c) A – (i), B – (iii), C – (v), D – (iv), E – (ii)
(d) A – (i), B – (v), C – (iii), D – (iv), E – (ii)

PRACTICE TEST PAPER

Chordates

1. Retrogressive metamorphosis is seen in:
   (a) Hemichordates
   (b) Urochordates
   (c) Cephalochordates
   (d) Vertebrates

2. The excretory organs of *Amphioxus* are:
   (a) Malpighian tubules
   (b) Protonephridia
   (c) Kidneys
   (d) Proboscis glands

3. Identify the correct statement:
   (a) All chordates are vertebrates
   (b) Vertebrates are called acraniata
   (c) Protochordates have a vertebral column
   (d) All vertebrates are chordates

4. A cartilaginous fish that produces electric current is:
   (a) *Torpedo* (b) *Trygon*
   (c) *Pristis* (d) *Scoliodon*

5. The teeth of shark are modified ______ scales and directed ______.
   Fill in the blanks with the correct option:
   (a) Placoid, forwardly
   (b) Ganoid, backwardly
   (c) Cycloid, backwardly
   (d) Placoid, backwardly

6. Which of the following is incorrect with respect to characters of *Chondrichthyes*?
   (a) Heterocercal tail
   (b) Absence of operculum
   (c) Terminal mouth
   (d) Absence of swim bladder

7. A larvivorous fish, used for biological control of malaria, is:
   (a) *Clarias* (b) *Gambusia*
   (c) *Betta* (d) *Labeo*

8. Which of the following matching pairs is wrong?
   (a) Silver fish – Arthropoda
   (b) Cuttlefish – Mollusca
   (c) Star fish – Echinodermata
   (d) Shell fish – Pisces

9. The sound producing organs in birds is called:
   (a) Larynx (b) Syrinx
   (c) Glottis (d) Trachea

10. The smallest flightless bird is:
    (a) *Struthio* (b) *Apteryx*
    (c) *Corvus* (d) *Neophron*

11. The venom of cobra attacks:
    (a) Digestive system
    (b) Circulatory system
    (c) Nervous system
    (d) Skeletal system

12. The limbless amphibian is:
    (a) *Rhacophorus* (b) *Ichthyophis*
    (c) *Hyla* (d) *Salamandra*

13. Which one of the following is not a homeothermic animal?
    (a) *Delphinus* (b) *Aptenodytes*
    (c) *Testudo* (d) *Neophron*

14. Elephant tusks are modified:
    (a) Canines (b) Incisors
    (c) Pre-molars (d) Molars

15. Which of the following characters is present in all mammals without exception?
    (a) Seven cervical vertebrae
    (b) Viviparity
    (c) Testis in scrotal sac
    (d) Mammary glands

16. In echolocation, the animal that produces high frequency sound is:
    (a) Monkey (b) Bat
    (c) Eagle (d) Butterfly

17. Vertebrates have:
    (a) Paired appendages that may be fins or limbs
    (b) Two-, three- or four-chambered muscular heart
    (c) Kidneys for excretion and osmoregulation
    (d) All of the above

18. Mark the odd one out with respect to amniotes:
    (a) Reptiles (b) Aves
    (c) Pisces (d) Mammals

19. Which among the following is the correct combination of aquatic mammals?
    (a) *Trygon, Delphinus, Torpedo*
    (b) Seal, Walrus, *Delphinus*
    (c) *Trygon*, Seal, Shark
    (d) Whale, Shark, Seal

20. Choose the correct statement:
    (a) All reptiles have a three-chambered heart
    (b) All pisces have gills covered with operculum
    (c) All mammals have hair and pinna
    (d) All cyclostomes do not possess jaws and paired fins

21. Which of the following is a mismatch?
(a) *Neophron* – Vulture
(b) *Columba livia* – Pigeon
(c) *Psittacula* – Peacock
(d) *Corvus splendens* – Crow

22. Identify the incorrect statement:
(a) *Varanus komodensis* is the largest lizard
(b) *Rhacophorus* is called flying frog
(c) Hind limbs and pelvic girdle are absent in dolphins
(d) Venom of viper is neurotoxic and paralyses respiratory muscles

23. Choose the incorrect statement:
(a) Air bladder in pisces regulates buoyancy
(b) In amphibians, tympanum represents the ear
(c) Birds of bone are pneumatic
(d) In *Chondrichthyes,* pectoral fins bear claspers

24. The group Amniota includes:
(a) Reptiles, aves, mammals
(b) Pisces, amphibians, reptiles
(c) Amphibians, reptiles, mammals
(d) Amphibians, pisces, mammals

25. Which of the following pairs of animals comprises "jawless fish"?
(a) Lampreys and eels
(b) Guppies and hag fish
(c) Lampreys and hag fish
(d) Mackerels and rohu

26. The characteristics of Reptilia are:
(a) Marine animals with cartilaginous endoskeleton
(b) Body covered by dry and cornified skin, having epidermal scales
(c) Freshwater animals with air bladder to regulate buoyancy
(d) Body devoid of scales and fins

27. Which of the following is correct with respect to the difference between frog and toad?
(a) Frog has a tail while toad does not
(b) Frog has cutaneous respiration while toad does not
(c) Frog is nocturnal while toad is diurnal
(d) Frog has no exoskeleton while toad has scales

28. Internal fertilisation is seen in:
(a) *Chondrichthyes* and amphibians
(b) *Osteichthyes* and reptiles
(c) *Chondrichthyes* and reptiles
(d) Amphibians and reptiles

29. Which of the following is an edible freshwater fish?
(a) *Exocoetus* (b) Bombay duck
(c) *Clarias* (d) Salmon

30. Identify the incorrect statement:
(a) Aves are homeothermic
(b) Skin of mammals is unique in possessing hair
(c) *Struthio* and *Apteryx* are flightless birds
(d) Neck and tail are absent in all amphibians

31. Which of the following is incorrect with respect to the characters of the chordate organism shown in the diagram?

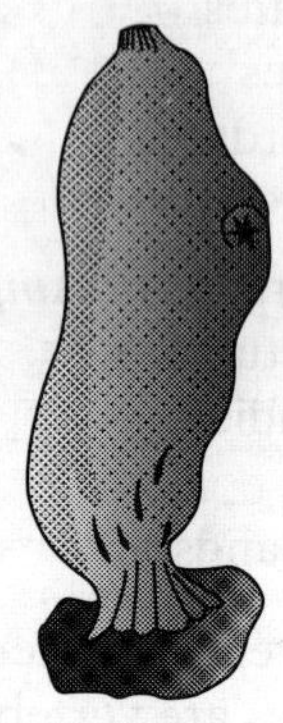

(a) The organism belongs to the subphylum Urochordata
(b) Circulatory system is of closed type
(c) It shows retrogressive metamorphosis
(d) It belongs to the genus *Ascidia*

32. Identify the organisms labelled A, B and C:

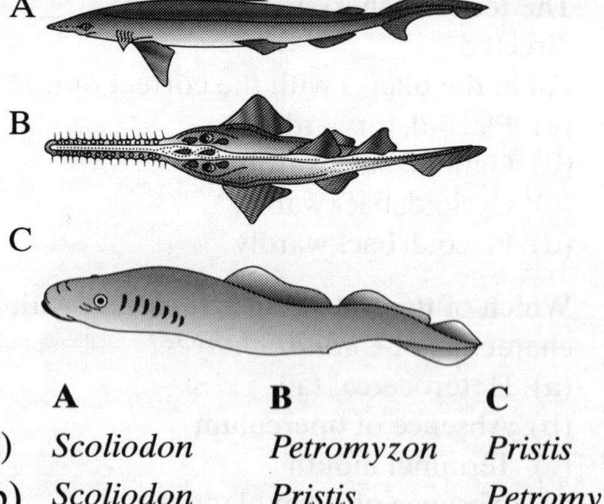

	A	B	C
(a)	*Scoliodon*	*Petromyzon*	*Pristis*
(b)	*Scoliodon*	*Pristis*	*Petromyzon*
(c)	*Pristis*	*Scoliodon*	*Petromyzon*
(d)	*Petromyzon*	*Pristis*	*Scoliodon*

33. Which of the following are examples of aquarium fish?
(a) *Labeo* and *Clarias*
(b) *Catla* and *Clarias*
(c) *Betta* and *Pterophyllum*
(d) *Scoliodon* and *Betta*

34. Which of the following is incorrect for *Rana*?
(a) Fertilisation is external
(b) Sexes are separate

(c) They lack cloaca
(d) They are oviparous and development is indirect

35. Which of the following is incorrect with respect to reptiles?
(a) Body covered with epidermal scales
(b) Homeothermic
(c) Monocondylic skull
(d) Internal fertilisation

36. Identify the incorrect statement:
(a) Birds possess preen glands or oil glands at the base of the tail to lubricate their feathers
(b) The most unique mammalian character is the presence of mammary glands
(c) Reptiles possess creeping or crawling mode of locomotion
(d) *Pavo* and *Neophron* are flightless birds

37. Read the following statements:
A. Members of Pisces have a two-chambered heart
B. Lizards and snakes shed their scales as skin casts
C. In all chordates, notochord is present throughout life
D. Urochordates and vertebrates are called protochordates
Which of the above statements are correct?
(a) (A) and (B) (b) (B) and (C)
(c) (C) and (D) (d) (A) and (D)

38. Which of the following is/are correct with respect to amphibians?
(a) Tympanum represents the ear
(b) The heart is three-chambered, with one auricle and two ventricles
(c) Alimentary canal, urinary and reproductive tracts open into a chamber called cloaca
(d) Both (a) and (c)

39. All vertebrates:
(a) Have two-, three- or four-chambered heart
(b) Are chordates and all chordates are vertebrates
(c) Have notochord persistent throughout life
(d) Lack kidneys

40. Select the incorrect statement:
(a) Cyclostomes have sucking and circular mouth without jaws
(b) Hagfish belongs to superclass Pisces
(c) *Petromyzon* has cartilaginous cranium and vertebral column
(d) Vertebrates have paired appendages which may be fins or limbs

41. Which of the following characteristics is not shared by birds and mammals?
(a) Viviparity
(b) Warm-blooded nature
(c) Breathing using lungs
(d) Kidneys for excretion

42. Which of the following animals is a reptile?
(a) *Salamandra* (b) Toad
(c) Tortoise (d) *Struthio*

43. Identify the incorrect statement with respect to *Chondrichthyes*:
(a) Mostly viviparous
(b) Heart is two-chambered
(c) Lacks jaws
(d) Fertilisation is mostly internal

44. Which of the following is a correct statement with no single exception with respect to the description stated?
(a) All mammals are viviparous
(b) All sponges are marine
(c) All reptiles have a three-chambered heart
(d) All bony fish have four pairs of gills and an operculum on each side

45. Identify the parts labelled X and Y in the diagram given below:

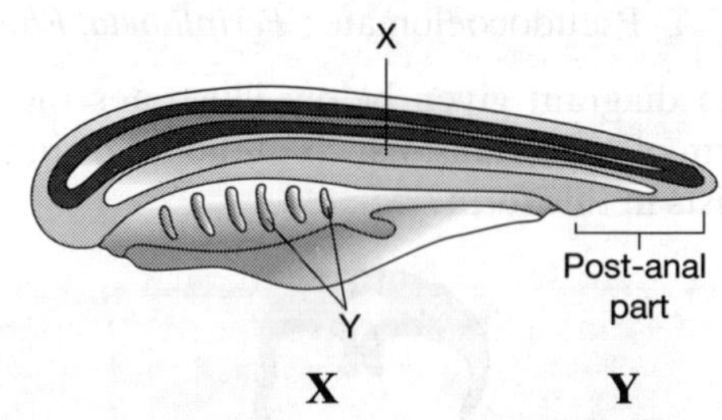

	X	Y
(a)	Nerve cord	Notochord
(b)	Notochord	Gill slits
(c)	Notochord	Nerve cords
(d)	Nerve cord	Heart

Non-chordates

46. Which of the following characters is exhibited by all members of kingdom Animalia?
(a) Cellular level of organisation
(b) Multicellularity
(c) Coelom
(d) Body segmentation

47. Which of the following are characters of cnidarians without any exception?
A. Marine habitat
B. Presence of cnidoblasts
C. Metagenesis
D. Both extracellular and intracellular digestion

(a) (A) and (B) (b) (B) and (C)
(c) (C) and (D) (d) (B) and (D)

48. X and Y are illustrations of the body cavity in animals. Study the diagrams and answer the following:

X.  Y. 

A. Which type of body cavity is represented by X and Y?

B. Following is a list of organisms. Select the organisms with X and Y type of body cavity: *Taenia, Ascaris, Ancylostoma, Periplaneta, Pheretima, Fasciola*

Choose the correct option:

(a) X- Pseudocoelom : *Ascaris, Ancylostoma*
Y- True coelom : *Periplaneta, Pheretima*
(b) X- True coelom : *Taenia, Fasciola*
Y- Pseudocoelom : *Periplaneta, Pheretima*
(c) X- True coelom : *Periplaneta, Pheretima*
Y- Pseudocoelom : *Ascaris, Ancylostoma*
(d) X- Acoelomate : *Fasciola, Taenia*
Y- Pseudocoelomate : *Periplaneta, Pheretima*

49. The diagram given below illustrates the medusa form of cnidarians. Which of the following animals exists in this form?

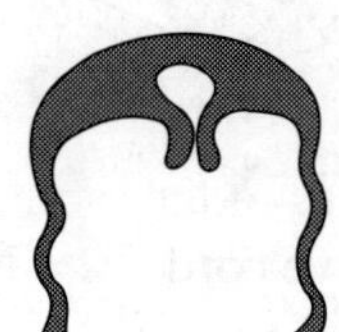

(a) *Hydra* (b) *Adamsia*
(c) *Aurelia* (d) *Meandrina*

50. Which of the following is not a character of members of Porifera?
(a) Cell aggregate body plan
(b) Intracellular digestion
(c) Mostly asymmetrical
(d) Exoskeleton of calcium carbonate

51. Identify the types of symmetry illustrated by figures A and B, respectively:

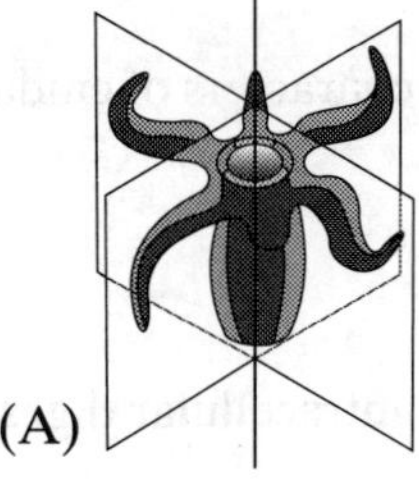
(A)

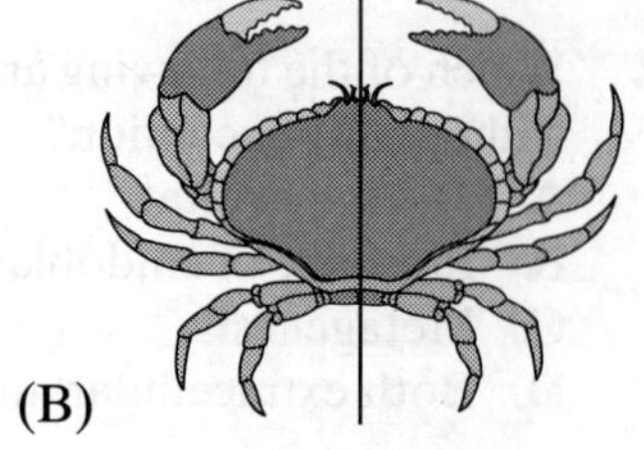
(B)

(a) A – Bilateral B – Radial
(b) A – Radial B – Bilateral
(c) A – Radial B – Asymmetrical
(d) A – Asymmetrical B – Radial

52. Identify the phylum and the corresponding example in the diagram given below:

(a) Porifera: *Euspongia*
(b) Cnidaria: *Adamsia*
(c) Platyhelminthes: *Fasciola*
(d) Aschelminthes: *Ascaris*

53. Which of the following statements is incorrect with respect to ctenophores?
(a) They are all marine
(b) They are commonly called comb jellies
(c) They have radial symmetry
(d) Sexes are separate

54. The only triploblastic acoelomate phylum is:
(a) Cnidaria (b) Ctenophora
(c) Platyhelminthes (d) Aschelminthes

55. Which of the following cnidarians exhibits metagenesis?
(a) *Hydra* (b) *Aurelia*
(c) *Obelia* (d) *Adamsia*

56. In which one of the following is there a solid mesoderm?
(a) Cnidaria (b) Platyhelminthes
(c) Annelida (d) Arthropoda

57. The open circulatory system is found in:
(a) Earthworm (b) Cockroach
(c) Snail (d) Both (b) and (c)

58. Mouth develops first in the embryo and anus is formed later in:
(a) Deuterostomes (b) Protostomes
(c) Echinoderms (d) Chordates

59. *Ascaris* is characterised by:
(a) Presence of true coelom but absence of metamerism
(b) Presence of true coelom and metamerism
(c) Presence of neither true coelom nor metamerism
(d) Absence of true coelom but presence of metamerism

60. Select the incorrect pair:
(a) Platyhelminthes – Tapeworm
(b) Insecta – Squid

(c) Arthropoda – Silver fish
(d) Mollusca – Apple snail

61. The mouth and alimentary canal are absent in:
(a) *Ascaris*
(b) *Fasciola*
(c) *Taenia solium*
(d) *Ancylostoma duodenale*

62. Which of the following is a free-living flatworm?
(a) Tapeworm (b) Roundworm
(c) *Planaria* (d) Liver fluke

63. Which of the following is a mismatched pair?
(a) Choanocytes: Porifera
(b) Parapodiae: Annelids
(c) Comb plates: Mollusca
(d) Jointed appendages: Arthropoda

64. Which of the following is the scientific name of pinworm?
(a) *Trichuris trichura*
(b) *Wuchereria bancrofti*
(c) *Enterobius vermicularis*
(d) *Trichinella spiralis*

65. Self-fertilisation is seen in:
(a) *Ascaris* (b) Cockroach
(c) Tapeworm (d) *Pheretima*

66. Which of the following features is not present in phylum Arthropoda?
(a) Parapodiae
(b) Jointed appendages
(c) Chitinous exoskeleton
(d) Metameric segmentation

67. Which of the following are examples of Arthropoda?
(a) Silver fish, star fish, prawn
(b) Clam worm, apple snail, honey bee
(c) Cockroach, scorpion, silver fish
(d) Tongue worm, scorpion, jelly fish

68. Book lungs are the respiratory organ of:
(a) Earthworm (b) Mollusca
(c) Scorpio (d) Echinodermata

69. Which of the following is not an exclusively marine phylum?
(a) Ctenophora (b) Echinodermata
(c) Cnidaria (d) Hemichordata

70. Water-vascular system is found in:
(a) Sea pen (b) Sea anemone
(c) Sea cucumber (d) Sea horse

71. In which of the following phylums does the larva have bilateral symmetry while the adult has radial symmetry?
(a) Cnidaria (b) Ctenophora
(c) Mollusca (d) Echinodermata

72. The proboscis glands of *Balanoglossus* are associated with:
(a) Digestion (b) Respiration
(c) Excretion (d) Circulation

73. Which of the following does not have an excretory system?
(a) *Pila* (b) *Fasciola*
(c) *Ascaris* (d) *Asterias*

74. An important characteristic that hemichordates share with chordates is:
(a) Absence of notochord
(b) Pharynx with gill slits
(c) Pharynx without gill slits
(d) Ventral tubular nerve cord

75. Haemocoelom is found in:
(a) Annelids and Arthropoda
(b) Arthropods and molluscs
(c) Roundworms and molluscs
(d) Flatworms and roundworms

76. Which of the following is not a character of Echinodermata?
(a) Pentamerous radial symmetry
(b) Sexes are separate
(c) External fertilisation
(d) Direct development

77. Identify the incorrect statement:
(a) Pseudocoelom is characteristic of round-worms
(b) Bioluminescence is seen in ctenophores
(c) Malpighian tubules are excretory organs of annelids
(d) Planarians are free-living flatworms

78. Filariasis is caused by:
(a) *Fasciola* (b) *Ascaris*
(c) *Entamoeba* (d) *Wuchereria*

79. Which of the following animals illustrated below are dioecious, with closed circulatory system and have parapodiae?

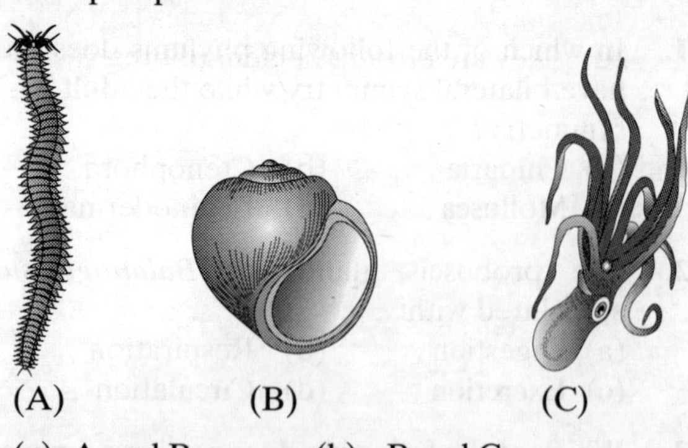

(a) A and B (b) B and C
(c) A only (d) C only

80. Among *Nereis, Pheretima, Hirudinaria, Ascaris, Wuchereria* and *Ancylostoma*, how many possess both longitudinal and circular muscles which help in locomotion?
(a) 2 (b) 3
(c) 4 (d) 5

81. Mark the incorrect statement about *Balanoglossus* and *Saccoglossus:*
(a) Circulatory system is closed
(b) Dioecious and has external fertilisation
(c) Exclusively marine
(d) Excretion by proboscis gland

82. Complete the following analogy:
Nephridium: excretion :: Statocyst: __________
(a) Respiration (b) Reproduction
(c) Balancing (d) Digestion

83. Select the incorrect statement:
(a) *Nereis* is dioecious while earthworm and leech are moneoecious
(b) Annelids have metameric segmentation and closed circulatory system
(c) Green glands are respiratory structures of prawns and crabs
(d) In molluscs, gills have respiratory and excretory functions

84. Identify the incorrect match:
(a) Liver fluke – Flame cells
(b) *Locusta* – Malpighian tubules
(c) *Pila* – Trachaeal respiration
(d) Sea urchin – Aristotle's lantern

85. Which of the following is called tusk shell?
(a) *Chiton* (b) *Physalia*
(c) *Antedon* (d) *Dentalium*

86. Identify the diagrams A and B given below:

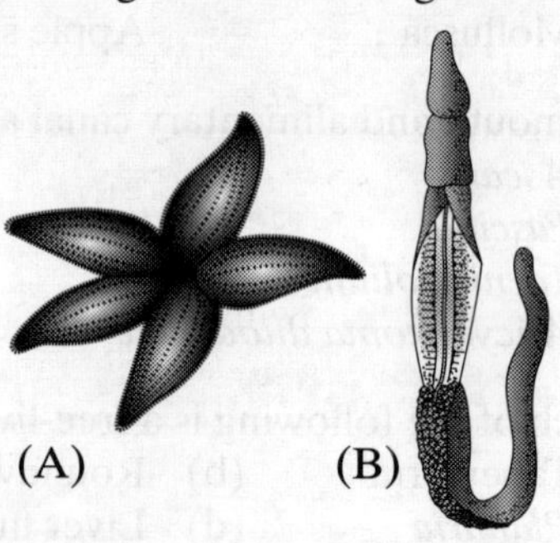

	A	B
(a)	*Ophiura*	*Balanoglossus*
(b)	*Asterias*	*Balanoglossus*
(c)	*Ophiura*	*Taenia solium*
(d)	*Asterias*	*Taenia solium*

87. Which of the following options includes correct examples for animals with chitinous exoskeleton and those covered by calcareous shell?
(a) *Locusta* and *Octopus*
(b) *Apis* and Scorpion
(c) *Pheretima* and *Dentalium*
(d) Butterfly and *Pila*

88. Which of the following is incorrect with respect to the diagram given below?

(a) Has book lungs
(b) Has jointed appendages
(c) Has closed circulatory system
(d) Has complete digestive tract

89. Which of the following statements is not correct about the organism depicted in the diagram given below?

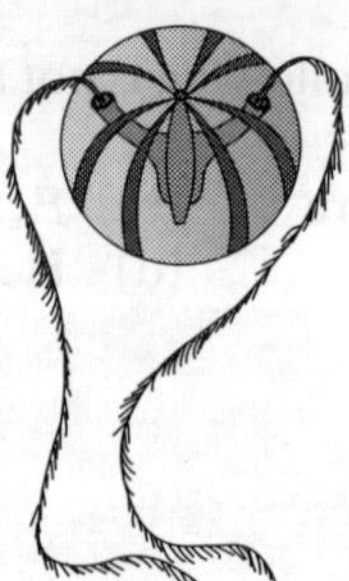

(a) It is an example of Ctenophora, *Pleurobrachia*
(b) It has eight flagellated comb plates which help in locomotion
(c) Reproduction takes place only by sexual means
(d) Digestion is both extracellar and intracellular

90. Which of the following is not a correct description of the cell depicted in the diagram?

(a) The diagram is of cnidoblast
(b) It is used for anchorage, defence and for capturing prey
(c) They are present on tentacles and body wall
(d) They are present in ctenophores

ANSWER KEYS

Exercises

1. (d)	**2.** (b)	**3.** (b)	**4.** (d)	**5.** (c)	**6.** (c)	**7.** (b)	**8.** (b)	**9.** (d)	**10.** (b)
11. (b)	**12.** (b)	**13.** (c)	**14.** (b)	**15.** (a)	**16.** (b)	**17.** (a)	**18.** (d)	**19.** (b)	**20.** (b)
21. (b)	**22.** (d)	**23.** (c)	**24.** (b)	**25.** (b)	**26.** (c)	**27.** (a)	**28.** (d)	**29.** (b)	**30.** (d)
31. (d)	**32.** (d)	**33.** (d)	**34.** (d)	**35.** (d)	**36.** (c)	**37.** (b)	**38.** (b)	**39.** (d)	**40.** (b)
41. (a)	**42.** (d)	**43.** (c)	**44.** (a)	**45.** (a)	**46.** (c)	**47.** (b)	**48.** (a)	**49.** (c)	**50.** (b)
51. (c)	**52.** (c)	**53.** (c)	**54.** (b)	**55.** (b)	**56.** (b)	**57.** (c)	**58.** (b)	**59.** (d)	**60.** (b)
61. (a)	**62.** (a)	**63.** (b)	**64.** (d)	**65.** (b)	**66.** (b)	**67.** (b)	**68.** (c)	**69.** (c)	**70.** (a)
71. (a)	**72.** (a)	**73.** (a)	**74.** (c)	**75.** (b)	**76.** (b)	**77.** (b)	**78.** (a)	**79.** (b)	**80.** (c)
81. (c)	**82.** (d)	**83.** (b)	**84.** (a)	**85.** (c)	**86.** (b)	**87.** (d)	**88.** (b)	**89.** (d)	**90.** (b)
91. (a)	**92.** (d)	**93.** (c)	**94.** (d)	**95.** (a)	**96.** (d)	**97.** (d)	**98.** (d)	**99.** (d)	**100.** (d)
101. (b)	**102.** (d)	**103.** (b)	**104.** (c)	**105.** (d)	**106.** (a)	**107.** (b)	**108.** (b)	**109.** (b)	**110.** (a)
111. (b)	**112.** (c)	**113.** (a)	**114.** (a)	**115.** (d)	**116.** (d)	**117.** (c)	**118.** (b)	**119.** (c)	**120.** (d)
121. (b)	**122.** (d)	**123.** (d)	**124.** (b)	**125.** (c)	**126.** (c)	**127.** (d)	**128.** (d)	**129.** (b)	**130.** (b)
131. (b)	**132.** (b)	**133.** (b)	**134.** (b)	**135.** (b)	**136.** (d)	**137.** (a)	**138.** (a)	**139.** (c)	**140.** (a)
141. (c)	**142.** (c)	**143.** (c)	**144.** (b)	**145.** (c)	**146.** (b)	**147.** (d)	**148.** (b)	**149.** (b)	**150.** (d)
151. (c)	**152.** (b)	**153.** (d)	**154.** (c)	**155.** (a)	**156.** (c)	**157.** (c)	**158.** (b)	**159.** (b)	**160.** (b)
161. (a)	**162.** (d)	**163.** (b)	**164.** (b)	**165.** (c)	**166.** (a)	**167.** (b)	**168.** (d)	**169.** (b)	**170.** (a)
171. (c)	**172.** (a)	**173.** (b)	**174.** (c)	**175.** (b)	**176.** (c)	**177.** (b)	**178.** (c)	**179.** (d)	**180.** (c)
181. (b)	**182.** (b)	**183.** (a)	**184.** (b)	**185.** (b)	**186.** (a)	**187.** (a)	**188.** (d)	**189.** (c)	**190.** (a)
191. (c)	**192.** (b)	**193.** (d)	**194.** (c)	**195.** (b)	**196.** (d)	**197.** (b)	**198.** (c)	**199.** (a)	**200.** (c)
201. (b)	**202.** (c)	**203.** (c)	**204.** (c)	**205.** (c)	**206.** (d)	**207.** (d)	**208.** (c)	**209.** (b)	**210.** (d)
211. (b)	**212.** (c)	**213.** (a)	**214.** (d)	**215.** (b)	**216.** (b)	**217.** (b)	**218.** (c)	**219.** (a)	**220.** (c)
221. (d)	**222.** (c)	**223.** (d)	**224.** (b)	**225.** (d)	**226.** (b)	**227.** (c)	**228.** (d)	**229.** (d)	**230.** (b)
231. (c)	**232.** (d)	**233.** (b)	**234.** (c)	**235.** (a)	**236.** (d)	**237.** (a)	**238.** (b)	**239.** (b)	**240.** (d)
241. (c)	**242.** (b)	**243.** (b)	**244.** (a)	**245.** (d)	**246.** (d)	**247.** (b)	**248.** (b)	**249.** (b)	**250.** (c)
251. (d)	**252.** (a)	**253.** (c)	**254.** (c)	**255.** (c)	**256.** (b)	**257.** (d)	**258.** (a)	**259.** (a)	**260.** (c)
261. (c)	**262.** (d)	**263.** (b)	**264.** (b)	**265.** (c)	**266.** (a)	**267.** (c)	**268.** (c)	**269.** (c)	**270.** (d)
271. (b)	**272.** (a)	**273.** (d)	**274.** (d)	**275.** (a)	**276.** (a)	**277.** (d)	**278.** (d)	**279.** (b)	**280.** (c)
281. (b)									

NCERT Exemplar Questions

1. (b)	**2.** (b)	**3.** (b)	**4.** (b)	**5.** (d)	**6.** (c)	**7.** (c)	**8.** (a)	**9.** (a)	**10.** (c)
11. (c)	**12.** (b)	**13.** (c)							

Practice Test Paper

1. (b)	**2.** (b)	**3.** (d)	**4.** (a)	**5.** (d)	**6.** (c)	**7.** (b)	**8.** (d)	**9.** (b)	**10.** (b)
11. (c)	**12.** (b)	**13.** (c)	**14.** (b)	**15.** (d)	**16.** (b)	**17.** (d)	**18.** (c)	**19.** (b)	**20.** (d)
21. (c)	**22.** (d)	**23.** (d)	**24.** (a)	**25.** (c)	**26.** (b)	**27.** (b)	**28.** (c)	**29.** (c)	**30.** (d)
31. (b)	**32.** (b)	**33.** (c)	**34.** (c)	**35.** (b)	**36.** (d)	**37.** (a)	**38.** (d)	**39.** (a)	**40.** (b)
41. (a)	**42.** (c)	**43.** (c)	**44.** (d)	**45.** (b)	**46.** (b)	**47.** (d)	**48.** (c)	**49.** (c)	**50.** (d)
51. (b)	**52.** (b)	**53.** (d)	**54.** (c)	**55.** (c)	**56.** (b)	**57.** (d)	**58.** (b)	**59.** (c)	**60.** (b)
61. (c)	**62.** (c)	**63.** (c)	**64.** (c)	**65.** (c)	**66.** (a)	**67.** (c)	**68.** (c)	**69.** (c)	**70.** (c)
71. (d)	**72.** (c)	**73.** (d)	**74.** (b)	**75.** (b)	**76.** (d)	**77.** (c)	**78.** (d)	**79.** (c)	**80.** (b)
81. (a)	**82.** (c)	**83.** (c)	**84.** (c)	**85.** (d)	**86.** (b)	**87.** (d)	**88.** (c)	**89.** (b)	**90.** (d)

SOLUTIONS

General Characters

1. All members of kingdom Animalia are multicellular.

2. No diploblastic animals have bilateral symmetry.

3. Option (b) is the correct answer.

4. Members of Coelenterata and Echinodermata have radial symmetry.

5. In members with bilateral symmetry, the body can be divided into identical right and left halves.

6. Members of Mollusca lack segmentation.

7. Notochord is a rod-like structure in chordates and derived from mesoderm.

8. In Aschelminthes, the mesoderm is present as scattered pouches in between ectoderm and endoderm.

9. Annelids have bilateral symmetry.

10. Tissue level is found in Coelenterata.

11. Platyhelminthes is the only triploblastic acoelomate phylum.

12. Open circulatory system is not found in annelids.

13. Cell aggregate body plan is found in sponges.

14. Haemocoelom is found in arthropods like cockroach.

15. Neurons are absent in members of Porifera.

16. Blind sac body plan is found in Coelenterata and Platyhelminthes.

17. The digestive system is termed complete if both mouth and anus are present.

18. Metamerism is characteristic of annelids.

19. Radial symmetry is exhibited by animals having benthos and sedentary nature.

20. In open circulatory system, blood moves slowly and at low pressure.

21. Annelids and arthropods have similar levels of organisation, body symmetry and coelom.

22. In annelids, blood is conveyed to the organs through capillaries.

23. *Hydra* has blind sac body plan and is acoelomate.

24. Non-chordates have double, ventral, solid nerve cords.

25. In animals with radial symmetry, any plane passing through the central axis of the body divides the organism into two identical halves.

26. Option (c) is the correct answer.

27. Option (a) is the correct answer.

28. Members of Aschelminthes have pseudocoelom and a tube within tube plan.

29. Mollusca is the second largest phylum of the animal kingdom.

30. Option (d) is the correct answer.

Porifera

31. Sponges lack a digestive tract and have intracellular digestion.

32. Members of Porifera have both asexual and sexual methods of reproduction.

33. In Porifera, the pathway of water through the water canal system does not help in food gathering, respiratory exchange and removal of waste.

34. In sponges, fertilisation is internal and development is indirect.

35. Choancytes are also called collar cells. They are present in the lining of spongocoel and the canals.

36. Sponges have both asexual and sexual methods of reproduction.

37. *Spongilla* is commonly called freshwater sponge.

38. The water current in the canals of sponges is maintained by choancytes.

39. 'Venus flower basket' is the name of *Euplectella*.

40. *Euspongia* is the common bath sponge.

41. *Cliona* is called the boring sponge and is detrimental for the pearl industry.

42. Archaeocytes are the totipotent cells of sponges.

43. Classification of sponges is based on spicules.

44. In sponges, digestion of food takes place in choancytes and amoebocytes.

45. Option (a) is the correct answer.

Coelenterata

46. *Obelia* exhibits metagenesis, that is, it exists in both polyp and medusa forms.

47. *Meandrina* and *Gorgonia* are members of the coral family and have a skeleton composed of calcium carbonate.

48. Members of Coelenterata are mostly marine and only a few are freshwater forms.

49. Cnidoblasts are present on both tentacles and body.

50. The gastrovascular cavity of cnidarians serves for both digestion and circulation.

51. The medusa form of Coelenterata is umbrella shaped and freely swimming/floating.

52. In members of Coelenterata, both polyp and medusa stages are diploid.

53. Cnidoblasts are used for capturing prey, defence and anchorage.

54. *Physalia* is called Portuguese Man-of-War.

55. Cnidarians have a blind sac body plan.

56. Cnidarians have tissue level of organisation.

57. Cnidarians have both intracellular and extracellular digestion.

58. *Gorgonia* is called sea fan.

59. Interstitial cells are the totipotent cells of cnidarians.

60. Planula larva stage is present in the life cycle of *Obelia*.

61. Polyp phase is absent in *Aurelia*.

62. Option (a) is the correct answer.

Ctenophora

63. Ctenophores are exclusively marine.

64. Ctenophores are also called sea walnuts/comb jellies.

65. In ctenophores, only sexual reproduction is seen.

66. Ctenophores do not have organ level of organisation.

67. The body of ctenophores has eight external rows of ciliated comb plates.

68. Digestion in ctenophores is both intracellular and extracellular.

69. The property of bioluminescence is well marked in ctenophores.

70. Members of Ctenophora are monoecious.

71. Cnidoblasts are absent in ctenophores.

72. *Pleurobrachia* and *Ctenoplana* are examples of Ctenophora.

73. The comb plates of ctenophores help in locomotion.

74. Ctenophores are exclusively marine and have cilia for locomotion.

75. Option (b) is the correct answer.

Platyhelminthes

76. Flatworms are triploblastic and acoelomate.

77. Platyhelminthes are mostly parasitic, and few are free living.

78. Tapeworms absorb nutrients from the host directly through their body surface.

79. In flatworms, specialised cells called flame cells help in osmoregulation and excretion.

80. Development is indirect in flatworms.

81. Anaerobic respiration is seen in tapeworms.

82. Liver fluke has a single opening that serves as mouth and anus.

83. Incomplete alimentary canal is present in *Fasciola*.

84. Digestive system is absent in tapeworms.

85. *Echinococcus granulosus* is commonly called dog tapeworm.

86. Transverse binary fission is seen in *Planaria*.

87. *Fasciola* has organ level of organisation.

88. Option (b) is the correct answer.

89. Metamerism is not seen in *Taenia solium*.

90. Solid mesoderm is seen in platyhelminthes.

91. *Planaria* has high regeneration ability.

92. *Obelia* is dipoblastic. *Ctenoplana* is marine while *Spongilla* is called freshwater sponge.

Aschelminthes

93. *Wuchereria* is found in lymph nodes and causes elephantiasis.

94. All roundworms are not parasitic.

95. The body of aschelminthes is circular in cross-section, and hence, they are called roundworms.

96. Option (d) is the correct answer.

97. In *Enterobius* (pinworm), mesoderm is present in the form of scattered pouches between the ectoderm and endoderm.

98. *Ascaris* (roundworm) has bilateral symmetry, pseudocoelom and muscular pharynx.

99. Option (d) is the correct answer.

100. Infection with pinworm is called enterobiasis/oxyuriasis.

101. *Ascaris* is a monogenetic parasite, that is, the life cycle is completed in only one host.

102. *Enterobius* is a parasite living in the caecum and colon of humans.

103. *Trichinella* is a viviparous parasite.

104. The food of *Ascaris* is intestinal chyme.

105. Elephantiasis is transmitted by the *Culex* mosquito.

106. Syncytial epidermis is seen in *Ascaris*.

107. In Aschelminthes, females are often longer than males.

108. *Ancylostoma* enters the human body through the skin.

109. Aschelminthes have a complete alimentary canal.

110. *Enterobius* (pinworm) is a monogenetic nematode parasite where adults live in the caecum and colon of humans.

111. *Rhabditis* is a free-living roundworm present in soil.

Annelida

112. Annelids are mostly free living and sometimes parasitic.

113. In annelids, locomotion takes place by contraction of longitudinal and circular muscles alternatively.

114. Aquatic annelids like *Nereis* use parapodiae for locomotion.

115. Annelids like earthworm and leech are momoecius while *Nereis* is dioecious.

116. In annelids, nephridia help in osmoregulation and excretion.

117. In annelids, respiration is cutaneous (moist skin).

118. True coelom appeared for the first time in annnelids.

119. Parapodiae are present in *Nereis*, but are not found in earthworm and leech. Setae are present in *Nereis* and earthworm, but not in leech.

120. Annelids reproduce only by sexual methods.

121. Botryoidal tissue is present in *Hirudinaria* (leech).

122. Option (d) is correct.

123. Option (d) is correct.

Arthropoda

124. Over two-thirds of all named species on earth are arthropods.

125. *Laccifer lacca* is an economically useful insect yielding lac.

126. Arthropods have exoskeleton and jointed appendages.

127. Both earthworms and cockroach have solid ventral nerve cords.

128. Antennal glands are used as excretory organs in aquatic arthropods like prawns.

129. Arhtopods have metameric segmentation and open circulatory system.

130. The exoskeleton of arthropods is made of chitin.

131. The excretory organs of crustaceans are antennal glands.

132. Sandfly is the vector for kala azar and oriental sore.

133. In arthropods, fertilisation is mostly internal.

134. The study of insects is called entomology.

135. *Apis, Bombyx* and *Laccifer* are economically useful insects.

136. The *Aedes* mosquito is the vector for dengue fever, yellow fever and chickungunya.

137. Members of class Insecta are also called hexapods.

138. Prawns use gills for respiration and green glands for excretion.

139. Housefly, sandfly and tsetse fly belong to class Insecta.

140. Option (a) is the correct answer.

141. *Lepisma* (silver fish) is a member of Arthropoda.

142. Jointed legs and chitinous exoskeleton are characters common to centipede, cockroach and crab.

143. *Anophelcs* is thc vector for malaria.

144. Open circulatory system is present in *Periplaneta* (cockroach).

145. Option (c) is the correct answer.

Mollusca

146. Segmentation is absent in Mollusca.

147. Members of Arthropoda and Mollusca have haemocoelom.

148. The body of Mollusca is divided into head, muscular foot and visceral hump.

149. In Mollusca, radula is a file-like rasping organ in the mouth.

150. The digestive system is complete in Mollusca.

151. Option (c) is the correct answer.

152. In Mollusca, respiratory organs are feather-like gills present in the mantle cavity.

153. *Aplysia* is commonly called sea hare.

154. *Chaetopleura* is commonly called chiton and is marine.

155. Visceral mass undergoes torsion in *Pila*.

156. Osphradium of *Pila* is meant for sensing the chemical nature of water.

157. *Dentalium* is commonly called tusk cell and is not a member of cephalopod.

158. Option (b) is correct.

Echinodermata

159. In echinoderms, the larvae are bilaterally symmetrical while adults are radially symmetrical.

160. Aristotle's lantern is a masticatory apparatus found in sea urchin.

161. Echinoderms are exclusively marine.

162. Water-vascular system in echinoderms helps in locomotion, respiration, capture and transport of food.

163. Excretory system is absent in echinoderms.

164. Only sexual reproduction is seen in echinoderms.

165. *Antedon* is commonly called sea lily/feather star.

166. Echinoderms have endoskeleton made of calcareous shells/ossicles.

167. Tube feet are the locomotary organs of echinoderms.

168. Echinoderms have mesodermal endoskeleton, enterocoelom and deuterostomous development.

169. Echinoderms are coelmate, spiny and exclusively marine.

170. Arms are absent in sea urchin.

171. Presence of enterocoelom and deuterostmous development have made echinoderms more advanced than remaining non-chordates.

172. Water-vascular system is found in echinoderms.

173. In sea lily, stalk is found for attachment with the substratum.

174. Radial symmetry is found in hydra and echinoderms.

175. Option (b) is correct.

176. Option (c) is correct.

Hemichordata

177. Hemichordates are marine.

178. The excretory organs of hemichordates are proboscis glands.

179. *Balanoglossus* and *Saccoglossus* are examples of hemichordates.

180. Hemichordates are now classified under non-chordates.

181. Circulatory system is open in hemichordates.

182. The body of hemichordates is composed of proboscis, collar and trunk.

183. In hemichordates, respiration takes place through gills.

184. Hemichordates are bilaterally symmetrical, coelomate and marine.

185. Notochord-like structure of hemichordates is called stomochord.

186. Option (a) is correct.

Chordata

187. Heart is ventral in chordates.

188. Hemichordates are a phylum of non-chordates.

189. *Branchiostoma* is a member of cephalochordate.

190. Urochordata is also called tunicate.

191. In cephalochordates, the notochord extends from head to tail and persists throughout life.

192. Subphyla, urochordata and cephalochordates are together called protochordates.

193. Option (d) is correct.

194. In urochordates, notochord is present in the larva's tail.

195. All protochordates are marine.

196. In cephalochordates, notochord is persistent throughout the life and extends from head to tail.

197. Cyclostomes are marine but migrate to freshwater bodies for spawning.

198. *Petromyzon* and *Myxine* are examples of jawless vertebrates.

199. Retrogressive metamorphosis is seen in urochordates like *Herdmania*.

200. In cephalochordates, excretion takes place with the help of flame cells/solenocytes.

201. Sea squirt is a member of tunicate.

202. Urochordates are example of chordates having open circulatory system.

203. In members of cyclostome, cranium and vertebral column are cartilaginous.

204. All living members of Agnatha are ectoparasites on some fishes.

205. Body of *Petromyzon* is devoid of scales and paired fins.

206. Members of Cephalochordata and Urochordata are together called Protochordata and also called Acraniata.

207. Notochord is absent in adult tunicate.

208. Development is indirect in urochordates.

209. Option (b) is correct.

210. In cyclostomes, the larvae migrate back from rivers to oceans after metamorphosis.

Pisces

211. *Carcharodon* is commonly called great white shark.

212. *Hippocampus* (sea horse) is an example of bony fish.

213. In *Chondrichthyes*, teeth are modified placoid scales and are backwardly directed.

214. Swim bladder is not present in cartilaginous fish.

215. *Hippocampus* (sea horse) has a prehensile tail.

216. *Trygon* is commonly called sting ray.

217. Among *Chondrichthyes*, the pelvic fins bear claspers.

218. *Labeo* (rohu) is an example of bony fish having four pairs of gills covered with operculum (gill cover).

219. *Exocoetus* is an edible marine fish.

220. *Gambusia* is a larvivorous fish used for biological control of malaria.

221. *Torpedo* is commonly called electric ray.

222. Isinglass is a gelatinous product obtained from the air bladder of bony fish like *Catla*.

223. Ichthyology is the branch of biology dealing with the study of pisces.

224. Placoid scales are present in cartilaginous fish.

225. *Betta* (fighting fish) and *Pterophylum* (angel fish) are examples of aquarium fish.

226. Shagreen is a kind of untanned leather obtained from the skin of shark.

227. Option (c) is the correct answer as the diagram is of *Labeo*, a member of *Osteichthyes* and which has external fertilisation.

228. Option (d) is the correct answer.

Amphibia

229. *Ichthyophis* is a limbless amphibian.

230. *Hyla* is commonly called tree frog.

231. The heart of amphibians is three-chambered, with two auricles and one ventricle.

232. Option (d) is the correct answer.

233. Frog is ureotelic, like most terrestrial amphibians.

234. Nucleated RBCs are present in amphibians.

235. Scales are absent in amphibians.

236. Tail is absent in members of order Anura of class Amphibia. Limbs are absent in members of order Apoda of class Amphibia.

237. Tadpole larva of frog respires through gills.

238. Amnion is absent in amphibians.

Reptilia

239. The class name Reptilia refers to the creeping or crawling mode of locomotion.

240. Thecodont dentition and four-chambered heart are common characters of humans and crocodile.

241. External ear openings are absent in reptiles.

242. *Naja* is commonly called cobra.

243. Snakes and lizards shed their scales as skin casts.

244. Reptiles, birds and mammals can be grouped under amniotes.

245. Fangs of snakes are modified maxillary teeth.

246. Fertilisation is internal in reptiles.

247. *Vipera* (viper) is a poisonous land snake.

248. Most lizards are oviparous. All reptiles do not possess three-chambered heart as crocodiles have four-chambered heart. Viper is a haemotoxic snake.

249. Jacobson's organs (vomeronasal organs) are present in the roof of the buccal cavity of snakes and lizards.

250. The largest living lizard in the world is *Varanus komodensis* (Komodo dragon).

251. Option (d) is correct.

252. *Draco* is commonly called flying lizard.

Aves

253. Air sacs provides buoyancy in aves. Skin of aves is dry without glands except the oil gland at the base of the tail. Heart of aves is four chambered.

254. In aves, forelimbs are modified into wings.

255. Option (c) is correct.

256. The digestive tract of birds has additional chambers, the crop and gizzard.

257. Teeth are absent in aves.

258. *Neophron* is commonly called vulture.

259. *Aptenodytes* (penguin) have forelimbs modified into flippers.

260. *Struthio* (ostrich) is the largest bird.

261. Hooded pitohui is the poisonous bird.

262. Quills are the feathers used for flying.

263. *Apteryx* (kiwi) is the smallest flightless bird.

264. The sound box of birds is called syrinx.

265. In birds, sternum is modified into keel.

266. In birds, two clavicles are fused to form a wishbone or furcula.

Mammalia

267. *Ornithorhynchus* (duck-billed platypus) is an egg laying mammal.

268. Corpus callosum is found in the brain of mammals, like rabbit.

269. The most unique feature of mammals is the presence of mammary glands.

270. Diaphragm is found in all mammals.

271. Mammals have 12 pairs of cranial nerves.

272. Tusks of elephants are modified incisors.

273. Dicondylic skull is found in amphibians and mammals.

274. Option (d) is the correct answer.

275. *Pteropus* is commonly called large bat/flying fox.

276. *Ornithorhynchus* is an egg laying mammal.

277. Both aves and mammals are homeothermic animals.

278. Option (d) is the correct answer.

279. Mammary glands are modified sweat glands.

280. *Delphinus* (dolphin) and *Balaenoptera* (blue whale) are aquatic mammals.

281. Double uteri and vaginae are present in *Macropus* (kangaroo).

5 Morphology of Flowering Plants

EXERCISES

The Root

1. Which of the following statements is incorrect with respect to the root?
(a) In mustard, primary roots and their branches form the taproot system
(b) In *Monstera*, roots arise from parts of the plant other than the radicle
(c) In pea, the primary root is short-lived and replaced by a large number of roots
(d) In wheat, the fibrous root system arises from the base of the stem

2. The root which arises from parts of the plant other than the radicle is called:
(a) Taproot (b) Adventitious root
(c) Radicle root (d) All except (b)

3. Which of the following plants has roots that originate from the base of the stem as the primary root is short-lived?
(a) Mustard plant (b) Mango plant
(c) Wheat plant (d) Pea plant

4. In the majority of dicotyledonous plants, the direct elongation of ____A____ leads to the formation of ____B____.
Select the correct option to fill in blanks A and B:

	A	B
(a)	Radicle	Root hairs
(b)	Plumule	Primary root
(c)	Root hair	Adventitious root
(d)	Radicle	Primary root

5. In which of the following plants is the root system depicted below found:
(a) Wheat (b) Pea
(c) Mustard (d) All except (a)

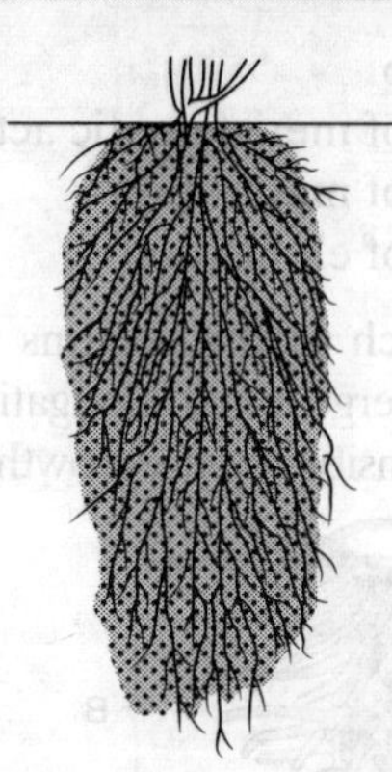

6. The supporting roots of the maize stem which arise from the lower nodes are called:
(a) Pneumatophores (b) Stilt roots
(c) Prop roots (d) Fibrous roots

7. Pneumatophores in a mangrove tree:
(a) Provide mechanical strength to heavy branches
(b) Act like the ropes of a tent
(c) Get oxygen for respiration
(d) Reduce the rate of transpiration

8. Which modification of the root does not store food?
(a) Tuberous root (b) Conical
(c) Prop root (d) Napiform

9. A fibrous root system is better adapted than the taproot system for:
(a) Food storage
(b) Anchorage of plant to soil
(c) Water and mineral absorption
(d) Translocation of water and organic food

10. In terrestrial plants, the thimble-like structure present at the apex of the root is:
(a) Root cap (b) Root hair
(c) Root pocket (d) Root apex

11. Which of the following is incorrectly matched?
(a) *Monstera* – Adventitious roots
(b) Photosynthetic root – *Trapa*

(c) Root tuber – *Ipomoea*
(d) *Haustoria* – Orchid

12. All the statements with respect to pneumatophores are correct, except:
(a) Seen in plants growing in swampy areas
(b) Come out of the ground and grow vertically upward
(c) Possess small pores which absorb moisture from air
(d) Help to get oxygen from air

13. Which of the following is the smallest region of the root?
(a) Root cap
(b) Region of meristematic activity
(c) Region of maturation
(d) Region of elongation

14. Cells of which of the regions shown in the given diagram, undergo rapid elongation and enlargement and are responsible for the growth of the root in length?

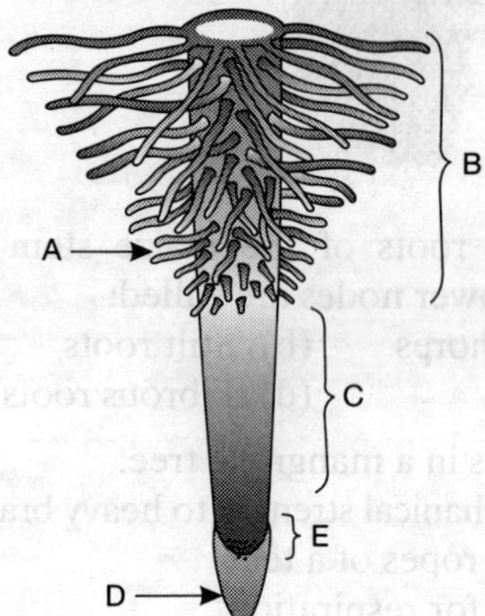

(a) B (b) D
(c) C (d) A

15. Cells of which region are very small, thin-walled and with dense protoplasm?
(a) Region of cell elongation
(b) Root hair
(c) Region of meristematic activity
(d) Region of maturation

16. In the root, the thread-like structure which increases the surface area for absorption of water and minerals is present in the region of:
(a) Maturation
(b) Proximal to region of elongation
(c) Meristematic activity
(d) All except (c)

17. Adventitious roots that become swollen to store food is:

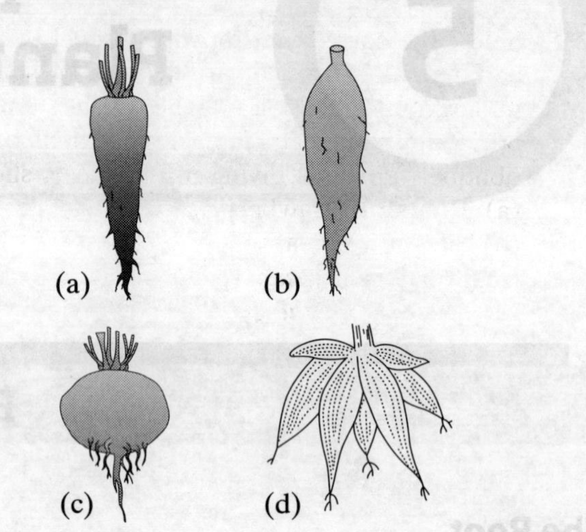

The Stem

18. Fill in the blanks correctly and select the correct option:
(i) Stem develops from ___**A**___ part of the embryo of a germinating seed
(ii) The regions of the stem where leaves are born are called ___**B**___ while ___**C**___ are the portions between two ___**D**___

	A	B	C	D
(a)	Plumule	Nodes	Internodes	Leaves
(b)	Radicle	Nodes	Internodes	Leaves
(c)	Plumule	Nodes	Internodes	Nodes
(d)	Plumule	Internodes	Nodes	Internodes

19. Complete the following analogy:
Jasmine:stolen:: *Pistia*: ______
(a) Runner (b) Corn
(c) Offset (d) Sucker

20. The main function of the stem is:
(a) Conduction of water, minerals and photosynthesis
(b) Spreading out branches bearing leaves, flowers and fruits
(c) Storage of food material
(d) Perennation and vegetative propagation

21. Sharp-pointed structures of *Bougainvillea* are:
(a) Modified leaves
(b) Modified axillary buds
(c) Called stipular spine
(d) Superficial outgrowth of stem

22. Which one is the correct match?
(a) Phylloclade: Australian acacia
(b) Stem thorn: *Opuntia*
(c) Stem tendril: Pea
(d) Corm: *Colocasia*

23. Select the correct option with respect to the sub-aerial weak stem that arises from the basal and underground portion of the main stem, grows horizontally beneath the soil and then comes out obliquely upwards, giving rise to a leafy shoot:
(a) Runner is strawberry
(b) Sucker is pineapple
(c) Stolen is *Chrysanthemum*
(d) Offset is *Eichhornia*

24. The type of sub-aerial stem modification depicted in the following diagram is present in:

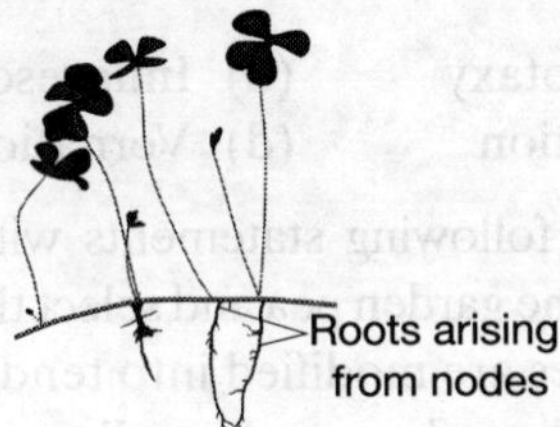

(a) *Eichhornia* (b) Jasmine
(c) *Oxalis* (d) Mint

25. The long, thin, thread-like, spirally coiled structure of the cucumber is the:
(a) Thorn
(b) Leaf tendril
(c) A modified axillary bud
(d) Both (a) and (c)

26. An underground stem which is highly reduced and disc-shaped and covered by fleshy leaves is called:
(a) Bulb (b) Rhizome
(c) Rhizophore (d) Corn

27. Select the correct option for A, B and C:

Structure	Modification	Example
Thorn	A	Citrus
B	Axillary bud	Grape vine
Rhizome	Stem	C

	A	B	C
(a)	Stem	Leaf tendril	Ginger
(b)	Axillary bud	Stem tendril	Turmeric
(c)	Leaf	Stem tendril	Zaminkand
(d)	Axillary bud	Leaf tendril	*Colocasia*

28. Which of the following is not a similarity between offsets and runners:
(a) Thin nodes bear adventitious roots
(b) Both are sub-aerial modifications of the stem
(c) Length of internodes
(d) All except (b)

29. The modified stem in grasses and strawberry is concerned with special functions:
A. Food storage
B. Spread to new niches
C. Perennation
D. Vegetative propagation
(a) Only (A) (b) (A), (C) and (D)
(c) (B), (C) and (D) (d) (A), (B), (C) and (D)

30. Type of modified stem meant for vegetative reproduction in *Chrysanthemum* is also present in:
(a) Strawberry (b) Pineapple
(c) Jasmine (d) All of the above

31. Choose the incorrect option with respect to the given diagram:

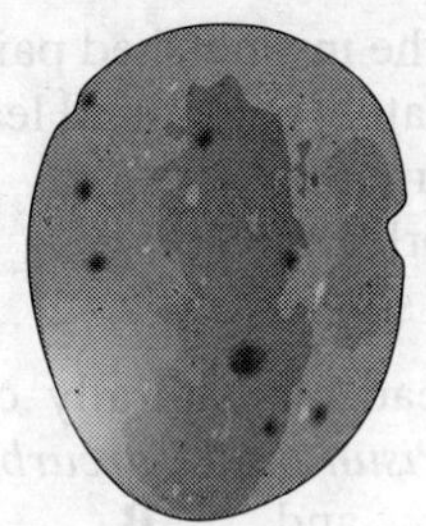

(a) Eyes represent nodes
(b) Arise at the tip of stolens
(c) Modification of stem for storage of food
(d) Underground stem meant for perennation only

32. In which feature is the stem similar to the root?
(a) Storing food material
(b) Ascending part of plant axis
(c) Spreading out branches with internodes
(d) Providing proper anchorage in soil

The Leaf

33. Select the incorrect statement with respect to the leaf:
(a) Lateral, generally flattened structure
(b) Develops at the node
(c) Bears a bud in its axile
(d) Arranged in basipetal succession

34. Which of the following structures is not a part of the leaf?
(a) Hyphopodium (b) Pedicel
(c) Petiole (d) Lamina

35. Pulvinus is:
(a) Swollen leaf base
(b) Modified petiole in Australian acacia
(c) Small lateral leaf-like structure at leaf base
(d) Cylindrical stalk that joins the leaf base with the lamina of the leaf

36. Leaves are generally responsible for:
A. Photosynthesis
B. Transpiration
C. Gaseous exchange

(a) Only (A) (b) Only (B)
(c) (A) and (B) only (d) (A), (B) and (C)

37. The leaf is attached to the stem by ___A___ and may bear two lateral leaf-like structures called ___B___:

	A	B
(a)	Petiole	Stipule
(b)	Leaf base	Stipule
(c)	Leaf base	Axillary bud
(d)	Petiole	Bracts

38. Choose the mismatched pair:
(a) Pinnately compound leaf – Neem
(b) Alternate phyllotaxy – Guava
(c) Whorled phyllotaxy – *Alstonia*
(d) Leaf spines – Cactus

39. The thread-like, spirally coiled, sensitive structure in *Pisum* and *Curcurbita* is a modification of ___A___ and ___B___, respectively:

	A	B
(a)	Leaf	Stem
(b)	Leaf	Leaflet
(c)	Stem	Leaf tip
(d)	Axillary bud	Leaf

40. A pair of leaves arise at each node and lie opposite to each other in:
(a) China rose and mustard
(b) Sunflower and mustard
(c) Guava and *Calotropis*
(d) *Alstonia* and *Nerium*

41. Arrangement of veins and veinlets in the lamina of the leaf is termed as:
(a) Venation (b) Vernation
(c) Ptysis (d) Phytaxy

42. Read the following statements and select the correct option:
A: In the leaf blade, the middle prominent vein is known as the mid-rib
B: Veins provide rigidity to the leaf blade and act as channels for the transport of water, minerals and food
(a) Only A is correct
(b) Only B is correct
(c) Both A and B are correct
(d) Both A and B are incorrect

43. Fill in the blanks by choosing the correct option for A, B and C:

Structure	Example	Function
Phyllode	A	Synthesis of food
B	Cactus	C

	A	B	C
(a)	*Euphorbia*	Leaf spine	Stored food
(b)	Australian acacia	Leaf spine	Defence
(c)	Australian acacia	Thorn	Defence
(d)	*Opuntia*	Thorn	Protective

44. In the pitcher plant, the pitcher is a modification of ______:
(a) Root (b) Stem
(c) Axillary bud (d) Leaf

45. A monocot plant can be distinguished from a dicot by:
(a) Phyllotaxy (b) Inflorescence
(c) Venation (d) Vernation

46. Read the following statements with respect to the leaves of the garden pea and select the correct option:
A. Leaves are modified into tendrils
B. Leaf base becomes swollen and is called the pulvinus
C. Shows reticulate venation
D. Stipules are absent at the base of the leaf
(a) Only (B) is correct
(b) (A), (B) and (C) are correct
(c) (B), (C) and (D) are correct
(d) (A), (C) and (D) are correct

47. The phenomenon observed in the monocot leaf, wherein the base is expanded into a sheath which covers the stem partially, is known as ______ leaf base:
(a) Shedding (b) Sheathing
(c) Swirling (d) Shelding

48. Palmately compound leaves:
(a) Have leaflets attached to a common point at the tip of the pedicel
(b) Are present in silk cotton
(c) Possess lateral leaflets on rachis
(d) All are correct except (c)

49. Examine the figures (A) and (B) and select the correct identification of the given plants:

A	B
(a) Whorl phyllotaxy	Pinnately compound leaf
(b) Pinnately compound leaf	Palmately compound leaf
(c) Whorl phyllotaxy	Palmately compound leaf
(d) Palmately compound leaf	Whorl phyllotaxy

50. The part of the leaf that allows the leaf blade to flutter in the wind, thereby cooling the leaf and bringing fresh air to the leaf surface is:
(a) Pulvinus (b) Petiole
(c) Leaf blade (d) Leaf base

51. Which of the following is not a modification of the leaf:
(a) Tendrils in pea
(b) Spines of *Opuntia*
(c) Thorns in *Bougainvillea*
(d) Fleshy leaves of onion

52. The given mode of leaf arrangement is shown by:

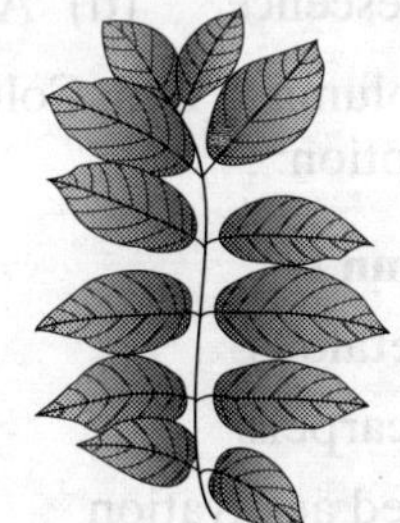

(a) China rose and *Calotropis*
(b) *Calotropis* and Guava
(c) *Alstonia* and *Nerium*
(d) Mustard and Sunflower

53. The labelled structures (A) and (B) given below represent modifications of the:

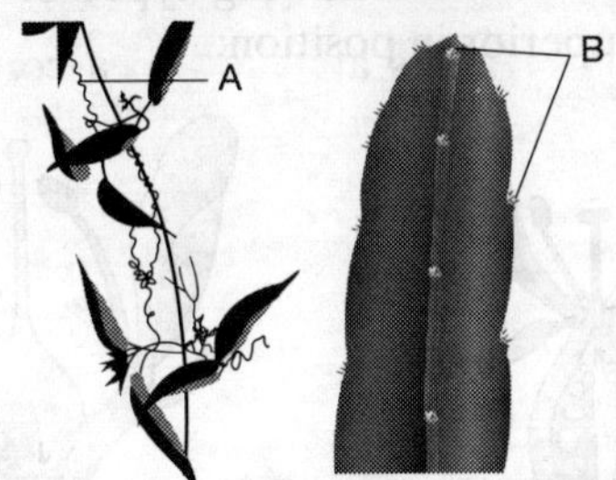

A	B
(a) Leaf	Leaf
(b) Stem	Leaf
(c) Leaf	Stem
(d) Leaf	Axillary bud

54. Compound leaf is similar to simple leaf in:
(a) Having a bud in the axile of the petiole
(b) Consisting of a partially incised lamina
(c) Having a bud in the axile of the leaflets
(d) Consisting of undivided lamina

The Inflorescence

55. Read the following statements and select the correct option:
A. A flower is a modified shoot wherein the shoot apical meristem changes to floral meristem
B. In the flower, the internodes are not formed and the axis is expended
(a) Only (A) is incorrect
(b) Only (B) is incorrect
(c) Both (A) and (B) are correct
(d) Both are incorrect

56. The arrangement of flowers on the floral axis is called:
(a) Aestivation (b) Placentation
(c) Vernation (d) Inflorescence

57. Look at the following figures and identify inflorescence A and B:

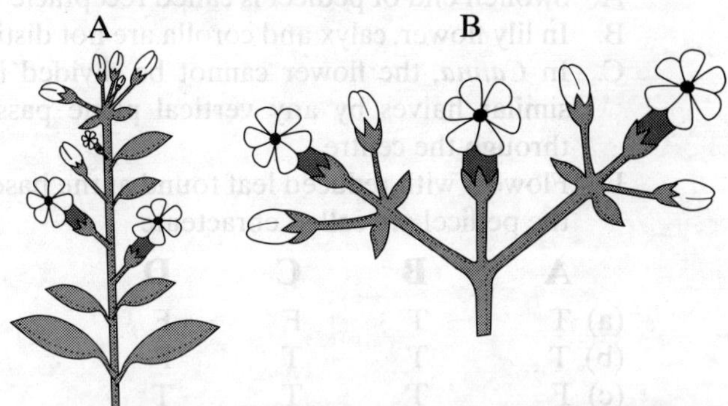

A	B
(a) Racemose	Cymose
(b) Racemose	Solitary
(c) Solitary	Cymose
(d) Cymose	Racemose

58. The axis of inflorescence is called:
(a) Rachis (b) Petiole
(c) Peduncle (d) All except (b)

59. In the cymose type of inflorescence:
A. The main axis terminates in a flower
B. The flowers are born in acropetal succession
C. The younger flowers are born at the base of the floral axis
Select the correct statements:
(a) Only (A) (b) (A) and (B)
(c) (A) and (C) (d) (A), (B), and (C)

60. The most highly evolved type of inflorescence is:
(a) Capitulum
(b) Spadix
(c) Hypanthodium
(d) Monochasial cyme

The Flower

61. Select the incorrect statement with respect to the flower:
(a) It is the reproductive unit in angiosperms
(b) It has four different kinds of whorls
(c) Floral whorls are arranged successively on the swollen end of the peduncle
(d) Calyx and corolla are accessory organs

62. Asymmetric flower is found in:
(a) *Canna* (b) *Datura*
(c) Chilli (d) Gulmohur

63. The reduced leaf found at the base of the pedicel is known as:
(a) Stipule (b) Bract
(c) Peduncle (d) Pulvinus

64. Identify the following statements as true (T) or false (F) and select the option accordingly:
A. Swollen end of pedicel is called receptacle
B. In lily flower, calyx and corolla are not distinct
C. In *Canna*, the flower cannot be divided into similar halves by any vertical plane passing through the centre
D. Flowers with reduced leaf found at the base of the pedicel are called ebracteate

	A	B	C	D
(a)	T	T	F	F
(b)	T	T	T	F
(c)	F	T	T	T
(d)	F	F	T	T

65. Select the incorrect match with respect to the symmetry of the flower:
(a) Mustard – Actinomorphic
(b) Chilli – Radial symmetry
(c) *Cassia* – Irregular
(d) Gulmohur – Zygomorphic

66. Match the following columns and select the correct option:

Column I	Column II
A. Actinomorphic flower	(i) Half inferior
B. Zygomorphic flower	(ii) Gynoecium occupies the highest position
C. Hypogynous flower	(iii) One-plane symmetry
D. Perigynous flower	(iv) Radial symmetry

(a) A – (iii), B – (iv), C – (i), D – (ii)
(b) A – (iii), B – (iv), C – (ii), D – (i)
(c) A – (iv), B – (iii), C – (ii), D – (i)
(d) A – (iv), B – (iii), C – (i), D – (ii)

67. In vexillary aestivation, standard represents:
(a) The largest posterior petal
(b) Two smaller lateral petals
(c) Two smallest anterior petals
(d) The largest anterior petal

68. If the margins of petals overlap one another, but not in a particular direction, then it is:
(a) Valvate in *Calotropis*
(b) Imbricate in *Cassia*
(c) Twisted in gulmohur
(d) Imbricate in china rose

69. The mode of arrangement of sepals or petals in a floral bud with respect to the other members of the same whorl is known as:
(a) Phyllotaxy (b) Venation
(c) Inflorescence (d) Aestivation

70. Match Column I and Column II and select the correct option:

Column I	Column II
A. Epipetalous	(i) Lotus
B. Free carpels	(ii) Brinjal
C. Twisted aestivation	(iii) Citrus
D. Polyadelphous condition	(iv) Cotton

(a) A – (ii), B – (i), C – (iv), D – (iii)
(b) A – (iv), B – (i), C – (iii), D – (ii)
(c) A – (ii), B – (iv), C – (iii), D – (i)
(d) A – (i), B – (iii), C – (ii), D – (iv)

71. In which of the following types of flowers is the ovary superior in position:

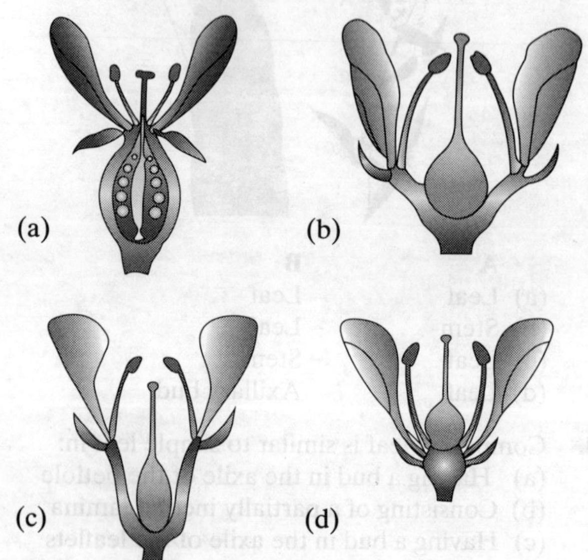

72. When the sepals or petals in a whorl just touch on another margin, without overlapping as in **A**, it is called **B** aestivation:

	A	B
(a)	*Cassia*	Axillary
(b)	Cotton	Twisted
(c)	*Calotropis*	Valvate
(d)	*Calotropis*	Imbricate

73. Identify the placentation shown below as well as the related plants of its occurrence and select the right option for the two together:

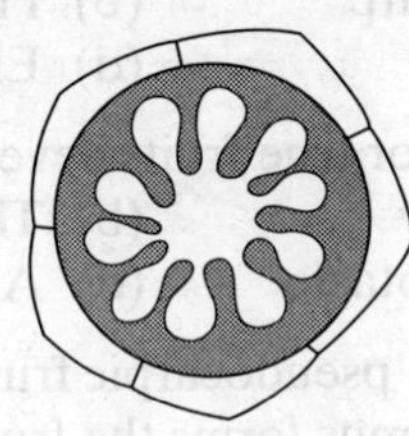

	Placentation	Plants
(a)	Parietal	Pea and bean
(b)	Basal	Sunflower and marigold
(c)	Axile	Primrose and pea
(d)	Free central	*Dianthus* and primrose

74. Consider the following statements A, B, C and D and select the option for two correct statements:

A. The ovary is said to be half-inferior in the perigynous flower

B. In axile placentation, the ovary is one chambered but it becomes two chambered due to the formation of false septum called replum

C. Inferior ovary is found in epigynous flower

D. When stamens are attached to tepals, they are epiphyllous in brinjal

The correct statements are:

(a) (C) and (D) (b) (A) and (B)
(c) (A) and (C) (d) (B) and (D)

75. Variation in the length of the filament within a flower occurs in:

(a) Mustard and *Salvia*
(b) Mustard and china rose
(c) China rose and pea
(d) *Salvia* and china rose

76. Butterfly-shaped corolla has/shows:

(a) Two lip-like structures and corolla mouth
(b) Vexillary aestivation
(c) Cross-like arrangement of petals
(d) All except (b)

77. Basal placentation develops when the ovary has:

(a) Single ovule
(b) Many ovules
(c) Single ovary in each locule
(d) Many locules

78. In **A** placentation, the ovary is one chambered but becomes two chambered due to formation of false septum called **B:**

	A	B
(a)	Marginal	Ridge
(b)	Parietal	Ventral suture
(c)	Parietal	Replum
(d)	Basal	Replum

79. Match the following columns and select the correct option:

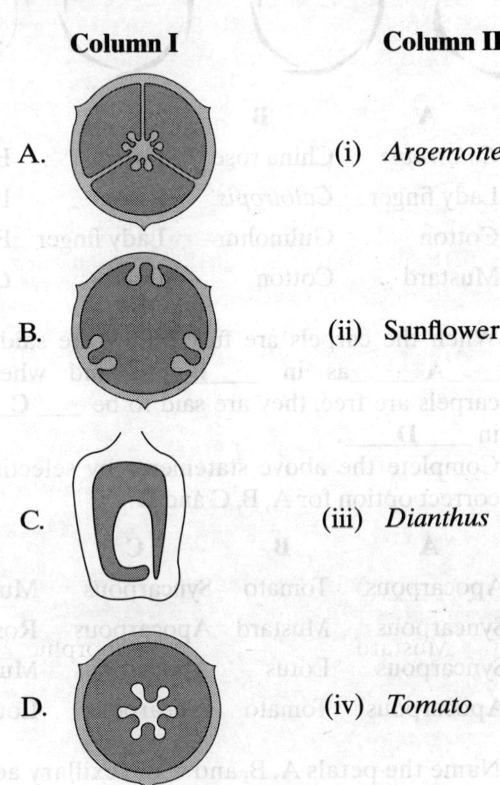

(a) A – (i), B – (iv), C – (iii), D – (ii)
(b) A – (iii), B – (i), C – (ii), D – (iv)
(c) A – (iv), B – (i), C – (ii), D – (iii)
(d) A – (ii), B – (iii), C – (iv), D – (i)

80. Long, filamentous threads protruding at the end of a young cob of maize are:

(a) Pappus
(b) Modified bract
(c) Epidermal outgrowth of seeds
(d) Styles

81. The given figure represents the cohesion of stamen. Identify the type of cohesion:

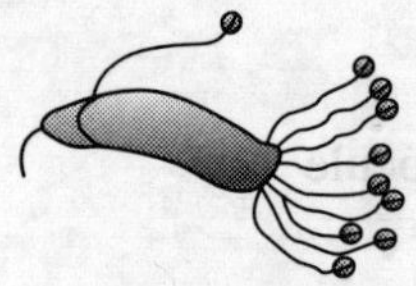

(a) Synandrous
(b) Monoadelphous
(c) Diadelphous
(d) Polyadelphous

82. Identify A, B, C and D (aestivation in petals) from the given figure and select the correct option with respect to plants in which it is found:

	A	B	C	D
(a)	*Calotropis*	China rose	*Cassia*	Bean
(b)	Lady finger	*Calotropis*	Bean	Pea
(c)	Cotton	Gulmohur	Lady finger	Pea
(d)	Mustard	Cotton	Bean	*Cassia*

83. When the carpels are fused, they are said to be ___A___ as in ___B___, and when the carpels are free, they are said to be ___C___ as in ___D___.
Complete the above statements by selecting the correct option for A, B, C and D:

	A	B	C	D
(a)	Apocarpous	Tomato	Syncarpous	Mustard
(b)	Syncarpous	Mustard	Apocarpous	Rose
(c)	Syncarpous	Lotus	Apocarpous	Mustard
(d)	Apocarpous	Tomato	Syncarpous	Lotus

84. Name the petals A, B, and C in vexillary aestivation shown in the given figure:

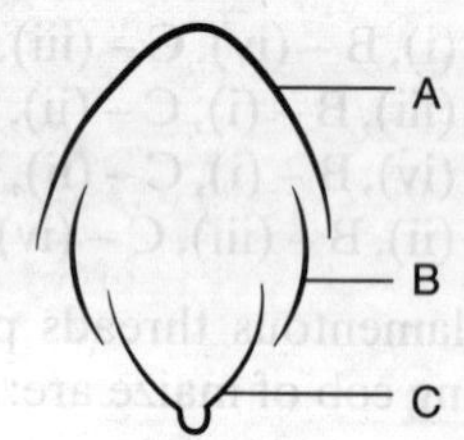

	A	B	C
(a)	Standard	Wing	Vexillum
(b)	Keel	Standard	Vexillum
(c)	Vexillum	Wings	Keel
(d)	Vexillum	Keel	Standard

85. Which of the following arrangement of petals is found in *Cassia* and gulmohur?

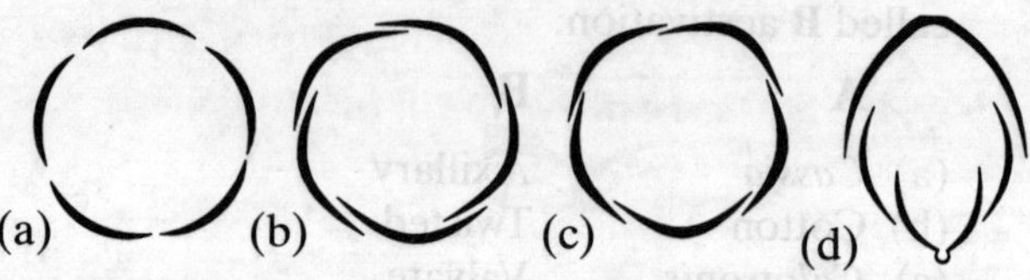

The Fruit

86. After fertilisation, the mature ovary is known as:
(a) Pericarp (b) Fruit
(c) Seed (d) Endosperm

87. Eucarpic or true fruits develop from:
(a) Ovary (b) Thalamus
(c) Receptacle (d) All of the above

88. Apple is a pseudocarpic fruit because:
(a) Thalamus forms the fruit along with ovary
(b) It develops from entire inflorescence
(c) Ovary proliferates without the process of fertilisation
(d) Pericarp is not differentiated into epicarp, mesocarp and endocarp

89. The inner layer of the pericarp is hard and stony in:
(a) Drupe (b) Pepo
(c) Amphisarca (d) Berry

90. Read the following statements and select the correct option:
A. When a fruit develops from the inflorescence, it is composite
B. Syconus is a composite fruit that develops from a hypanthodium inflorescence and is found in the *Ficus* species
(a) Only (A) is correct
(b) Only (B) is correct
(c) Both (A) and (B) are correct
(d) Both (A) and (B) are incorrect

91. Read the following statements and choose the correct option:
A. The drupe fruit is characterised by the presence of stony endocarp
B. The mango fruit develops from a monocarpellary superior ovary, with fibrous mesocarp
(a) Only (A) is correct
(b) Only (B) is correct
(c) Both (A) and (B) are correct
(d) Both (A) and (B) are incorrect

92. Which part of the coconut produces coir?
(a) Seed coat (b) Mesocarp
(c) Stem (d) Epicarp

93. _________ is the edible part of the apple:
(a) Mesocarp (b) Fleshy thalamus
(c) Pericarp (d) Mesocarp and endocarp

94. Identify the structure labelled X:

(a) Epicarp (b) Endocarp
(c) Mesocarp (d) Seed

95. Match the following:

A.	Mango	(i)	Juicy placental hairs
B.	Tomato	(ii)	Pericarp and placenta
C.	Litchi	(iii)	Fleshy mesocarp
D.	Orange	(iv)	Fleshy aril

(a) A – (ii), B – (i), C – (iv), D – (iii)
(b) A – (iii), B – (iv), C – (i), D – (ii)
(c) A – (iii), B – (ii), C – (iv), D – (i)
(d) A – (ii), B – (iii), C – (iv), D – (i)

96. Identify the figure and select the incorrect statements:

(a) The fruit is known as drupe
(b) These fruits develop from monocarpellary superior ovaries
(c) One-seeded simple fruit
(d) In coconut, pericarp is not differentiated and it is fibrous

97. Consider the following statements and state true (T) or false (F):
A. Most of the fruits are true fruits
B. Fruits which are developed without fertilisation are called parthenocarpic
C. Parthenocarpic fruits are seedless
D. Aggregate fruits develop from multicarpellary apocarpus gynoecium

	A	B	C	D
(a)	T	T	T	T
(b)	T	T	F	F
(c)	T	T	F	T
(d)	F	F	T	T

98. A fruit is said to be fleshy if:
(a) It is formed without fertilisation
(b) It develops from monocarpellary superior ovary
(c) Its pericarp is differentiated
(d) Its pericarp is not differentiated

99. The fruit that develops from hypanthodium inflorescence is called:
(a) Sorosis (b) Syconus
(c) Drupe (d) Siliqua

100. Select the incorrectly matched pair:
(a) Aggregate fruit – Develops from multicarpellary syncarpous gynoecium
(b) Parthenocarpic fruit – Seedless fruit
(c) Geocarpic fruit – Groundnut
(d) Composite fruit – Develops from entire inflorescence

The Seed

101. After fertilisation, the mature _____ develop into seeds:
(a) Ovules (b) Ovary
(c) Loculus (d) Placenta

102. Consider the following statements and select the option that correctly fills in the blanks:
(i) The outermost covering of a seed is called seed coat. It has two coverings, the outer ___A___ and inner ___B___.
(ii) ___C___ is the scar on the seed coat through which the developing seed is attached to the fruit

	A	B	C
(a)	Tegmen	Testa	Hilum
(b)	Testa	Tegmen	Hilum
(c)	Testa	Tegmen	Funicle
(d)	Tegmen	Testa	Caruncle

103. The thin membranous sheath that encloses the radicle in a monocot embryo is:
(a) Scutellum (b) Aleurone layer
(c) Coleoptile (d) Coleorhiza

104. At the two ends of the embryonal axis, ______ are present in dicot seeds:
(a) Radicle and plumule
(b) Coleorhiza and coleoptile
(c) Micropyle and hilum
(d) Endosperm and embryo

105. Select the incorrect statement with respect to the maize seed:
(a) Seed coat is membranous and generally fused with the fruit wall

(b) Endosperm is bulky and stores starch
(c) Embryo is small and situated in the groove at one end of the endosperm
(d) Coleoptile covers both radicle and plumule

106. In seed, perisperm is:
(a) Third integument
(b) Persistent nucleus after fertilisation
(c) Persistent endosperm in monocot seeds
(d) A part of the embryo

107. Select the odd one out with respect to the endospermic seed:
(a) Wheat (b) Coconut
(c) Orchid (d) Castor

108. In the maize seed, the outer covering of the endosperm is separated from the embryo by a proteinaceous layer called:
(a) Coleorhiza (b) Scutellum
(c) Testa (d) Aleurone layer

109. In a cereal grain, the single cotyledon of embryo is represented by:
(a) Scutellum (b) Testa
(c) Tegmen (d) Perisperm

110. Endosperm is a product of double fertilisation in angiosperm. Endosperm is not present in the seeds of:
(a) Castor (b) Pea
(c) Maize (d) Coconut

111. Identify A, B, C and D in the given figure:

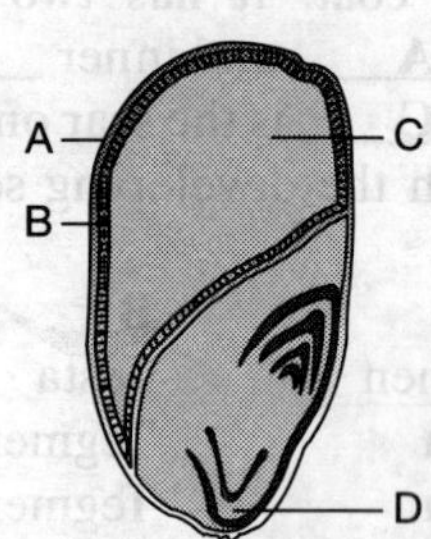

	A	B	C	D
(a)	Pericarp	Aleurone layer	Endosperm	Colcorhiza
(b)	Fruit wall	Pericarp	Endosperm	Radicle
(c)	Fruit wall	Aleurone layer	Perisperm	Coleoptile
(d)	Pericarp	Seed coat	Scutellum	Coleorhiza

112. Embryo is made up of:
A. A radicle
B. Embryonal axis
C. A plumule
D. One or two cotyledons
(a) (A) and (C) only
(b) (B) and (D) only
(c) (A), (B) and (C) only
(d) (A), (B), (C) and (D)

113. The plumule and radicle in monocot seed are enclosed in sheaths, which are called ___**A**___ and ___**B**___, respectively:

	A	B
(a)	Coleoptile	Scutellum
(b)	Coleoptile	Coleorhiza
(c)	Coleorhiza	Coleoptile
(d)	Scutellum	Coleorhiza

114. Identify A, B, and C in a dicotyledonous seed:

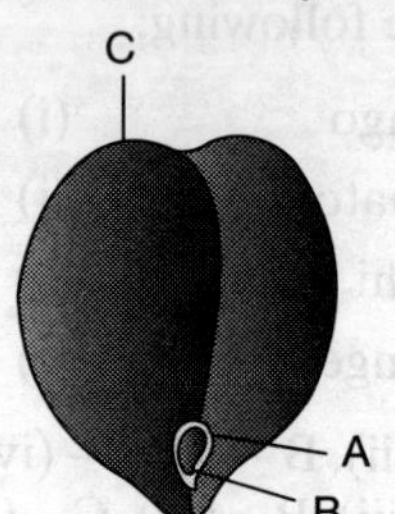

	A	B	C
(a)	Hilum	Micropyle	Pericarp
(b)	Micropyle	Hilum	Seed coat
(c)	Hilum	Micropyle	Seed coat
(d)	Seed coat	Micropyle	Micropyle

115. Identify A, B and C in this figure:

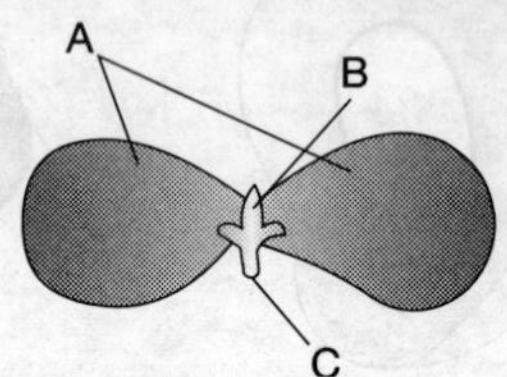

	A	B	C
(a)	Cotyledon	Plumule	Radicle
(b)	Cotyledon	Radicle	Plumule
(c)	Radicle	Plumule	Cotyledon
(d)	Plumule	Radicle	Cotyledon

Description of Some Important Families

116. Select the mismatched pair
(a) $\oplus$ Actinomorphic flower
(b) $\overset{\frown}{C\ \ A}$ Epiphyllous condition
(c) $K_{(n)}$ Gamosepalous calyx
(d) $\overline{G}_n$ Inferior ovary

117. Identify the plant family with the following features:
A. Perennial herbs with underground bulbs, corns or rhizome

B. There are six stamens with epitepalous condition

C. Tricarpellary syncarpous ovary with axile placentation

The floral diagram of the plant family is:

(a) Liliaceae (b) Solanaceae
(c) Fabaceae (d) Brassicaceae

118. The side of a flower facing the mother axis is called:

(a) Anterior side (b) Posterior side
(c) Dorsal side (d) Ventral side

119. Which of the following characters cannot be expressed by floral formula?

(a) Symmetry of flower
(b) Sexuality of flower
(c) Cohesion and adhesion of floral parts
(d) Aestivation

120. Which of the following is not described in the floral diagram of a typical flowering plant?

(a) Mother axis
(b) Number of loculus in the ovary
(c) Position of calyx, corolla and androecium with respect to the ovary in the thalamus
(d) Placentation

121. The $\widehat{C_{(5)} A_5}$ is found in the floral formula of:

(a) *Solanum* (b) *Trifolium*
(c) *Gloriosa* (d) *Brassica*

122. Select the correct match:

(a) Vexillary aestivation – Solanaceae family
(b) Epiphyllous condition – Liliaceae family
(c) Marginal placentation – Brassicaceae family
(d) Unilocular ovary with false septum – Fabaceae family

123. Tricarpellary, syncarpous, tricolour superior ovary with axile placentation are characteristic features of:

(a) Potato family (b) Pea family
(c) Mustard family (d) Lily family

124. Identify the plant and its family having the following features:

A. Androecium – Diadelphous condition
B. Flower – Zygomorphic and hypogynous
C. Gynoecium – Monocarpellary, unilocular, superior ovary
D. Corolla – Vexillary aestivation

	Plant	Family
(a)	*Trifolium*	Fabaceae
(b)	Lupin	Liliaceae
(c)	Ashwagandha	Fabaceae
(d)	*Gloriosa*	Solanaceae

125. The floral diagram of a plant family is given below. By identifying the figure, select the true statement about the family:

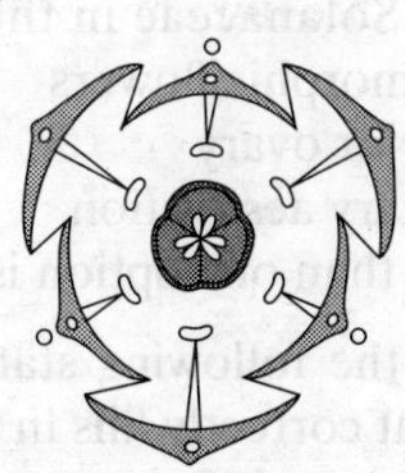

(a) Bisexual and zygomorphic flower
(b) Bicarpellary superior ovary with marginal placentation
(c) Valvate aestivation of sepals and petals
(d) Five stamen with epiphyllous condition

126. A diagnostic feature for the identification of the mustard family is/are:

(a) Cruciferon corolla
(b) Tetradynamous condition
(c) Bicarpellary syncarpous superior ovary with parietal placentation
(d) All of the above

127. Identify the floral diagram shown below as well as the related plant of its occurrence and select the correct option of the two together:

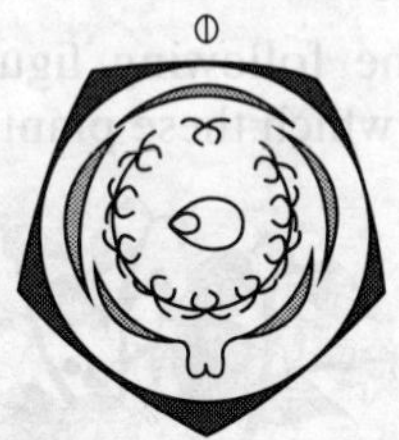

	Family	Plant
(a)	Fabaceae	*Indigofera*
(b)	Fabaceae	*Petunia*
(c)	Solanaceae	Chilli
(d)	Solanaceae	Lupin

128. How many of the given features are true for the Solanaceae family?

A. Exstipulate leaves with reticulate venation
B. Five united sepals, persistent, valvate aestivation
C. Placenta swollen with many ovules
D. Many non-endospermous seeds
E. Tricarpellary syncarpous ovary with axile placentation

(a) (D) (b) (C)
(c) (E) (d) (B)

129. What is the outermost whorl of the flower of a tulip?
(a) Calyx (b) Corolla
(c) Androecium (d) Perianth

130. Members of family Fabaceae differ from members of family Solanaceae in the presence of:
(a) Zygomorphic flowers
(b) Superior ovary
(c) Vexillary aestivation
(d) More than one option is correct

131. Consider the following statements and select the option that correctly fills in the blanks:
(i) ___A___ inflorescence is found in Solanum
(ii) *Colchium autumnale* is a member of family ___B___
(iii) *Trifolium* is a ___C___ plant belonging to family ___D___

	A	B	C	D
(a)	Cymose	Liliaceae	Fodder	Fabaceae
(b)	Racemose	Liliaceae	Ornamental	Solanaceae
(c)	Cymose	Solanaceae	Ornamental	Fabaceae
(d)	Solitary	Liliaceae	Medicinal	Solanaceae

132. Which of the following characteristics is common for families Liliaceae and Solanaceae?
(a) Perianth (b) Epiphyllous condition
(c) Fruit type (d) Placentation

133. Look at the following figures and identify the families to which these plant species belong:

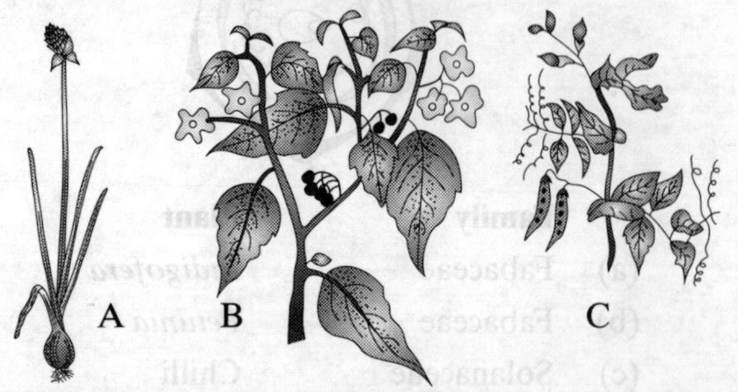

	A	B	C
(a)	Liliaceae	Solanaceae	Fabaceae
(b)	Liliaceae	Fabaceae	Solanaceae
(c)	Fabaceae	Solanaceae	Liliaceae
(d)	Lilaceae	Malvaceae	Fabaceae

134. Which of the following plants is used to extract blue dye?
(a) Petunia (b) *Indigofera*
(c) *Gloriosa* (d) Tulip

135. Which of the following members of family Solanaceae is fumigatory?
(a) Belladonna (b) *Datura*
(c) *Nicotiana* (d) *Atropa*

136. A family delimited by the type of inflorescence is:
(a) Fabaceae (b) Asteraceae
(c) Solanaceae (d) Liliaceae

137. The number and arrangement of stamens in family Fabaceae is:
(a) $A_{(10)+1}$ (b) A_{2+4}
(c) $A_{(9)+1}$ (d) $\widehat{C\,A_{(5)}}$

138. The position of the ______ with respect to the flower is represented by a dot on the top of the floral diagram:
(a) Ovary (b) Stamen
(c) Petals (d) Mother axis

139. The given floral formula is represented by which group of plant family?
⚥ $K_{(5)}\ C_{1+2+(2)}\ A_{(9)+1}\ G_1$
(a) Lupin, sweet pea, *Trifolium*
(b) Ashwagandha, petunia, chilli
(c) *Asparagus*, tulip, *Gloriosa*
(d) Muliathi, belladonna, colchicine

140. Flowers are trimerous in:
(a) Plants having stamens with epitepalous condition
(b) *Asparagus* and colchicine family
(c) Plants with underground bulbs/corms/rhizomes
(d) All of the above

NCERT EXEMPLAR QUESTIONS

1. Rearrange the following zones seen in the root in the vertical section and choose the correct option:
A. Root hair zone
B. Zone of meristems
C. Root cap zone
D. Zone of maturation
E. Zone of elongation

(a) (C), (B), (E), (A), (D)
(b) (A), (B), (C), (D), (E)
(c) (D), (E), (A), (C), (B)
(d) (E), (D), (C), (B), (A)

2. In an inflorescence where flowers are borne laterally in acropetal succession, the position of the youngest floral bud shall be:

(a) Proximal (b) Distal
(c) Intercalary (d) Anywhere

3. The mature seeds of plants such as gram and peas possess no endosperm because:
(a) These plants are not angiosperms
(b) There is no double fertilisation in them
(c) Endosperm is not formed in them
(d) Endosperm is used up by the developing embryo during seed development

4. Roots developed from parts of the plant other than the radicle are called:
(a) Taproots (b) Fibrous roots
(c) Adventitious roots (d) Nodular roots

5. Venation is a term used to describe the pattern of arrangement of:
(a) Floral organs
(b) Flower in inflorescence
(c) Veins and veinlets in a lamina
(d) All of the above

6. Endosperm, a product of double fertilisation, is absent in the seeds of:
(a) Coconut (b) Orchids
(c) Maize (d) Castor

7. Many pulses of daily use belong to one of the families below:
(a) Solanaceae (b) Fabaceae
(c) Liliaceae (d) Poaceae

8. The placenta is attached to the developing seed near the:
(a) Testa (b) Hilum
(c) Micropyle (d) Chalaza

9. Which of the following plants is used to extract blue dye?
(a) *Trifolium* (b) *Indigofera*
(c) Lupin (d) *Cassia*

10. Match the following and choose the correct option:

	Column I	Column II
A.	Aleurone layer	(i) Without fertilisation
B.	Parthenocarpic fruit	(ii) Nutrition
C.	Ovule	(iii) Double fertilisation
D.	Endosperm	(iv) Seed

(a) A – (i), B – (ii), C – (iii), D – (iv)
(b) A – (ii), B – (i), C – (iv), D – (iii)
(c) A – (iv), B – (ii), C – (i), D – (iii)
(d) A – (ii), B – (iv), C – (i), D – (iii)

PRACTICE TEST PAPER

1. Axillary buds of the stem may also be modified into woody straight and pointed structures called ______:
(a) Prickles (b) Stipules
(c) Thorns (d) Spines

2. Root hairs are formed from the epidermal cells of the:
(a) Root apex
(b) Region of meristematic activity
(c) Region of elongation
(d) Region of maturation

3. Underground stem of potato, ginger and turmeric:
A Are modified to store food in them
B Act as organs of perennation to tide over conditions unfavourable for growth
(a) Only A is correct
(b) Only B is correct
(c) Both A and B are correct
(d) Both A and B are incorrect

4. Supporting roots arising from the nodes of the stem are called ____**A**____ and found in ____**B**____:

	A	B
(a)	Prop root	Maize
(b)	Stilt root	Sugarcane
(c)	Stilt root	Banyan tree
(d)	Pneumatophores	*Rhizophora*

5. Whorl phyllotaxy is found in:
(a) *Calotropis* (b) Silk cotton
(c) *Alstonia* (d) *Brassica*

6. Match the following:
A. Stem tendril (i) *Allium*
B. Bulb (ii) *Euphorbia*
C. Phylloclade (iii) Pea
D. Leaf tendril (iv) Cucumber
(a) A – (iv), B – (i), C – (ii), D – (iii)
(b) A – (i), B – (iv), C – (ii), D – (iii)
(c) A – (iii), B – (i), C – (ii), D – (iv)
(d) A – (iii), B – (ii), C – (i), D – (iv)

7. In Australian acacia:
(a) Leaves are small and long lived
(b) Petioles expand, become green and synthesize food
(c) There is a swollen leaf base called phyllode
(d) More than one option is correct

8. Select the mismatched pair:
(a) Tuber – Potato
(b) Offset – *Eichhornia*
(c) Stolen – *Chrysanthemum*
(d) Rhizome – Banana

9. Which of the following is not the main function of the root system:
(a) Proper anchorage to the plant
(b) Absorption of water and minerals
(c) Nitrogen fixation in nodules
(d) Synthesis of plant growth regulators

10. Roots in *Rhizophora*:
(a) Grow vertically upward
(b) Help in moisture conservation
(c) Are assimilatory
(d) All except (c)

11. Actinomorphic flowers are found in all of the following families, except in:
(a) Fabaceae (b) Brassicaceae
(c) Solanaceae (d) Liliaceae

12. In maize, the outer covering of the endosperm:
(a) Is a proteinaceous layer
(b) Stores oils in its elaioplasts
(c) Is called aleurone layer
(d) All except (b)

13. A hypogynous flower:
(a) Has a superior ovary
(b) Is always unisexual
(c) Has an inferior ovary
(d) Does not produce seeds

14. On the basis of the position of floral parts on the thalamus, select the odd one out:
(a) Guava (b) Cucumber
(c) Mustard (d) Ray florets of sunflower

15. In the floral formula of which of the following plants is androecium is represented as A_{2+4}:
(a) *Solanum* (b) Lily
(c) Pea (d) Mustard

16. In the plants belonging to family Fabaceae, fruits are:
(a) Drupe (b) Legume
(c) Berry (d) Capsule

17. The number of petals in the keel, wings and standard in a flower having vexillary aestivation are, respectively:
(a) 1, 2 and 1 (b) 2, 1 and 2
(c) 2, 2 and 1 (d) 1, 2 and 2

18. In a floral formula, gamosepalous condition is represented as:
(a) (K) (b) $K_{(n)}$
(c) C_n (d) $C_{(n)}$

19. Match the following columns and select the correct option:

Column I	Column II
A. *Trifolium*	(i) An ornamental plant
B. Lupin	(ii) Grown as fodder
C. Belladonna	(iii) Source of dye
D. *Indigofera*	(iv) A medicinal plant

(a) A – (ii), B – (i), C – (iv), D – (iii)
(b) A – (ii), B – (iv), C – (i), D – (iii)
(c) A – (iv), B – (i), C – (iii), D – (ii)
(d) A – (iv), B – (iii), C – (ii), D – (i)

20. Non-endospermic seeds are found in:
(a) Groundnut (b) Brinjal
(c) Makoi (d) Onion

21. *Salvia* and mustard flowers are characterised by the presence of:
(a) Cruciform corolla
(b) Filaments of stamens united in two bundles
(c) Variation in length of staminal filament
(d) Verticillaster inflorescence

22. Identify the following statements as true (T) or false (F) and choose the option accordingly:
A. In monoadelphous condition, stamens are united in a single bundle
B. In *Salvia* and mustard, there may be variation in length within a flower
C. Each anther is usually bilobed and each lobe has four chambers called pollen sacs
D. *Citrus* and pea show the same type of stamen cohesion

	A	B	C	D
(a)	T	T	F	T
(b)	T	T	F	F
(c)	F	F	T	T
(d)	F	T	F	F

23. Select the incorrect match:
(a) Actinomorphic flower – Mustard
(b) Bilateral symmetry – *Cassia*
(c) Half inferior ovary – Plum
(d) Apocarpous ovary – Tomato

24. The floral formula of a flower is:

$$⚥\ K_{(5)}\ C_{1+2+(2)}\ A_{(9)+1}\ G_1$$

On the basis of this information, we cannot say that:
(a) Ovary is superior or inferior
(b) Gynoecium is apocarpous or syncarpous
(c) Placentation of ovary
(d) Cohesion and adhesion within parts of whorls and between whorls

25. Which of the following statements is/are correct?
A. Fruits are mature ripened ovaries

B. Parthenocarpic fruits are seedless
C. Drupe fruit develops from monocarpellary inferior ovaries and are one-seeded
D. Endocarp of mango is fibrous and edible
(a) (A) and (B) (b) (A), (B) and (C)
(c) (A) and (C) (d) (A), (B), (C) and (D)

26. Find the incorrect statement with respect to racemose inflorescence:
(a) Main axis has unlimited growth
(b) It occurs in *Solanum*
(c) Flowers are born in aeropetal manner
(d) Youngest flowers towards apex

27. Modified stem offset as observed in *Pistia* and *Eichhornia* is characterised by all, except:
(a) A lateral branch with long internode
(b) Each node bearing a rosette of leaves
(c) A tuft of roots at the base
(d) Petiole of leaves become swollen

28. Pinnately compound leaf differs from palmately compound leaf in:
(a) Possessing partially incised lamina
(b) Presence of bud in the axile of the petiole
(c) Having leaflets at the top of the petiole
(d) Possessing a number of leaflets laterally on a common axis called rachis

29. In ___**A**___ flower, the margin of petals overlaps each other in a particular direction. This condition is called ___**B**___ aestivation.
Select the correct option to fill in the blanks (A) and (B):

	A	B
(a)	*Calotropis*	Twisted
(b)	*Calotropis*	Valvate
(c)	China rose	Twisted
(d)	*Cassia*	Imbricate

30. Find the incorrect match:
(a) Ashwagandha – Fabaceae
(b) Colchicine yielding family – Liliaceae
(c) Petunia – Ornamental plant of Solanaceae
(d) Eggplant – Brassicaceae

31. Which of the following families has sepals, petals stamens and carpels in the ratio of 2: 2: 2:1?
(a) Solanaceae (b) Brassicaceae
(c) Liliaceae (d) Composite

32. Identify the type of aestivation for the corolla in the given floral diagram:

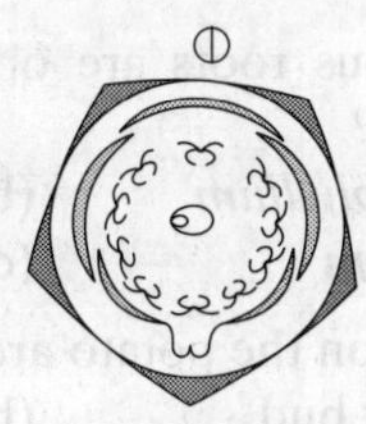

(a) Vexillary (b) Twisted
(c) Imbricate (d) Valvate

33. A particular type of placentation is shown below:

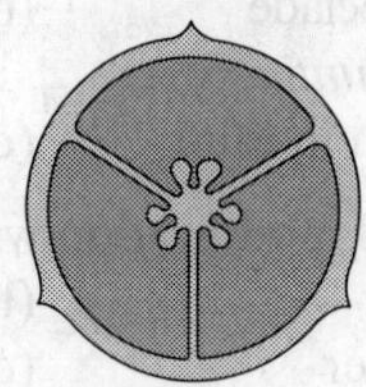

Select the option in which all the plants have this type of placentation:
(a) China rose, tomato, lemon
(b) *Argemone, Calotropis*, mustard
(c) *Dianthus*, primrose, *Argemone*
(d) Cucumber, lily, mustard

34. Which of the following features is not associated with family Lilaceae?
(a) Axile placentation
(b) Non-endospermous seeds
(c) Calyx and corolla are not distinct
(d) Epipetalous condition of stamen

35. The medicinal plant of family Liliaceae is:
(a) Ashwagandha (b) Belladonna
(c) Muliathi (d) Aloe

36. Select the correct statements:
A. The root is covered at the apex by a thimble-like unicellular structure called the root cap
B. Root cap protects the tender root apex
C. A few millimetres above the root cap is the region of meristematic activity
(a) (A) and (B) (b) Only (B)
(c) (B) and (C) (d) (A), (B) and (C)

37. Read the statements given below:
(i) Testa and tegmen are a part of the ovule
(ii) The hilum is a scar on the pericarp
(iii) Coleoptile represents the protective sheath of plumule in a dicot seed
(iv) In castor, the endosperm is a food storing tissue
How many of the above statements are correct?
(a) 2 (b) 1
(c) 4 (d) 3

38. Epiphyllous roots are observed in which of the following?
(a) *Bryophyllum* (b) *Kalanchoe*
(c) *Nerium* (d) Both (a) and (b)

39. The eyes on the potato are modified:
(a) Shoot bud (b) Radicle bud
(c) Axillary bud (d) Accessory bud

40. Select the odd one out with respect to the modification of a stem:
(a) Phylloclade in *Opuntia* (b) Thorn in *Bougainvillea*
(c) Stolen in mint (d) Spine in cactus

41. Sub-aerial modification with long internodes is:
(a) Tuber (b) Phyllode
(c) Runner (d) Stolen

42. In which type of placentation are ovules present on the central axis?
(a) Axile (b) Parietal
(c) Free central (d) Both (a) and (c)

43. How many of the following features are associated with the mustard plant?
(i) Alternate phyllotaxy
(ii) Radial symmetry of flower
(iii) Valvate aestivation of corolla
(iv) Tetradynamous condition of stamens
(v) Parietal placentation
(a) 2 (b) 3
(c) 4 (d) 5

44. A floral diagram is given below that belongs to:

(a) Liliaceae (b) Poaceae
(c) Fabaceae (d) Solanaceae

45. A diagnostic feature of the angiosperm family that includes petunia and belladonna is:
(a) Diadelphous condition
(b) Marginal placentation
(c) Bicarpellary obligately placed superior ovary
(d) Underground, bulb, corn, rhizome

ANSWER KEYS

Exercises

1. (c)	**2.** (b)	**3.** (c)	**4.** (d)	**5.** (a)	**6.** (b)	**7.** (c)	**8.** (c)	**9.** (b)	**10.** (a)
11. (d)	**12.** (c)	**13.** (a)	**14.** (c)	**15.** (c)	**16.** (d)	**17.** (d)	**18.** (c)	**19.** (c)	**20.** (b)
21. (b)	**22.** (d)	**23.** (b)	**24.** (c)	**25.** (c)	**26.** (a)	**27.** (b)	**28.** (c)	**29.** (c)	**30.** (b)
31. (d)	**32.** (a)	**33.** (d)	**34.** (b)	**35.** (a)	**36.** (d)	**37.** (b)	**38.** (b)	**39.** (a)	**40.** (c)
41. (a)	**42.** (c)	**43.** (b)	**44.** (d)	**45.** (c)	**46.** (b)	**47.** (b)	**48.** (b)	**49.** (c)	**50.** (b)
51. (c)	**52.** (b)	**53.** (a)	**54.** (a)	**55.** (b)	**56.** (d)	**57.** (a)	**58.** (d)	**59.** (c)	**60.** (a)
61. (c)	**62.** (a)	**63.** (b)	**64.** (b)	**65.** (c)	**66.** (c)	**67.** (a)	**68.** (b)	**69.** (d)	**70.** (a)
71. (d)	**72.** (c)	**73.** (d)	**74.** (c)	**75.** (a)	**76.** (b)	**77.** (a)	**78.** (c)	**79.** (c)	**80.** (d)
81. (c)	**82.** (a)	**83.** (b)	**84.** (c)	**85.** (c)	**86.** (b)	**87.** (a)	**88.** (a)	**89.** (a)	**90.** (c)
91. (a)	**92.** (b)	**93.** (b)	**94.** (d)	**95.** (c)	**96.** (d)	**97.** (a)	**98.** (c)	**99.** (b)	**100.** (a)
101. (a)	**102.** (b)	**103.** (d)	**104.** (a)	**105.** (d)	**106.** (b)	**107.** (c)	**108.** (d)	**109.** (a)	**110.** (b)
111. (a)	**112.** (d)	**113.** (b)	**114.** (c)	**115.** (a)	**116.** (b)	**117.** (a)	**118.** (b)	**119.** (d)	**120.** (c)
121. (a)	**122.** (b)	**123.** (d)	**124.** (a)	**125.** (c)	**126.** (d)	**127.** (a)	**128.** (b)	**129.** (d)	**130.** (d)
131. (a)	**132.** (d)	**133.** (a)	**134.** (b)	**135.** (c)	**136.** (b)	**137.** (c)	**138.** (d)	**139.** (a)	**140.** (a)

NCERT Exemplar Questions

1. (a)	**2.** (b)	**3.** (d)	**4.** (c)	**5.** (c)	**6.** (b)	**7.** (b)	**8.** (d)	**9.** (b)	**10.** (b)

Practice Test Paper

1. (c)	**2.** (d)	**3.** (c)	**4.** (b)	**5.** (c)	**6.** (a)	**7.** (b)	**8.** (c)	**9.** (c)	**10.** (a)
11. (a)	**12.** (d)	**13.** (a)	**14.** (c)	**15.** (d)	**16.** (b)	**17.** (c)	**18.** (b)	**19.** (a)	**20.** (a)
21. (c)	**22.** (b)	**23.** (d)	**24.** (c)	**25.** (a)	**26.** (b)	**27.** (a)	**28.** (d)	**29.** (c)	**30.** (d)
31. (b)	**32.** (a)	**33.** (a)	**34.** (b)	**35.** (d)	**36.** (c)	**37.** (b)	**38.** (d)	**39.** (c)	**40.** (d)
41. (c)	**42.** (a)	**43.** (d)	**44.** (a)	**45.** (c)					

SOLUTIONS

The Root

1. In pea, the primary root develops from the radicle. It bears lateral roots of several orders that are referred to as secondary roots, tertiary roots, etc.

2. Adventitious roots arise from any part of the plant body other than the radicle.

3. In wheat plant, primary root is short-lived and is replaced by a large number of roots, called fibrous roots.

4. A – Radicle, B – Primary root

5. Wheat (Fibrous root)

6. Stilt roots arise obliquely from the lower nodes of the stem and provide mechanical support.

7. Pnuematophores are found in mangrove plants. The branches of the taproot grow vertically upward (negative geotrophic), and possess minute pores to absorb oxygen from the atmosphere – hence called respiratory roots.

8. Prop root is meant for mechanical support.

9. Anchorage of plant to soil

10. The root is covered at the apex by a thimble-like structure called the root cap. It protects the root tip.

11. Haustorial roots are parasitic roots found in *Cuscuta*.

12. Minute pores present on pneumatophores help in gaseous exchange.

13. Root cap

14. The region C that lies proximal to the meristematic region (E) is the region of cell elongation. Cells of this region undergo rapid elongation and enlargement and are responsible for the growth of the root in length.

15. The cells of the region of meristematic activity are very small, thin-walled with dense cytoplasm and have a prominent nucleus.

16. Root hairs are present in the region of cell maturation or differentiation. This region lies proximal to the elongation zone.

17. Zaminkand is a modified stem for storage of food material.

The Stem

18. A – Plumule, B – Nodes, C – Internodes, D – Nodes

19. Offset

20. The main function of the stem is spreading out branches, bearing leaves, flowers and fruits.

21. In *Bougainvillea*, the axillary bud is modified into woody, pointed straight structures called thorns, which protect the plants from browsing animals.

22. Phylloclade:*Opuntia*, Stem tendril: Gourds, Grape vines, Stem thorn: *Bougainvillea, Citrus*

23. Sucker present in banana, pineapple and *Chrysanthemum*

24. *Oxalis*

25. In cucurbits, axillary buds are modified into slender and spirally coiled structures called stem tendrils.

26. Bulb: Onion, Garlic

27. A – Axillary bud, B – Stem tendril, C – Turmeric

28. In runner, the sub-aerial branch is elongated, prostrate with long internodes and adventitious roots at the node. In offset, the sub-aerial branch has short internodes and each node bears a rosette of leaves and tuft of adventitious roots.

29. The modified stem in grasses and strawberry is concerned with functions such as spread to new niches when older parts die, perennation and vegetative propagation.

30. Pineapple and Banana

31. Potato tuber is an underground stem modified to perform different functions: storage of food, perennation and vegetative propagation.

32. Stem is similar to root in storing food material.

The Leaf

33. Leaves originate from shoot apical meristem and are arranged in basipetal manner (larger and older towards base, smaller and younger towards apex).

34. Pedicel is the stalk of a flower; it is not a part of the leaf.

35. In some leguminous plants, the leaf base become swollen, which is called the pulvinus.

36. (A), (B) and (C) are correct.

37. A – Leaf base, B – Stipule

38. In alternative phyllotaxy, a single leaf arises at each node; for example, china rose, mustard, sunflower.

39. A – Leaf tendril, B – Stem tendril

40. Opposite phyllotaxy is seen in guava and *Calotropis*.

41. Venation is the arrangement of veins and veinlets in the lamina of the leaf.

42. Both (A) and (B) are correct.

43. A – Australian acacia, B – Leaf spine, C – Defence

44. Leaves of certain insectivorous plants such as pitcher plant and Venus flytrap are modified leaves.

45. Dicot leaves show reticulate venation while monocot leaves show parallel venation.

46. Only A, B and C are correct. In the garden pea, stipulate leaves are present.

47. Sheathing

48. In palmately compound leaves, leaflets are attached to a common point, that is, the tip of the petiole, as in silk cotton.

49. A – Whorl phyllotaxy, B – Palmately compound leaf

50. Petiole

51. Thorns in *Bougainvillea* are modified axillary buds.

52. The given figure represents opposite phyllotaxy as found in *Calotropis* and guava.

53. A – Leaf tendril, B – Leaf spine

54. Compound leaf is similar to simple leaf in having a bud in the axile of the petiole.

The Inflorescence

55. Only B is incorrect.

56. Inflorescence

57. A – Racemose, B – Cymose

58. The inflorescence axis is called rachis or peduncle.

59. Only (A) and (C) are correct. In cymose, inflorescence is seen in basipetal succession.

60. Capitulum is considered as the most advanced type of inflorescence because the aggregation of florets results in pollination of many flowers by a single visit of insects.

The Flower

61. Floral whorls are arranged successively on the swollen end of the pedicel called thalamus or receptacle.

62. A flower is asymmetric if it cannot be divided into two similar halves by any vertical plane passing through the centre, as in *Canna*.

63. Bracts

64. Only D is false. Flowers with reduced leaf found at the base of the pedicel are called bracts. A flower with bract is called bracteate and without it is called ebractiate.

65. In *Cassia*, floral symmetry is zygomorphic.

66. Actinomorphic flower – Radial symmetry, Zygomorphic flower – One-plane symmetry, Hypogynous flower – Superior ovary, Perigynous flower – Half inferior ovary

67. In vexillary aestivation, there are five petals. The posterior one is the largest and called standard or vexillum and almost covers the two lateral petals, called wings, which in turn cover the two lateral petals called keel.

68. Imbricate aestivation in *Cassia*

69. Aestivation

70. Epipetalous – Brinjal, Free Carpels – Lotus, Twisted Aestivation – Cotton, Polyadelphous condition – *Citrus*

71. Gynoecium occupies the highest position while the other parts are situated below it. The ovary in such flowers is said to be superior.

72. A – *Calotropis*, B – Valvate

73. The given placentation represents free central placentation and it is found in *Dianthus* and primrose.

74. Only (A) and (C) are correct. B is related to parietal placentation. The epiphyllus condition is seen in lily.

75. There may be variation in the length of filaments within a flower, as in *Salvia* and mustard.

76. Vexillary aestivation

77. In basal placentation, the ovary is unicellular and the placenta develops at the base of the ovary on the thalamus and bears a single ovule.

78. A – Parietal, B – Replum

79. A (Axile) – Tomato, B (Parietal) – *Argemone*, C (Basal) – Sunflower, D (Free central) – *Dianthus*

80. Styles

81. Diadelphous condition

82. A (Valvate) – *Calotropis*, B (Twisted) – China rose, Lady finger, C (Imbricate) – *Cassia*, Gulmohur, D (Vexillary) – Pea and bean flowers

83. A – Syncarpous, B – Mustard, C – Apocarpous, D – Bean

84. A – Vexillum, B – Wings, C – Keel

85. Imbricate aestivation

The Fruit

86. Fertilised and ripened ovary is called fruit and the fruit wall is called the pericarp.

87. The fruit thats develops from the ovary is called a true fruit or eucarpic fruit. Most fruits are true fruit.

88. If any other floral part along with the ovary takes part in fruit formation, it is called false fruit or pseudocarpic fruit. Apple is a pseudocarpic fruit because the thalamus forms the fruit along with the ovary.

89. Drupe fruits are characterised by the presence of stony endocarp.

90. Both (A) and (B) are correct.

91. Only (A) is correct. Mango fruit develops from monocarpellary superior ovary with fleshy mesocarp.

92. Coir is a natural fibre extracted from the mesocarp of the coconut fruit.

93. Fleshy thalamus is the edible part in apple and pear.

94. Seed

95. Mango: Fleshy mesocarp, Tomato: Pericarp and Placenta, Litchi: Fleshy aril, Orange: Juicy placental hairs

96. In coconut, the pericarp is differentiated with fibrous mesocarp and stony endocarp.

97. (A), (B), (C) and (D) are true.

98. A fruit is said to be fleshy if its pericarp is differentiated into epicarp, mesocarp and endocarp.

99. Syconus fruit develops from hypanthodium inflorescence. For example, the *Ficus* species.

100. Aggregate fruits are formed by multicarpellary ovary of a single flower. Each carpel develops into a fruitlet and all the fruitlets together form an aggregate fruit.

The Seed

101. The ovule after fertilisation develops into the seed.

102. A – Testa, B – Tegmen, C – Hilum

103. In monocot seed, the radicle enclosed in a thin membranous sheath is called coleorhiza.

104. Radicle and Plumule

105. Coleoptile covers the plumule only.

106. The remains of the nucellus in the seed is called perisperm.

107. Generally, monocotyledonous seeds are endospermic seed or albuminous seed but some are non-endospermic like orchids.

108. The outer covering of the endosperm is separated from the embryo by a proteinaceous layer called the aleurone layer.

109. Monocot seeds consist of one large and shield-shaped cotyledon called scutellum.

110. Dicot seeds are non-endospermic or ex-albuminous like bean, gram and pea, but some are endospermic, like castor.

111. A – Pericarp, B – Aleurone layer, C – Endosperm, D – Coleorhiza

112. Embryo is made up of a radicle, embryonal axis (tigellum), a plumule and one or two cotyledons.

113. A – Coleoptile, B – Coleorhiza

114. A – Hilum, B – Micropyle, C – Seed coat

115. A – Cotyledon, B – Plumule, C – Radicle

Description of Some Important Families

116. Epipetalous condition

117. Family Liliaceae

118. The side of a flower facing the mother axis is called posterior side.

119. Aestivation of sepals or petals cannot be expressed by floral formula.

120. Posterior of calyx, corolla and androecium with respect to ovary or thalamus cannot be expressed by floral diagram.

121. Family Solanaceae

122. Vexillary aestivation: Family Fabaceae, Marginal placentation: Family Fabaceae, Unilocular ovary with false septum: Family Fabaceae

123. Lily family (Liliaceae)

124. *Trifolium* (Family Fabaceae)

125. Valvate aestivation of sepals and petals

126. All are correct.

127. Family Fabaceae, Plant *Indigofera*

128. Three statements (i, ii and iii) are correct. In family Solanaceae, seeds are many and endospermic and the ovary is bicarpellary and syncarpous with axile placentation.

129. In some flowers like lily and tulip (family Liliaceae), the calyx and corolla are not distinct and are termed as perianth.

130. Members of family Solanaceae are characterised by the presence of actinomorphic flowers and valvate aestivation.

131. A – Cymose, B – Liliaceae, C – Fodder, D – Fabaceae

132. Members of familes Solanaceae and Fabaceae show axile placentation.

133. A – Liliaceae, B – Solanaceae, C – Fabaceae

134. *Indigofera*

135. Tobacco (*Nicotiana*)

136. Aesteraceae

137. $A_{(9+1)}$

138. Mother axis

139. Lupin, sweet pea, *Trifolium* (family Fabaceae)

140. All statements are related to family Liliaceae.

Anatomy of Flowering Plants

EXERCISES

The Tissues (Meristematic and Permanent)

1. A tissue is a group of cells:
   (a) Similar in origin but dissimilar in form and function
   (b) Similar in origin and form but usually dissimilar in function
   (c) Similar in origin and form and function
   (d) Similar or dissimilar in origin but always similar in function

2. Meristematic cells are characterised by all, except
   (a) High nucleocytoplasmic ratio
   (b) Low surface area to volume ratio
   (c) Presence of thin cellulosic cell wall
   (d) Absence of large central vacuole

3. Apical meristem, intercalary meristem and lateral meristem are differentiated on the basis of:
   (a) Development (b) Function
   (c) Origin (d) Location

4. Which one is a secondary lateral meristem?
   (a) Cork cambium
   (b) Intrafascicular cambium
   (c) Intercalary meristem
   (d) Both (a) and (b)

5. Cells in a continuous state of division also show the presence of:
   (a) Plastids
   (b) Large-sized prominent nucleus
   (c) Thick cellulosic cell wall
   (d) Large-sized vacuoles

6. Which meristem occurs in grasses and regenerates parts removed by grazing herbivores?
   (a) Intercalary meristem (b) Epical meristem
   (c) Lateral meristem (d) Phellogen

7. Label correctly A, B, C and D in the following diagram:

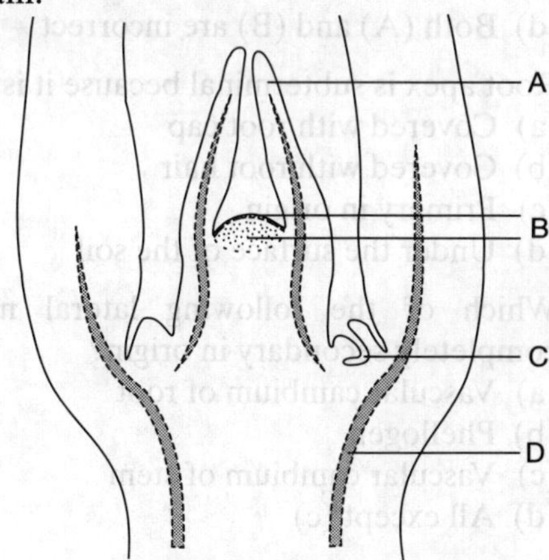

	A	B	C	D
(a)	Leaf primordium	Shoot apical meristem	Axillary bud	Protoderm
(b)	Leaf primordium	Shoot apical meristem	Axillary bud	Differentiating vascular tissue
(c)	Protoderm	Radical bud	Lateral bud	Shoot apical meristem
(d)	Lateral bud	Procambium	Ground meristem	Protoderm

8. During formation of leaves and elongation of stems, some cells left behind from the shoot apical meristem constitute________:
   (a) Axillary bud (b) Phellogen
   (c) Intrafascicular cambium (d) Periderm

9. All are lateral meristems, except:
   (a) Intrafascicular cambium
   (b) Vascular cambium
   (c) Phellogen
   (d) Intercalary meristem

10. Select the incorrectly matched pair:
(a) Lateral meristem – Cylindrical meristem
(b) Aerenchyma – Large air spaces
(c) Bast fibres – Xylem fibres
(d) Non-porous food – Wood of gymnosperm

11. Read the following statements and select the correct option:
A. Apical meristem and intercalary meristem are primary meristems
B. Primary meristems appear early in the life of plants and contribute to the formation of the primary plant body
(a) Only (A) is correct
(b) Only (B) is correct
(c) Both (A) and (B) are correct
(d) Both (A) and (B) are incorrect

12. Root apex is subterminal because it is:
(a) Covered with root cap
(b) Covered with root hair
(c) Primary in origin
(d) Under the surface of the soil

13. Which of the following lateral meristems is completely secondary in origin?
(a) Vascular cambium of root
(b) Phellogen
(c) Vascular cambium of stem
(d) All except (c)

14. Collenchyma cells are characterised by:
A. Presence below the epidermis in most dicotyledonous plants
B. Presence as homogenous layer or in patches
C. Presence of cells which are much thickened at corners due to the deposition of cellulose, hemicellulose and pectin
D. Chloroplast as constant feature
(a) Only (A) and (C) are correct
(b) (A), (C) and (D) are correct
(c) (A), (B) and (C) are correct
(d) (A), (B), (C) and (D) are correct

15. Sclerenchyma differs from collenchyma as the former:
(a) Is dead mechanical tissue
(b) Can take part in the synthesis of food
(c) Has pits in the cell wall
(d) Both (a) and (c) are correct

16. Secondary meristems:
(a) Are responsible for increase in the girth of plant organs
(b) Arise from dedifferentiation of permanent cells
(c) Are not present from the very beginning
(d) All of the above

17. Fascicular, interfascicular and extrastelar cambium together constitute:
(a) Lateral meristem (b) Apical meristem
(c) Secondary meristem (d) Ground meristem

18. Read the following statements with respect to intercalary meristem and state whether they are true (T) or false (F):
A. These are intercalated between permanent tissues
B. They are present at the base of the node or at the base of the internode
C. These are a part of the lateral meristem
D. It helps in the growth of internodes

	A	B	C	D
(a)	T	T	F	F
(b)	F	F	T	F
(c)	T	T	F	T
(d)	T	F	T	T

19. Clowes proposed the quiescent centre theory, which is related to:
(a) Root apex (b) Shoot apex
(c) Endodermis (d) Epidermis

20. Living mechanical tissue is:
(a) Parenchyma (b) Collenchyma
(c) Sclerenchyma (d) Chlorenchyma

21. Which is not a function of the parenchyma?
(a) Photosynthesis
(b) Translocation of food
(c) Storage of food material
(d) Mechanical support

22. On the basis of the given features of simple permanent tissue, identify its type and choose the correct option:
A. Cells are usually living, with vacuolated chloroplast
B. Cells are much thickened at corners due to cellulose, hemicellulose and pectin
C. When they contain chloroplast, the cells perform photosynthesis
(a) Chlorenchyma (b) Parenchyma
(c) Sclerenchyma (d) Collenchyma

23. Lignin is an important constituent in the cell wall of:
(a) Apical meristem
(b) Parenchymatous tissue
(c) Sclerenchymatous tissue
(d) Collenchymatous tissue

24. Parenchymatous cells which help in photosynthesis are called:

(a) Idioblast (b) Chlorenchyma
(c) Prosenchyma (d) Collenchyma

25. The parenchymatous tissue, which can store secretory substances such as tannins, resins and gums, is known as:
(a) Prosenchyma
(b) Aerenchyma
(c) Complementary cells
(d) Storage parenchyma

26. Select the incorrect statement with respect to sclerenchyma:
(a) Provide mechanical support
(b) Thick lignified cell walls having few or numerous pits
(c) Provide flexibility and elasticity
(d) Cells are usually dead and without protoplast

27. Read the following statements and select the correct option:
A. Intercalary meristem is found between permanent cells
B. Intrafascicular cambium represents primary meristem
C. Lateral meristems are responsible for producing secondary tissue
D. Quiescent centre is an inverted cup-like structure in the shoot apex
(a) Only (A) and (B) are correct
(b) (A), (C) and (D) are correct
(c) Only (A), (B) and (C) are correct
(d) All are correct

28. Sclereids are found in:
A. Pulp of fruits like guava
B. Fruit wall of nuts
C. Seed coat of legumes
D. Leaves of tea
(a) Only (A)
(b) Only (B) and (C)
(c) Only (A), (C) and (D)
(d) All are correct

29. The element of xylem which is responsible for radial conduction of water is:
(a) Tracheid (b) Vessel
(c) Xylem fibre (d) Xylem parenchyma

30. Which of the following is absent in the phloem of monocots?
(a) Sieve tube element (b) Phloem parenchyma
(c) Companion cells (d) Phloem fibres

31. In which of the following cells does the nucleus disappear at maturity?
(a) Sieve tubes (b) Companion cells
(c) Phloem parenchyma (d) Sclerenchyma

32. All the components of the xylem are dead, except:
(a) Tracheids (b) Vessels
(c) Xylem parenchyma (d) Xylem fibres

33. All are features of xylem fibres, except:
(a) Highly thickened cell walls
(b) Septate or aseptate
(c) Obliterated central lumen
(d) Non-lignified cell wall

34. The multicellular, dead, long, cylindrical tube-like conducting element of the xylem present in the mustard plant is:
(a) Vessels (b) Tracheids
(c) Fibres (d) Xylem parenchyma

35. Read the following statements and select the correct option:
A. The first formed primary xylem elements are called protoxylem and the later formed primary xylem element is called metaxylem
B. In stems, the protoxylem lies towards the centre
(a) Only (A) is correct
(b) Only (B) is correct
(c) Both (A) and (B) are correct
(d) Both (A) and (B) are incorrect

36. Label correctly A, B, C and D in the following diagram:

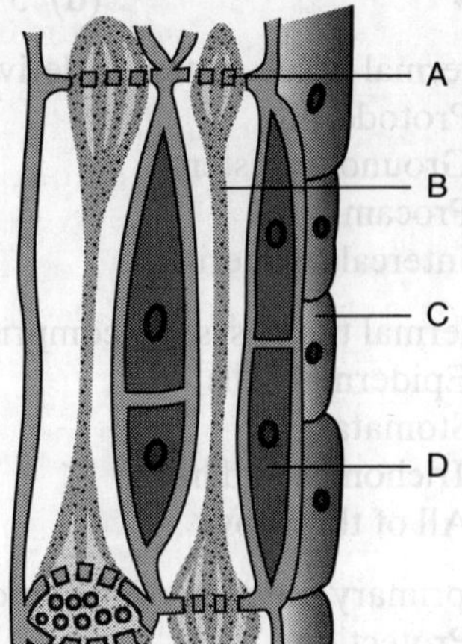

	A	B	C	D
(a)	Sieve pore	Sieve tube	Companion cell	Phloem parenchyma
(b)	Sieve pore	Sieve tube	Phloem parenchyma	Companion cell

(c) Sieve tube	Companion cell	Phloem parenchyma	Phloem fibre
(d) Sieve plate	Sieve plate	Phloem fibre	Companion cell

37. All are phloem fibres obtained from stems and branches, except:
(a) Jute (b) Flax
(c) Cotton (d) Hemp

38. Find the incorrectly matched pair:
(a) Companion cells – Maintain pressure gradient in the sieve tubes
(b) Sieve tube – Peripheral cytoplasm with large central vacuole and prominent nucleus
(c) Phloem parenchyma – Stores food materials and other substances like resin, latex and mucilage
(d) Phloem fibres – Absent in primary phloem but present in secondary phloem

39. Protophloem is generally devoid of:
(a) Phloem fibres (b) Narrow sieve tubes
(c) Bigger sieve tubes (d) Both (a) and (c)

The Tissue System

40. On the basis of location and function, how many types of tissue systems are present in vascular plants?
(a) 2 (b) 3
(c) 4 (d) 5

41. Epidermal tissue system is derived from:
(a) Protoderm
(b) Ground meristem
(c) Procambium
(d) Intercalary meristem

42. Epidermal tissue system comprises:
(a) Epidermal cells
(b) Stomata
(c) Trichomes and hairs
(d) All of the above

43. The primary function of the epidermis is:
(a) Protection
(b) Food assimilation
(c) Conduction of organic and inorganic solutes
(d) Mechanical support

44. Epidermal cells are ___A___ cells and are usually ___B___:

	A	B
(a)	Parenchymatous	Multi-layered
(b)	Collenchymatous	Single-layered
(c)	Sclerenchyma	Multi-layered
(d)	Parenchyma	Single-layered

45. The outside of the epidermis is often covered with a waxy thick layer, called:
(a) Periblem (b) Cuticle
(c) Epiblema (d) Trichome

46. Trichome is soft or stiff hair present on the stem of a plant and its function is:
(a) Protection
(b) Secretion
(c) Preventing water loss
(d) All of the above

47. Guard cells are a special kind of epidermal cell. These cells:
(a) Have chloroplast (b) Thick inner wall
(c) Thin outer wall (d) All of the above

48. All tissues on the ___A___ of the ___B___ such as pericycle, vascular bundles and pith in a dicot root together constitute ___C___:

	A	B	C
(a)	Inner side	Endodermis	Stele
(b)	Outer side	Endodermis	Stele
(c)	Inner side	Hypodermis	Cambium
(d)	Outer side	Secondary cortex	Periderm

49. Well-developed pith is found in:
(a) Monocot root and dicot stem
(b) Dicot root and dicot stem
(c) Monocot stem and dicot root
(d) Monocot stem and monocot root

50. Endodermal cells are:
(a) Barrel-shaped
(b) Without any intercellular spaces
(c) Suberised in the roots
(d) All of the above

51. Primary vascular tissue are formed by:
(a) Procambium (b) Protoderm
(c) Phellogen (d) Ground meristem

52. Identify the given diagram and choose the correct statement:

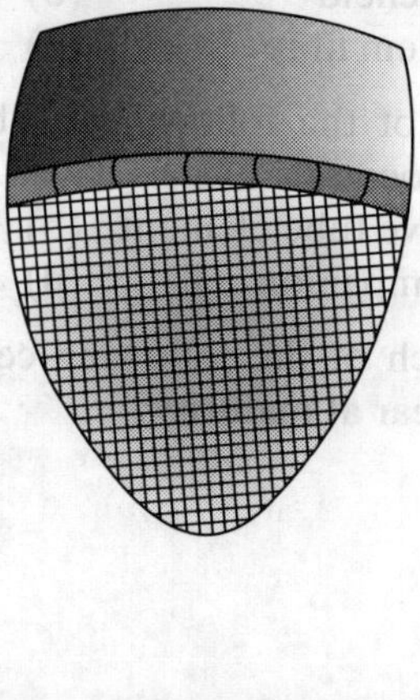

(a) Xylem is centrifugal
(b) Common in dicot stem
(c) Phloem is usually located on the outer side of xylem
(d) All of the above

53. The type of vascular bundles in which xylem and phloem are situated at the same radius, is present in:
A. Root
B. Stem
C. Leaf
(a) (A) and (B) (b) (B) only
(c) (A) and (C) (d) (B) and (C)

54. Given below is the transverse section of a plant part:

Identify the appropriate answer:
(a) Radial vascular bundles of monocot root
(b) Depicts tetrarch condition of dicot root
(c) Depicts conjoint collateral open vascular bundles of dicot stem
(d) Radial vascular bundle with centripetal xylem differentiation

55. Open vascular bundles are:
(a) Always radial
(b) Possess intrafascicular cambium between primary xylem and primary phloem
(c) Shows xylem and phloem are in direct contact
(d) All of the above

56. Stomata:
A. Are structures present in leaf endodermis
B. Consist of two bean-shaped guard cells in monocots
C. Regulate the process of transpiration and gaseous exchange
D. Their guard cell contains chloroplasts
(a) (B) and (D) (b) (B) and (C)
(c) (A) and (B) (d) (C) and (D)

57. Fill in the blanks and select the correct option:
(i) Cuticle is absent in ___A___
(ii) The specialised cells present in the vicinity of guard cells are known as ___B___
(iii) ___C___ helps in preventing water loss due to transpiration

	A	B	C
(a)	Roots	Subsidiary cells	Trichomes
(b)	Stem	Epidermal cell	Trichomes
(c)	Leaf	Epidermal cells	Stem hairs
(d)	Roots	Subsidiary cells	Root hairs

58. Open vascular bundles:
(a) Are found in roots
(b) Lack cambium
(c) Possess the ability to form secondary tissue
(d) Have xylem and phloem on different radii

59. The first formed primary xylem element in stem and roots is towards (respectively):
(a) Pith, Periphery (b) Periphery, Pith
(c) Periphery, Periphery (d) Pith, Pith

60. Ground tissue system:
(a) Includes tissue other than epidermis and vascular tissue
(b) Is composed of simple permanent tissue
(c) Forms outermost covering of the whole plant body
(d) More than one option is correct

61. Biological checkpost in the root of angiosperm is:
(a) Starch sheath (b) Bulliform cells
(c) Endodermis (d) Epidermis

62. The condition where the first formed xylem element faces the pericycle is called:
(a) Endarch (b) Exarch
(c) Mesarch (d) Diarch

63. Procambium forms:
A. Primary xylem and primary phloem
B. Intrafascicular cambium
C. Interfascicular cambium
D. Secondary xylem and secondary phloem
(a) Only (A) (b) (A) and (B)
(c) Only (D) (d) (B) and (D)

Anatomy of Dicotyledonous and Monocotyledonous Plants

64. Exarch and polyarch vascular bundles are seen in:
(a) Dicot stem (b) Dicot root
(c) Monocot root (d) Monocot leaf

65. Fill in the blanks and select the correct option:
A. Large central pith is seen in ___A___
B. Cortex and pith are not distinguished in ___B___
C. Collenchymatous hypodermis is characteristic of ___C___
D. Mesophyll is differentiated into palisade tissue and spongy parenchyma in ___D___

	A	B	C	D
(a)	Monocot root	Monocot stem	Dicot stem	Dicot leaf
(b)	Monocot root	Dicot stem	Monocot stem	Monocot leaf
(c)	Dicot root	Monocot root	Monocot stem	Dicot leaf
(d)	Monocot stem	Dicot stem	Monocot root	Monocot leaf

66. The cells that lie between the xylem and phloem in a dicot root are:
(a) Intrafascicular cambium
(b) Interfascicular cambium
(c) Conjunctive tissue
(d) Medullary rays

67. Match the columns and select the correct option:

	Column I		Column II
A.	Collenchymatous hypodermis	(i)	Root endodermis
B.	Starch sheath	(ii)	Monocot stem
C.	Casparian strip	(iii)	Dicot stem
D.	Undifferentiated ground tissue	(iv)	Endodermis of dicot stem

(a) A – (iii), B – (iv), C – (i), D – (ii)
(b) A – (ii), B – (iv), C – (i), D – (iii)
(c) A – (iii), B – (i), C – (iv), D – (ii)
(d) A – (ii), B – (i), C – (iv), D – (iii)

68. The figure depicted below is the transverse section of:

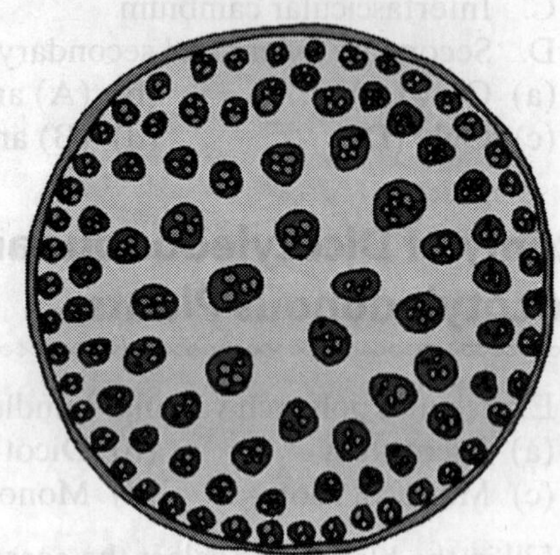

(a) Monocot stem (b) Monocot root
(c) Dicot root (d) Dicot stem

69. Sclerenchymatous bundle sheath is present in the vascular bundle of:
(a) Dicot root (b) Monocot root
(c) Dicot stem (d) Monocot stem

70. Conjoint and closed vascular bundles are not found in:
(a) Mustard leaf (b) Wheat leaf
(c) Wheat stem (d) Mustard stem

71. Isobilateral leaf differs from dorsiventral leaf in:
(a) Presence of stomata on the adaxial epidermis
(b) Absence of differentiated mesophyll
(c) Presence of large, empty, colourless cells on the adaxial epidermis
(d) More than one option is correct

72. Which of the following statements are correct?
A. Radial vascular bundles are present in dicot and monocot roots
B. Conjoint and closed vascular bundles are present in monocot stem
C. Vascular cambium is present in dicot stem
(a) Only (A) (b) (A) and (C) only
(c) (B) and (C) only (d) (A), (B) and (C)

73. Pith and cortex is not differentiated in:
(a) Dicot stem (b) Monocot root
(c) Monocot stem (d) Dicot root

74. Stem with open vascular bundles arranged in a ring contains:
(a) Water containing cavities within the vascular bundles
(b) Large and well-developed pith present in the centre
(c) Absence of phloem parenchyma in the vascular bundles
(d) Vascular bundles are surrounded by sclerenchymatous bundle sheath

75. Bulliform cells are present in:
(a) Adaxial epidermis of grass leaf
(b) Abaxial epidermis of grass leaf
(c) Adaxial epidermis of mustard leaf
(d) Abaxial epidermis of mustard leaf

76. Vascular bundles present in dicot leaves are:
(a) Conjoint, open
(b) Conjoint, closed
(c) Radial, open
(d) Bicollateral and open

77. The tangential as well as radial walls of the endodermal cells have a deposition of water-impermeable, waxy material ___A___ in the form of ___B___:

	A	B
(a)	Suberin	Starch sheath
(b)	Cutin	Stile
(c)	Suberin	Casparian strip
(d)	Lignin	Pericycle

78. Vascular bundles in isobilateral leaves are:
(a) Radial, closed (b) Conjoint, closed
(c) Conjoint, open (d) Concentric, closed

79. Find the incorrect statement with respect to the diagram given below:

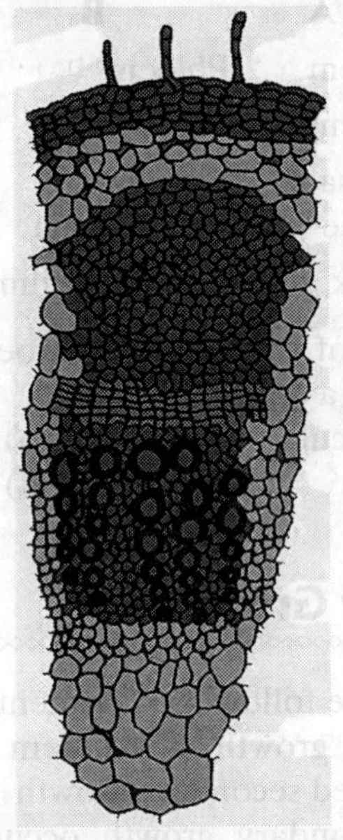

(a) Collenchymatous hypodermis present below the epidermis
(b) Semilunar patches of sclerenchyma present above the phloem
(c) Vascular bundles are conjoint, collateral and open
(d) Water-containing cavities are present within the vascular bundles

80. The following anatomical features (A–C) are related with:
A. Polyarchy condition of vascular bundles
B. Absence of cambium
C. Presence of large and well-developed pith in centre
(a) Dicot stem (b) Monocot root
(c) Monocot root (d) Dicot root

81. Label correctly A, B and C in the following diagram:

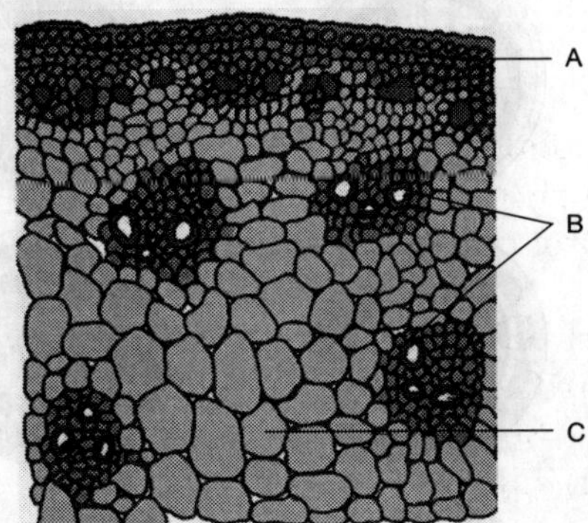

	A	B	C
(a)	Hypodermis	Vascular bundles	Pith
(b)	Hypodermis	Vascular bundles	Ground tissue
(c)	Epidermis	Vascular bundles	Pith
(d)	Cortex	Vascular bundles	Ground tissue

82. Identify the given anatomy and choose the correct statements:

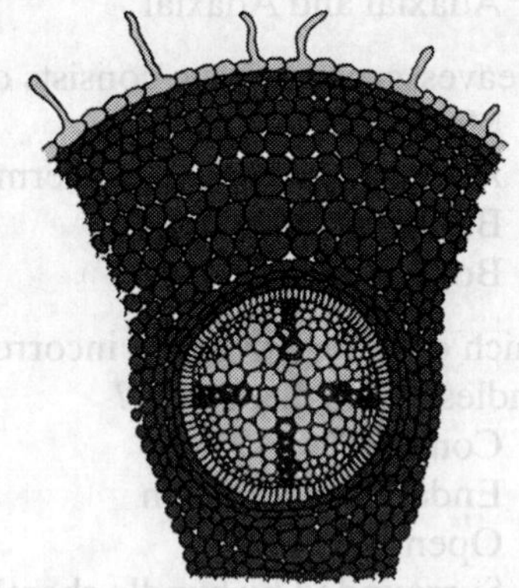

(a) It has semilunar patches of sclerenchyma just above the medullary rays
(b) The innermost layer of cortex is rich in starch sheath
(c) The tangential as well as radial walls of endodermal cells have depositions of suberin
(d) Phloem parenchyma is absent

83. The figures (a) and (b) show the internal structures of certain plant parts. Select the correct answer with respect to a common characteristic of these structures:

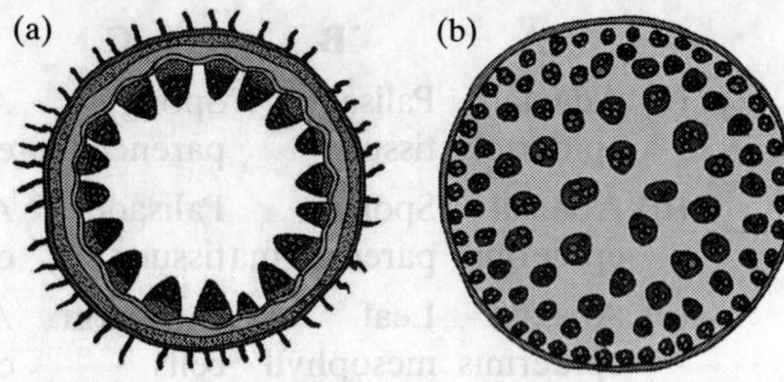

(a) Presence of undifferentiated ground tissue
(b) Pith is large and well developed
(c) Presence of endarch primary xylem
(d) Presence of trichomes

84. During water stress, the bulliform cells:
(a) Become turgid
(b) Become flaccid
(c) Make the leaf curl inwards
(d) All except (a)

85. Dicot stem differs from dicot root in:
(a) Presence of cortex
(b) Position of protoxylem

(c) Absence of hypodermis
(d) Presence of Casparian strip in endodermis

86. In dorsiventral leaf, the locations of the phloem and palisade tissue, respectively, are:
(a) Abaxial and Adaxial
(b) Adaxial and Abaxial
(c) Abaxial and Abaxial
(d) Adaxial and Adaxial

87. In leaves, ground tissue consists of:
(a) Mesophyll
(b) Adaxial and abaxial epidermis
(c) Bulliform cells
(d) Both (a) and (c)

88. Which of the following is incorrect about vascular bundles of the dicot stem?
(a) Conjoint
(b) Endarch protoxylem
(c) Open
(d) Surrounded by bundle sheath

89. Label correctly A, B, C and D in the given figure:

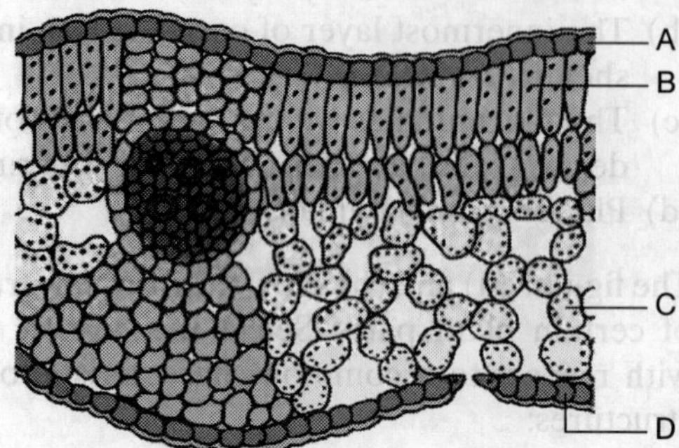

	A	B	C	D
(a)	Adaxial epidermis	Palisade tissue	Spongy parenchyma	Abaxial epidermis
(b)	Abaxial epidermis	Spongy parenchyma	Palisade tissue	Adaxial epidermis
(c)	Abaxial epidermis	Leaf mesophyll	Bulliform cells	Adaxial epidermis
(d)	Adaxial epidermis	Leaf mesophyll	Bulliform cells	Abaxial epidermis

90. In dorsiventral leaf, the size of the vascular bundles is dependent on the:
(a) Structure of lamina
(b) Size of veins and veinlets
(c) Season
(d) Length of petiole

91. A dorsiventral leaf is characterised by all, except:
(a) Amphistomatic
(b) Differentiated leaf mesophyll
(c) Absence of bulliform cells
(d) Reticulate venation

92. Conjunctive tissue is present between ____A____ and ____B____ in ____C____:

	A	B	C
(a)	Xylem	Phloem	Monocot stem
(b)	Xylem	Phloem	Dicot root
(c)	Palisade tissue	Spongy parenchyma	Dorsiventral leaf
(d)	Cork	Cork cambium	Dicot stem

93. Which of the following does not constitute the stele of a dicot root?
(a) Vascular bundles
(b) Pericycle
(c) Pith
(d) Endodermis

Secondary Growth

94. Read the following statements:
A. The growth of the stem and roots in girth is called secondary growth
B. Secondary growth occurs with the help of lateral meristem
C. Secondary growth occurs due to joint activity of the vascular cambium and cork cambium
D. Secondary growth does not occur in monocots due to lack of vascular cambium

Correct statements is/are:
(a) Only (A) and (B)
(b) Only (A) and (C)
(c) Only (A), (B) and (C)
(d) (A), (B), (C) and (D)

95. The activity of the cambium is under the control of:
(a) Physiological factors
(b) Environmental factors
(c) Both (a) and (b)
(d) None of the above

96. Find the correct sequence of events during secondary growth in the dicot stem:

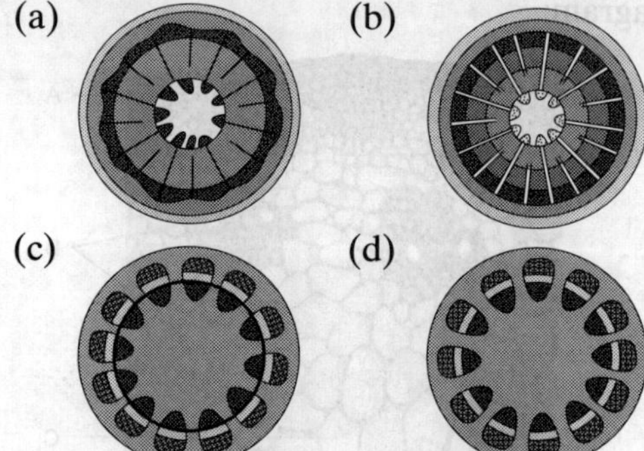

(a) DCBA (b) DCAB
(c) CDAB (d) CDBA

97. During the period of secondary growth in dicot roots, which cells divide and give rise to part of the vascular cambium?
(a) Parenchymatous cells of the cortex
(b) Parenchymatous cells of the pericycle
(c) Parenchymatous cells of pith
(d) Collenchymatous cells of the hypodermis

98. Periderm includes:
(a) Cork cambium and cork only
(b) Cork cambium and secondary cortex only
(c) Cork and secondary cortex only
(d) Cork, cork cambium and secondary cortex

99. Annual rings are well distinguished in trees growing in:
A. Uniform climatic conditions
B. Temperate regions
C. Tropical regions
(a) Only (A) (b) Only (C)
(c) (A) and (B) (d) (A) and (C)

100. Cambium ring in dicot stem is:
(a) Primary in origin
(b) Secondary in origin
(c) Half primary and half secondary in origin
(d) Promeristem in origin

101. Tyloses are:
(a) Balloon-like outgrowths of parenchymatous cells in the lumen of vessels
(b) Present in sieve tubes
(c) Present in secondary xylem and make it non-functional
(d) More than one option is correct

102. The impervious nature of cork for water is due to deposition of ______
(a) Cutin (b) Lignin
(c) Suberin (d) Silica

103. After secondary growth, what is the actual future of primary xylem?
(a) Converts into secondary xylem
(b) Remains more or less intact
(c) Is gradually crushed
(d) Transform into pith

104. Autumn wood characteristics include all, except:
(a) Narrow lumen in vessels
(b) Higher density
(c) Darker in colour
(d) Larger number of xylary elements

105. Early wood differs from late wood in:
(a) Dark colour wood with higher density
(b) Light colour wood with lower density
(c) Vessels with wider cavities
(d) All except (a)

106. Choose the correct option with respect to alburnum or sapwood:
(a) Peripheral and dead
(b) Central and physiological active
(c) Light-coloured and peripheral
(d) Provides mechanical support to the stem

107. Secondary growth in the extrastelar region is due to the activity of:
(a) Cork cambium
(b) Vascular cambium
(c) Phellogen
(d) More than one option is correct

108. Sapwood includes:
A. Peripheral secondary xylem
B. Xylary elements are filled with tannins, resins, oil, gum, aromatic compounds, etc.
C. Functional secondary xylem
D. Highly lignified non-functional xylary element
(a) (A) and(C) (b) (A) and (B)
(c) (A) and (D) (d) (B) and (D)

109. Bark includes all, except:
(a) Primary cortex (b) Primary phloem
(c) Periderm (d) Primary xylem

110. Label correctly A, B and C in the following diagram:

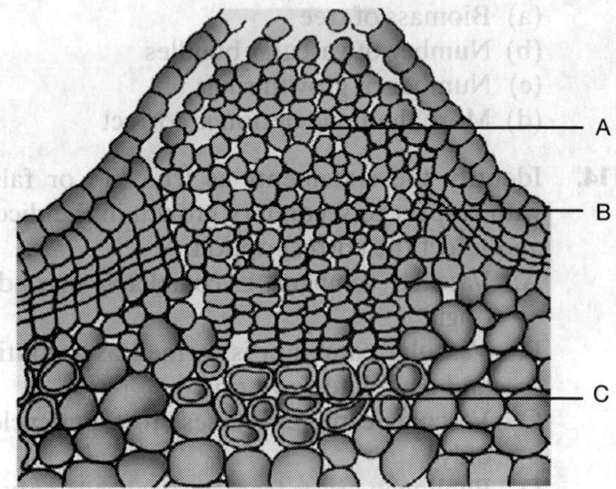

	A	B	C
(a)	Cork	Cork cambium	Secondary cortex
(b)	Complementary cells	Phellogen	Phelloderm
(c)	Cortex	Cork cambium	Cork
(d)	Phellem	Phellogen	Phelloderm

111. Read the following statements and state whether true (T) or false (F):
A. Annual rings are distinct in plants growing in tropical areas
B. Lenticels occur in most woody trees and permit gaseous exchange
C. In older trees, the greater part of secondary xylem is dark brown and does not conduct water
D. All tissues outside the cork cambium constitute the bark

	A	B	C	D
(a)	T	T	T	F
(b)	T	T	F	F
(c)	F	F	T	F
(d)	T	F	T	T

112. Which of the given sets of tissues can be observed in bark?
(a) Secondary xylem, secondary phloem and secondary cortex
(b) Secondary phloem, secondary cortex, cork cambium and cork
(c) Secondary xylem, secondary cortex, phellem and pericycle
(d) Patches of sclerenchymatous pericycle, primary and secondary xylem

113. The age of a tree can be estimated by:
(a) Biomass of tree
(b) Number of vascular bundles
(c) Number of growth rings
(d) More than one option is correct

114. Identify the statements as true (T) or false (F) with respect to secondary growth in the dicot root and select the correct option:
A. Vascular cambium is completely secondary in origin
B. Vascular cambium is formed as a continuous and complete wavy ring
C. Vascular cambium arises from pericycle cells only
D. Phellogen arises from cortical cells only

	A	B	C	D
(a)	F	F	T	T
(b)	T	T	F	F
(c)	T	F	T	T
(d)	T	T	T	F

115. Choose the option that correctly differentiates between secondary growth of the dicot root and dicot stem:

	Dicot root	Dicot stem
(a)	Vascular cambium is completely secondary in origin	Vascular cambium is completely primary in origin
(b)	Pericycle forms vascular cambium	Conjunctive tissue forms vascular cambium
(c)	Cambium activity is uniform throughout year	Cambium activity is not uniform throughout the year
(d)	Cork cambium forms periderm	Cork cambium does not form periderm

116. Which of the following are formed through redifferentiation?
A. Secondary xylem
B. Secondary phloem
C. Phellem
D. Phellogen
E. Phelloderm
F. Vascular cambium
(a) Only (A) and (B)
(b) (A), (B), (C) and (E)
(c) Only (D) and (F)
(d) (C), (D) and (E)

117. Which of the following functions is not performed by heartwood?
(a) Mechanical support
(b) Durability
(c) Pathogen resistant
(d) Conduction of sap

118. Which of the following is incorrect during secondary growth in the dicot stem?
(a) Vascular cambium formed by intra- and interfascicular cambium
(b) The cambium is generally more active on the outer side than on the inner side
(c) Secondary medullary rays pass through the secondary xylem and secondary phloem in the radial direction
(d) The primary xylem remains more or less intact in or around the centre

NCERT EXEMPLAR QUESTIONS

1. A transverse section of the stem is stained first with safranin and then with fast green following the usual schedule of double staining for the preparation of a permanent slide. What would be the colour of the stained xylem and phloem?
(a) Red and green
(b) Green and red
(c) Orange and yellow
(d) Purple and orange

2. Match the following and choose the correct option from below:

Column I	Column II
A. Meristem	(i) Photosynthesis
B. Parenchyma	(ii) Mechanical support
C. Collenchyma	(iii) Actively dividing cells
D. Sclerenchyma	(iv) Stomata
E. Epidermal tissue	(v) Sclereids

(a) A – (i), B – (iii), C – (v), D – (ii), E – (iv)
(b) A – (iii), B – (i), C – (ii), D – (v), E – (iv)
(c) A – (ii), B – (iv), C – (v), D – (i), E – (iii)
(d) A – (v), B – (iv), C – (iii), D – (ii), E – (i)

3. Match the following and choose the correct option from below:

Column I	Column II
A. Cuticle	(i) Guard cells
B. Bulliform cells	(ii) Single layer
C. Stomata	(iii) Waxy layer
D. Epidermis	(iv) Empty colourless cell

(a) A – (iii), B – (iv), C – (i), D – (ii)
(b) A – (i), B – (ii), C – (iii), D – (iv)
(c) A – (iii), B – (ii), C – (iv), D – (i)
(d) A – (iii), B – (ii), C – (i), D – (iv)

4. Identify the simple tissue from among the following:
(a) Parenchyma (b) Xylem
(c) Epidermis (d) Phloem

5. The cells of this root are living and show angular wall thickening. They also provide mechanical support. The tissue is:
(a) Xylem (b) Sclerenchyma
(c) Collenchyma (d) Epidermis

6. Epiblema of roots is equivalent to:
(a) Pericycle (b) Endodermis
(c) Epidermis (d) Stele

7. A conjoint and open vascular bundle will be observed in the transverse section of:
(a) Monocot root (b) Monocot stem
(c) Dicot root (d) Dicot stem

8. Interfascicular cambium and cork cambium are formed due to:
(a) Cell division
(b) Cell differentiation
(c) Cell dedifferentiation
(d) Redifferentiation

9. Phellogen and phellem, respectively, denote:
(a) Cork and cork cambium
(b) Cork cambium and cork
(c) Secondary cortex and cork
(d) Cork and secondary cortex

10. In which of the following pairs of parts of a flowering plant is the epidermis absent?
(a) Root tip and shoot tip
(b) Shoot bud and floral bud
(c) Ovule and seed
(d) Petiole and pedicel

11. How many shoot apical meristems are likely to be present in a twig of a plant possessing 4 branches and 26 leaves?
(a) 26 (b) 1
(c) 5 (d) 30

12. A piece of wood having no vessels (trachea) must belong to:
(a) Teak (b) Mango
(c) Pine (d) Palm

13. A plant tissue, when stained, showed the presence of hemicellulose and pectin in the cell wall of its cells. The tissue represents:
(a) Collenchyma (b) Sclerenchyma
(c) Xylem (d) Meristem

14. In conifers, fibres are likely to be absent in:
(a) Secondary phloem (b) Secondary xylem
(c) Primary phloem (d) Leaves

15. When we peel the skin of a potato tuber, we remove:
(a) Periderm (b) Epidermis
(c) Cuticle (d) Sapwood

16. A vesselless piece of stem possessing prominent sieve tubes would belong to:
(a) *Pinus* (b) *Eucalyptus*
(c) Grass (d) *Trochodendron*

17. Which one of the following cell types always divides by anticlinal cell division?
(a) Fusiform initial cells (b) Root cap
(c) Protoderm (d) Phellogen

18. What is the fate of primary xylem in a dicot showing extensive secondary growth?
(a) It is retained in the centre of the axis
(b) It gets crushed
(c) May or may not get crushed
(d) It gets surrounded by primary phloem

PRACTICE TEST PAPER

1. Which of the following is not a feature of meristematic tissue?
(a) Cells are immature and living
(b) Presence of thin cellulosic cell wall
(c) Cells are metabolically less active
(d) Presence of dense cytoplasm and large prominent nucleus

2. Which of the following meristems is primary in origin but responsible for secondary growth?
(a) Intercalary meristem
(b) Intrafascicular cambium
(c) Interfascicular cambium
(d) Cork cambium

3. Consider the following statements:
(i) It occurs in between mature tissue
(ii) It is a primary meristem
(iii) It occurs in grasses and regenerates parts removed by grazing herbivores
The above characteristics are attributed to:
(a) Axillary meristem
(b) Intrafascicular cambium
(c) Lateral meristem
(d) Intercalary meristem

4. Permanent tissue consists of cells having no power of cell division. Such tissue are formed by:
(a) Primary meristem (b) Lateral meristem
(c) Vascular cambium (d) Phellogen

5. Highly thick-walled lignified cells generally formed in the fruit wall of nuts and pulp of fruits are:
(a) Fibres
(b) Albuminous cells
(c) Sclereids
(d) Complementary cells

6. Stomata are usually more in number on the lower epidermis in:
(a) Dorsiventral leaf (b) Isobilateral leaf
(c) Monocot leaf (d) Both (b) and (c)

7. All of the following are absent in monocot stem, except:
(a) Endodermis (b) Pericycle
(c) Pith (d) Hypodermis

8. Which of the following is not a function of collenchyma?
(a) Provides mechanical support to young stems
(b) They assimilate food when they contain chloroplast
(c) Provides mechanical support to the petiole of a leaf
(d) Storage of food material and secretion

9. Read the following statements and select the correct option:
A. Gymnosperm lack sieve tubes and companion cells
B. Phloem fibres are generally absent in primary phloem but are found in the secondary phloem
C. First formed primary phloem has bigger sieve tubes and is referred to as protoplasm
(a) Only (A) and (B) are correct
(b) Only (B) is correct
(c) (A), (B) and (C) are correct
(d) Only (B) and (C) are correct

10. Phloem fibres are:
(a) Jute, Flax, Coir (b) Jute, Flax, Hemp
(c) Cotton, Jute, Coir (d) Hemp, Cotton, Flax

11. Distinct endodermis with deposition of suberin can be observed in:
(a) Dicot roots (b) Dicot stem
(c) Monocot stem (d) Both (a) and (b)

12. Fill in the blanks and choose the correct option:
(i) Metabolically active tissue made up of thin-walled isodiametric cells without intercellular space is **A**
(ii) Gymnosperms lack **B** in their xylem

(iii) Phloem parenchyma is absent in most of **C**
(iv) In stem, the primary xylem is **D**

	A	B	C	D
(a)	Parenchyma	Vessels	Monocots	Endarch
(b)	Meristematic tissue	Vessels	Monocots	Endarch
(c)	Parenchyma	Tracheids	Dicots	Exarch
(d)	Meristematic tissue	Vessels	Dicots	Exarch

13. The region of ground tissue system present just below the epidermis in monocot stem:
(a) Parenchyma (b) Collenchyma
(c) Chlorenchyma (d) Sclerenchyma

14. A TS shows conjoint, collateral, endarch and open vascular bundles arranged in a ring. It is:
(a) Dicot stem (b) Monocot stem
(c) Dicot root (d) Monocot root

15. Dicot root is characterised by all of the following features, except:
(a) Vascular bundles are radial and hexarch
(b) Pith is well developed
(c) Pericycle gives rise to cork cambium
(d) Centrifugal xylem development

16. Which of the following components of the xylem is not present in all vascular plants?
(a) Vessels
(b) Tracheids
(c) Xylem parenchyma
(d) Xylem fibres

17. The meristematic tissue which develops between two vascular bundles is called:
(a) Phellogen
(b) Intrafascicular cambium
(c) Intercalary meristem
(d) Interfascicular cambium

18. Lenticel is produced by the activity of:
(a) Vascular cambium
(b) Phellogen
(c) Primary meristem
(d) Intercalary meristem

19. Read the following statements and select the correct option:
A. Ground tissue consists of simple tissues such as parenchyma, collenchyma and sclerenchyma
B. In leaves, ground tissue consists of chloroplast-containing cells called mesophyll
(a) Only (A) is correct
(b) Only (B) is correct
(c) Both (A) and (B) are incorrect
(d) Both (A) and (B) are correct

20. Which is true about sapwood/laburnum?
A. It is darker in colour
B. It is hard, durable and resistant to attachment of microorganisms and insects
C. It is a peripheral part of secondary xylem
D. It is involved in conduction of water and minerals from root to leaf
(a) (A) and (B) (b) (A) and (D)
(c) (B) and (C) (d) (C) and (D)

21. Vascular tissue is derived from:
(a) Protoderm (b) Phellogen
(c) Procambium (d) Promeristem

22. Bark includes:
(a) All tissues exterior to vascular cambium
(b) Both living and dead tissues
(c) Periderm and secondary phloem
(d) All of the above

23. The cork formed from cork cambium is impervious to water deposition because of:
(a) Suberin (b) Cutin
(c) Lignin (d) Pectin

24. Spring wood:
A. Is lighter in colour
B. Has low density
C. Has wider lumen of vessels
(a) Only (A) (b) Only (A) and (C)
(c) Only (B) and (C) (d) (A), (B) and (C)

25. Identify the following statements as true (T) or false (F) and choose the option accordingly:
A. In monocot root, polyarchy condition is usually seen in their vascular bundles
B. In isobilateral leaf palisade, tissue is present towards the adaxial epidermis
C. In monocot stem, peripheral vascular bundles are generally smaller than the centrally located ones
D. In collateral vascular bundles, phloem is situated towards the inner side

	A	B	C	D
(a)	T	T	F	F
(b)	T	F	T	F
(c)	F	T	F	T
(d)	F	F	T	T

26. Which of the following figures is not related to dicot?

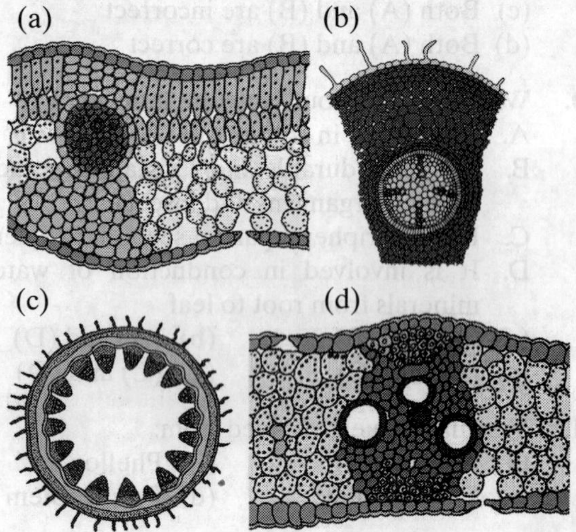

27. Bulliform cells:
(a) Are large emptily colourless cells
(b) Are found in dicot leaf
(c) Are modified abaxial epidermal cells
(d) All except (b)

28. Why are vascular bundles closed in monocots?
(a) Xylem and phloem are derived from primary meristem
(b) Xylem and phloem occur in separate bundles
(c) Vascular cambium is present between xylem and phloem
(d) Vascular cambium is not present

29. Aerenchyma provides:
(a) Flexibility to plants
(b) Mechanical strength
(c) Buoyancy to hydrophytic plants
(d) All except (c)

30. Tyloses are:
(a) Tracheal plugs
(b) Found in heartwood
(c) Balloon-like structures formed by xylem parenchyma in the lumen of vessels
(d) All of the above

31. Select the incorrectly matched pair:
(a) Y-shaped arrangement of xylem vessels – Monocot stem
(b) Differentiated leaf mesophyll – Monocot leaf
(c) Vascular bundles are surrounded by bundle sheath – Dicot stem
(d) Large and well-developed pith – Monocot root

32. Cork cambium is also called **A** and generally develops in **B** region:

	A	B
(a)	Phellogen	Medullary rays
(b)	Phellogen	Cortex
(c)	Phelloderm	Cortex
(d)	Phellem	Pith

33. Read the following statements with respect to secondary growth in dicot stem:
A. Cambium is more active on the inner side than on the outer side
B. Due to the continued formation and accumulation of secondary xylem, secondary and primary phloem are obliterated
C. The primary xylem remains more or less intact in or around the centre
D. Activity of cambium ring is not uniform in trees growing in temperate regions

Select the correct statements:
(a) Only (A)
(b) Only (A), (B) and (D)
(c) (A), (B) and (C)
(d) (A), (B), (C) and (D)

34. Biological checkpost in the roots of angiosperm is:
(a) Pericycle (b) Endodermis
(c) Hypodermis (d) Epiblema

35. Select the incorrect statement for cork cambium:
(a) It is lateral meristem
(b) Completely secondary in origin
(c) Couple of layers thick
(d) Narrow, thick-walled, nearly rectangular cells

36. Specialised epidermal cells surrounding the guard cells are called:
(a) Bulliform cells
(b) Lenticels
(c) Complementary cells
(d) Subsidiary cells

37. Which of the following anatomical features will you use to distinguish between dicot stem and dicot root?
(a) Cortical cell
(b) Position of protoxylem
(c) Absence of phloem parenchyma in dicot root
(d) Absence of pith in dicot stem

38. Select the incorrect statement for secondary growth in dicot root:
(a) Cambium ring is completely secondary in origin
(b) A portion of the pericycle tissue above the protoxylem contributes to cambium ring formation

(c) Annual rings are clearly distinguished as in dicot stem
(d) Cork cambium originates from the pericycle

39. Vessels differ from tracheids in:
(a) Presence of lignified cell wall
(b) Being interconnected through perforations on their transverse walls
(c) Being tube-like cells with tapering ends
(d) Being dead and without protoplast

40. Radial conduction of water takes place by:
(a) Vessels
(b) Vessels and tracheids
(c) Phloem
(d) Ray parenchyma cells

41. Read the following statements:
A. Collenchymatous hypodermis
B. Sclerenchymatous pericycle above the phloem in the form of semi-lunar patches
C. Well-developed pith

The above characteristics are attributed to:
(a) Dicot root (b) Monocot root
(c) Monocot stem (d) Dicot stem

42. Xylem parenchyma stores:
(a) Starch (b) Fat
(c) Tannins (d) All of the above

43. Trichomes, root hairs and stomata are included under:
(a) Epidermal tissue system
(b) Ground tissue system
(c) Vascular tissue system
(d) Non-living component

44. Which of the following is absent in monocots?
(a) Cambium (b) Sieve tubes
(c) Companion cells (d) Phloem fibres

45. Starch sheath is endodermis rich in starch grains and found in:
(a) Dicot root (b) Monocot root
(c) Dicot stem (d) Monocot stem

ANSWER KEYS

Exercises

1. (c) **2.** (b) **3.** (d) **4.** (a) **5.** (b) **6.** (a) **7.** (b) **8.** (a) **9.** (d) **10.** (c)
11. (c) **12.** (a) **13.** (d) **14.** (c) **15.** (d) **16.** (d) **17.** (a) **18.** (c) **19.** (a) **20.** (b)
21. (d) **22.** (d) **23.** (c) **24.** (b) **25.** (a) **26.** (c) **27.** (d) **28.** (d) **29.** (d) **30.** (b)
31. (a) **32.** (c) **33.** (d) **34.** (a) **35.** (c) **36.** (b) **37.** (c) **38.** (b) **39.** (d) **40.** (b)
41. (a) **42.** (d) **43.** (a) **44.** (d) **45.** (b) **46.** (d) **47.** (d) **48.** (a) **49.** (a) **50.** (d)
51. (a) **52.** (d) **53.** (d) **54.** (b) **55.** (b) **56.** (d) **57.** (a) **58.** (c) **59.** (a) **60.** (d)
61. (c) **62.** (b) **63.** (b) **64.** (c) **65.** (a) **66.** (c) **67.** (a) **68.** (a) **69.** (d) **70.** (d)
71. (d) **72.** (d) **73.** (c) **74.** (b) **75.** (a) **76.** (b) **77.** (c) **78.** (b) **79.** (d) **80.** (b)
81. (b) **82.** (c) **83.** (c) **84.** (d) **85.** (b) **86.** (a) **87.** (a) **88.** (d) **89.** (a) **90.** (b)
91. (a) **92.** (b) **93.** (d) **94.** (d) **95.** (c) **96.** (b) **97.** (b) **98.** (d) **99.** (b) **100.** (c)
101. (d) **102.** (c) **103.** (b) **104.** (d) **105.** (d) **106.** (c) **107.** (d) **108.** (a) **109.** (d) **110.** (b)
111. (a) **112.** (b) **113.** (c) **114.** (b) **115.** (c) **116.** (b) **117.** (d) **118.** (b)

NCERT Exemplar Questions

1. (b) **2.** (b) **3.** (a) **4.** (a) **5.** (c) **6.** (c) **7.** (d) **8.** (c) **9.** (b) **10.** (a)
11. (c) **12.** (c) **13.** (a) **14.** (b) **15.** (a) **16.** (d) **17.** (c) **18.** (a)

Practice Test Paper

1. (c) **2.** (b) **3.** (d) **4.** (a) **5.** (c) **6.** (a) **7.** (d) **8.** (d) **9.** (a) **10.** (b)
11. (a) **12.** (b) **13.** (d) **14.** (a) **15.** (c) **16.** (a) **17.** (d) **18.** (b) **19.** (d) **20.** (d)
21. (c) **22.** (d) **23.** (a) **24.** (d) **25.** (b) **26.** (d) **27.** (a) **28.** (d) **29.** (c) **30.** (d)
31. (c) **32.** (b) **33.** (c) **34.** (b) **35.** (d) **36.** (d) **37.** (b) **38.** (c) **39.** (b) **40.** (d)
41. (d) **42.** (d) **43.** (a) **44.** (a) **45.** (c)

SOLUTIONS

The Tissues (Meristematic and Permanent)

1. A group of similar or dissimilar cells that perform a common function and have a common origin is called tissue.
2. Meristematic cells are characterised by the presence of high surface area and volume ratio.
3. Location
4. Cork cambium is a secondary lateral meristem. Intrafascicular cambium is also a lateral meristem, but primary in origin.
5. Meristematic cells are characterised by the presence of a large prominent nucleus.
6. Intercalary meristem
7. A – Leaf primordium, B – Shoot apical meristem, C – Axillary bud, D – Differentiated vascular tissue
8. Axillary bud
9. Intercalary meristem is a primary meristem and present at the base of the internode or at the base of the node of a stem or leaf. It helps in elongation and growth of the plant at the node and internode.
10. Phloem fibres are also called bast fibres.
11. Both (A) and (B) are correct.
12. Root apex is subterminal due to the presence of root cap at its tip.
13. Phellogen (cork cambium) and vascular cambium of the roots are completely secondary in origin. They develop by dedifferentiation of permanent tissue at the time of secondary growth.
14. (A), (B) and (C) are correct. Presence of chloroplast is not a constant feature.
15. Sclerenchyma is dead mechanical tissue, with highly thickened lignocellulosic cell wall and without protoplast. These cells have simple pits in their cell wall.
16. All the statements are correct.
17. Lateral meristem is present on the lateral side of stems and roots. For example, fascicular, interfascicular and extrastelar cambium.
18. C is false. Intercalary meristem is a primary meristem and responsible for primary growth.
19. Quiescent centre is present at the centre of the root apex. Cells of this region show very low metabolic activity due to synthesis of a little amount of DNA, RNA and protein.
20. Collenchyma is living mechanical tissue.
21. Parenchyma performs various functions like photosynthesis, storage and secretion.
22. Collenchyma
23. Sclerenchyma consists of long, narrow cells with thick and lignified cell walls.
24. The parenchyma containing chloroplast is called chlorenchyma.
25. Prosenchyma
26. Collenchyma cells provide flexibility and elasticity.
27. (A), (B), (C) and (D) are correct.
28. Option (d) is correct.
29. The radial conduction of water takes place by ray parenchyma or xylem parenchyma.
30. Phloem parenchyma is absent in most monocotyledons.
31. Sieve tube elements are long, tube-like structures that possess a peripheral cytoplasm and a large vacuole but lack a nucleus.
32. Xylem parenchyma
33. Xylem fibres are sclerenchymatous fibres. Cells are elongated with pointed ends. Cell wall is highly lignified, having simple pits.
34. Vessels
35. Both (A) and (B) are correct. In the stem, protoxylem lies towards the centre and metaxylem lies towards the periphery.

36. A – Sieve pore, B – Sieve tube, C – Companion cell, D – Phloem parenchyma

37. Cotton fibres are exogenous outgrowths of seed coats and made up of pure cellulose.

38. Sieve tubes lack a nucleus.

39. Phloem fibres and bigger sieve tubes

The Tissue System

40. Three types of tissue systems are present in vascular plants: epidermal, ground and vascular.

41. Proderm

42. Epidermal tissue system comprises epidermal cells, stomata and epidermal outgrowth (trichomes, root hairs).

43. Epidermis is protective in function and forms the outermost covering of the whole plant body.

44. A – Parenchymatous, B – Single layered

45. Cuticle

46. Trichomes help in preventing water loss due to transpiration. They may even be secretory.

47. Guard cells possess chloroplast. The outer wall of the guard cells are thin and the inner walls (towards the aperture) are thick and elastic.

48. A – Inner side, B – Endodermis, C – Stele

49. Pith occupies the central part of stele. Well-developed pith is found in monocot root and dicot stem.

50. All of the above.

51. Procambium forms the primary xylem, primary phloem and intrafascicular cambium, if present.

52. The given figure represents conjoint, collateral and open vascular bundles.

53. Roots have radial vascular bundles.

54. Tetrarch condition of dicot root

55. Vascular bundle is open if intrafascicular cambium is present between xylem and phloem.

56. (A) and (B) are incorrect. Stomata are present in the leaf epidermis. Each stomata is composed of two bean-shaped cells known as guard cells, which enclose the stomatal aperture. In grasses (monocots), guard cells are dumbbell-shaped.

57. A – Roots, B – Subsidiary cells, C – Trichomes

58. In open vascular bundles, a layer of cambium will be present between the xylem and phloem. Thus, they show secondary growth.

59. The first formed primary xylem element is called protoxylem. In the stem, it lies towards the pith and in the roots, it lies towards the periphery.

60. (A) and (B) are correct.

61. Endodermis acts as a biological checkpost.

62. Exarch condition is seen in both monocot and dicot roots, where the protoxylem lies on the outer side (towards the pericycle) and the metaxylem lies on the inner side.

63. Procambium forms the primary xylem, primary phloem and intrafascicular cambium, if present.

Anatomy of Dicotyledonous and Monocotyledonous Plants

64. In monocot roots, vascular bundles are polyarch (more than 6) and exarch.

65. A – Monocot root, B – Monocot stem, C – Dicot stem, D – Dicot leaf

66. The parenchymatous cells present between the xylem and phloem in the dicot root are called conjuctive tissue.

67. Collenchymatous hypodermis – Dicot stem, Starch sheath – Endodermis of dicot stem, Casparian strip – Root endodermis, Undifferentiated ground tissue – Monocot stem

68. Monocot stem

69. In monocot stem, vascular bundles are surrounded by sclerenchymatous bundle sheath.

70. In mustard stem (dicot stem), vascular bundles are conjoint, collateral and open.

71. Isobilateral leaves are amphistomatic. They possess bulliform cells (large, empty, colourless cells) on its adaxial surface and their leaf mesophyll is not differentiated into palisade tissue and spongy parenchyma.

72. (A), (B) and (C) are correct.

73. In monocot stem, the ground tissue is not differentiated into pith and cortex.

74. In dicot stem, conjoint, collateral and open vascular bundles are arranged in a ring with a large, well-developed parenchymatous pith at the centre.

75. Bulliform cells are present in the adaxial epidermis of grass leaves.

76. Vascular bundles in leaves are conjoint and closed.

77. A – Suberin, B – Casparian strip

78. Conjoint and closed

79. In monocot stem, water containing cavities are present within the vascular bundles.

80. Monocot roots

81. A – Hypodermis, B – Vascular bundles, C – Ground tissue

82. The given diagram represents a TS of a dicot root. In dicot root, the tangential and radial walls of the endodermal cells have deposition of suberin in the form of the Casparian strip.

83. Both dicot and monocot stem show endarch condition of primary xylem.

84. During water stress, bulliform cells become flaccid and they make the leaf curl inwards to minimise water loss.

85. Position of protoxylem (exarch or endarch condition)

86. In a dorsiventral leaf, the palisade tissue lies towards the adaxial epidermis and the phloem lies towards the abaxial epidermis.

87. In leaves, ground tissue consists of leaf mesophyll, which is often differentiated into palisade tissue and spongy parenchyma.

88. In dicot stem, vascular bundles are not surrounded by bundle sheath.

89. A – Adaxial epidermis, B – Palisade tissue, C – Spongy parenchyma, D – Abaxial epidermis

90. The size of the vascular bundles is dependent on the size of the vein.

91. In a dorsiventral leaf, the abaxial epidermis generally bears more stomata (hypostomatic).

92. A – Xylem, B – Phloem, C – Dicot root

93. Stele = Pericycle + Vascular bundles + Non-vascular pith

Secondary Growth

94. (A), (B), (C) and (D) are correct.

95. The activity of cambium is under the control of many physiological and environmental factors.

96. DCAB

97. In dicot roots, the vascular cambium originates from the tissue located just below the phloem bundles and a portion of the pericycle tissue, above the protoxylem.

98. Phellogen (cork cambium), phellem (cork) and phelloderm (secondary cortex) are collectively known as periderm.

99. Growth rings are distinct or sharply demarcated in the plants of the temperate climate.

100. In dicot stem, vascular cambium is half primary and half secondary in origin.

101. Tyloses are balloon-like structures and develop from xylem parenchyma. These block the passage of xylem vessels and are also called tracheal plug.

102. Suberin

103. Primary xylem remains more or less intact in or around the centre.

104. In the autumn, cambium is less active and forms fewer xylary elements that have narrow vessels.

105. Early wood is higher in colour with lower density and has vessels with wider lumen.

106. Sapwood or albernum is a peripheral part, light in colour and physiologically active.

107. Secondary growth in the extrastelar region is due to the activity of cork cambium or phellogen. It develops from the outer layer of the cortex.

108. (A) and (C) are correct.

109. Bark includes all the dead and living tissues outside the vascular cambium.

110. A – Complementary cells, B – Phellogen, C – Phelloderm

111. Only D is false. All tissues outside the vascular cambium constitute the bark.

112. Secondary phloem, secondary cortex, cork cambium cork

113. By counting the number of annual rings in the main stem at the base, we can determine the age of a tree.

114. (C) and (D) are false. In dicot root, vascular cambium arises from the tissue located just below the phloem bundles and a portion of the pericycle just above the protoxylem. Phellogen or cork cambium arises from the outer layers of the pericycle cells.

115. In roots, cambium activity is uniform throughout the year due to no change in the climatic condition of the soil.

116. The cells derived from dedifferentiated tissue are again redifferentiated. These are secondary xylem, secondary phloem, phellen and phelloderm. Cork cambium and vascular cambium are dedifferentiated tissues.

117. Heartwood is not physiologically active.

118. The cambium is generally more active on the inner side than on the outer side.

Structural Organisation in Animals

EXERCISES

Epithelial Tissue

1. The cells of epithelial tissues are:
(a) Compactly packed with abundant intercellular matrix
(b) Loosely packed with little intercellular matrix
(c) Compactly packed with little intercellular matrix
(d) Loosely packed with abundant intercellular matrix

2. Simple cuboidal epithelium is found in:
(a) Air sacs of lungs
(b) Walls of blood vessels
(c) Tubular parts of nephron
(d) Lining of intestine

3. ________ epithelium has a single layer of flattened cells with irregular boundaries:
(a) Cuboidal (b) Squamous
(c) Columnar (d) Pseudostratified

4. Which of the following is incorrect with respect to columnar epithelium?
(a) It is a single layer of tall and slender cells
(b) Their nuclei are located at the base
(c) Free surface may have microvilli
(d) It is involved in functions like forming a diffusion boundary

5. Which of the following is not a property of epithelial tissue?
(a) Cells are held together by cell junctions
(b) The tissue has a free surface, which faces either a body fluid or the outside environment
(c) It provides a covering or a lining for some part of the body
(d) It is highly vascular

6. Which of the following is not applicable to compound epithelium?
(a) It consists of two or more cell layers
(b) It has protective function
(c) Both (a) and (b)
(d) It is composed of a single layer of cells and functions as a lining for body cavities, ducts and tubes

7. Squamous epithelium is not found in the lining of:
(a) Walls of blood vessels
(b) Air sacs of lungs
(c) Bowman's capsule
(d) Ducts of glands

8. Glandular epithelium is not possible for:
(a) Squamous epithelium
(b) Cuboidal epithelium
(c) Columnar epithelium
(d) Both (b) and (c)

9. Stratified and non-keratinised epithelium is found in:
(a) Vagina (b) Buccal cavity
(c) Intestine (d) Both (a) and (b)

10. Ciliated epithelium is found in:
(a) Vagina (b) Fallopian tube
(c) Bronchioles (d) Both (b) and (c)

11. Microvilli of epithelial cells:
(a) Engulf the foreign particles
(b) Increase surface area for absorption
(c) Push mucus
(d) More than one option is correct

12. Select the incorrect statement:
(a) Squamous epithelium is called germinal epithelium
(b) Tessellated epithelium lines the walls of alveoli
(c) The main function of compound epithelium is protection
(d) Gap junction allows rapid transfer of ions

13. Brush-bordered cuboidal epithelium with microvilli is found in the lining of:
(a) Blood vessels
(b) Proximal convoluted tubule
(c) Fallopian tube
(d) Alveoli

14. Which of the following is an incorrect statement?
(a) Ciliated epithelium lines the inner surface of hollow organs like bronchioles
(b) Cuboidal epithelium is composed of a single layer of cube-like cells
(c) Columnar epithelium has nuclei located near the base
(d) The function of cilia is to move particles or mucus non-specifically in any direction over the epithelium

15. The cells of which epithelium are involved in functions like forming a diffusion boundary?
(a) Cuboidal epithelium
(b) Columnar epithelium
(c) Squamous epithelium
(d) Ciliated epithelium

16. Mark the odd one out with respect to endocrine glands:
(a) Thyroid gland (b) Adrenal gland
(c) Pituitary gland (d) Mammary gland

17. The cell junctions that perform cementing to keep neighbouring cells together is:
(a) Desmosome
(b) Gap junction
(c) Adhering junction
(d) Both (a) and (c)

18. Select the incorrect statement:
(a) Some of the cuboidal or columnar cells are specialised for secretion and are called glandular epithelium
(b) Goblet cells are examples of unicellular glands
(c) The products of exocrine glands are hormones and are secreted directly into the fluid bathing the gland
(d) Compound epithelium is multi-layered and thus has a limited role in secretion and absorption

19. ______ help to stop substances from leaking across a tissue:
(a) Adhering junctions (b) Tight junctions
(c) Gap junctions (d) Macula occludens

20. Which of the following is incorrect with respect to compound epithelium?
(a) It has more than one layer of cells
(b) Its main function is to protect against chemical and mechanical stresses
(c) It covers the dry surface of the skin, the moist surface of the buccal cavity, pharynx, inner lining of ducts, salivary glands and pancreatic ducts
(d) None of the above

21. Ciliated pseudostratified epithelium is found in:
(a) Fallopian tube (b) Pharynx
(c) Trachea (d) Skin

22. Which of the following is a correct set of glands having both exocrine and endocrine parts?
(a) Thyroid, Testis, Adrenal
(b) Pituitary, Adrenal, Thymus
(c) Mammary gland, Thyroid, Pancreas
(d) Pancreas, Testis, Ovary

23. Consider the following diagram. Choose the correct option with respect to the location where this type of epithelium is found:

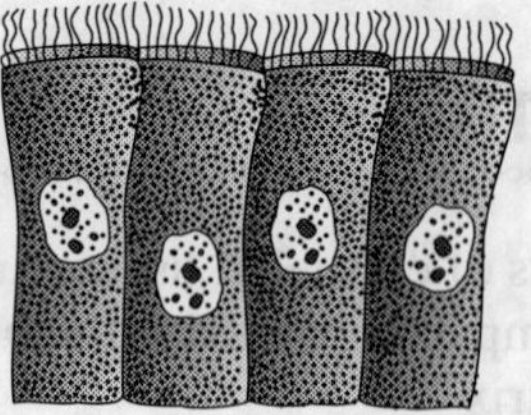

(a) Fallopian tube (b) Bronchioles
(c) Both (a) and (b) (d) Air sacs of lungs

24. Identify A, B and C in the diagram given below:

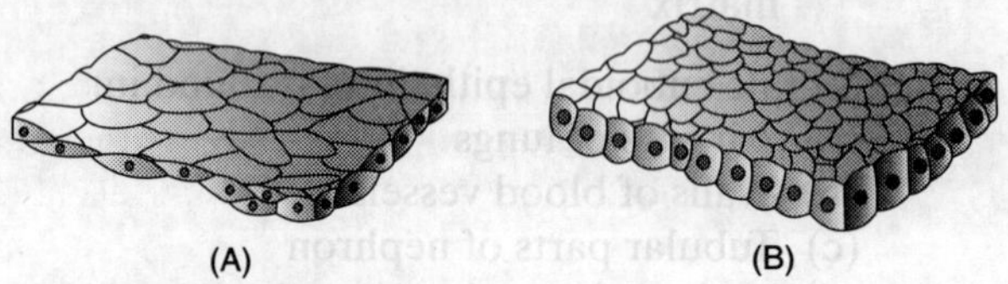

(A) (B)

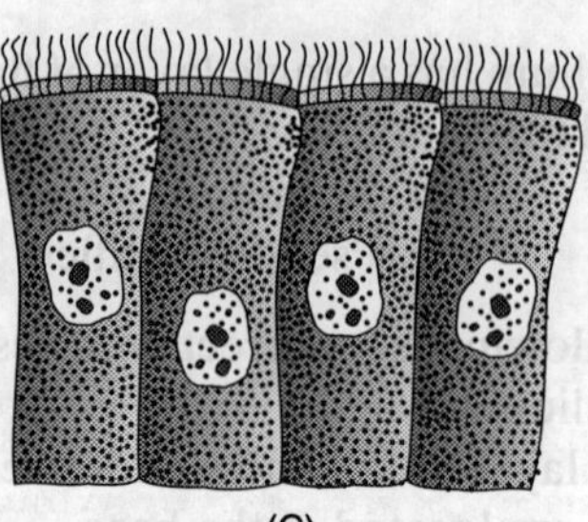

(C)

	A	B	C
(a)	Simple columnar	Simple squamous	Simple cuboidal
(b)	Simple cuboidal	Simple squamous	Simple ciliated
(c)	Simple squamous	Simple cuboidal	Simple ciliated
(d)	Simple cuboidal	Simple columnar	Simple squamous

25. Identify A, B and C in the diagram given below:

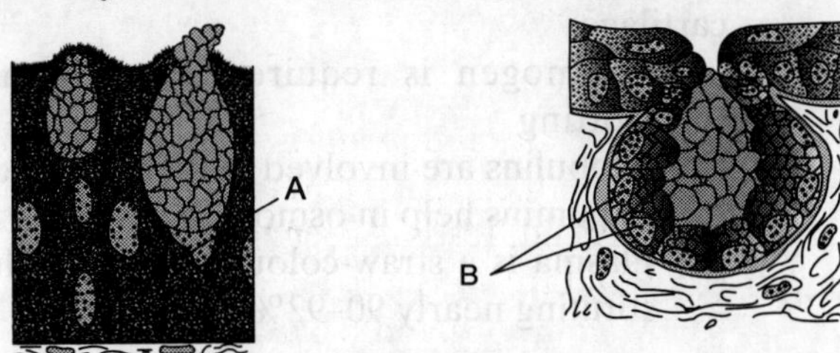

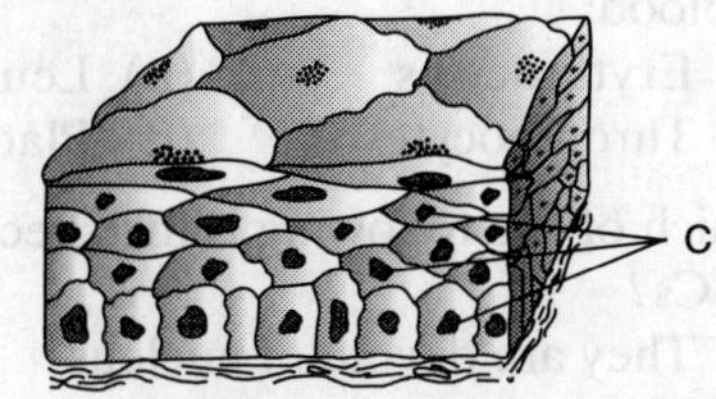

	A	B	C
(a)	Unicellular gland	Multicellular gland	Simple epithelium
(b)	Multicellular gland	Unicellular gland	Stratified epithelium
(c)	Unicellular gland	Multicellular gland	Stratified epithelium
(d)	Multicellular gland	Unicellular gland	Simple epithelium

Connective Tissue

26. Which of the following is/are examples of dense regular connective tissue?
(a) Tendon (b) Ligament
(c) Skin (d) Both (a) and (b)

27. Tendons connect:
(a) Muscle to muscle (b) Bone to muscle
(c) Bone to bone (d) Nerve to muscle

28. The main function of connective tissue is to:
(a) Store nutrients
(b) Facilitate diffusion
(c) Hold other tissues together
(d) Conduct impulses

29. The matrix or ground substance of the connective tissue is secreted by:
(a) Mast cells (b) Fibroblasts
(c) Plasma cells (d) Histiocytes

30. Areolar tissue connects:
(a) Bone to bone (b) Muscle to muscle
(c) Skin with muscle (d) Bone to muscle

31. Which of the following are examples of loose connective tissue?
A. Areolar tissue
B. Adipose tissue
C. Tendon
D. Ligament
E. Bone
F. Cartilage
(a) (A) and (C) (b) (D) and (F)
(c) (B) and (E) (d) (A) and (B)

32. ________ tissues are most abundant and widely distributed in the body of complex animals:
(a) Epithelial (b) Connective
(c) Neural (d) Muscular

33. Identify the example of connective tissue in which cells do not secrete fibres of structural proteins called collagen or elastin:
(a) Cartilage (b) Bone
(c) Adipose tissue (d) Blood

34. Which of the following is incorrect with respect to areolar tissue?
(a) It is present beneath the skin
(b) It serves as a support framework for epithelium
(c) It contains fibroblasts, macrophages and mast cells
(d) Most cells of this tissue store excess nutrients as fat

35. In which of the following types of dense connective tissue are the fibroblasts and many fibres (mostly collagen) oriented differently?
(a) Tendon (b) Ligament
(c) Skin (d) Cartilage

36. Identify examples of specialised connective tissue:
A. Bone
B. Tendon
C. Cartilage
D. Areolar tissue
E. Blood
F. Ligament
(a) (A), (C) and (F) (b) (A), (C) and (E)
(c) (B), (D) and (E) (d) (A), (E) and (F)

37. Connective tissue has fibres of structural proteins, collagen or elastin. These fibres do not provide:
(a) Strength (b) Flexibility
(c) Elasticity (d) Conductivity

38. The intercellular material of the cartilage is:
(a) Solid and non-pliable
(b) Hard and non-pliable
(c) Solid and pliable
(d) Rich in calcium salts

39. Which of the following is incorrect?
(a) The intercellular material of cartilage resists compression
(b) The cells of the cartilage are enclosed in small cavities within the matrix secreted by them
(c) Most of the cartilage in vertebrate embryos are replaced by bones in adults
(d) Cartilage is the site of production of blood cells

40. Select the incorrect statement with respect to bones:
(a) Bones support and protect softer tissues and organs
(b) It is the main tissue that provides a structural framework to the body
(c) The bone cells are present in spaces called lacunae
(d) They interact with smooth muscles attached to bones to bring about movement

41. Bones have a ____**I**____ and ____**II**____ ground substance rich in ____**III**____ and ____**IV**____ which give bones their strength:

	I	II	III	IV
(a)	Soft	Non-pliable	Magnesium salts	Elastin fibres
(b)	Hard	Non-pliable	Calcium salts	Collagen fibres
(c)	Hard	Pliable	Calcium salts	Elastin fibres
(d)	Soft	Pliable	Calcium salts	Collagen fibres

42. Identify an example of sesamoid bone:
(a) Humerus (b) Tibia
(c) Patella (d) Coxal bone

43. The most abundant cartilage in the human body is:
(a) Calcified (b) Hyaline
(c) Fibrous (d) Elastic

44. Elastic cartilage is not found in:
(a) Trachea (b) Epiglottis
(c) Pinna (d) Tip of nose

45. Blood platelets are formed by the breakdown of ________ in the bone marrow:
(a) Megakaryocytes (b) Erythroblasts
(c) Lymphoblasts (d) Myeloblasts

46. Bone and cartilage can be differentiated on the basis of:
(a) Lacunae (b) Matrix
(c) Haversian canals (d) Both (a) and (b)

47. Identify the incorrect statement with respect to cartilage:
(a) Fibrinogen is required for normal blood clotting
(b) Globulins are involved in defence mechanism
(c) Albumins help in osmotic balance
(d) Plasma is a straw-coloured, viscous fluid constituting nearly 90–92% of blood

48. ________ are the most abundant of all the cells in blood:
(a) Erythrocytes (b) Leucocytes
(c) Thrombocytes (d) Platelets

49. Which of the following is incorrect with respect to RBCs?
(a) They are devoid of nucleus
(b) They are biconcave in shape
(c) They have iron-containing complex proteins called haemoglobin
(d) Their number averages 6000–8000 mm^{-3} of blood

50. The over-stretching of ligaments is called
(a) Tetanus (b) Tetany
(c) Sprain (d) Fatigue

51. Read the following statements:
A. Hyaline cartilage is present in the nasal septum and costal ends of ribs
B. Spongy bone has marrow cavity and is formed in the shaft of the long bone
C. The longitudinal Haversian canals are interconnected by transverse Volkmann's canal
D. The type of cartilage present in our ear pinna is also present in the intervertebral disc
Which of the above statements are correct?
(a) (A) and (B) (b) (B) and (C)
(c) (C) and (D) (d) (A) and (C)

52. White fibrous cartilage is:
(a) Strongest cartilage
(b) Found in pubic symphysis and intervertebral disc
(c) Glass-like, translucent and found on joint surfaces
(d) Both (a) and (b)

53. Identify the structures labelled A, B and C in the diagrams given below:

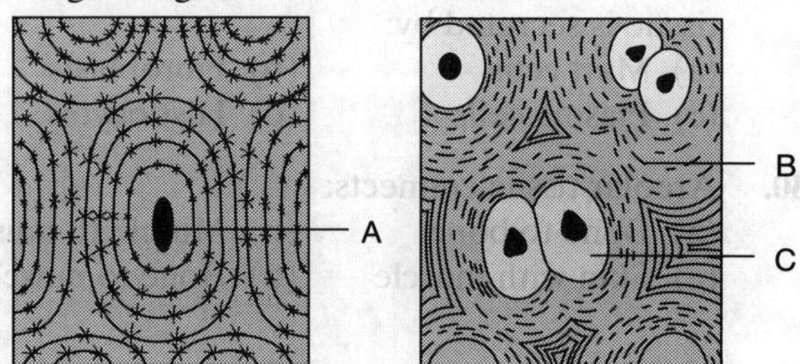

	A	B	C
(a)	Haversian canal	Collagen fibres	Osteocyte
(b)	Haversian canal	Collagen fibres	Chondrocyte
(c)	Volkmann's canal	Elastin fibres	Chondrocyte
(d)	Volkmann's canal	Collagen fibres	Osteocyte

54. Choose the correct information about tissues A and B given below:

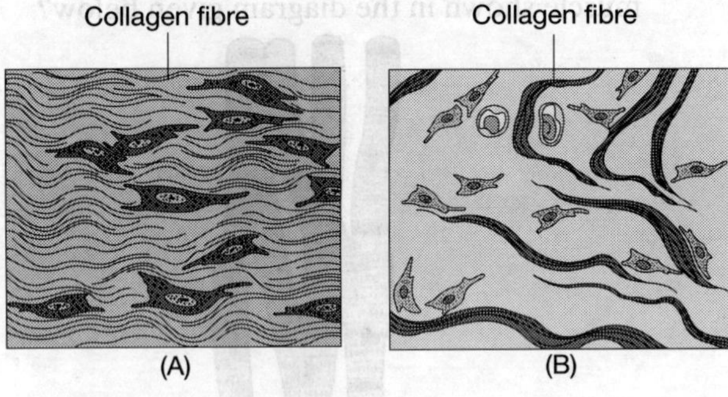

(a) A = Dense irregular connective tissue, B = Dense regular connective tissue
(b) A = Found in tendons, B = Found in ligaments
(c) A = This tissue is present in the skin, B = Found in tendons and ligaments
(d) A = Found in tendons and ligaments, B = This tissue is present in the skin

55. Given below is a diagrammatic sketch of areolar tissue. Identify the parts labelled A, B, C and D and select the correct option describing them:

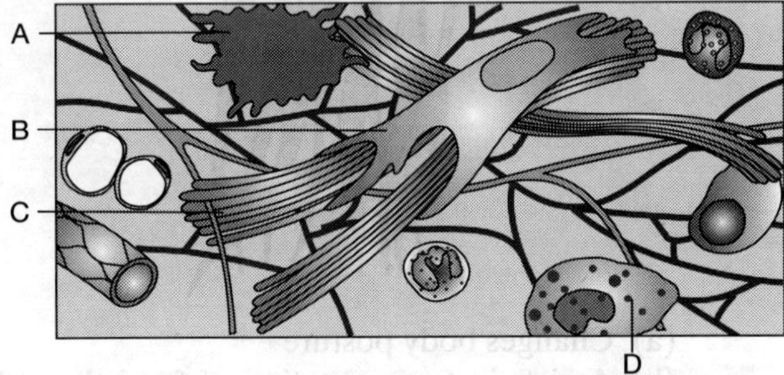

	A	B	C	D
(a)	Macrophage	Fibroblast	Collagen fibres	Mast cell
(b)	Mast cell	Fibroblast	Collagen fibres	Macrophage
(c)	Mast cell	Macrophage	Elastin fibres	Fibroblast
(d)	Macrophage	Mast cell	Collagen fibres	Fibroblast

Muscle Tissue

56. Smooth muscles:
(a) Are present in the heart
(b) Taper at both ends
(c) Show striations
(d) Are attached to skeletal bones

57. Which set clearly defines skeletal muscles?
(a) Cylindrical, striped and branched
(b) Cylindrical, syncytial and unbranched
(c) Fusiform, nucleated and branched
(d) Cylindrical, unstriped and unbranched

58. Oblique bridges, intercalated discs, cylindrical and branched fibres are the features of:
(a) Muscles of the stomach and intestine
(b) Biceps
(c) Cardiac muscles
(d) Muscles of the iris

59. In cardiac muscles, ________ at some fusion points allow the cells to contract as a unit:
(a) Desmosomes
(b) Tight junctions
(c) Intercalated discs
(d) Adhering junction

60. Smooth muscles are ____I____ as their functioning ____II____ be directly controlled:

	I	II
(a)	Voluntary	Can
(b)	Involuntary	Cannot
(c)	Involuntary	Can
(d)	Voluntary	Cannot

61. During muscle contraction:
(a) Mechanical energy is changed into chemical energy
(b) Chemical energy is changed into mechanical energy
(c) Mechanical energy is changed into potential energy
(d) Chemical energy is changed into electrical energy

62. Which of the following is an incorrect statement?
(a) Each muscle is made up of many long, cylindrical fibres arranged in parallel arrays
(b) These fibres are composed of numerous fine fibrils called myofibrils
(c) Muscles move the body to adjust to changes in the environment and the position of various parts of the body
(d) Muscles lengthen in response to stimulation, then shorten to their uncontracted state in a coordinated fashion

63. The muscles surrounding the pupil of the eye are:
(a) Striated and involuntary
(b) Unstriated and voluntary
(c) Unstriated and involuntary
(d) Branched and striated

64. Select the incorrect statement:
(a) Skeletal muscles attached to our bones are responsible for locomotion
(b) Smooth muscles are present in the wall of the alimentary canal and blood vessels
(c) Cardiac muscles are striped and multinucleate
(d) Smooth muscles are fusiform and do not show striations

65. Which of the following statements is not applicable to muscles?
(a) Muscles receive and respond to stimuli
(b) Muscles can return to their original shape after contraction and extension
(c) Smooth muscles are located in the inner walls of hollow visceral organs
(d) Myofibrils show light and dark bands in smooth muscles

66. At the centre of the A-band, a comparatively less dark zone called ____ is present:
(a) Z-line
(b) Krause's membrane
(c) Isotropic band
(d) Hensen zone

67. Identify the correct answer:
(a) Striated muscles never get fatigued
(b) Sarcoplasm is the cytoplasm of skeletal muscle fibre
(c) Peristalsis occurs in alimentary canal due to the action of skeletal muscles
(d) Myocytes of skeletal muscle are uninucleate

68. Which of the following is correct with respect to cardiac muscles?
(a) Myofibrils lack dark and light bands
(b) They may get fatigued
(c) They lack cell junctions
(d) They contract as a unit; when one cell receives a signal to contract, its neighbours are also stimulated to contract

69. Select the incorrect statement:
(a) All voluntary muscles are striated, but all striated muscles are not voluntary
(b) All smooth muscles are involuntary, but all involuntary muscles are not smooth
(c) About 80–90% of the body weight of a human adult is contributed by muscles
(d) Muscles have special properties like excitability, contractility, elasticity and extensibility

70. ________ muscles are primarily involved in locomotory action and changes in body posture:
(a) Smooth (b) Non-striated
(c) Cardiac (d) Skeletal

71. Which of the following is correct with respect to the muscle shown in the diagram given below?

(a) It is syncytial, voluntary and has cylindrical muscle fibres
(b) It is multinucleate, involuntary and fusiform
(c) It is uninucleate, involuntary and has intercalated discs
(d) It is uninucleate, voluntary and unbranched

72. Identify the functions of the muscles given below:

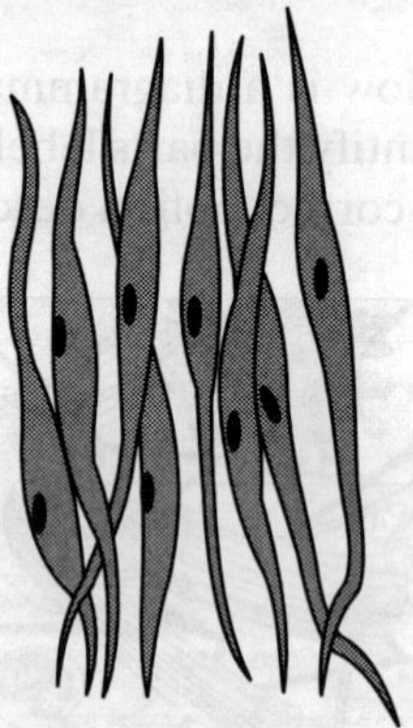

(a) Changes body posture
(b) Assists in transportation of food through the digestive tract
(c) Assists in transportation of gametes through the genital tract
(d) Both (b) and (c)

73. Each organised skeletal muscle in our body is made up of a number of muscle bundles or fascicles held together by a common collagenous connective tissue layer called:
(a) Endomysium (b) Fascia
(c) Perimysium (d) Epimysium

74. Identify the statement that is not applicable to the diagram given below:

(a) Each muscle fibre has a large number of filaments arranged in parallel in the sarcoplasm called myofilaments
(b) Each myofibril has alternating light and dark bands
(c) These muscles are closely associated with the skeletal components of the body
(d) They do not exhibit striations and are smooth in appearance

75. Which of the types of tissue correctly match with their location?

Tissue	Location
(a) Cuboidal epithelium	Lining of the stomach
(b) Smooth muscle	Wall of intestine
(c) Areolar tissue	Tendons
(d) Transitional epithelium	Tip of the nose

76. Which one of the following is the correct pairing of body part and the type of muscle tissue that moves it?
(a) Heart wall – Involuntary unstriated muscle
(b) Iris – Involuntary smooth muscle
(c) Biceps of upper arm – Smooth muscle fibres
(d) Alimentary canal – Syncytial voluntary muscle

Neural Tissue

77. The area where the medullary sheath is absent in the nerve fibre is called:
(a) Schwann node (b) Nissl granules
(c) Node of Ranvier (d) Neuroglia

78. Which of the following is not a type of neuroglial cell?
(a) Astrocyte (b) Microglia
(c) Oligodendrocyte (d) Chondrocyte

79. Blood–brain barrier is formed by:
(a) Ependymal cells
(b) Microglia cells
(c) Oligodendrocyte
(d) Astrocyte

80. Neural tissue originates in:
(a) Endoderm (b) Mesoderm
(c) Ectoderm (d) All of the above

81. Centrioles are absent in:
(a) Epithelial cells (b) Muscle cells
(c) Nerve cells (d) Germinal cells

82. Neuroglial cells are:
(a) Sensory cells
(b) Secretory cells
(c) Sensory and supporting cells
(d) Non-sensory and supporting cells

83. Which of the following is incorrect with respect to neurons?
(a) They cannot regenerate or divide
(b) They are excitable
(c) They may be sensory or motor
(d) None of the above

84. ________ tissue exerts the greatest control over the body's responsiveness to changing conditions:
(a) Epithelial (b) Neural
(c) Connective (d) Muscle

85. Neuroglia make up:
(a) Less than one-half of the volume of neural tissue
(b) More than one-half of the volume of neural tissue
(c) Less than one-fourth of the volume of neural tissue
(d) More than three-fourths of the volume of neural tissue

86. Axons serve to:
(a) Bring impulses towards the cell body of the neuron
(b) Take impulses away from the cell body of the neuron
(c) Bring blood to the heart
(d) Protect and support the neuron

87. Most neurons in the cerebral cortex are:
(a) Unipolar (b) Bipolar
(c) Multipolar (d) Pseudounipolar

88. Multipolar neuron has:
(a) One dendrite and one axon
(b) Many dendrites and one axon
(c) Many dendrites and many axons
(d) One dendrite and many axons

89. Nissl granules are absent in:
(a) Axon (b) Dendrite
(c) Cyton (d) Both (b) and (c)

90. Bipolar nerve cells are present in:
(a) Dorsal root ganglia
(b) Retina
(c) Cerebral cortex
(d) Embryo stages

91. Neuroglia:
(a) Are excitable cells
(b) Support and protect neurons
(c) Receive and conduct impulses
(d) Have Nissl granules

92. When a neuron is suitably stimulated, an electrical disturbance is generated, which swiftly travels:
(a) Along the cytoplasm
(b) Along its plasma membrane
(c) Along the axoplasm
(d) Both (a) and (c)

93. In the central neural system, myelin is secreted by:
(a) Oligodendrocyte (b) Microglia
(c) Schwann cell (d) Astrocyte

94. The longest cells in the human body are:
(a) Muscle cells (b) Sperms
(c) Neurons (d) Bone cells

95. Nissl granules are:
(a) Centrioles (b) Ribosomes
(c) Mitochondria (d) Golgi bodies

96. The dendrite carries impulses:
(a) Towards cell body
(b) Away from cell body
(c) From one neuron to another
(d) From neuron to muscle

97. Given below is a type of neuron which is found in:

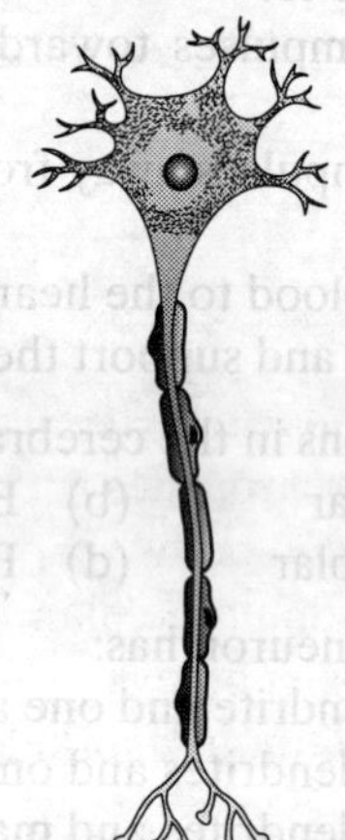

(a) Retina (b) Hydra
(c) Cerebral cortex (d) Embryonic stage

Morphology of Cockroach

98. Cockroach is:
(a) Omnivorous, nocturnal, fossorial
(b) Insectivorous, diurnal, cursorial
(c) Omnivorous, nocturnal, cursorial
(d) Herbivorous, diurnal, cursorial

99. Which of the following is an incorrect statement regarding the morphology of cockroach?
(a) Its size ranges from ¼ inch to 3 inches
(b) Presence of bright yellow, red and green colour of the body
(c) Body is covered with a thick, hard chitinous cuticle
(d) The abdomen in both males and females consists of 12 segments

100. The chitinous plates or sclerites of the exoskeleton in cockroach are attached to each other by:
(a) Basilar membrane
(b) Schneiderian membrane
(c) Arthrodial membrane
(d) Reissner's membrane

101. The body of cockroach is differentiated into:
(a) Cephalothorax and abdomen
(b) Head and trunk
(c) Head, thorax and abdomen
(d) Prosoma and opisthosoma

102. Male cockroach can be differentiated from female cockroach by:
(a) Long and narrow abdomen
(b) Presence of larger antennae
(c) Presence of anal style
(d) All of the above

103. Identify the correct statement with respect to cockroach:
(a) Alary muscles help in respiration
(b) Mouth parts are siphoning type
(c) Excretion occurs through green glands
(d) Gizzard helps in grinding the food particles

104. In cockroach, the respiratory system consists of a network of trachea that open out through ________ of small holes called spiracles:
(a) 8 pairs (b) 10 pairs
(c) 4 pairs (d) 20 pairs

105. Which abdominal sternite in the female cockroach is boat shaped?
(a) 7th sternum (b) 9th sternum
(c) 4th sternum (d) 6th sternum

106. Which of the following statements is incorrect for hepatic or gastric caecae of cockroach?
(a) They are present at the junction of the foregut and midgut
(b) They secrete digestive juices
(c) They are part of foregut
(d) They help in absorption

107. Which of the following are used for flying in cockroach?
(a) Forewings (b) Tegmina
(c) Elytra (d) Hindwings

108. In cockroach, the attachment of the head with the rest of the longitudinal body axis is termed as:
(a) Prognathous (b) Orthognathous
(c) Hypognathous (d) Agnathous

109. The length of *Periplaneta americana* is about ____ mm:
(a) 20–25 (b) 34–53
(c) 60–70 (d) 74–93

110. Which of the following is incorrect with respect to the head of cockroach?
(a) Head is triangular in shape
(b) Head lies at right angles to the longitudinal body axis
(c) It is formed by fusion of eight segments
(d) Head shows great mobility in all directions due to flexible neck

111. Which of the following is incorrect with respect to the wings of cockroach?
(a) Forewings are called tegmina
(b) Forewings are dark, opaque and leathery
(c) Hindwings are transparent, membranous and used in flight
(d) The first pair of wings arises from the prothorax and the second pair arises from the mesothorax

112. In both sexes, the __A__ segment bears a pair of jointed filamentous structures called __B__. Fill in the blanks A and B with correct options given below:
(a) 9th, anal cerci (b) 10th, anal cerci
(c) 9th, anal style (d) 10th, anal style

113. In female cockroach, the genital pouch is formed by:
(a) 7th, 8th and 9th sternites
(b) 6th, 7th and 8th sternites
(c) 9th and 10th tergites dorsally and 9th sternite ventrally
(d) 10th tergite dorsally and 8th, 9th sternites ventrally

114. In the following figure, A, B and C are ______, ______ and ______, respectively:

A. B. C.

(a) Maxilla, labium and mandible
(b) Labium, maxilla and mandible
(c) Maxilla, labrum and hypopharynx
(d) Labrum, maxilla and mandible

115. Heart of cockroach is an elongated muscular tube which is made up of:
(a) 10 chambers (b) 13 chambers
(c) 5 chambers (d) 20 chambers

116. Select the figure labelled incorrectly:

(a) Hypopharynx (b) Mandible
(c) Maxilla (d) Labrum

117. In cockroach, each compound eye is made up of 2000 units called:
(a) Arolium (b) Ommatidia
(c) Phaosome (d) Ommatophore

118. The most important character for the evolutionary success of cockroaches is:
(a) Segmented body
(b) Tube within tube plan
(c) Three pairs of legs
(d) Hard chitinous exoskeleton

119. Haemolymph of cockroach is devoid of:
(a) Haemocytes
(b) Plasma
(c) Respiratory pigment
(d) Electrolytes

120. Select the incorrect statement with respect to the excretory system of cockroach:
(a) Cockroach is uricotelic
(b) Malpighian tubules are present at the junction at foregut and midgut
(c) Each tubule is linked by glandular and ciliated cells
(d) Urecose glands are located at the periphery of mushroom glands in males

121. In male cockroach:
(a) Testes are present from 4th to 6th abdominal segments
(b) Mushroom glands are present in 6th, 7th abdominal segment
(c) External genitalia consist of three small irregular chitinous plates called phallomeres
(d) All are correct

122. The double ventral nerve cords of cockroach are present along the mid-ventral line of the thorax and abdomen and have 9 ganglia, of which:
(a) 4 are present in the thorax and 5 in the abdomen
(b) 3 are present in the thorax and 6 in the abdomen
(c) 6 are present in the thorax and 3 in the abdomen
(d) 2 are present in the thorax and 7 in the abdomen

123. Which of the following structures is not associated with the reproductive system of male cockroach?
(a) Seminal vesicles (b) Phallic gland
(c) Vasa efferentia (d) Ejaculatory duct

124. The development in *Periplaneta americana* is:
(a) Hemimetabolus (b) Holometabolus
(c) Ametabolus (d) Paurometabolus

125. How many gonapophyses are present in female cockroach?
(a) 3 (b) 3 pairs
(c) 2 (d) 2 pairs

126. Mosaic or apposition vision of cockroach has:
(a) More sensitivity and less resolution
(b) Less sensitivity and more resolution
(c) More sensitivity and more resolution
(d) Less sensitivity and less resolution

127. Female cockroach has a pair of ovaries in:
(a) 2nd to 6th abdominal segments laterally
(b) 4th to 8th abdominal segments laterally
(c) 2nd to 6th abdominal segments dorsally
(d) 4th to 8th abdominal segments dorsally

128. The longest part in the leg of the cockroach is:
(a) Coxa (b) Trochanter
(c) Femur (d) Tibia

129. Fat bodies of cockroach are considered analogous with:
(a) Skin (b) Lungs
(c) Liver (d) Pancreas

130. Ootheca is secreted in females by:
(a) Phallic gland (b) Collaterial gland
(c) Conglobate gland (d) Utricular gland

131. Select the incorrect statement with respect to the reproductive system of the female *Periplaneta americana*:
(a) Each ovary is formed of a group of eight ovarian tubules or ovarioles
(b) Oviducts of each ovary unite into a single median oviduct
(c) A pair of spermatheca is present in the 4th segment which opens into a genital chamber
(d) A pair of branched collaterial glands is present in females

132. On an average, females produce ____ ootheca, each containing ____ eggs:
(a) 9–10, 14–16 (b) 14–16, 9–10
(c) 18–20, 9–10 (d) 20–25, 15–20

133. Mark the incorrect option regarding the number of a particular structure in cockroach:
(a) 6–8 gastric caeca
(b) 100–150 Malpighian tubules
(c) 12 chitinous plates called teeth in gizzard
(d) 10 segments in abdomen

134. Which of the following is incorrect with respect to the digestive system in cockroach?
(a) Oesophagus opens into sac-like structure called crop used for storing food
(b) Gizzard helps in grinding food particles
(c) The entire midgut is lined by cuticle
(d) The hindgut is broader than midgut and is differentiated into ileum, colon and rectum

135. The blood vascular system of cockroach is of ____I____ type and blood vessels are ____II____:

	I	II
(a)	Closed type	Well developed
(b)	Open type	Well developed
(c)	Closed type	Poorly developed
(d)	Open type	Poorly developed

136. Identify the incorrect statement with respect to cockroach:
(a) Thin, branched tracheal tubes carry oxygen from air to all the body parts
(b) Visceral organs in haemocoel are bathed in blood

(c) The haemolymph is composed of coloured plasma and haemocytes
(d) Heart has funnel-shaped chambers with ostia on either side

137. The nymph of cockroach grows by moulting about ______ times to reach the adult form:
(a) 20 (b) 13
(c) 7 (d) 18

138. Select the incorrect statement with respect to cockroach:
(a) The second-last nymphal stage has wing pads but only adult cockroaches have wings
(b) Cockroaches are pests because they destroy food and contaminate it with their smelly excreta
(c) They can transmit a variety of bacterial diseases by contaminating food material
(d) They are diurnal herbivore animals living in dry places throughout the world

139. Given below is a diagram of the female reproductive system of cockroach. Choose the correct option with respect to the function of the labelled structure:
(a) Secretion of wall of ootheca
(b) Regulates moulting in cockroach
(c) Storage of sperm after sexual mating
(d) Secretes juvenile hormone which retains nymphal character

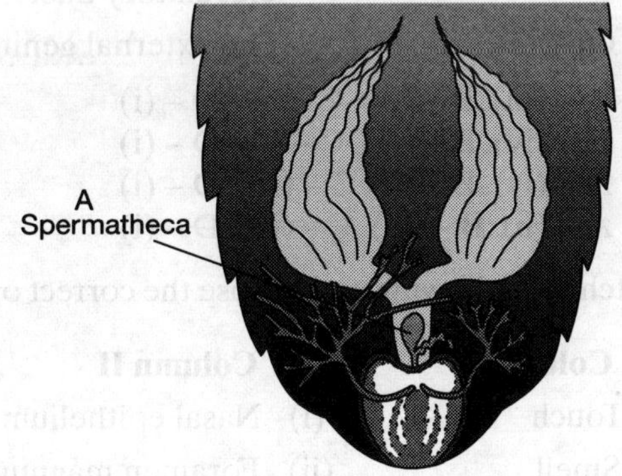

140. The haemolymph of cockroach will show the presence of all of the following, except:
(a) Haemocytes (b) Haemocyanin
(c) Plasma (d) Trehalose

141. In female cockroach, which of the following are paired structures?
A. Maxilla
B. Labrum
C. Anal cerci
D. Salivary glands
E. Anal style
(a) (B), (C) and (E) only
(b) (A), (B) and (C) only
(c) (A), (B), (C) and (D)
(d) (A), (C), (D) and (E)

142. Phallomeres are:
(a) Mouth parts of cockroach
(b) Vestigial wings in female cockroach
(c) External genitalia in cockroach
(d) Structures present in the head of cockroach

143. Which of the following is not associated with the reproductive system of male cockroach?
(a) Conglobate gland (b) Collaterial gland
(c) Utricular gland (d) Seminal vesicle

144. Which of the following is an incorrect statement?
(a) Heart of cockroach is dorsal, tubular and neurogenic
(b) Cockroach has 10 pairs of spiracles on the lateral side
(c) Malpighian tubules help in removal of excretory products in the form of urea
(d) Prothoracic gland secretes the moulting hormone

NCERT EXEMPLAR QUESTIONS

1. Which one of the following types of cells is involved in the making of the inner walls of blood vessels?
(a) Cuboidal epithelium
(b) Columnar epithelium
(c) Squamous epithelium
(d) Stratified epithelium

2. To which one of the following categories does adipose tissue belong?
(a) Epithelial (b) Connective
(c) Muscular (d) Neural

3. Which one of the following is not a connective tissue?
(a) Bone (b) Cartilage
(c) Blood (d) Muscles

4. Setae help in locomotion in earthworm but are not uniformly present in all the segments. They are present in:
(a) 1st segment
(b) Last segment
(c) Clitellar segment
(d) 20th–22nd segment

5. Which one of the following statements is true for cockroach?
(a) The number of ovarioles in each ovary is 10
(b) The larval stage is called caterpillar
(c) Anal styles are absent in females
(d) They are ureotelic

6. Match the following and choose the correct option:

Column I	Column II
A. Adipose tissue	(i) Nose
B. Stratified epithelium	(ii) Blood
C. Hyaline cartilage	(iii) Skin
D. Fluid connective tissue	(iv) Fat storage

(a) A – (i), B – (ii), C – (iii), D – (iv)
(b) A – (iv), B – (iii), C – (i), D – (ii)
(c) A – (iii), B – (i), C – (iv), D – (ii)
(d) A – (ii), B – (i), C – (iv), D – (iii)

7. Match the following and choose the correct option:

Column I	Column II
A. Hermaphrodite	(i) Produces blood cells and haemoglobin
B. Direct development	(ii) Testis and ovary in the same animal
C. Chemoreceptor	(iii) Larval form absent
D. Blood gland in earthworm	(iv) Sense of chemical substances

(a) A – (ii), B – (iii), C – (iv), D – (i)
(b) A – (iii), B – (ii), C – (iv), D – (i)
(c) A – (i), B – (iii), C – (ii), D – (i)
(d) A – (ii), B – (iv), C – (iii), D – (i)

8. Match the following with reference to cockroach and choose the correct option:

Column I	Column II
A. Phallomere	(i) Chain of developing ova
B. Gonopore	(ii) Bundles of sperm
C. Spermatophore	(iii) Opening of the ejaculatory duct
D. Ovarioles	(iv) The external genitalia

(a) A – (iii), B – (iv), C – (ii), D – (i)
(b) A – (iv), B – (iii), C – (ii), D – (i)
(c) A – (iv), B – (ii), C – (iii), D – (i)
(d) A – (ii), B – (iv), C – (iii), D – (i)

9. Match the following and choose the correct option:

Column I	Column II
A. Touch	(i) Nasal epithelium
B. Smell	(ii) Foramen magnum
C. Cranial nerves	(iii) Sensory papillae
D. Medulla oblongata	(iv) Peripheral nervous system

(a) A – (iii), B – (i), C – (ii), D – (iv)
(b) A – (ii), B – (i), C – (iv), D – (iii)
(c) A – (iii), B – (iv), C – (ii), D – (i)
(d) A – (iii), B – (i), C – (iv), D – (ii)

PRACTICE TEST PAPER

1. Simple squamous epithelium:
(a) Is present in the wall of Bowman's capsule
(b) Is found in the walls of blood vessels
(c) Has role in diffusion of gases
(d) All are correct

2. Select the correct statement:
(a) Goblet cells are multicellular glands
(b) Tight junction helps to stop substances from leaking across tissue
(c) Compound epithelium is found in the lining of the trachea
(d) Cells of columnar epithelium are flat and assist in forming diffusion boundary

3. Choose the correct statement with respect to epithelial tissue:
(a) Cells are compactly packed with little intercellular matrix
(b) The free surface of the epithelium faces the body fluid or outside environment
(c) It may be simple epithelium with two or more layers or compound epithelium with a single layer of cells
(d) Both (a) and (b) are correct

4. The inner surface of hollow organs like the fallopian tube or bronchioles have:
(a) Compound epithelium
(b) Ciliated epithelium
(c) Cuboidal epithelium
(d) Squamous epithelium

5. Choose the correct statement:
(a) Thyroid gland has ducts
(b) Microvilli of epithelial cells increase the absorptive area
(c) The moist surface of buccal cavity and pharynx have simple epithelium
(d) Cuboidal epithelium is composed of a single layer of tall and slender cells

6. Identify the diagrams of epithelial tissue:

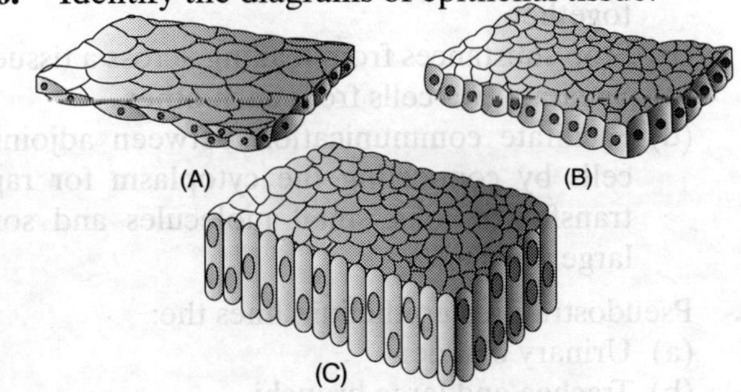

	A	B	C
(a)	Simple cuboidal	Simple squamous	Simple columnar
(b)	Simple squamous	Simple columnar	Simple cuboidal
(c)	Simple squamous	Simple cuboidal	Simple columnar
(d)	Simple columnar	Simple squamous	Simple cuboidal

7. Exocrine glands:
(a) Are ductless
(b) Secrete hormones
(c) Secrete saliva, milk, ear wax, mucus and digestive enzymes
(d) More than one option is correct

8. Identify the epithelium given below and its corresponding location:

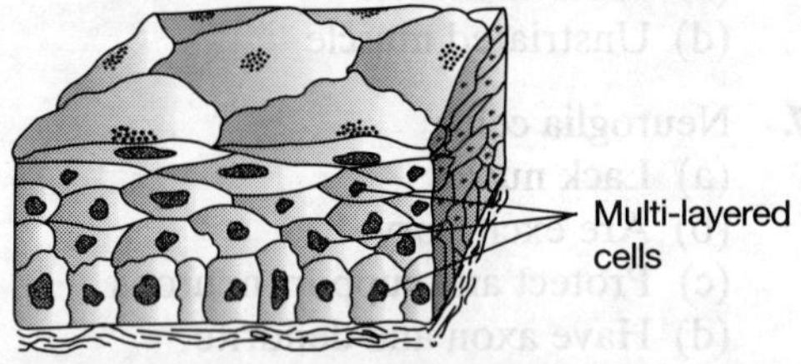

(a) Compound epithelium: walls of blood vessels
(b) Simple epithelium: ducts of glands
(c) Simple epithelium: tubular parts of nephron
(d) Compound epithelium: buccal cavity

9. Select the incorrect statement:
(a) Gap junctions are abundant in the cardiac muscles
(b) Fibroblasts secrete collagen fibres
(c) Tendons are an example of dense irregular connective tissue
(d) Antibodies are secreted by plasma cells

10. Which of the following is/are examples of loose connective tissue?
(a) Areolar tissue (b) Adipose tissue
(c) Both (a) and (b) (d) Ligament

11. Select the odd one out with respect to examples of specialised connective tissue:
(a) Blood (b) Bone
(c) Cartilage (d) Adipose tissue

12. Elastic cartilage is found in:
(a) Tip of the nose
(b) Trachea

(c) Eustachian tube
(d) More than one option is correct

13. The intercellular material of cartilage is ________ and ________ and resists compression:
(a) Solid, non-pliable
(b) Hard, non-pliable
(c) Solid, pliable
(d) Liquid, pliable

14. Which of the following describes sarcomere?
(a) Part between two Z-lines
(b) Part between two I-bands
(c) Part between two A-bands
(d) Part between two M-lines

15. Smooth muscle is:
(a) Branched and uninucleate
(b) Syncytial and unbranched
(c) Fusiform and uninucleate
(d) Striated and uninucleate

16. Intercalated discs are found in:
(a) Smooth muscle
(b) Cardiac muscle
(c) Skeletal muscle
(d) Unstriated muscle

17. Neuroglia cells:
(a) Lack nuclei
(b) Are excitatory
(c) Protect and support neurons
(d) Have axon and dendrite

18. Choose the correct statement:
(a) Smooth muscles are attached to the skeleton and assist locomotion
(b) Dendrites carry impulses away from the cell body
(c) Nissl granules are made of ribosomes
(d) Cardiac muscles are voluntary and non-striated

19. Cardiac muscles:
(a) Have light and dark bands
(b) Do not fatigue
(c) Are branched and uninucleate
(d) All are correct

20. Select the incorrect statement:
(a) Smooth muscle fibres taper at both ends
(b) Bone cells are formed in lacunae
(c) The ground substance of bone is hard and non-pliable
(d) Adipocytes mediate inflammatory reactions by secreting histamine and serotonin

21. The phagocytic cells of areolar tissue are:
(a) Fibrocytes (b) Mast cells
(c) Histiocytes (d) Plasma cells

22. Select the correct example in the diagram given below:

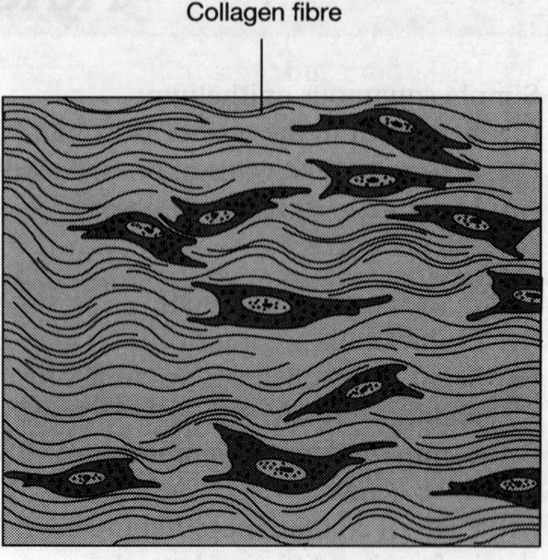

(a) Cartilage (b) Bone
(c) Adipose tissue (d) Tendon

23. Identify the incorrect statement with respect to the diagram given below and its label A:

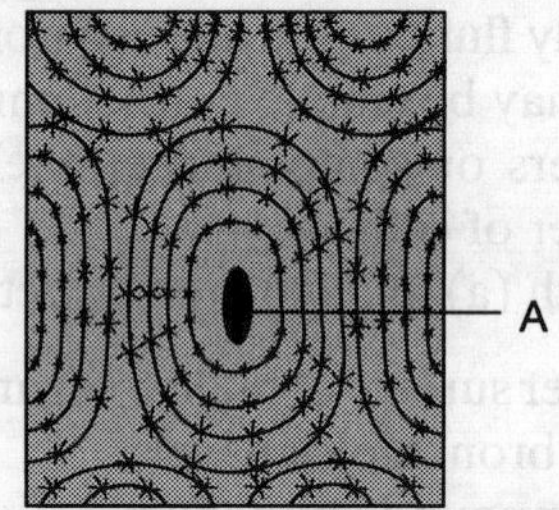

(a) It has hard and non-pliable ground substance
(b) 'A' is the Haversian canal
(c) Cells of this tissue are called chondrocytes
(d) The tissue supports and protects softer tissues and organs

24. The extracellular matrix is maximum in:
(a) Dense, regular connective tissue
(b) Areolar tissue
(c) Neural tissue
(d) Skeletal tissue

25. The function of the gap junction is to:
(a) Perform cementing to keep neighbouring cells together
(b) Stop substances from leaking across a tissue
(c) Separate two cells from each other
(d) Facilitate communication between adjoining cells by connecting the cytoplasm for rapid transfer of ions, small molecules and some large molecules

26. Pseudostratified epithelium lines the:
(a) Urinary bladder
(b) Trachea and large bronchi

(c) Air sacs of lungs
(d) Proximal convoluted tubule

27. The mast cells of connective tissue secrete:
(a) Histamine and melanin
(b) Heparin and histamine
(c) Heparin and calcitonin
(d) Serotonin and melanin

28. Volkmann's canal connects:
(a) Two Haversian canals
(b) Haversian canal with matrix
(c) Marrow cavity with Haversian canal
(d) Two lacunae

29. Bone is mainly composed of:
(a) Calcium and sodium
(b) Sodium and magnesium
(c) Calcium and phosphorous
(d) Magnesium and sulphur

30. White fibrous cartilage is:
(a) Found in Eustachian tube
(b) Found in intervertebral disc and pubic symphysis
(c) Rich in yellow fibres made of elastin
(d) Fibreless and glass-like

31. Segmented exoskeleton of the body of cockroach is formed by sclerites which are joined by:
(a) Enamel
(b) Osteocyte
(c) Basilar membrane
(d) Arthrodial membrane

32. Choose the correct statement with respect to *Periplaneta americana*:
(a) Hypopharynx is called the upper lip
(b) Forewings are tegmina which are used in flight
(c) Malpighian tubules are excretory organs projecting out from the rectum
(d) Male bears anal styles which are not present in females

33. Match Column I and Column II with respect to the cockroach. Choose the correct option:

Column I	Column II
A. Hepatic caecae	(i) 6–8 in number
B. Malpighian tubules	(ii) 100–150 in number
C. Spiracles	(iii) 20 in number
D. Ommatidia	(iv) 2000 in number

(a) A – (i), B – (iii), C – (ii), D – (iv)
(b) A – (i), B – (ii), C – (iii), D – (iv)
(c) A – (ii), B – (i), C (iii), D – (iv)
(d) A – (iv), B – (ii), C – (iii), D – (i)

34. On an average, the female cockroach produces ______ ootheca, each containing ______ eggs. Choose the options which fill in the blanks correctly:
(a) 9–10, 14–16
(b) 14–16, 9–10
(c) 12–16, 20–22
(d) 2–4, 6–8

35. In cockroach, the mouth part that has the grinding and incising region is:
(a) Labrum
(b) Labium
(c) Mandible
(d) Maxillae

36. Choose the correct odd one out with respect to the male reproductive system of cockroach:
(a) Collaterial gland
(b) Conglobate gland
(c) Mushroom gland
(d) Seminal vesicle

37. The largest part of the cockroach foregut is:
(a) Gizzard
(b) Oesophagus
(c) Crop
(d) Hepatic caecae

38. Which gland secretes ecdysone to control moulting of the cockroach nymph?
(a) Corpora allata
(b) Phallic gland
(c) Prothoracic gland
(d) Collaterial gland

39. How many ganglia are present in the nervous system of the cockroach in the region of the thorax and abdomen?
(a) 9
(b) 6
(c) 3
(d) 12

40. Choose the incorrect statement with respect to the cockroach:
(a) Wings are attached to the thorax
(b) Three pairs of legs are attached to the abdomen
(c) Head lies anteriorly at right angles to the longitudinal body axis
(d) A pair of thread-like antennae arise from membranous sockets lying in front of the eyes

41. Which of the following is incorrect with respect to the gizzard of the alimentary canal in cockroach?
(a) It is also called proventriculus
(b) It has an outer layer of thick circular muscles
(c) It has a thick inner cuticle forming 12 highly chitinous plates called teeth
(d) It helps in grinding food particles

42. Which of the following is incorrect with respect to the circulatory system of cockroach?
(a) Blood vascular system is of open type
(b) Blood vessels are poorly developed and open into space (haemocoel)
(c) Haemolymph is composed of colourless plasma and haemocytes

(d) Heart of cockroach consists of an elongated muscular tube lying along the mid-ventral line of the abdomen only

43. In the male reproductive system of cockroach, a characteristic mushroom-shaped gland is present in the _______ abdominal segments which functions as an accessory reproductive gland:
(a) 4th–5th (b) 6th–7th
(c) 8th–9th (d) 9th–10th

44. The mosaic vision of cockroach has:
(a) More resolution and less sensitivity
(b) More sensitivity and less resolution
(c) Very little sensitivity
(d) Very high resolution

45. Choose the option which correctly identifies 'A' and 'B' in the reproductive system of cockroaches:

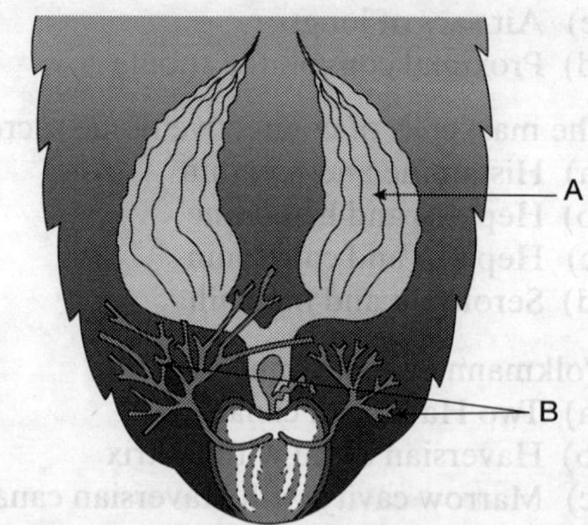

	A	B
(a)	Oviduct	Ovary
(b)	Ovary	Collaterial glands
(c)	Ovary	Oviduct
(d)	Genital chamber	Ovary

ANSWER KEYS

Exercises

1. (c)	**2.** (c)	**3.** (b)	**4.** (d)	**5.** (d)	**6.** (d)	**7.** (d)	**8.** (a)	**9.** (d)	**10.** (d)
11. (b)	**12.** (a)	**13.** (b)	**14.** (d)	**15.** (c)	**16.** (d)	**17.** (d)	**18.** (c)	**19.** (b)	**20.** (d)
21. (c)	**22.** (d)	**23.** (c)	**24.** (c)	**25.** (c)	**26.** (d)	**27.** (b)	**28.** (c)	**29.** (b)	**30.** (c)
31. (d)	**32.** (b)	**33.** (d)	**34.** (d)	**35.** (c)	**36.** (b)	**37.** (d)	**38.** (c)	**39.** (d)	**40.** (d)
41. (b)	**42.** (c)	**43.** (b)	**44.** (a)	**45.** (a)	**46.** (c)	**47.** (d)	**48.** (a)	**49.** (d)	**50.** (c)
51. (d)	**52.** (d)	**53.** (b)	**54.** (d)	**55.** (a)	**56.** (b)	**57.** (b)	**58.** (c)	**59.** (c)	**60.** (b)
61. (b)	**62.** (d)	**63.** (c)	**64.** (c)	**65.** (d)	**66.** (d)	**67.** (b)	**68.** (d)	**69.** (c)	**70.** (d)
71. (c)	**72.** (d)	**73.** (b)	**74.** (d)	**75.** (b)	**76.** (b)	**77.** (c)	**78.** (d)	**79.** (d)	**80.** (c)
81. (c)	**82.** (d)	**83.** (d)	**84.** (b)	**85.** (b)	**86.** (b)	**87.** (c)	**88.** (b)	**89.** (b)	**90.** (b)
91. (b)	**92.** (b)	**93.** (a)	**94.** (c)	**95.** (b)	**96.** (a)	**97.** (c)	**98.** (c)	**99.** (d)	**100.** (c)
101. (c)	**102.** (d)	**103.** (d)	**104.** (b)	**105.** (a)	**106.** (c)	**107.** (d)	**108.** (c)	**109.** (b)	**110.** (c)
111. (d)	**112.** (b)	**113.** (a)	**114.** (b)	**115.** (b)	**116.** (d)	**117.** (b)	**118.** (b)	**119.** (c)	**120.** (b)
121. (d)	**122.** (b)	**123.** (c)	**124.** (d)	**125.** (b)	**126.** (a)	**127.** (a)	**128.** (d)	**129.** (c)	**130.** (b)
131. (c)	**132.** (a)	**133.** (c)	**134.** (c)	**135.** (d)	**136.** (c)	**137.** (b)	**138.** (d)	**139.** (c)	**140.** (b)
141. (d)	**142.** (c)	**143.** (b)	**144.** (c)						

NCERT Exemplar Questions

1. (c)	**2.** (b)	**3.** (d)	**4.** (d)	**5.** (c)	**6.** (b)	**7.** (a)	**8.** (b)	**9.** (d)

Practice Test Paper

1. (d)	**2.** (b)	**3.** (d)	**4.** (b)	**5.** (b)	**6.** (c)	**7.** (c)	**8.** (d)	**9.** (c)	**10.** (c)
11. (d)	**12.** (d)	**13.** (c)	**14.** (a)	**15.** (c)	**16.** (b)	**17.** (c)	**18.** (c)	**19.** (d)	**20.** (d)
21. (c)	**22.** (d)	**23.** (c)	**24.** (b)	**25.** (d)	**26.** (b)	**27.** (b)	**28.** (a)	**29.** (c)	**30.** (b)
31. (d)	**32.** (d)	**33.** (b)	**34.** (a)	**35.** (c)	**36.** (a)	**37.** (c)	**38.** (c)	**39.** (a)	**40.** (b)
41. (c)	**42.** (d)	**43.** (b)	**44.** (b)	**45.** (b)					

SOLUTIONS

Epithelial Tissue

1. The cells of epithelial tissue are compactly packed with little intercellular matrix.
2. Simple cuboidal epithelium is found in the tubular parts of the nephron.
3. Squamous epithelium has a single layer of flattened cells with irregular boundaries.
4. Squamous epithelium is involved in functions such as forming a diffusion boundary.
5. Epithelial tissue is avascular.
6. Compound epithelium has two or more layers of cells.
7. Cuboidal epithelium is found in the lining of the ducts of glands.
8. Glandular epithelium is not possible for squamous epithelium.
9. Vagina and buccal cavity are lined with non-keratinised stratified squamous epithelium.
10. Fallopian tube and bronchioles are lined with ciliated epithelium.
11. Microvilli of epithelial cells increase the surface area for absorption.
12. Cuboidal epithelium is called germinal epithelium.
13. Brush-bordered cuboidal epithelium with microvilli is found in the lining of PCT.
14. The function of cilia is to move particles or mucus in a specific direction over the epithelium.
15. Squamous epithelium forms diffusion boundary.
16. Mammary glands are exocrine glands.
17. Desmosomes or adhering junctions perform cementing to keep neighbouring cells together.
18. The products of the endocrine glands are hormones and they are secreted directly into the fluid bathing the gland.
19. Tight junctions help to stop substances from leaking across the tissues.
20. Option (d) is correct.
21. The trachea is lined with ciliated pseudostratified epithelium.
22. Pancreas, testis and ovary have both exocrine and endocrine parts.
23. Option (c) is correct.
24. Option (c) is correct.
25. Option (c) is correct.

Connective Tissue

26. Both tendon and ligament have dense regular connective tissue.
27. Tendons connect bones to muscles.
28. The main function of connective tissue is to hold other tissues together.
29. Fibroblasts secrete matrix or ground substance of the connective tissue.
30. Areolar tissue connects skin with muscle.
31. Areolar tissue and adipose tissue are examples of loose connective tissue.
32. Connective tissue are most abundant and widely distributed in the body of complex animals.
33. Blood is an example of connective tissue in which cells do not secrete fibres of structural proteins like collagen or elastin.
34. The cells of adipose tissue store excess nutrients as fat.
35. In the skin, fibroblasts and fibres are oriented differently.
36. Blood, lymph, bone and cartilage are examples of specialised connective tissue.
37. The fibres of structural proteins provide strength, flexibility and elasticity to the connective tissue.
38. The intercellular material of the cartilage is solid and pliable.
39. The bone marrow is the site of production of blood cells.

40. Bones interact with skeletal muscles to bring about movement.

41. Option (b) is correct.

42. Patella is an example of the sesamoid bone.

43. Hyaline cartilage is the most abundant cartilage in our body.

44. The trachea has cartilaginous rings made up of hyaline cartilage.

45. Platelets are formed by the breakdown of megakaryocytes in the bone marrow.

46. Haversian canals are present only in the diaphysis of the long bones in mammals.

47. Plasma is a straw-coloured viscous fluid constituting nearly 55–60% of blood.

48. Erythrocytes are the most abundant cells in blood.

49. A healthy adult man has, on an average, 5 to 5.5 million RBCs per mm^{-3} of blood.

50. The over-stretching of ligaments is called sprain.

51. Spongy bone has a marrow cavity and is found at the ends of the long bone. The ear pinna has elastic cartilage, while the intervertebral disc has fibrous cartilage.

52. White fibrous cartilage is rich in white fibres made of collagen and is the strongest cartilage. It is found in pubic symphysis and intervertebral disc.

53. Option (b) is correct.

54. Option (d) is correct.

55. Option (a) is correct.

Muscle Tissue

56. The smooth muscle fibres taper at both ends (fusiform) and do not show striations.

57. Skeletal muscle fibres are cylindrical, syncytial and unbranched.

58. Cardiac muscle fibres have oblique bridges, intercalated discs, and are branched and cylindrical.

59. Communication junctions/intercalated discs at some fusion points allow the cardiac muscle fibres to contract as a unit.

60. Smooth muscles are involuntary as their functioning cannot be directly controlled.

61. During muscle contraction, chemical energy is converted into mechanical energy.

62. Muscle fibres contract (shorten) in response to stimulation, then relax (lengthen) and return to their uncontrolled state in a coordinated fashion.

63. The muscles surrounding the pupil of the eye are unstriated and involuntary.

64. Cardiac muscles are striped and uninucleate.

65. The smooth muscle fibres are non-striated and do not show dark and light bands.

66. In skeletal muscle, at the centre of the A-band, a comparatively less dark zone called the Hensen zone is present.

67. Striated muscles can undergo fatigue. Peristalsis occurs in alimentary canal due to the action of smooth muscles. Myocytes of skeletal muscles are multinucleate.

68. Cardiac muscles have light and dark bands and do not undergo fatigue. Cell junctions fuse the plasma membranes of the cardiac muscle cells and make them stick together.

69. About 40–50% of the body weight of a human adult is contributed by muscles.

70. Skeletal muscles are primarily involved in locomotory action and changes in body posture.

71. Option (c) is the correct answer.

72. Option (d) is the correct answer.

73. Skeletal muscles in our body are made up of a number of muscle bundles or fascicles held together by a common collagenous tissue layer called fascia.

74. Option (d) is the correct answer.

75. Option (b) is the correct answer.

76. Heart wall has cardiac muscles, biceps of the upper arm have skeletal muscle, while alimentary canal has smooth muscles.

Neural Tissue

77. Nodes of Ranvier are the areas where the myelin sheath is absent on the myelinated neurons.

78. Option (d) is the correct answer.

79. The blood–brain barrier is formed by astrocytes.

80. The neural tissues originate in the ectoderm.

81. Centrioles are absent in nerve cells.

82. Neuroglia cells are non-sensory and supporting cells.

83. Option (d) is correct.

84. Neural tissue exerts the greatest control over the body's responsiveness to changing conditions.

85. Neuroglia make up more than one-half of the volume of neural tissue.

86. Axons serve to take impulses away from the cell body of the neuron.

87. Most neurons in the cerebral cortex are multipolar.

88. Multipolar neuron has many dendrites and one axon.

89. Nissl granules are absent in neurons.

90. Bipolar nerve cells are present in the retina.

91. Neuroglia are non-excitable cells that support and protect neurons.

92. When a neuron is suitably stimulated, an electrical disturbance is generated which swiftly travels along its plasma membrane.

93. In the CNS, myelin is secreted by oligodendrocytes.

94. Neurons are the longest cells in the body.

95. Nissl granules are masses of ribosomes.

96. The dendrite carries impulses towards the cell body.

97. Option (c) is correct.

Morphology of Cockroach

98. Cockroach is omnivorous, nocturnal and cursorial.

99. The abdomen in both male and female cockroach has 10 segments.

100. The chitinous plates or sclerites of the exoskeleton in cockroach are attached to each other by an arthrodial membrane.

101. The body of the cockroach is differentiated into head, thorax and abdomen.

102. Option (d) is correct.

103. Alary muscles help in pushing the blood forward in the heart. Mouth parts of cockroach are of biting and chewing type. Excretion occurs through malpighian tubules in cockroach.

104. Cockroach has 10 pairs of spiracles.

105. In the female, the 7th sternite is boat shaped.

106. The hepatic or gastric caeca are part of midgut.

107. In cockroach, hindwings are used for flying. The forewings are modified into wing covers/ tegmina/elytra.

108. The attachment of head in cockroach is hypognathous: the head is attached perpendicular to the long axis of the body.

109. The length of *Periplaneta americana* is about 34–53 mm.

110. Head of cockroach is formed by fusion of 6 segments.

111. The first pair of wings arises from the mesothorax while the second pair of wings arises from the metathorax.

112. In both sexes, the 10th segment bears a pair of jointed filamentous structures called anal cerci.

113. In female cockroach, the genital pouch is formed by the 7th, 8th and 9th sternites.

114. Option (b) is correct.

115. Heart of cockroach is an elongated muscular tube which is made up of 13 chambers.

116. Option (d) is correct.

117. In cockroach, each compound eye is made up of 2000 units called ommatidia.

118. Hard chitinous exoskeleton is the most important character for the evolutionary success of cockroaches.

119. Haemolymph of cockroach is devoid of respiratory pigment.

120. In cockroach, the malpighian tubules are present at the junction of the midgut and hindgut.

121. Option (d) is correct.

122. In cockroach, there are 9 ganglia attached to the nerve cords, 3 of which are present in the thorax while 6 are present in the abdomen.

123. Option (c) is correct.

124. The development of *Periplaneta americana* is paurometabolous.

125. The female cockroach has 3 pairs of gonapophyses.

126. Mosaic or apposition vision of cockroach has more sensitivity and less resolution.

127. Female cockroach has a pair of ovaries in the 2nd to 6th abdominal segments laterally.

128. The longest part in the leg of the cockroach is the tibia.

129. Fat bodies of cockroach are analogous with the liver.

130. In female cockroach, ootheca is secreted by colleterial glands.

131. In female *Periplaneta americana*, a pair of spermatheca is present in the 6th segment which opens into the genital chamber.

132. On an average, female produces 9–10 ootheca, each containing 14–16 eggs.

133. Gizzard of cockroach has chitinous teeth for crushing food.

134. Only foregut and hindgut are lined by cuticle. Midgut is thus the only part that absorbs food.

135. The blood vascular system is of open type in cockroach and blood vessels are poorly developed.

136. The haemolymph is composed of colourless plasma and haemocytes.

137. The nymph of cockroach grows by moulting about 13 times to reach adult form.

138. Cockroaches are nocturnal, omnivorous animals living in damp places throughout the world.

139. Option (c) is correct.

140. The blood of cockroach (haemolymph) does not have any respiratory pigment.

141. Labrum is the only unpaired structure among them.

142. Phallomeres are external genitalia in males. They are three chitinous asymmetrical structures.

143. Colleterial glands are present in female cockroach and secrete ootheca.

144. Malpighian tubules help in the removal of excretory products in the form of uric acid.

Cell: The Unit of Life

EXERCISES

What is a Cell?, Cell Theory, An Overview of the Cell

1. Read the following statements and select the correct option:
   A. Anything less than a complete structure of a cell does not ensure independent existence
   B. Unicellular organisms are capable of independent existence and of the essential functions of life
   (a) Only (A) is correct
   (b) Only (B) is correct
   (c) Both (A) and (B) are correct
   (d) Both (A) and (B) are incorrect
2. Who amongst the following scientists is credited with the discovery of the cell?
   (a) Robert Brown (b) Robert Hook
   (c) Schleiden (d) Schwann
3. ________ first showed and described a live cell:
   (a) Anton van Leeuwenhook
   (b) Robert Brown
   (c) Huxley
   (d) Robert Hook
4. The nucleus was first observed in the cells of orchid roots in 1831 by:
   (a) Robert Brown (b) Schleiden
   (c) Purkinje (d) Corti
5. Who stated that "Protoplasm is the physical basis of life"?
   (a) Purkinje (b) Huxley
   (c) Rudolf Virchow (d) Schwann
6. ________ first explained that cells divide and new cells are formed from pre-existing cells:
   (a) Schleiden (b) Rudolf Virchow
   (c) Schwann (d) Robert Brown
7. Which of the following is a true conclusion of Matthias Schleiden?
   (a) Animals and plants are composed of cells and are the product of cells
   (b) Cells have a thin outer layer which is today known as the plasma membrane
   (c) The presence of cell wall is a unique character of plant cells
   (d) Plants are composed of different kinds of cells which form the tissue of the plant
8. Cell theory was put forward by:
   (a) Schleiden and Schwann
   (b) Sutton and Boveri
   (c) Watson and Crick
   (d) Darwin and Wallace
9. Who concluded, based on his studies on plant tissues, that the presence of cell wall is a unique character of the plant cell?
   (a) A German scientist
   (b) A British zoologist
   (c) A British botanist
   (d) A German zoologist
10. The average size of bacteria is:
    (a) 3–5 micrometre (b) 30–50 micrometre
    (c) 0.3 micrometre (d) 10 micrometre
11. The statement "Omnis cellulae cellula" was given by:
    (a) Schleiden and Schwann
    (b) Rudolf Virchow
    (c) Palade
    (d) Von Mohl
12. The smallest cell organelles are:
    (a) Lysosomes (b) Spherosomes
    (c) Peroxisomes (d) Ribosomes
13. The largest isolated single cell is:
    (a) Human RBC (b) Egg of an ostrich
    (c) Nerve cell (d) PPLO
14. Which of the following is an exception to cell theory?
    (a) Fungi (b) Lichen
    (c) Virus (d) Animals

15. Modified cell theory differs from Schleiden and Schwann's cell theory in which of the following statements?
(a) Cell is the structural and functional unit of life
(b) All living organisms are composed of cells and are the product of cells
(c) Metabolically, all cells are alike
(d) All cells arise from pre-existing cells

16. Match the following columns and select the correct option:

Column I	Column II
A. Red blood cells	(i) Elongated
B. Tracheid	(ii) Long and narrow
C. Mesophyll cells	(iii) Round and biconcave
D. Columnar epithelial cells	(iv) Round and oval

(a) A – (iii), B – (i), C – (iv), D – (ii)
(b) A – (iii), B – (iv), C – (i), D – (ii)
(c) A – (iv), B – (i), C – (ii), D – (iii)
(d) A – (iv), B – (iii), C – (ii), D – (i)

17. Which of the following organelles lack membrane?
A. Ribosome
B. Flagella
C. Centriole
D. Nucleolus
E. Microbodies
F. Lysosomes
(a) Only (A)
(b) (A), (C) and (B)
(c) (A), (C) and (D)
(d) (C), (E) and (F)

18. Double membrane is absent in:
(a) Mitochondria (b) Chloroplast
(c) Nucleus (d) Lysosomes

19. Choose the correct option to fill in the blanks:
(i) Various chemical reactions occurring in ____A____ keep the cell in the living state
(ii) ____B____ is a membrane-less organelle which is absent in higher plants

	A	B
(a)	Nucleus	Centriole
(b)	Cytoplasm	Centriole
(c)	Mitochondria	Contractile vacuole
(d)	Cytoplasm	Flagella

20. Which of the following is present in both plant and animal cells?
(a) Centriole (b) Primary cell wall
(c) Sap vacuole (d) Plasma membrane

21. What is the main arena of all the cellular activities of a cell?
(a) Nucleus (b) Cytoplasm
(c) Mitochondria (d) Vacuole

22. Read the following features:
A. Cells are held together by various types of junctions
B. An organelle with cartwheel appearance is present

Identify the cell type on the basis of the above features:
(a) Prokaryotic cell (b) Animal cell
(c) Mesokaryotic cell (d) Plant cell

23. Non–membrane-bound cell organelle found in all living cells is:
(a) Centriole (b) Ribosomes
(c) Lysosomes (d) Both (a) and (b)

24. How many of the following structures are found in plant cells?

Cell wall, Plasmids, Large centriole, Vacuoles, Centriole, Nucleus, Inclusion bodies
(a) 3 (b) 2
(c) 4 (d) 5

25. Identify the smallest cell among the following:
(a) Human RBC (b) Mycoplasma
(c) Egg of an ostrich (d) *Bacillus*

26. Human red blood cells are about:
(a) 7 μm in diameter
(b) 70 μm in diameter
(c) 3.5 μm in diameter
(d) 300–500 μm

27. The size of which of the following ranges between 3 and 5 μm?
(a) Typical eukaryotic cell
(b) PPLO
(c) Typical bacteria
(d) Viruses

28. Read the following statements carefully:
A. Ribosomes are non–membrane-bound organelles found in all cells
B. A non–membrane-bound cell organelle helps in the division of animal and plant cells
C. The semifluid matrix which occupies the volume of the cell is cytoplasm
D. The shape of the cell may vary with the function it performs

How many statements are correct?
(a) One (b) Three
(c) Two (d) Four

29. Match Column I (diagrams) with Column II (type of cells) and select the correct option:

Column I		Column II
A.	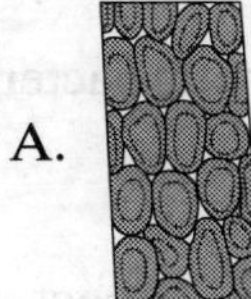	(i) Columnar epithelium
B.	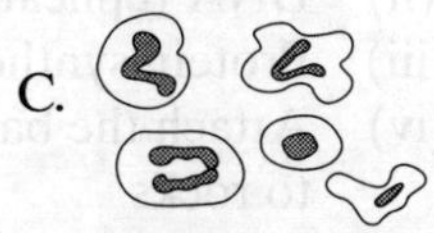	(ii) Red blood cells
C.		(iii) Mesophyll cells
D.		(iv) White blood cells

(a) A – (iii), B – (i), C – (iv), D – (ii)
(b) A – (ii), B – (iii), C – (iv), D – (i)
(c) A – (ii), B – (i), C – (ii), D – (iv)
(d) A – (iii), B – (iv), C – (ii), D – (i)

30. Select the incorrectly matched pair with respect to the size of various cells/cellular particles:

Cells/Cellular particles	Size
(a) PPLO	30 μm
(b) Typical eukaryotic cell	10–20 μm
(c) Typical bacteria	1–2 μm
(d) Viruses	0.02–0.2 μm

31. Which of the following features is incorrect regarding eukaryotic cell?
(a) Presence of double envelope system
(b) Presence of membrane-bound organelle
(c) Presence of plasmids
(d) Occurrence of both 70S and 80S ribosomes

32. Identify the organelles labelled A, B and C:

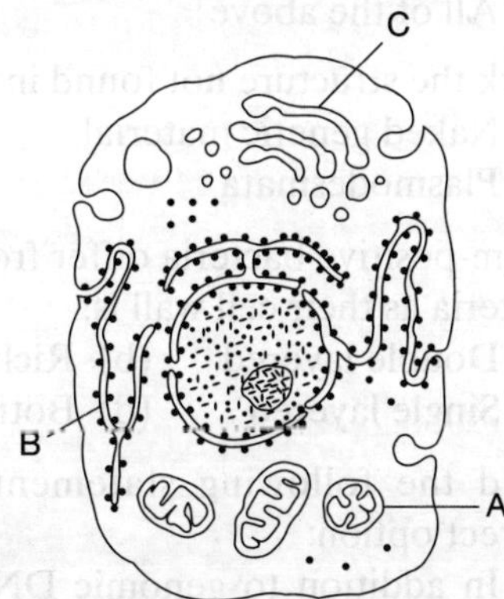

	A	B	C
(a)	Mitochondria	RER	SER
(b)	Mitochondria	RER	Golgi apparatus
(c)	Ribosomes	SER	Golgi apparatus
(d)	Lysosomes	Microtubules	SER

33. Read the following statements carefully and choose the correct ones:
A. All eukaryotic cells are identical in structure
B. There is extensive compartmentalisation of the cytoplasm through the presence of membrane-bound organelles in a eukaryotic cell

(a) Only (A) is correct
(b) Only (B) is correct
(c) Both (A) and (B) are correct
(d) Both (A) and (B) are incorrect

34. Theodore Schwann reported plasma membrane in:
(a) Plant cell (b) Animal cell
(c) Bacterial cell (d) Both (a) and (b)

35. Some features of cells are listed below. Categorise these features with respect to prokaryotic and eukaryotic cells and choose the option accordingly:
A. Nuclear region is poorly defined
B. Extra chromosomal genetic material is present
C. Nuclear material is not in direct contact with cytoplasm
D. Nucleolus is present
E. Presence of 80S ribosomes in cytoplasm
F. Thylakoid if present, lies freely in the cytoplasm

	Prokaryotic cell	Eukaryotic cell
(a)	(A), (B) and (F)	(C), (D) and (E)
(b)	(A), (B), (E) and (F)	(C) and (D)
(c)	(A), (B) and (C)	(D), (E) and (F)
(d)	(C), (E) and (F)	(A), (B) and (D)

Prokaryotic Cells

36. Organisation of prokaryotic cells is fundamentally similar and has genetic material without nuclear membrane and:
A. Mesosomes
B. 70S ribosomes
C. Microbodies
D. Pili and fimbriae

(a) Only (A) and (B) are correct
(b) (A), (B) and (D) are correct
(c) (B), (C) and (D) are correct
(d) (A), (B), (C) and (D) are correct

37. There are _____ basic shapes of bacteria:
(a) Four (b) Two
(c) Three (d) Five

38. Match the following columns and select the correct option:

Column I	Column II
A. Coccus	(i) Comma shaped
B. Bacillus	(ii) Spherical
C. Vibrio	(iii) Spiral shaped
D. Spirillum	(iv) Rod like

(a) A – (ii), B – (iv), C – (i), D – (iii)
(b) A – (iii), B – (iv), C – (i), D – (ii)
(c) A – (ii), B – (iii), C – (i), D – (ii)
(d) A – (iii), B – (i), C – (ii), D – (iv)

39. Read the following statements and select the correct option:
A. Cell envelope in bacteria is tightly linked in a three-layer structure
B. Each layer of the envelope performs distinct functions but acts together as a single protection unit
(a) Only (A) is correct
(b) Only (B) is correct
(c) Both (A) and (B) are correct
(d) Both (A) and (B) are incorrect

40. Which of the following structures determines the shape of the bacterial cell and provides strong structural support to prevent the bacterium from bursting?
(a) Glycocalyx (b) Capsule
(c) Slime layer (d) Cell wall

41. A bacterial cell structure which functionally resembles the mitochondria of eukaryotes is:
(a) Plasmid (b) Polysome
(c) Mesosome (d) Chromatophore

42. The layer of the cell envelope which differs in composition and thickness among bacteria and could be a loose sheath is:
(a) Cell membrane (b) Cell wall
(c) Glycocalyx (d) Chromatophore

43. State true (T) or false (F) for the following statements with respect to prokaryotic cells:
A. Prokaryotic cells are generally smaller and multiply more rapidly than eukaryotic cells
B. The plasma membrane is semifluid in nature and interacts with the outside world
C. Mesosome is formed by extensions of plasma membrane in the form of vesicles, tubules and lamella
D. Bacterial cells may be motile or non-motile

	A	B	C	D
(a)	T	T	T	T
(b)	T	F	F	T
(c)	T	F	T	T
(d)	F	T	F	F

44. Gas vacuoles are found in:
(a) Blue-green algae
(b) Purple and green photosynthetic bacteria
(c) Red algae and brown algae
(d) All except (c)

45. Select the correct match with respect to the function of cell organelles in prokaryotic cells:

Column I	Column II
A. Mesosome	(i) Inclusion body
B. Polysome	(ii) DNA replication
C. Fimbriae	(iii) Protein synthesis
D. Cyanophycean granules	(iv) Attach the bacteria to rocks

(a) A – (ii). B – (iii), C – (iv), D – (i)
(b) A – (iv), B – (iii), C – (i), D – (ii)
(c) A – (iv), B – (iii), C – (i), D – (ii)
(d) A – (ii), B – (i), C – (iii), D – (iv)

46. The prokaryotic cell structure exemplified by statements A and B is:
A. They are associated with the plasma membrane of the cell
B. They are about 15–20 nm in size
(a) Ribosomes
(b) Mesosomes
(c) Inclusion bodies
(d) Chromatophores

47. Prokaryotic cells differ from eukaryotic cells in the absence of:
(a) DNA (b) Basic protein
(c) Histones (d) Both (b) and (c)

48. Which of the following structures represents membranous extensions into the cytoplasm of a prokaryotic cell?
(a) Mesosome, Pili, Fimbriae
(b) Chromatophores
(c) Flagellum and Mesosomes
(d) All of the above

49. Mark the structure not found in a bacterial cell:
(a) Naked genetic material (b) Plasmid
(c) Plasmodesmata (d) Polysomes

50. Gram-positive bacteria differ from Gram-negative bacteria as their cell wall is:
(a) Double layered (b) Rich in peptidoglycan
(c) Single layered (d) Both (b) and (c)

51. Read the following statements and select the correct option:
A. In addition to genomic DNA, many bacteria have smaller circular DNA outside the genomic DNA
B. The plasmid confers certain unique phenotypic characters to bacteria

C. Plasmid DNA is used to monitor bacterial transformation with foreign DNA
(a) Only (A) is correct
(b) Only (B) is correct
(c) Only (A) and (C) are correct
(d) (A), (B) and (C) are correct

52. Match the following columns and select the correct option:

	Column I		Column II
A.	Plasmid	(i)	Helps in DNA distribution to daughter cell
B.	Fimbriae	(ii)	Prevents the bacterium from collapsing
C.	Mesosome	(iii)	Confers unique phenotypic characters to bacteria
D.	Cell wall	(iv)	Helps in attachment of bacteria to rocks and to the host tissues

(a) A – (iii), B – (iv), C – (i), D – (ii)
(b) A – (i), B – (iii), C – (ii), D – (iv)
(c) A – (iii), B – (i), C – (iv), D – (ii)
(d) A – (iv), B – (ii), C – (i), D – (iii)

53. The longest portion of bacterial flagella which extends from the cell surface to the outside is:
(a) Basal body (b) Hook
(c) Axoneme (d) Filament

54. Which of the given structures of prokaryotes resemble in composition that of eukaryotes?
(a) Cell wall (b) Cell membrane
(c) Ribosomes (d) Both (b) and (c)

55. Select the incorrect match:

Structure	Function
(a) Mesosome	Cell wall formation and secretion
(b) Polysome	Translate mRNA into protein
(c) Nucleoid	Resistance to antibiotics
(d) Fimbriae	Attachment of bacteria to host tissue

56. Out of the given features, how many are correctly related to the genetic material of bacteria?
A. Naked
B. Circular
C. Single-stranded DNA
D. Nucleoid
E. Larger than plasmid DNA
(a) Only (A), (B), (D) and (E)
(b) Only (A), (B), (C) and (D)
(c) Only (A), (B) and (C)
(d) Only (B), (C) and (D)

57. Mesosomes are extensions of the cell membrane in the form of:
(a) Vesicles (b) Lamellae
(c) Tubules (d) All of the above

58. Select the incorrect statement with respect to inclusion body:
(a) It is the site of food storage
(b) They lie freely in cytoplasm
(c) They are single-membrane-bound structures
(d) They are found only in prokaryotic cells

59. Plasmids have the following characters:
A. They contain fewer genes than the nucleoid DNA
B. They have double-stranded circular DNA
C. They are used to monitor bacterial transformation with foreign DNA
D. They provide resistance to certain antibiotics
The correct statements are:
(a) Only (A) and (B)
(b) Only (B) and (D)
(c) Only (B) and (C)
(d) All (A), (B), (C) and (D)

60. Bacterial cell wall is composed of:
(a) Cellulose (b) Chitin
(c) Lignin (d) Peptidoglycan

61. The Gram stain used for differentiating Gram-positive and Gram-negative forms of bacteria is:
(a) Crystal violet and iodine
(b) Safranin and iodine
(c) Acetocarmine and iodine
(d) Safranin and gentian violet

62. The characteristic cell wall material peptidoglycan has another covering of lipopolysaccharide. This specialised condition is found in:
(a) Gram+ve bacteria (b) Gram–ve bacteria
(c) *Mycoplasma* (d) Both (a) and (b)

63. Mesosomes help in all of the following cellular activities, except:
(a) Synthesis of protein (b) DNA replication
(c) Secretion process (d) Cell wall formation

64. Which of the following is not an inclusion body in prokaryotes?
(a) Phosphate granules
(b) Cyanophycean granules
(c) Gas vacuoles
(d) Plasmid

65. Read the following statements:
A. Storage or reserve food material in prokaryotes
B. Not bound by any membrane system
C. Lie freely in cytoplasm
D. Involved in translation of mRNA into proteins
Which of the given statements is/are associated with inclusion bodies in bacteria?
(a) Only (A) and (D)

(b) Only (B) and (C)
(c) Only (A) and (B)
(d) Only (A), (B) and (C)

66. Slime layer and capsule in bacteria are modifications of the:
(a) Glycocalyx (b) Cell wall
(c) Cell membrane (d) Tonoplast

67. Which of the following does not differ in *E.coli* and *Chlamydomonas*?
(a) Ribosomes
(b) Chromosomal organisation
(c) Cell membrane
(d) Cell wall

Eukaryotic Cells (Cell Membrane, Cell Wall)

68. A detailed structure of the membrane was studied after the advent of electron microscope in:
(a) 1930s (b) 1950s
(c) 1960s (d) 1990s

69. The universally accepted model of plasma membrane is:
(a) Unit membrane model
(b) Trilamellar model
(c) Fluid mosaic model
(d) Robertson model

70. Fluid mosaic model of plasma membrane was given by:
(a) SJ Singer and GL Nicolson
(b) SS Singer and GL Nicolson
(c) SJ Singer and HL Nicolson
(d) SS Singer and JL Nicolson

71. Biochemical investigation clearly revealed that the cell membrane:
(a) Mainly consists of phospholipids and sugars
(b) Possesses protein, lipid and carbohydrates
(c) Possesses lipid bilayer with polar heads towards the inner side
(d) Both (b) and (c) are correct

72. Which of the following features is not shown by lipids of the plasma membrane?
(a) Quasi-fluid
(b) Flip-flop movement
(c) Act as permeability barrier
(d) Circulation

73. Cell wall shows:
(a) Complete permeability
(b) Semi-permeability
(c) Differential permeability
(d) Impermeability

74. Cell wall:
A. Helps in cell to cell interaction
B. Provides a barrier to undesirable macromolecules
C. Protects the cell from mechanical damage and infection
(a) Only (A) and (B) are correct
(b) Only (C) is correct
(c) Only (A) and (C) are correct
(d) (A), (B) and (C) are correct

75. Middle lamella is made up of:
(a) Microfibrils
(b) Lignin
(c) Calcium and magnesium pectates
(d) Suberin

76. Complete the given statements with the correct option:
(i) ____**A**____ acts as cementing layer during tissue formation
(ii) The ____**B**____ and ____**C**____ may be traversed by plasmodesmata

	A	B	C
(a)	Middle lamella	Cell wall	Middle lamella
(b)	Middle lamella	Cell wall	Cell membrane
(c)	Primary cell wall	Secondary cell wall	Middle lamella
(d)	Cell membrane	Secondary cell wall	Primary cell wall

77. Outermost layer of cell wall is:
(a) Plasma lemma (b) Secondary cell wall
(c) Middle lamella (d) Primary cell wall

78. Cell wall layer having suberin deposition is:
(a) Middle lamella (b) Secondary wall
(c) Tertiary wall (d) Primary wall

79. Smallest unit in the plant cell wall is:
(a) Micelle (b) Micro fibril
(c) Fibril (d) Microtubules

80. In the fluid mosaic model, the terms fluid and mosaic signify, respectively:
(a) Protein and lipids
(b) Lipids and proteins
(c) Protein and carbohydrates
(d) Lipids and carbohydrates

81. In humans, the membrane of erythrocyte has approximately ____**A**____ per cent protein and ____**B**____ per cent lipids:

	A	B
(a)	52	40
(b)	40	52
(c)	20	20
(d)	20	80

82. Membrane proteins can be classified as integral or peripheral depending on the:
(a) Ease of extraction (b) Type of lipids
(c) Quality of proteins (d) Both (a) and (c)

83. Membrane proteins are held in the bilayer mainly by:
(a) Covalent bond
(b) Ionic bond
(c) Hydrophobic attraction
(d) Hydrophilic attraction

84. Plasma membrane is asymmetric because:
(a) Lipids present in the outer and inner sides of the bilayer are different
(b) Extrinsic proteins are more abundant on the inner surface than on the outer surface
(c) Oligosaccharides are attached only to the external surface of lipids and proteins of a bio membrane
(d) All of the above

85. The ability of lipids to move within a membrane is measured as its fluidity. The fluid nature of a membrane is important for:
A. Cell growth
B. Formation of intercellular junction
C. Endocytosis
D. Cell division
E. Secretion
(a) Only (A), (C) and (E)
(b) Only (B) and (D)
(c) Only (B), (D) and (E)
(d) (A), (B), (C) and (D)

86. The ability of the protein to move within the cell membrane is due to the membrane's:
(a) Rigidity (b) Permeability
(c) Elasticity (d) Fluidity

87. Algal cell wall is made up of:
(a) Cellulose, Hemicellulose, Pectin and Proteins
(b) Hemicellulose, Chitin, Galactans
(c) Cellulose, Galactans, Mannans and Calcium Carbonate
(d) Cellulose, Hemicellulose, Calcium and Magnesium Pectate and Chitin

88. Cell membrane is:
(a) Permeable
(b) Impermeable
(c) Semi-permeable
(d) Differentially permeable

89. In *Amoeba*, whenever soluble organic substances and salts are absorbed, the mode of nutrition is called:
(a) Holozoic (b) Pinocytosis
(c) Holophytic (d) Exocytosis

90. Engulfing of solid materials by the cell is called:
(a) Pinocytosis (b) Active transport
(c) Phagocytosis (d) Autolysis

91. Read the following statements with respect to the plasma membrane:
A. Asymmetric nature
B. Oligosaccharides are found on the inner side of the plasma membrane
C. Fluidity is due to quasi-fluid nature of lipid
(a) Only (C) is correct
(b) (A) and (C) are correct
(c) Only (C) is incorrect
(d) (A), (B), (C) are correct

92. Which of these transport methods between cells involves a specific carrier protein and cellular energy?
(a) Na^+K^+ pump in animal cells
(b) Simple diffusion
(c) Facilitated diffusion
(d) Both (a) and (c)

93. Select the mismatched pair:
(a) Cell wall - Polymer of NAG in Fungi
(b) Middle lamella - Cementing layer of cell
(c) Secondary cell wall - Present on the outer side of the middle lamella
(d) Plasmodesmata - Connect the cytoplasm of neighbouring cells

94. Mark the incorrect match:
(a) Neutral solute - Simple diffusion
(b) Water - Osmosis
(c) Non-polar molecules - Facilitated diffusion
(d) ATP expenditure - Active transport

95. The fluidity of the plasma membrane is due to:
(a) Lipids (b) Proteins
(c) Carbohydrates (d) Both (a) and (b)

96. Carrier molecules facilitating transport across cell membranes are:
(a) Proteinaceous (b) Fatty acids
(c) Starch (d) Alkaloid

97. Cell wall in higher plants is made up of:
(a) Cellulose + Lignin
(b) Cellulose + Pectin + Suberin
(c) Cellulose + Hemicellulose + Pectin + Proteins
(d) Cellulose + Lipid

98. The strength and rigidity of a cell wall is duc to substances known as:
(a) Suberin (b) Lignin
(c) Cellulose (d) Pectin

99. The fluid mosaic model explains:
(a) Only structural aspects of the cell membrane
(b) Only functional aspects of the cell membrane

(c) Both structural and functional aspects of the cell membrane
(d) Only fluidity of the membrane

100. State true (T) or false (F) for the following statements with respect to the plasma membrane:
A. Non-polar tails of phospholipids are towards the cytoplasm
B. Quasi-fluid nature of lipids enables lateral movement of proteins within the overall bilayer
C. Peripheral proteins are partially or totally buried
D. Proteins make up 60–70% of the cell membrane

	A	B	C	D
(a)	F	T	F	F
(b)	F	T	F	T
(c)	T	T	T	F
(d)	T	F	T	F

101. Plasma membrane in eukaryotes may also contain sterols of various kinds. Sterol molecules are more rigid than the phospholipid and:
(a) Confer stability on the eukaryotic membrane
(b) Increase fluidity of the membrane
(c) Increase conductivity of the membrane
(d) All of the above

102. One of the most important functions of the plasma membrane is the transport of molecules. Which of the following is correct with respect to this statement?
(a) Polar membrane can pass easily through the lipid bilayer
(b) The membrane does not support passive transport
(c) Neutral solutes require a carrier protein of membrane for transport
(d) The membrane is selectively permeable to some molecules present on either side of it

103. Protoplast does not include:
(a) Plasma membrane (b) Cell wall
(c) Nucleus (d) Cytoplasm

104. According to modern ideas, the cell membrane is:
(a) Solid (b) Fluid
(c) Quasi-fluid (d) Solidified sheath

105. The fluid nature of the membrane is important from the point of view of functions like:
(a) Cell growth and formation of intracellular junctions
(b) Secretion and endocytosis
(c) Cell division
(d) All of the above

106. Glycocalyx, which is exclusively present on the outer side of the membrane and functions as a cell recognition centre, is made up of:
(a) Proteins
(b) Lipids
(c) Protein and lipids
(d) Glycoprotein and glycolipids

107. Why are the tails of lipids in the membrane towards the inner part?
(a) The tail has non-polar hydrocarbon and so is protected within an aqueous environment
(b) The tail has polar hydrocarbon and so is protected from the aqueous environment
(c) The non-polar or hydrophobic hydrocarbon tails of lipid, being on the inner side, ensures their protection from the aqueous environment
(d) The tail is hydrophilic, so it tends to be located on the aqueous inner side of the membrane

108. Which of the following features is not shown by lipids of the plasma membrane?
(a) Quasi-fluid nature
(b) Flip-flop movement
(c) Acts as permeability barrier
(d) Circulation

109. Cell wall layer having suberin deposition is:
(a) Middle lamella (b) Secondary wall
(c) Tertiary wall (d) Primary wall

110. Molecules in the cell membrane that function as receptors are:
(a) Proteins (b) Glycerol
(c) Cholesterol (d) Phospholipid

Endomembrane System (Endoplasmic Reticulum, Golgi Apparatus, Lysosomes, Vacuoles)

111. Many cell organelles are considered as an endomembrane system because:
(a) They contain ds circular DNA
(b) They perform photosynthesis
(c) Their functions are coordinated
(d) They are double-membrane-bound structures

112. The endomembrane system comprises:
A. Endoplasmic reticulum
B. Golgi apparatus
C. Lysosome
D. Vacuole
E. Microbodies
(a) Only (A), (B) and (C)
(b) Only (A), (C) and (D)

(c) Only (B), (C) and (E)
(d) All except (E)

113. This function is not related to the endomembrane system:
(a) Steroid hormone synthesis
(b) Formation of glycoprotein and glycolipid
(c) Intracellular digestion
(d) Phosphorylation

114. ________ are called suicidal bags:
(a) Gas vacuoles (b) Lysosomes
(c) Nucleus (d) Food vacuole

115. Which of the following cell organelles is single-membrane-bound but does not coordinate with the organelles of the endomembrane system?
(a) ER (b) Golgi apparatus
(c) Vacuole (d) Peroxisome

116. Which one is apparato reticolare?
(a) Golgi apparatus (b) Endoplasmic reticulum
(c) Microfilament (d) Microtubules

117. Select the answer which incorrectly matches the organelle and its function:

Organelle	**Function**
(a) RER	Protein synthesis and secretion
(b) SER	Synthesis of steroidal hormones
(c) Golgi apparatus	Synthesis of glycoprotein and glycolipid
(d) Contractile vacuole	Phagocytosis

118. Select the correct set of statements from those given below:
A. The endomembrane system does not include semiautonomous organelles
B. Concave face of Golgi bodies is maturing face
C. RER takes part in detoxification of toxic chemicals in the body
D. Proteins synthesized by ribosomes are modified in the cisternae of the Golgi apparatus
(a) Only (B) and (C)
(b) Only (A) and (D)
(c) Only (A), (B) and (D)
(d) All (A), (B), (C) and (D)

119. Diameter of the Golgi apparatus is
(a) 0.5 μm–1.0 μm (b) 0.2 μm–1.0 μm
(c) 5.0 μm–10 μm (d) 1.0 μm–5 μm

120. Water-soluble pigment present in sap vacuole is:
(a) Anthocyanin (b) Carotene
(c) Xanthophyll (d) Chlorophyll

121. Membrane-bound vesicular structures formed by the process of packaging in the Golgi apparatus and filled with hydrolytic enzymes are called:
(a) Vacuoles (b) Lysosomes
(c) Microbodies (d) Centrosome

122. Disappearance of tadpole tail during metamorphosis is brought about by:
(a) Golgi bodies (b) Peroxisomes
(c) Endoplasmic reticulum (d) Lysosomes

123. The term lipocondria was suggested for:
(a) Mitochondria (b) Golgi complex
(c) ER (d) All of these

124. Detoxification of lipid-soluble drugs and other harmful compounds in ER is carried out by cytochrome:
(a) $a–a_3$ (b) c
(c) $b–f_6$ (d) P450

125. Organelle connected with glycosidation of proteins and glycosidation of lipids is:
(a) Golgi apparatus (b) ER
(c) Glyoxisomes (d) Lysomes

126. In the plant cell, the vacuole contains:
(a) Metabolic gases
(b) Water, sap, excretory product
(c) Materials not useful for cell
(d) All except (a)

127. In *Amoeba*, contractile vacuole is important for:
(a) Storing food
(b) Providing buoyancy
(c) Osmoregulation and excretion
(d) Engulfing food particles

128. Which of the following is not concerned with the diagram given below?

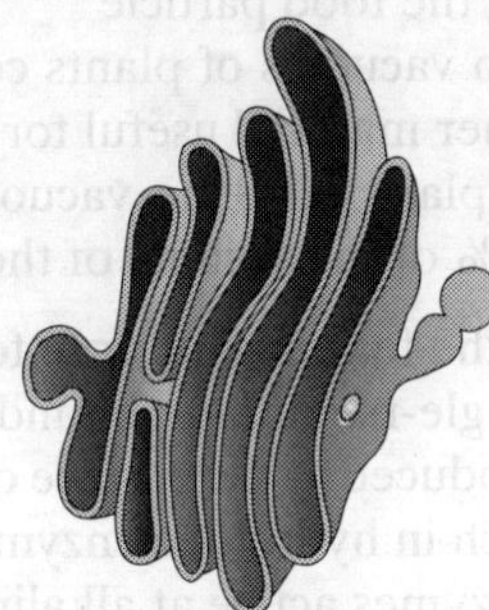

(a) Remains in close association with ER
(b) Helps in the formation of plant cell wall
(c) Performs the function of packaging particles
(d) Phosphorylation with the help of oxygen

129. Find the incorrect statement with respect to lysosomes:
(a) Are bounded by single membrane
(b) Contain hydrolytic enzyme synthesized in the Golgi apparatus
(c) Enzymes are optimally active at acidic pH
(d) Clean up the cell by digesting foreign materials as well as worn-out cell organelles

130. Read the following statements and select the correct option:
A. Neutral or non-polar solutes may move across the membrane by the process of simple diffusion
B. Some solutes are transported across the plasma membrane from lower concentration to higher concentration through a transporter protein without expenditure of energy
(a) Only (A) is correct
(b) Only (B) is correct
(c) Both (A) and (B) are correct
(d) Both (A) and (B) are incorrect

131. The membrane covering the vacuole is known as:
(a) Desmoneme (b) Tonoplast
(c) Glycocalyx (d) Slime layer

132. Which of the following is non-living cell inclusion?
(a) Golgi apparatus (b) Centrosome
(c) Vacuole (d) Ribosome

133. The bright colour of petals is due to the presence of:
(a) Plastocyanin (b) Anthocyanin
(c) Xanthophyll (d) Chloroplast

134. Ribosomes are formed attached to RER by:
(a) Ribophorins (b) Glycolipids
(c) Porins (d) Lipids

135. Select the incorrect statement from the following:
(a) In *Amoeba*, the contractile vacuole is important for excretion
(b) In protists, food vacuoles are formed by engulfing the food particle
(c) Sap vacuoles of plants contain water, sap and other material useful for the cell
(d) In plant cells, the vacuoles can occupy up to 90% of the volume of the cell

136. Select the features not true for lysosome:
A. Single-membrane-bound vesicular structure
B. Produced at the cis face of the Golgi apparatus
C. Rich in hydrolytic enzymes
D. Enzymes active at alkaline pH
(a) (A) and (B) (b) (B) and (D)
(c) (B) and (C) (d) (C) and (D)

137. Identify the correctly matched pair with respect to the given diagram:

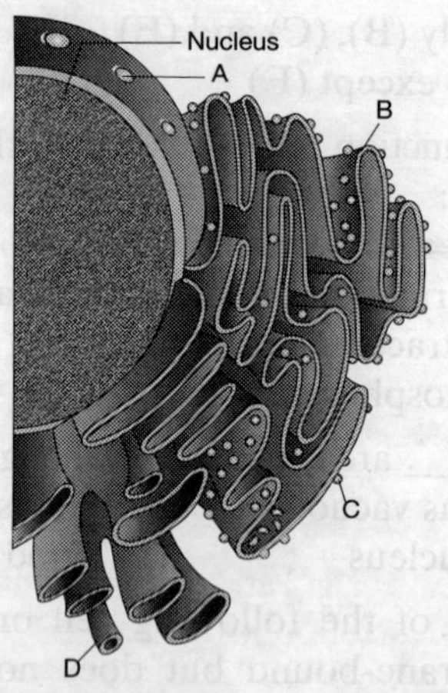

(a) A – Controls bidirectional movement of RNA and protein
(b) B – Major site of synthesis of lipid
(c) C – 70S ribosomes attached by its larger subunit
(d) D – Frequently observed in cell activity involved in protein synthesis and secretion

138. Which of the following organelles is rich in acid phosphatases?
(a) Lysosomes (b) Golgi complex
(c) RER (d) Nucleolus

139. The rough ER is so named because it has an abundance of _______ on it:
(a) Inclusion bodies (b) Lysosomes
(c) Palade particles (d) Protein granules

140. Ribosomes attach to ER by its:
(a) 50S subunit (b) 60S subunit
(c) 40S subunit (d) 30S subunit

141. Which one is not true with respect to lysosomes?
(a) Very rich in hydrolytic enzymes
(b) Active at acidic pH
(c) Single-membranous vesicular structures
(d) Formation of secretory vesicles in the cis face

142. Which organelle is present in higher numbers in secretory cells?
(a) Dicytosome (b) ER
(c) Lysosome (d) Vacuole

Mitochondria, Plastids, Ribosomes

143. Mitochondria do not occur in:
(a) Human liver cell (b) Human nerve cell
(c) Human erythrocyte (d) Frog liver cell

144. What is mitoplast?
(a) Membrane-less mitochondria
(b) Mitochondria without inner membrane

(c) Another name for mitochondria
(d) Mitochondria without outer membrane

145. The size of mitochondria in the plant cell is:
(a) 3–4 micrometre (b) 2–4 micrometre
(c) 1–4 micrometre (d) 0.1–1.0 micrometre

146. Mitochondria are the site of:
(a) Photophosphorylation
(b) Oxidative phosphorylation
(c) Macromolecule degradation
(d) Intracellular digestion

147. Identify A, B, C and D in the given figure:

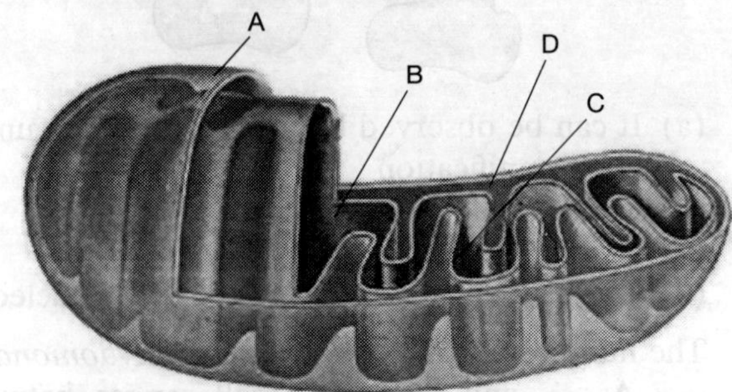

	A	B	C	D
(a)	Outer membrane	Crista	Matrix	Intermembrane space
(b)	Inner membrane	Crista	Matrix	Inter membrane space
(c)	Inner membrane	Thylakoid	Crista	Peri-mitochondrial space
(d)	Outer membrane	Hub	Spoke	Peri-mitochondrial space

148. The matrix of mitochondria does not contain:
(a) 70S ribosomes and ds circular DNA
(b) Components required for protein synthesis
(c) Enzymes for Krebs cycle
(d) Electron carriers for ATP synthesis

149. Electron microscope has revealed the presence of:
(a) Ribosomes (b) Chromosomes
(c) Chloroplast (d) Leucoplast

150. F_0–F_1 particles are present on the surface of:
(a) Outer membrane of mitochondria
(b) Inner membrane of mitochondria
(c) Cristae
(d) Both (b) and (c)

151. Mitochondria and chloroplast are similar in the presence of:
A. ds circular DNA
B. 70S ribosome
C. Electron transport system
D. Porin protein in their membrane
E. Phosphorylation
(a) Only (A) and (B)
(b) Only (A), (B) and (E)
(c) All except (D)
(d) (A), (B), (C), (D) and (E)

152. The stroma of the chloroplast possesses all, except:
(a) Flattened membranous sacs
(b) Enzymes required for carbohydrates and proteins
(c) Photosynthetic pigments
(d) DNA molecules and ribosomes

153. Mitochondria and chloroplasts are considered to be endosymbionts of cells because they:
(a) Do not arise de novo
(b) Possess their own nucleic acids
(c) Have membranes similar to those of bacteria
(d) All of the above

154. Match the following:

Column I	Column II
A. Thylakoid	(i) Infoldings of inner membrane in mitochondria
B. Cristae	(ii) Site of carbon reduction
C. Stroma	(iii) Oxidative phosphorylation
D. Oxysome	(iv) Flattened membranous sac

(a) A – (ii), B – (iii), C – (iv), D – (i)
(b) A – (ii), B – (iii), C – (i), D – (iv)
(c) A – (iv), B – (i), C – (ii), D – (iii)
(d) A – (iv), B – (ii), C – (i), D – (iii)

155. In rRNA, 'r' stands for:
(a) Ribophorins (b) Ribosomal
(c) Ribozyme (d) Ribonuclease

156. Read the following statements and select the correct option:
A. In chromoplasts, fat-soluble carotenoid pigments like carotene and xanthophylls are present
B. The leucoplasts are colourless plastids of varied shape and size which store nutrients
C. Majority of the chloroplast of green plants is found in the mesophyll cells of the leaves
(a) Only (B) and (C) (b) Only (C)
(c) Only (A) and (C) (d) (A), (B) and (C)

157. Which of the following statements is incorrect for ribosomes:
(a) Composed of ribonucleic acid and proteins
(b) Synthesize protein in cytoplasm and nucleus
(c) Not surrounded by any membrane
(d) Organelle within organelle

158. Which is not a true organelle?
(a) Lysosome (b) Mitochondria
(c) Ribosome (d) Chloroplast

159. Non-secretory proteins are synthesized on:
(a) Ribosomes attached to nuclear envelope
(b) Ribosomes attached to RER
(c) Ribosomes present freely in the cytoplasm
(d) Both (a) and (b)

160. Read the following statements and select the correct option:
A. The ribosomes of chloroplasts are smaller than cytoplasmic ribosomes
B. The stroma of chloroplast contain enzymes required for the synthesis of carbohydrates and proteins
(a) Only (A) is correct
(b) Only (B) is correct
(c) Both (A) and (B) are correct
(d) Both (A) and (B) are incorrect

161. The sedimentation coefficient of ribosomes is an indirect measure of its:
(a) Density and size (b) Size and shape
(c) Density and mass (d) Density and volume

162. Classification of plastids into chloroplast, chromoplast and leucoplast is based on:
(a) Stored food (b) Pigments
(c) Structure (d) Size

163. Percentage of mitochondrial DNA in the cell is ________ of total cellular DNA:
(a) 10% (b) 1%
(c) 30% (d) 25%

164. The figure given below shows the structure of a chloroplast with its four parts labelled as A, B, C and D. Select the part incorrectly matched with its function:

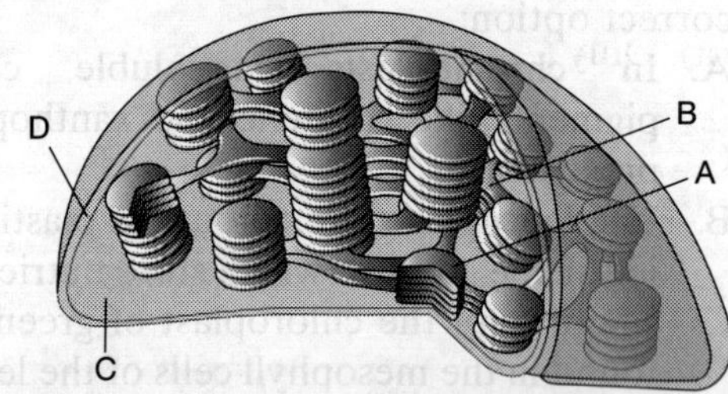

(a) A. Thylakoid: Contains pigment required for light reaction
(b) B. Granum: Stacks of thylakoid
(c) C. Stroma: Contains enzymes required for photophosphorylation and CO_2 fixation
(d) D. Inner membrane: Relatively less permeable than outer membrane

165. Chromoplast contains:
(a) Fat-soluble carotenoid pigments
(b) Water-soluble carotenoid pigments
(c) Water-soluble anthocyanin pigment
(d) Stores nutrients like starch, oil or protein

166. 80S ribosomes break up into two subunits whose sedimentation constants are:
(a) 50S and 30S (b) 60S and 20S
(c) 60S and 40S (d) 50S and 40S

167. Identify the following figure and select the incorrect statement:

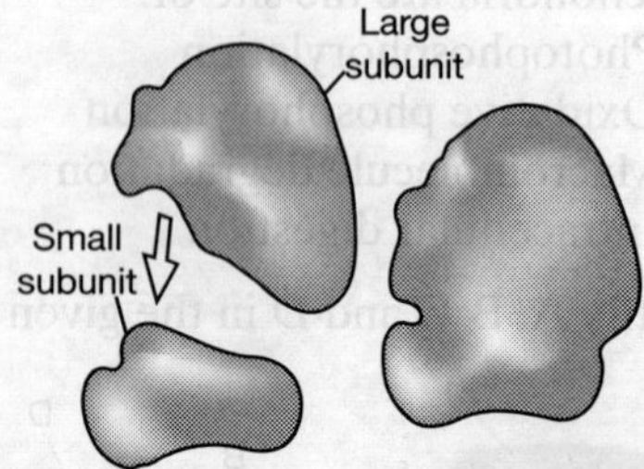

(a) It can be observed by light microscope under low magnification
(b) Made up of two subunits
(c) Also called palade particles
(d) Their subunits are synthesized in the nucleolus

168. The number of chloroplasts in *Chlamydomonas* is ___**A**___ and mesophyll cell ranges between ___**B**___:

	A	B
(a)	1	2 and 4
(b)	1	20 and 40
(c)	2–6	10 and 20
(d)	2–4	20 and 40

169. In 80S ribosomes, S stands for
(a) Svedberg unit (b) Shape
(c) Subunit (d) All of the above

170. In a cell, the maximum number of enzymes are present in:
(a) Nucleus (b) Mitochondria
(c) Chloroplast (d) ER

171. Which of the following human cells does not contain mitochondria?
(a) Nerve cell (b) Liver cell
(c) Red blood cell (d) White blood cell

172. Amyloblast, elaioblast and aleuroblast store:
(a) Starch, fat and protein, respectively
(b) Nutrients and pigments
(c) Fat-soluble pigments and nutrients
(d) Starch, protein and fat, respectively

173. The mitochondria divide by:
(a) Multiple fission (b) Binary fission
(c) Fragmentation (d) Budding

174. How many of the given features are associated with chloroplasts?

Double membrane, Thylakoid, Cristae, 80S ribosome, Stroma, Circular ss-DNA

(a) 5 (b) 6
(c) 4 (d) 3

175. Read the following statements about mitochondria and select the correct option:
A. Site of aerobic and anaerobic respiration
B. Produce cellular energy in the form of ATP
C. Also called powerhouse of the cell
D. The matrix of mitochondria possesses many circular DNA molecules
(a) Only (A) and (B) are correct
(b) (B), (C) and (D) are correct
(c) Only (D) is incorrect
(d) Only (B) and (C) are correct

176. All plastids are similar to each other in:
(a) Storing starch, fats and proteins
(b) Possessing photosynthetic pigments
(c) Being found in plants cells
(d) Lacking membrane

Cytoskeleton, Cilia and Flagella, Centrosome and Centriole

177. Mark the following statements as true (T) or false (F) and select the correct option:
A. Cilia are small structures which work like oars, causing the movement of either the cell or the surrounding fluid
B. Bacterial flagella are structurally similar to that of eukaryotic flagella
C. Centrioles are absent in plants including algae
D. Centrioles form the basal body of cilia or flagella

	A	B	C	D
(a)	T	F	T	T
(b)	T	F	F	T
(c)	T	F	F	F
(d)	F	T	T	F

178. Study the following diagram and select the incorrect option:

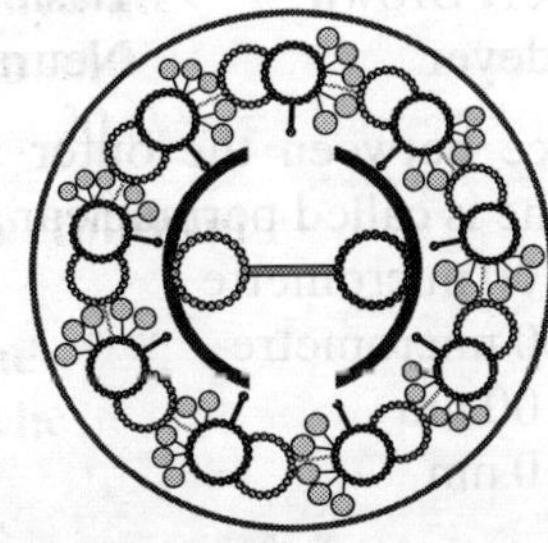

(a) Peripheral microtubules doublets – 9
(b) Central microtubule – 2
(c) Radial spoke – 9
(d) It forms the basal body of cilia and flagella

179. The core of the flagellum, which is composed entirely of microtubules, is:
(a) Axoneme (b) Hub
(c) Protofilament (d) Blepharoplasty

180. Cytoskeleton is made up of:
(a) Microtubules and ER
(b) Microtubules and microfilament
(c) Cytoplasm and ER
(d) Cytoplasm with a network of microtubules and microfilament

181. 9+2 organisation of microtubules is present in:
(a) Prokaryotic flagella
(b) Centrioles
(c) Basal body of eukaryotic flagella
(d) Flagella and cilia of a eukaryotic cell

182. Read the following statements with respect to the given figure and select the correct option:

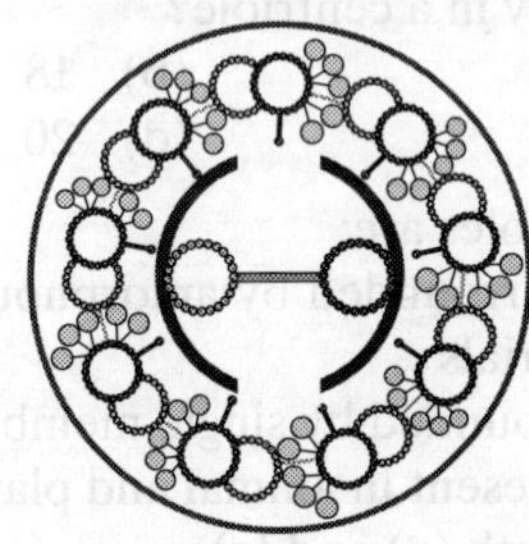

A. The core of the structure given in the diagram has (9+0) arrangement of microtubules
B. Peripheral doublets of microtubules are connected to each other by linkers
C. It emerges from a centriole-like structure
D. It is a non–membrane-bound structure

The correct statements are:
(a) Only (A) and (B)
(b) Only (B) and (C)
(c) Only (A), (B) and (C)
(d) All (A), (B), (C) and (D)

183. Select the incorrect statements about centrioles:
(a) They have an organisation like a cartwheel
(b) They are made up of nine evenly spaced peripheral fibrils of tubulin protein
(c) They show (9+2) microtubular organisation
(d) Both (a) and (b)

184. Movement of cilia and flagella is due to the presence of:
(a) Central sheath (b) Dynein
(c) Radial spokes (d) Singlet tubules

185. A: They are made up of nine evenly spaced peripheral fibrils in triplet, no centrally located microtubule

B: They have nine radial spokes attached to peripheral triplet microtubules

Identify the cell organelle, on the basis of the above features:
(a) Centriole
(b) Eukaryotic flagella
(c) Prokaryotic flagella
(d) Basal body of bacterial flagella

186. Select the incorrect match:
(a) Centriole – Shows cartwheel organisation with (9+0) microtubular arrangement
(b) Cilium – Central doublet microtubules are connected by central bridge
(c) Flagellum – Axoneme usually has nine pairs of doublets of radially arranged peripheral microtubules
(d) Centrosome – Hub and radial spokes are proteinaceous

187. How many peripheral microtubules are arranged radially in a centriole?
(a) 9 (b) 18
(c) 27 (d) 20

188. Centrioles are:
(a) Surrounded by amorphous pericentriolar materials
(b) Bounded by single membrane
(c) Present in animal and plant cells
(d) Both (a) and (c)

189. Choose the odd one out with respect to the components/features of centrioles:
(a) Tubulin, peripheral triplets
(b) 9+0 arrangement of microtubules
(c) Central sheath and central bridge
(d) Hub and radial spokes

190. The solid linear cytoskeletal element, having a diameter of 6 nm and made up of a single type of monomer, is known as:
(a) Microtubule (b) Microfilament
(c) Intermediate filament (d) Lamin

191. Microtubules are ___**A**___ unbranched cylinders generally about ___**B**___ in diameter:

	A	B
(a)	Hollow	25 nm
(b)	Solid	25 Å
(c)	Solid	10 Å
(d)	Hollow	10 nm

192. The cytoskeleton includes all of the following, except:
(a) Microtubule (b) Microfilament
(c) Myosin filament (d) Intermediate filament

193. Microfilaments perform all of the following functions, except:
(a) Provide support to plasma membrane
(b) Involved in cyclosis
(c) Help in cell plate formation during cell division
(d) Help in pseudopodia formation

194. Centrosome is an organelle usually containing two cylindrical structures called:
(a) Basal body
(b) Diplosome
(c) Centriole
(d) Pericentriolar material

195. Which is not a function of the cytoskeleton?
(a) Mechanical support
(b) Motility
(c) Maintenance of the shape of the cell
(d) Secretion

196. Constituents of the cytoskeleton are composed of:
(a) Cellulose
(b) Proteinaceous filament
(c) Chitin
(d) Cellulose and hemicellulose

Nucleus, Chromosomes, Microbodies

197. *Acetabularia* used in Hammerling's nucleocytoplasmic experiments is:
(a) Unicellular fungus
(b) Multicellular uninucelated fungus
(c) Unicellular uninucelated green algae
(d) Multinucleate differentiated green algae

198. Nucleus is absent in:
(a) Mature mammalian RBC
(b) Mature sieve tube cells
(c) *Paramecium*
(d) All except (c)

199. The term chromatin was given by ___**A**___ after staining of nucleus with ___**B**___ dyes:

	A	B
(a)	Flemming	Basic
(b)	Flemming	Acidic
(c)	Robert Brown	Basic
(d)	Waldeyer	Neural

200. The space between the outer and inner nuclear membrane is called perinuclear space, which is:
(a) 50–100 micrometre
(b) 10–50 micrometre
(c) 50–100 nm
(d) 10–50 nm

201. A single human cell has approximately ______ long thread of DNA distributed among its ______ chromosomes:
(a) 2 cm, 46 (b) 2 metre, 46
(c) 2 cm, 23 (d) 2 metre, 23

202. All the statements are correct with respect to the nucleolus, except:
(a) A membrane-bound structure in the nucleus
(b) A site for active rRNA synthesis
(c) Large and many in number in cells which are actively involved in protein synthesis
(d) It is produced from and is associated with a specific nucleolar organising region of certain chromosomes

203. Nuclear pores are the regions in the nuclear envelope:
A. Where both membranes are fused
B. Which allows unidirectional flow of RNA and proteins
(a) Only (A) is correct
(b) Only (B) is correct
(c) Both (A) and (B) are correct
(d) Both (A) and (B) are incorrect

204. How many of the following statements is/are correct for chromatin material of eukaryotes:
A. Composed of DNA, histone protein, some non-histone proteins and also RNA
B. At interphase, it occurs as highly extended and elaborate nucleoprotein fibres
C. It can be stained by acidic dyes
D. Heterochromatin is the densely packed darky stained and transcriptionally inactive region
(a) 1 (b) 3
(c) 4 (d) 2

205. Which of the following is a linear DNA molecule?
(a) Plasmid DNA
(b) Eukaryotic genomic DNA
(c) Mitochondrial DNA
(d) Both (b) and (c)

206. Chromosomes present in prolonged prophase in the salivary gland of *Drosophila* are:
(a) Polytene
(b) Lamp brush
(c) B-chromosome
(d) Supernumerary chromosomes

207. Ribosomal subunits are synthesized in a eukaryotic cell inside the:
(a) Nucleus (b) Nucleolus
(c) Nucleosome (d) Cytoplasm

208. Spherical structure present in the nucleoplasm:
A. Has contents continuous with rest of the nucleoplasm
B. Are larger and more numerous in cells actively involved in lipid and steroid synthesis
(a) Only (A) is correct
(b) Only (B) is correct
(c) Both (A) and (B) are correct
(d) Both (A) and (B) are incorrect

209. In humans, allosomes are:
(a) SAT chromosomes
(b) Sex chromosomes
(c) Marker chromosomes
(d) Other than sex chromosomes

210. The darkly stained region of chromatin is:
(a) Called heterochromatin
(b) Densely packed
(c) Transcriptionally inactive
(d) All of the above

211. The best stage to observe the shape, size and number of chromosomes is:
(a) Interphase (b) Metaphase
(c) Prophase (d) Telophase

212. The chromosome in which the centromere is situated close to its one end, forming one extremely short arm and one long arm, is:
(a) Telocentric (b) Submetacentric
(c) Metacentric (d) Acrocentric

213. Telomeres:
(a) Initiate RNA synthesis
(b) Help chromatids move towards the poles
(c) Seal the ends of the chromosomes
(d) Identify the correct members of homologous pairs of chromosomes

214. The amount of DNA (in picograms) present in a genome denotes:
(a) C value
(b) Complementary DNA
(c) Paradox value
(d) Circular DNA of mitochondria and plastids

215. Somatic pairing is a characteristic feature of:
(a) Lamp brush chromosomes
(b) Polytene chromosomes
(c) SAT chromosomes
(d) All of the above

216. A diagrammatic representation of the karyotype of a species is called:
(a) Dendrogram (b) Cladogram
(c) Karyogram (d) Idiogram

217. Find the mismatched pair:
(a) Kinetochore Disc shaped structure on both sides of the chromosome
(b) Metacentric chromosome – Heterobrachial
(c) Chromatin – Nucleoprotein fibres
(d) Satellite – Non-staining secondary constrictions at a constant location

218. Euchromatin is:

A. Lightly stained region of chromatin
B. Tightly packed region of DNA
C. The part of the DNA that undergo transcription
(a) Only (A) (b) Only (A) and (B)
(c) Only (A) and (C) (d) (A), (B) and (C)

219. The long and short arms of chromosomes are designated, respectively, as
(a) p and q (b) q and p
(c) α and β (d) m and p

220. Polytene chromosomes were discovered in:
(a) *Chironomus* (b) *Drosophila*
(c) *Musca* (d) *Culex*

221. In idiogram, the chromosome of an organism is arranged according to:
(a) Increasing size
(b) Decreasing size
(c) Position of centromere
(d) Number of chromosomes

222. The point at which polytene chromosomes appear attached is:
(a) Centriole (b) Chromocenter
(c) Chromosome (d) Centromere

223. Balbiani rings are the site of:
(a) DNA replication
(b) RNA and protein synthesis
(c) Synthesis of polysaccharides
(d) Synthesis of lipids

224. Lamp brush chromosomes are observed in:
(a) Meiotic metaphase (b) Mitotic metaphase
(c) Mitotic prophase (d) Meiotic prophase

225. Polytene chromosomes are formed due to:
(a) Endoduplication
(b) Duplication without separation
(c) Replication of DNA without cell division
(d) All of the above

226. Which possess giant chromosomes?
(a) *Drosophila* (b) *Xenopsylla*
(c) *Branchiomyces* (d) Mouse

227. Lamp brush chromosomes occur inside:
(a) Nucleus of human cells
(b) Oocyte of frog/amphibian
(c) Salivary gland of *Drosophila*
(d) Salivary gland of silk moth

228. Chromosomes can be classified into four types based on the position of:
(a) Satellite
(b) Primary constriction
(c) Secondary constriction
(d) Telomere

229. The name of the microbody which is usually found in close association with chloroplast and mitochondria in the cells where photorespiration occurs is:
(a) Gas vacuole (b) Peroxisome
(c) Spherosome (d) Glyoxysome

230. Enzymes associated with converting fats to carbohydrates are located in:
(a) Liposomes (b) Golgi bodies
(c) Glyoxysomes (d) Microsomes

231. Match the following columns and select the correct option with respect to the shape of the chromosomes:

	Column I		Column II
A.	Metacentric	(i)	I-shaped
B.	Telocentric	(ii)	V-shaped
C.	Acrocentric	(iii)	L-shaped
D.	Submetacentric	(iv)	J-shaped

(a) A – (iii), B – (i), C – (iv), D – (ii)
(b) A – (ii), B – (i), C – (iv), D – (iii)
(c) A – (iii), B – (iv), C – (i), D – (ii)
(d) A – (iii), B – (ii), C – (i), D – (iv)

232. Nuclear membrane is derivative of:
(a) Smooth ER
(b) RER
(c) Microtubules
(d) Membrane of Golgi body

233. Which of the following is not a heterobrachial chromosome?
(a) Metacentric (b) Submetacentric
(c) Telocentric (d) Acrocentric

234. Recent research shows that peroxisomes have ______ origin:
(a) Cyanobacterial (b) Fusobacterial
(c) Proteobacterial (d) Actinobacterial

235. An oxidative organelle is:
(a) Ribosome (b) Golgi body
(c) Peroxisome (d) ER

236. The non-staining secondary constriction at a constant location that forms a small fragment in few chromosomes is called:
(a) Satellite (b) Centromere
(c) Primary constriction (d) Kinetochore

237. Microbodies are present in:
(a) *Mycoplasma*
(b) Cyanobacteria
(c) All prokaryotic and eukaryotic cells
(d) Both plant and animal cells

238. Identify the structures labelled A, B, C and D in the given figure:

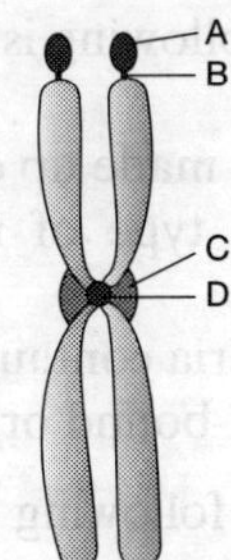

A	B	C	D
(a) Satellite	Secondary constriction	Centromere	Kinetochore
(b) Satellite	Secondary constriction	Primary constriction	Centromere
(c) Telomere	Primary constriction	Centromere	Kinetochore
(d) Telomere	Primary constriction	Secondary constriction	Centromere

239. The given diagram shows a chromosome, in which X represents:

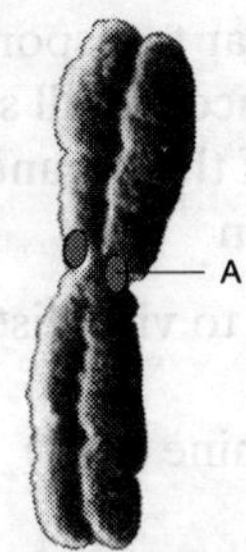

(a) Centromere (b) Telomere
(c) Kinetochore (d) Primary constriction

240. Which of the following chromosomes are SAT chromosomes (having satellite) and acrocentric chromosomes, respectively?

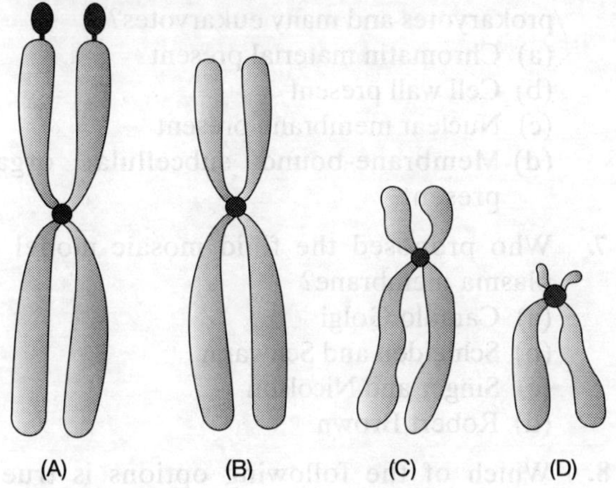

	SAT chromosome	Acrocentric chromosome
(a)	D only	B only
(b)	A only	D only
(c)	A only	C and D
(d)	C and D	A only

241. Microbodies are minute vesicles and are present in:
(a) Plant cells only
(b) Animal cells only
(c) Prokaryotic cells only
(d) Both plants and animal cells

242. Proteins required for the functioning of the nucleus are formed in:
(a) Nucleolus (b) RER
(c) Cytoplasm (d) Nucleoplasm

NCERT EXEMPLAR QUESTIONS

1. A common characteristic feature of plant sieve tube cells and most mammalian erythrocytes is:
(a) Absence of mitochondria
(b) Presence of cell wall
(c) Presence of haemoglobin
(d) Absence of nucleus

2. Select the one which is not true for ribosomes:
(a) Madc up of two subunits
(b) Form polysome
(c) May attach to mRNA
(d) Have no role in protein synthesis

3. Which one of these is not a eukaryote?
(a) *Euglena* (b) *Anabaena*
(c) *Spirogyra* (d) *Agaricus*

4. Which of the following is not used for staining chromosomes?
(a) Basic fuchsin
(b) Safranin
(c) Methylene green
(d) Carmine

5. Different cells have different sizes. Arrange the following cells in ascending order of their size and select the correct option:
A. *Mycoplasma*
B. Ostrich eggs
C. Human RBCs
D. Bacteria
(a) (A)-(D)-(C)-(B)
(b) (A)-(C)-(D)-(B)

(c) (B)-(A)-(C)-(D)
(d) (C)-(B)-(A)-(D)

6. Which of the following features is common to prokaryotes and many eukaryotes?
(a) Chromatin material present
(b) Cell wall present
(c) Nuclear membrane present
(d) Membrane-bound subcellular organelles present

7. Who proposed the fluid mosaic model of the plasma membrane?
(a) Camillo Golgi
(b) Schleiden and Schwann
(c) Singer and Nicolson
(d) Robert Brown

8. Which of the following options is true for a secreted cell?
(a) Golgi apparatus is absent
(b) RER is easily observed in the cell
(c) Only SER is present
(d) Secondary granules are formed in the nucleus

9. What is a tonoplast?
(a) Outer membrane of mitochondria
(b) Inner membrane of chloroplast
(c) Membrane boundary of the vacuole of plant cells
(d) Cell membrane of a plant cell

10. Which of the following is not true for a eukaryotic cell?
(a) Cell wall is made up of peptidoglycan
(b) It has 80S type of ribosome present in the cytoplasm
(c) Mitochondria contains circular DNA
(d) Membrane-bound organelles are present

11. Which of the following is not true for the cell membrane?
(a) It is present in both plant and animal cells
(b) Lipids are present in it as bilayer
(c) Proteins may be peripheral or integral in it
(d) Carbohydrates are never found in it

12. Plastids differ from mitochondria on the basis of which of the following features?
(a) Presence of two layers of membrane
(b) Presence of ribosome
(c) Presence of thylakoids
(d) Presence of DNA

13. Which of the following is not_a function of the cytoskeleton in a cell?
(a) Intracellular transport
(b) Maintenance of cell shape and structure
(c) Support of the organelles
(d) Respiration

14. The stain used to visualise mitochondria is:
(a) Fast green (b) Safranin
(c) Acetocarmine (d) Janus green

PRACTICE TEST PAPER

1. Which of the following non–membrane-bound structures is found in prokaryotic cells?
(a) Vacuole (b) Ribosome
(c) Centriole (d) All except (a)

2. Pigment containing membrane extensions are present in the cytoplasm of:
(a) Virus (b) Cyanobacteria
(c) Fungi (d) Animal cell

3. Choose the correct statement regarding eukaryotic cells:
A. Presence of extensive compartmentalisation of cytoplasm
B. Presence of an organised nucleus having a nuclear envelope
C. Locomotory and cytoskeleton structures are absent
D. The genetic material is organised into chromosomes
(a) A and B only (b) B and C only
(c) A, B and D only (d) A, B, C and D

4. Read the following statements carefully and select the incorrect statement:
(a) Nucleoid is equivalent to a single chromosome
(b) Plasmid occurs in all eukaryotic cells
(c) Gas vacuoles are commonly found in BGA purple and green sulphur bacteria
(d) Prokaryotic cells lack membrane-bound cell organelles

5. Choose the feature which is common to mitochondria, chloroplast and prokaryotic cell?
(a) Presence of 70S ribosomes
(b) ds circular DNA
(c) Surrounded by rigid cell wall
(d) All except (c)

6. Contractile vacuole is important for excretion in:
(a) *Chlamydomonas*
(b) *Amoeba*
(c) Fungi
(d) Prokaryotic cells

7. The enzymes of lysosomes are optimally active at:
(a) Acidic pH
(b) Highly basic pH
(c) Moderate basic pH
(d) Neutral pH

8. Fill in the blanks and choose the correct option:
(i) ___A___ is the single membrane around the sap vacuole
(ii) ___B___ is related to packaging of material
(iii) Synthesis of steroidal hormones occurs in ___C___

	A	B	C
(a)	Tonoplast	Golgi apparatus	RER
(b)	Tonoplast	Golgi apparatus	SER
(c)	Cell membrane	Lysosome	RER
(d)	Cell membrane	RER	SER

9. How many of the following features are associated with mitochondria?

Single Circular DNA, 70S ribosomes, RNA aerobic respiration, Carbon assimilation, Double membranous cell organelle
(a) 4 (b) 5
(c) 3 (d) 6

10. Phosphoglycerides are the main constituent of:
(a) Lipid component of cell membrane
(b) Protein component of cell membrane
(c) Lipid component of cell wall
(d) Protein component of cell wall

11. How many microtubules are found in the transverse section of prokaryotic flagella?
(a) 18 (b) 20
(c) 27 (d) 0

12. Read the following statements and select the correct option:
A. Plastids are found in all plant cells and in euglenoids
B. Aleuroplasts, amyloplasts and elaioplasts all are types of plastids having photosynthetic pigments
(a) Only A is correct
(b) Only B is correct
(c) Both A and B are incorrect
(d) Both A and B are correct

13. Find the mismatched pair:
(a) Kinetochores: Site of spindle fibre attachment
(b) Nucleolus: Single-membrane–bound spherical bodies in nuclear matrix
(c) Metacentric chromosome: Isobrachial
(d) Chromatin: Nucleoprotein fibres

14. Chromosome in which centromere is present at the terminal:
(a) Is called telocentric chromosome
(b) Is I-shaped
(c) Appears to have only one arm
(d) All of the above

15. Many membrane-bound minute vesicles called microbodies that contain various enzymes are present in both:
(a) Plant and animal cells
(b) Prokaryotic and eukaryotic cells
(c) Bacterial and cyanobacterial cells
(d) Plant and bacterial cells

16. The structure that works like oars, causing movement of cell or surrounding fluid, is
(a) Flagella (b) Cilia
(c) Central hub (d) Massule

17. How many centrioles are present in the centrosome?
(a) 1 (b) 2
(c) 3 (d) 0

18. The comma-shaped bacterial cell is called:
(a) Bacillus (b) Vibrio
(c) Coccus (d) Spirillum

19. Constituents of the cytoskeleton are composed of:
(a) Cellulose
(b) Nucleoprotein
(c) Proteinaceous filaments
(d) Cellulose and hemicellulose

20. The term chromatin was coined by:
(a) Hofmeister (b) Waldeyer
(c) Flemming (d) Robert Brown

21. Read the following features of a cell organelle:
A. Membrane-bound vesicular structures
B. Formed by the process of packaging in the Golgi apparatus
C. Filled with hydrolytic enzymes which are optimally active at acidic pH

Identify the organelle on the basis of the above features:
(a) Peroxisome (b) Lysosome
(c) Vacuole (d) Oxysome

22. Which of the given structures of prokaryotes resembles in composition that of eukaryotes?
(a) Cell wall (b) Ribosomes
(c) Cell membrane (d) Both (b) and (c)

23. The living connection of two adjacent plant cells is formed by:
(a) Middle lamella (b) Carrier proteins
(c) Gap junction (d) Plasmodesmata

24. Identify the inclusion bodies found in BGA, purple and green photosynthetic bacteria:
(a) Cyanophycean granules
(b) Polysome
(c) Gas vacuoles
(d) All except (b)

25. Which of the given structures of bacteria has functional resemblance with the mitochondria?
(a) Mesosome (b) Chromatophore
(c) Polysome (d) Oxysome

26. Identify the incorrectly matched pair with respect to the given diagram:

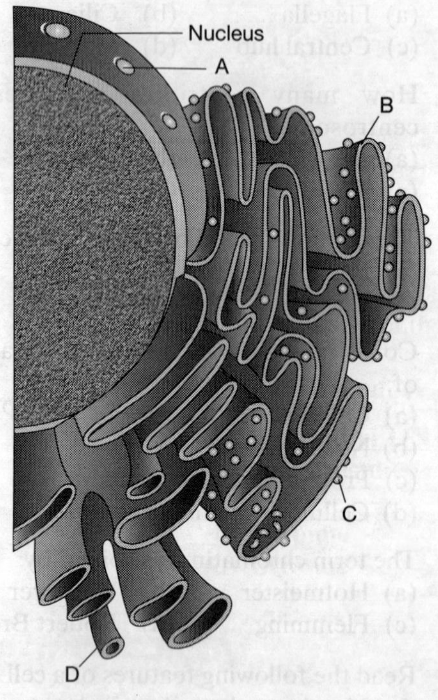

(a) A – Controls bidirectional movement of RNA and proteins
(b) B – Major site for protein synthesis
(c) C – 70S ribosomes
(d) D – Frequently observed in cells which are actively involved in lipid synthesis

27. Which of the following statements is not true for prokaryotic cells?
(a) All prokaryotic cells are fundamentally identical
(b) Ribosomes are associated with the plasma membrane
(c) Mesosome helps in cell wall synthesis, DNA replication and distribution to daughter cells
(d) Gas vacuoles are unit membrane structures

28. Find the correct match:

	Column I		Column II
A.	Tonoplast	(i)	Lipid molecule
B.	Glycocalyx	(ii)	Vacuole
C.	Flip-flop movement	(iii)	Chloroplast
D.	Thylakoid	(iv)	Recognition site of the cell

(a) A – (ii), B – (iv), C – (i), D – (iii)
(b) A – (ii), B – (i), C – (iv), D – (iii)
(c) A – (iii), B – (i), C – (iv), D – (ii)
(d) A – (iv), B – (iii), C – (ii), D – (i)

29. Microtubules are not the constituent of:
(a) Centriole
(b) Spindle apparatus
(c) Basal granule
(d) Bacterial flagella

30. The size of a typical eukaryotic cell is:
(a) 1–2 micrometre
(b) 10–20 micrometre
(c) 0.1 micrometre
(d) 0.02–0.2 micrometre

31. Bacteria differ from PPLO in having:
(a) Nucleoid (b) Cell membrane
(c) Cell wall (d) 70S ribosomes

32. Select the mismatched pair with respect to the function of the structure:
(a) Plasmid – Resistance to antibiotics
(b) Fimbriae – Help in attachment of bacteria to host cell
(c) Polysomes – Translate the mRNA into proteins
(d) Pili – Help bacteria in motility

33. All of the following can be justified, except:
(a) The body of all living organisms is made up of cells and their product
(b) All cells arise from pre-existing cells
(c) Presence of cell wall is a unique character of plant cells
(d) Mammalian RBC and sieve tube cells continue to live without nucleus

34. A protoplast is:

(a) A cell organelle
(b) A cell without cell wall
(c) A cell without nucleus
(d) Capsule of bacterial cell

35. The quasi-fluid nature of ___A___ enables ___B___ movement of proteins within the overall bilayer of a plasma membrane:

	A	B
(a)	Protein	Lateral
(b)	Protein	Flip-flop
(c)	Lipid	Lateral
(d)	Carbohydrate	Flip-flop

36. Ribosomes of chloroplast are:
(a) 80S type
(b) Larger than cytoplasmic ribosomes
(c) Smaller than cytoplasmic ribosomes
(d) Both (a) and (c)

37. Identify the following statements as true (T) or false (F) and choose the option accordingly:
A. Reserve materials in prokaryotic cells are stored in cytoplasm in the form of inclusion body
B. The cytoplasm is the main arena of cellular activities in both plant and animal cells
C. Secondary lysosomes are formed which phagosomes fuse with Golgi bodies
D. The axoneme of eukaryotic flagellum is made up of nine peripheral triplets and two centrally located singlet microtubules

	A	B	C	D
(a)	T	T	F	F
(b)	T	T	T	F
(c)	T	T	F	T
(d)	F	F	T	T

38. The unit of measurement of sedimentation of ribosome is:
(a) Angstrom (Å)
(b) Micron (μ)
(c) Svedberg unit (s)
(d) Nanometre (nm)

39. Non-membranous organelle without DNA with duplicating nature is:
(a) Plasmid (b) Centriole
(c) Ribosome (d) Plastid

40. Second constriction differs from primary constriction in being:
(a) Present at the chromosomal end in all chromosomes
(b) Euchromatin
(c) Attachment site for spindle fibres
(d) Transcriptionally inactive chromatin

41. A living plant cell which becomes enucleated at maturity is:
(a) Sieve tubes (b) Vessel
(c) Tracheid (d) Companion cell

42. Which of the following chromosomes has non-staining secondary constriction?

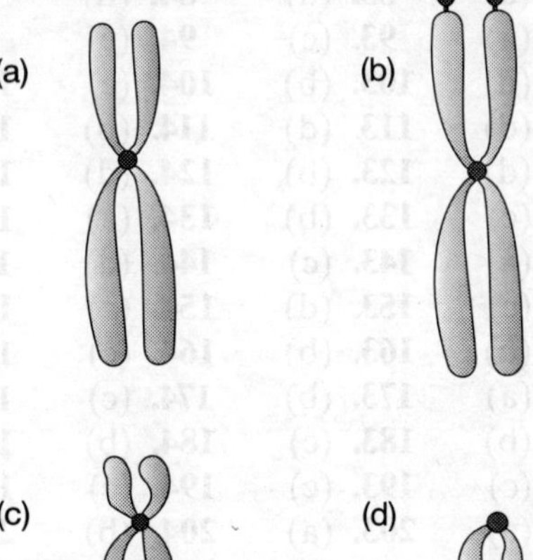

43. Which of the following organelles is the cell's garbage disposal system?
(a) Peroxisomes (b) Lysosomes
(c) Phagosomes (d) Golgi apparatus

44. The function of the nucleolus in the cell is:
(a) Secretory
(b) Synthesis of DNA
(c) Synthesis of RNA and ribosomal subunits
(d) Synthesis of protein and enzymes

45. What is common between chloroplast, chromoplast and leucoplast?
(a) Presence of pigments
(b) Presence of thylakoid in grana
(c) Ability to multiply by a fission-like process
(d) Storage of starch protein and lipids

ANSWER KEYS

Exercises

1. (c) 2. (b) 3. (a) 4. (a) 5. (b) 6. (b) 7. (d) 8. (a) 9. (b) 10. (a)
11. (b) 12. (d) 13. (b) 14. (c) 15. (d) 16. (a) 17. (c) 18. (d) 19. (b) 20. (d)
21. (b) 22. (b) 23. (b) 24. (a) 25. (b) 26. (a) 27. (c) 28. (b) 29. (a) 30. (a)
31. (c) 32. (b) 33. (b) 34. (b) 35. (a) 36. (b) 37. (a) 38. (a) 39. (c) 40. (d)
41. (c) 42. (c) 43. (a) 44. (d) 45. (a) 46. (a) 47. (c) 48. (b) 49. (c) 50. (d)
51. (d) 52. (a) 53. (d) 54. (b) 55. (c) 56. (a) 57. (d) 58. (c) 59. (d) 60. (d)
61. (a) 62. (b) 63. (a) 64. (d) 65. (d) 66. (a) 67. (c) 68. (b) 69. (c) 70. (a)
71. (b) 72. (d) 73. (a) 74. (d) 75. (d) 76. (c) 77. (c) 78. (b) 79. (a) 80. (b)
81. (a) 82. (a) 83. (a) 84. (d) 85. (d) 86. (d) 87. (c) 88. (d) 89. (b) 90. (c)
91. (b) 92. (a) 93. (c) 94. (c) 95. (a) 96. (a) 97. (c) 98. (b) 99. (c) 100. (a)
101. (a) 102. (d) 103. (b) 104. (c) 105. (d) 106. (d) 107. (c) 108. (d) 109. (b) 110. (a)
111. (c) 112. (d) 113. (d) 114. (b) 115. (d) 116. (a) 117. (d) 118. (c) 119. (a) 120. (a)
121. (b) 122. (d) 123. (b) 124. (d) 125. (a) 126. (d) 127. (c) 128. (d) 129. (b) 130. (a)
131. (b) 132. (c) 133. (b) 134. (a) 135. (c) 136. (b) 137. (a) 138. (a) 139. (c) 140. (b)
141. (d) 142. (a) 143. (c) 144. (d) 145. (c) 146. (b) 147. (a) 148. (d) 149. (a) 150. (c)
151. (d) 152. (c) 153. (d) 154. (c) 155. (b) 156. (d) 157. (b) 158. (c) 159. (c) 160. (c)
161. (a) 162. (b) 163. (b) 164. (c) 165. (a) 166. (c) 167. (a) 168. (b) 169. (a) 170. (b)
171. (c) 172. (a) 173. (b) 174. (c) 175. (d) 176. (c) 177. (b) 178. (d) 179. (a) 180. (b)
181. (d) 182. (b) 183. (c) 184. (b) 185. (a) 186. (b) 187. (c) 188. (a) 189. (c) 190. (b)
191. (a) 192. (c) 193. (c) 194. (c) 195. (d) 196. (b) 197. (c) 198. (d) 199. (a) 200. (d)
201. (b) 202. (a) 203. (a) 204. (b) 205. (b) 206. (a) 207. (b) 208. (a) 209. (b) 210. (d)
211. (b) 212. (d) 213. (c) 214. (a) 215. (b) 216. (d) 217. (b) 218. (c) 219. (a) 220. (a)
221. (b) 222. (b) 223. (b) 224. (d) 225. (d) 226. (a) 227. (b) 228. (b) 229. (b) 230. (c)
231. (b) 232. (b) 233. (a) 234. (d) 235. (c) 236. (a) 237. (d) 238. (a) 239. (c) 240. (b)
241. (d) 242. (c)

NCERT Exemplar Questions

1. (d) 2. (d) 3. (b) 4. (b) 5. (a) 6. (b) 7. (c) 8. (b) 9. (c) 10. (a)
11. (d) 12. (c) 13. (d) 14. (d)

Practice Test Paper

1. (b) 2. (b) 3. (d) 4. (b) 5. (d) 6. (b) 7. (a) 8. (b) 9. (b) 10. (a)
11. (d) 12. (a) 13. (b) 14. (d) 15. (a) 16. (b) 17. (b) 18. (b) 19. (c) 20. (c)
21. (b) 22. (d) 23. (d) 24. (d) 25. (a) 26. (c) 27. (d) 28. (a) 29. (d) 30. (b)
31. (c) 32. (d) 33. (d) 34. (b) 35. (c) 36. (c) 37. (a) 38. (c) 39. (b) 40. (d)
41. (a) 42. (b) 43. (b) 44. (c) 45. (c)

SOLUTIONS

What is a Cell?, Cell Theory, An Overview of the Cell

1. Both (A) and (B) are correct.
2. Robert Hook
3. Anton van Leeuwenhook first saw and described a live cell.
4. Robert Brown discovered the nucleus.
5. Huxley stated that protoplasm is the physical basis of life.
6. Rudoff Virchow first explained that cells divide and new cells are formed from pre-existing cells; that is, omnis cellulae cellula.
7. Option (d) is correct.
8. Schleiden and Schwann
9. Schwann (British zoologist) concluded, based on his studies on plant tissues, that the presence of the cell wall is a unique character of plant cells.
10. 3–5 micrometre
11. Rudolf Virchow
12. Ribosomes are the smallest cell organelles.
13. The largest isolated single cell is the egg of an ostrich.
14. Viruses are acellular structures.
15. All cells arise from pre-existing cells.
16. Red blood cells – Round and biconcave, Tracheids – Elongated, Mesophyll cells – Round and oval, Columnar epithelial cells – Long and narrow
17. Ribosomes, centriole and nucleolus are non-membrane-bound organelles.
18. Lysosome is a single-membrane–bound cell organelle.
19. A – Cytoplasm, B – Centriole
20. Plasma membrane
21. Cytoplasm is the main arena of cellular activities.
22. Animal cell
23. Ribosomes are non–membrane-bound cell organelles found in all cells.
24. Cell wall, large central vacuole, nucleus
25. Mycoplasma, the smallest cells, are only 0.3 micrometre in length.
26. Human RBCs are about 7.0 micrometre in diameter.
27. A typical bacteria could be 3–5 micrometre.
28. Three statements (i, iii and iv) are correct. Centriole is only present in animal cells and helps in cell division.
29. Option (a) is correct.
30. PPLO (Mycoplasma) – 0.3 micrometre
31. Plasmids are present in prokaryotic cells.
32. A – Mitochondria, B – RER, C – Golgi apparatus
33. Only (B) is correct. Cells differ greatly in size, shape and activity.
34. Schwann studied different types of animal cells and reported that cells had a thin outer layer known as the plasma membrane.
35. Prokaryotic cell – a, b and f, Eukaryotic cell – c, d and e

Prokaryotic Cells

36. Microbodies are single-membrane–bound structures that contain various enzymes and are present in both plant and animal cells.
37. There are four basic shapes of bacteria – bacillus, coccus, vibrio, spirillum.
38. Coccus – Spherical, Bacillus – Rod-like, Vibrio – Comma shaped, Spirillum – Spiral shaped
39. Both (A) and (B) are correct. Cell envelope includes glycocalyx, cell wall and cell membrane.
40. The cell wall determines the shape of the cell and provides strong structural support and prevents the bacterium from bursting and collapsing.

41. Mesosomes perform the function of cellular respiration in bacteria.

42. Glycocalyx could be a loose sheath (slime layer) or it may be thick and tough (capsule).

43. (A), (B), (C) and (D) are true.

44. Gas vacuoles are formed in blue-green and purple and green photosynthetic bacteria.

45. Mesosomes – DNA replication, Polysome Protein synthesis, Fimbriae – Attach the bacteria to rocks, Cyanophycean granules – Inclusion body

46. In prokaryotes, ribosomes are associated with the plasma membrane of the cell and they are about 15–20 nm in size.

47. Histone protein

48. In some prokaryotes like BGA, there are membranous extensions into the cytoplasm called chromatophores, which contain pigments.

49. Plasmodesmata

50. In Gram-negative bacteria, the cell wall is single-layered, thick and made up of peptidoglycan. In Gram-negative bacteria, the cell wall is double-layered; the outer layer is thick and made up of lipopolysaccharide, while the inner cell wall is thin and made up of peptidoglycan.

51. (A), (B) and (C) are correct.

52. Option (a) is correct.

53. Bacterium flagellum is composed of three parts: filament, hook and basal body. The filament is the longest portion and extends from the cell surface to the outside.

54. Cell membrane is made up of lipoprotein.

55. Nucleoid is the genetic material of a prokaryotic cell. It is naked, ds circular DNA without histone protein.

56. A, B, D and E are correct.

57. Mesosomes are extensions of the cell membrane in the form of vessels, tubules and lamellae.

58. Inclusion bodies are not bounded by any membrane system and lie free in the cytoplasm.

59. (A), (B) (C) and (D) are correct.

60. Bacterial cell wall is made up of peptidoglycan or murein or mucopeptide.

61. Crystal violet and iodine solution

62. Gram-negative bacteria

63. Synthesis of protein is a function of ribosomes.

64. Plasmids are extra chromosomal self-replicating genetic material.

65. The ribosomes of a polysome translate the mRNA into proteins.

66. Glycocalyx

67. Cell membrane

Eukaryotic Cells (Cell Membrane, Cell Wall)

68. 1950s

69. The most accepted model of the plasma membrane is the fluid mosaic model.

70. SJ Singer and GL Nicolson

71. Cell membrane possesses protein, lipid and carbohydrates. Lipids are arranged in a bilayer, with the polar head towards the outer side and hydrophobic tails towards the inner side.

72. Circulation

73. Cell wall shows complete permeability.

74. (A), (B) and (C) are correct.

75. Middle lamella is the cementing layer between the primary walls of adjacent cells and it is made up of calcium and magnesium pectate.

76. A – Middle lamella, B – Cell wall, C – Middle lamella

77. Middle lamella

78. Secondary cell wall

79. Cell wall of plants contains:
(i) Matrix (Hemicellulose, protein, pectin water)
(ii) Fibrils
(iii) Depositions

Fibrils are made up of cellulose. One macrofibril is composed of 250 microfibrils. Each microfibril is composed of 20 elementary fibrils or micelles. Each micelle is composed of 100 cellulose chains.

80. Plasma membrane is mosaic in nature due to the presence of embedded protein. The protein has been compared to icebergs floating in the sea of phospholipid bilayer.

81. A – 52%, B – 40%

82. Ease of extraction

83. Membrane proteins are held in the bilayer mainly by covalent bonds

84. All the given statements are correct.

85. The fluid nature of the membrane is also important from the point of view of functions like cell growth, formation of intercellular junctions, secretion, endocytosis and cell division.

86. Fluidity

87. Algal cell wall is made up of cellulose, galactans, mannans and minerals like calcium carbonate.

88. Cell membrane is differentially or selectively permeable in nature.

89. Pinocytosis or cell drinking

90. Engulfing of solid materials by a cell is called phagocytosis or cell eating.

91. (A) and (C) are correct.

At some place, extrinsic protein and the lipid of the cell membrane bear oligosaccharides, forming glycoprotein and glucolipid, and they constitute the glycocalyx.

92. Active transport requires special carrier proteins and energy expenditure. For example, Na^+/K^+ pump.

93. Secondary cell wall is formed on the inner side of the cell (towards the plasma membrane).

94. The molecules which are lipid soluble (non-polar) are easily transported through the membrane. They do not need proteins for transport and can diffuse across quickly.

95. The fluidity of the membrane is due to lipids.

96. Intrinsic or integral proteins act as carrier proteins.

97. Cell wall in higher plants is made up of cellulose + hemicellulose + pectin + proteins.

98. Lignin

99. The fluid mosaic model explains both structural and functional aspects of the cell membrane.

100. (A), (C) and (D) are false. Non-polar tail of phospholipids is buried in the membrane. Peripheral proteins lie on the surface. Proteins make up 52% of the cell membrane.

101. Cholesterol helps in stability of the plasma membrane.

102. Membrane is selectively permeable to some molecules present on either side of it.

103. Protoplasts are cells which have had their cell wall removed, usually by digestion with enzymes.

104. Cell membrane is quasi-fluid or semi-fluid in nature.

105. All are correct.

106. Glycocalyx is made up of glycoprotein and glycolipids.

107. Hydrophobic tails of lipids, being on the inner side, ensure their protection from the aqueous environment.

108. Circulation

109. Secondary cell wall

110. Options (a) is correct.

Endomembrane System (Endoplasmic Reticulum, Golgi Apparatus, Lysosomes, Vacuoles)

111. Endomembrane system includes those membranous cell organelles whose formation and functions are coordinated.

112. Endomembrane constitutes ER, Golgi apparatus, lysosomes and vacuoles.

113. Phosphorylation is a function of mitochondria.

114. Lysosomes

115. Peroxysomes are not a part of the endomembrane system.

116. Golgi apparatus

117. Contractile vacuole performs the function of osmoregulation and excretion.

118. (A) (B) and (D) are correct. SER takes part in detoxification of toxic chemicals in the body.

119. 0.5–1.0 micrometre

120. Anthocyanin is a water-soluble pigment present in sap vacuoles. It provides colouration to different flowers.

121. Lysosomes are newly pinched off vesicles from the trans face of the Golgi apparatus.

122. Lysosomal enzymes

123. Lipochondria perform temporary storage of lipids found in the Golgi apparatus.

124. SER performs detoxification of drugs/xenobiotics by using cytochrome P450.

125. Golgi apparatus is involved in the synthesis of glycoprotein and glycolipids.

126. Vacuoles contain water, sap, excretory products and other materials not useful for the cells.

127. Osmoregulation and excretion

128. Oxidative phosphorylation is a function of mitochondria.

129. A number of proteins are synthesised by RER and modified in the cisternae of Golgi apparatus.

130. Only (A) is correct. Solutes are transported from low to high concentration with the expenditure of energy.

131. Vacuole is bound by a single-membrane tonoplast.

132. Vacuole

133. Anthocyanin provides colouration to different flowers.

134. Ribosomes remain attached to ER with the help of two glycoproteins, ribophorin I and II.

135. Sap vacuoles contain water, sap and other materials not useful for the cell.

136 (b) and (d) are correct. Lysosomes are produced at the trans face or concave face of the Golgi apparatus. Lysosomes contain 40–50 types of hydrolytic enzymes which are active at acidic pH.

137. A is the correctly matched pair. B is rough endoplasmic reticulum which is the major site of protein synthesis. C is 80S ribosomes attached by larger subunits. D is SE, which are frequently observed in cells which are actively involved in lipid synthesis.

138. Lysosomes are rich in acid hydrolases which work best at acidic pH.

139. Ribosomes (Palade particles)

140. 80S ribosomes attach to ER by their larger subunit, 60S.

141. Secretory vesicles are produced at the trans face of the Golgi apparatus.

142. Plant cells usually contain smaller arrays of Golgi-type vesicles called dictyosomes.

Mitochondria, Plastids, Ribosomes

143. Human RBCs lack mitochondria.

144. Mitochondria without outer membrane is called mitoplast.

145. Typically, mitochondria is sausage-shaped, having a diameter of 0.2–1.0 micrometre and length 1.0–4.1 micrometre.

146. Oxidative phosphorylation is the synthesis of ATP by phosphorylation of ADP for which energy is obtained by elution transport and which takes place in mitochondria during aerobic respiration.

147. A – Outer membrane, B – Crista, C – Matrix, D – Inter membrane space

148. Electron transport chain is present in the inner mitochondrial membrane.

149. Ribosomes are the granular structures first observed under the electron microscope.

150. The cristae and the inner face of the inner membrane are studded with numerous spherical or knob-like structures called elementary particles.

151. (a), (b), (c), (d) and (e)

152. Photosynthetic pigments are present in the thylakoid membrane.

153. All statements are correct.

154. a(iv), b(i), c(ii), d(iii)

155. In rRNA, r strands for ribosomal.

156. (A) (B) and (C) are correct.

157. Ribosomes synthesize protein in the cytoplasm and on the RER.

158. Ribosomes are not membrane enclosed instead they are macromolecules made up of both RNA and proteins.

159. Non-secretory proteins are synthesized by cytoplasmic ribosomes.

160. Both (A) and (B) are correct.

161. The sedimentation coefficient is indirectly a measure of density and size.

162. Depending upon their colour and pigments, plastids are of three types: leucoplasts, chromoplasts and chloroplasts.

163. Mitochondrial DNA constitutes 1% of total cellular DNA.

164. Photophosphorylation reaction occurs in the thylakoid membrane.

165. In chromoplasts, fat-soluble carotenoid pigments like carotene, xanthophylls and others are present.

166. The two subunits of 80S ribosomes are 40S and 60S.

167. Ribosomes were first observed under electron microscope.

168. A – 1, B – 20 to 40

169. Svedberg unit

170. Mitochondria

171. RBCs 32.

172. Starch, fat and protein

173. Mitochondria divide by binary fission.

174. Four features are associated with chloroplasts: double membranous, thylakoids, stroma, RuBisCO. In chloroplasts, 70S ribosomes and ds circular DNA are present.

175. Only (B) and (C) are the mitochondria sites of aerobic respiration. The matrix of the mitochondria possesses a single circular DNA molecule, a few RNA molecules and 70S ribosomes.

176. Plastids are found in plant cells.

Cytoskeleton, Cilia, Flagella, Centrosome and Centriole

177. B and C are false. Bacterial flagella do not show 9+2 pattern of microtubules.

178. Centriole forms the basal body of eukaryotic flagella.

179. The core of bacterial flagellum is axoneme.

180. Cytoskeleton is an elaborate network of filamentous proteinaceous structures present in the cytoplasm of eukaryotic cells. They are made up of microtubules and microfilament.

181. Eukaryotic cilia or flagella show (9+2) arrangement of microtubules. Nine microtubule doublets arranged in a circle around two central microtubules.

182. Only (B) and (C) are correct. Eukaryotic cilia or flagella are hair-like outgrowths of the cell membrane. Its basal body is similar to centriole and shows 9+0 organisation of microtubules.

183. 9+0 organisation of microtubules

184. ATPase activity is present in dynein protein

185. Centriole

186. In cilia or flagella, two central singlet microtubules are connected by a central bridge.

187. 27 (9 triplet microtubules)

188. Centrioles are surrounded by amorphous pericentriolar materials for growth of microtubules during formation of new centrioles.

189. Central sheath and central bridge are part of flagellum.

190. Microfilament

191. A – Hollow, B – 25 nm

192. Cytoskeleton includes microtubules, microfilament and intermediate filament.

193. Phragnoplast helps in cell plate formation.

194. In a centrosome, two centrioles lie at approximately 90° to each other.

195. Cytoskeleton in a cell is involved in many functions such as mechanical support, motility, and maintenance of the shape of the cell.

196. Cytoskeleton is an elaborate network of filamentous proteinaceous structures.

Nucleus, Chromosomes, Microbodies

197. *Acetobacteria* is unicellular, uni-nucelated, differentiated green algae.

198. Mature mammalian RBC and mature sieve tube cells lack nucleus.

199. A – Flemming, B – Basic dyes

200. 10–50 nm

201. 2 metre, 46 chromosomes

202. Nucleolus is a naked (non–membrane-bound) rounded structure present in the nucleoplasm.

203. Only (A) is correct. Nuclear pores are a passage through which movement of RNA and protein takes place in both directions between the nucleus and cytoplasm.

204. Three statements (a, b and d) are correct. Chromatin material can be stained by basic dyes.

205. Eukaryotic genomic DNA is a double-stranded linear molecule.

206. Polytene chromosomes are found in the salivary glands of insects.

207. Nucleolus acts as a site of synthesis of rRNA (ribosomal subunits).

208. Only (A) is correct. Larger and more numerous nucleoli are present in cells actively carrying out protein synthesis.

209. Allosomes are called sex chromosomes.

210. All statements are correct.

211. Metaphase is the best stage to study the morphology of chromosomes.

212. In aerocentric chromosomes, the centromere is situated close to its end, forming one extremely short and one very long arm.

213. Telomeres are the ends of chromosomes and seal the ends of chromosomes.

214. C – Value

215. Polytene chromosomes show somatic pairing.

216. Idiogram is the diagrammatic representation of karyotype, where the somatic metaphase of chromosomes is arranged in decreasing order of their size.

217. Metacentric chromosomes are isobrachial chromosomes, that is, they have two equal arms because the centromere is median.

218. (a) and (c) are correct. Euchromatin is a loosely packed region of DNA.

219. Short arm is designated as p-arm and long is q-arm.

220. Polytene chromosomes were first reported by EG Balbiani in the larval salivary glands of *Chironomus* (an insect of order Diptera).

221. In idiogram, chromosomes of an organism are arranged according to decreasing size.

222. Centromere

223. Polytene chromosomes bear a number of dark bands and chromosomal puffs (Balbiani rings). These Balbiani rings are active sites of RNA and protein synthesis.

224. Lamp brush chromosomes are observed in the diplotene phase of Meiosis I.

225. All are correct.

226. *Drosophila* (diplotene chromosome)

227. Lamp brush chromosomes are found in the primary oocyte nucleus of vertebrates and invertebrates.

228. Depending upon the position of the centromere (primary constriction), chromosomes are of four types: metacentric, submetacentric, aerocentric and telocentric.

229. Peroxysomes

230. Glyoxysomes (glyoxylate cycle)

231. Metacentric – V shaped, Telocentric – I shaped, Aerocentric – J shaped, Submetacentric – L shaped

232. Nuclear membrane is derivative of RER.

233. Metacentric chromosomes are isobrachial (equal arms).

234. Actinobacterial origin

235. Peroxysomes contain oxidative enzymes.

236. Satellite

237. Microbodies are membrane-bound minute vesicles, present in both plant and animal cells.

238. A – Satellite, B – Secondary constriction, C – Centromere, D – Kinetochore

239. Kinetochore

240. A – Sat chromosome, D – Acrocentric chromosome

241. Option (d) is correct.

242. Proteins are synthesized in cell cytoplasm by ribosomes.

Biomolecules

EXERCISES

The Cellular Pool, Primary and Secondary Metabolites

1. If we observe the average composition of the cell, then:
   (a) Percentage of lipids is more than carbohydrates
   (b) Percentage of nucleic acid is more than proteins
   (c) Water constitutes 70–90% of a living cell
   (d) All of the above are correct

2. Biomicromolecules:
   (a) Are large sized
   (b) Have molecular weight of more than 10,000 daltons
   (c) Are found in acid soluble fraction
   (d) Are always polymeric in nature

3. ______ are considered as macromolecules though their molecular weight is less than 800 daltons:
   (a) Proteins (b) Lipids
   (c) Nucleic acids (d) Polysaccharides

4. Identify the incorrect match with respect to the average composition of a living cell:
   (a) Nucleic acid – 5–7%
   (b) Carbohydrates – 3%
   (c) Protein – 10–15%
   (d) Lipids – 20–22%

5. All of the following are primary metabolites, except:
   (a) Glycine (b) Fructose
   (c) Adenylic acid (d) Cellulose

6. Select the incorrect match with respect to secondary metabolites:
   (a) Alkaloid – Codeine
   (b) Drug – Vinblastine
   (c) Toxin – Lemongrass oil
   (d) Lectin – Concanavalin A

7. Identity the molecule which will be found in the retentate fraction after treating tissue with trichloroacetic acid:
   (a) Serine (b) Maltose
   (c) Triglyceride (d) Guanosine

8. Which of the following secondary metabolites are polymers?
   (a) Abrin, Ricin
   (b) Morphine, Codeine
   (c) Vinblastine, Curcumin
   (d) Rubber, Cellulose

9. Except ______, other macromolecules are formed by polymerisation of subunits called monomers:
   (a) Polysaccharides (b) Lipids
   (c) Nucleic acids (d) Proteins

10. Which of the following is not found in acid insoluble fractions?
    (a) Collagen (b) Ribonucleic acid
    (c) Cellulose (d) Serine

11. Inorganic compounds like sulphate and phosphate, which become available in the filtrate after grinding the living tissue in trichloroacetic acid, represent:
    (a) Acid soluble pool
    (b) Acid insoluble pool
    (c) Alkali soluble pool
    (d) Alkali insoluble pool

12. The inorganic compounds and elements in living tissue are analysed by:
    (a) Treating the tissue with alkali
    (b) Treating the tissue with chloroform
    (c) Treating the tissue with trichloroacetic acid
    (d) Drying and burning the tissue so that all carbon compounds are oxidised to gaseous form and the remaining ash contains inorganic compounds

13. Which of the following substances is present in minimum percentage of the total cellular mass?
(a) Proteins (b) Ions
(c) Lipids (d) Carbohydrates

14. Arrange the following elements in ascending order of the percentage weight of the human body:
(a) Hydrogen < Calcium < Nitrogen < Silicon < Oxygen
(b) Silicon < Nitrogen < Oxygen < Hydrogen < Calcium
(c) Hydrogen < Calcium < Nitrogen < Carbon < Oxygen
(d) Hydrogen < Nitrogen < Oxygen < Carbon < Calcium

15. During chemical analysis of a cell:
(a) The acid soluble fraction is called filtrate
(b) The acid insoluble fraction is called retentate
(c) Biomicromolecules have molecular weight less than 1000 daltons
(d) All of the above are correct

16. Select the incorrect statement:
(a) The relative abundance of carbon and hydrogen with respect to other elements is higher in living organisms than the earth's crust
(b) The percentage weight of silicon in the earth's crust is very high in comparison to the human body
(c) The percentage weight of nitrogen in the earth's crust is very high in comparison to the human body
(d) All the elements present in a sample of the earth's crust are also present in a sample of living tissue

17. When we homogenise any tissue in acid, the acid soluble pool represents:
(a) Cytoplasm (b) Cell membrane
(c) Mitchondria (d) Nucleus

18. Identify the incorrect statement:
(a) The molecules in the acid soluble pool have molecular weight ranging from 18 to 800 daltons
(b) The molecular weight of lipids is less than 800 daltons
(c) When we grind a tissue, the cell membranes are broken into pieces that are water soluble
(d) The macromolecules from cytoplasm and organelles become the acid insoluble fraction

19. Read the following statements and choose the correct option:
I: Primary metabolites have identifiable functions and play known roles in normal physiological processes
II : The role or function of all secondary metabolites is known to us
(a) Both statements are correct
(b) Both statements are incorrect
(c) Only I is correct
(d) Only II is correct

20. Which of the following secondary metabolites are pigments?
(i) Morphine
(ii) Abrin
(iii) Anthocyanin
(iv) Carotenoids
(v) Gums
(a) (i) and (ii) (b) (iii) and (iv)
(c) (iv) and (v) (d) (i) and (v)

Carbohydrates

21. Which of the following is not correct with respect to glycogen?
(a) Homopolysaccharide
(b) Heteropolysaccharide
(c) Stored in liver and muscle
(d) Branched chain molecule

22. Select the incorrect statement with respect to sucrose:
(a) It is a disaccharide
(b) It is non-reducing
(c) It is composed of fructose and galactose
(d) It accumulates in the cytoplasm

23. Which of the following is a ketone sugar?
(a) Glucose (b) Fructose
(c) Ribose (d) Galactose

24. ______ is a natural polymer found in insects and fungi:
(a) Cellulose (b) Pectin
(c) Suberin (d) Chitin

25. Which of the following sugars cannot be hydrolysed further to yield simple sugar?
(a) Maltose (b) Sucrose
(c) Lactose (d) Mannose

26. The monosaccharides are linked by ______ in a polysaccharide. Select the option to fill in the blank correctly:
(a) Peptide bond
(b) Glycosidic bond
(c) Hydrogen bond
(d) Phosphodiester bond

27. The monomer unit of cellulose is:
(a) α-D glucose (b) β-D glucose
(c) α-D galactose (d) β-D fructose

28. Select the incorrect statement:
(a) Mannitol is a sugar alcohol
(b) Chitin is a nitrogen containing polysaccharide
(c) The right end of a polysaccharide chain is non-reducing
(d) Sucrose is commonly called cane sugar

29. Lactose is made up of:
(a) Glucose–glucose
(b) Glucose–galactose
(c) Glucose–fructose
(d) Fructose–galactose

30. Select the incorrect statement:
(a) Starch is a polymer of α-glucose
(b) Starch is made up of amylose and amylopectin
(c) Amylose is a helical chain of glucose molecules arranged linearly by 1,4-glycosidic bonds
(d) Amylopectin is a straight chain with several glucose residues joined only by 1,4-glycosidic linkages

31. Which of the following is/are not applicable to inulin?
(a) It is a heteropolymer
(b) It is used to estimate glomerular filtration rate
(c) It has fructose molecules linked by glycosidic bond
(d) Both (b) and (c)

32. Which of the following polysaccharides is called animal starch?
(a) Glycogen (b) Chitin
(c) Collagen (d) Agar

33. Which of the following derivatives of cellulose is called artificial silk?
(a) Cellulose xanthate
(b) Cellulose acetate
(c) Cellulose nitrate
(d) Carboxymethyl cellulose

34. Ribose is different from deoxyribose in having:
(a) Ketone group
(b) No oxygen
(c) One extra oxygen at the second carbon
(d) Aldehyde group

35. Select the incorrect match:
(a) Grape Sugar – Glucose
(b) Milk Sugar – Lactose
(c) Brain Sugar – Galactose
(d) Malt Sugar – Sucrose

36. The mucopolysaccharide used for treatment of constipation is:
(a) Hyaluronic acid
(b) Agar-agar
(c) Chondroitin sulphate
(d) Husk of *Plantago ovata*

37. Identify the correct set of non-reducing disaccharides:
(a) Sucrose and maltose
(b) Sucrose and trehalose
(c) Lactose and maltose
(d) Maltose and trehalose

38. Which of the following is not a pentose sugar?
(a) Ribose (b) Xylose
(c) Arabinose (d) Erythrose

39. Cellulose nitrate is present in:
(a) Rayon and cellophane
(b) Cigarette filters
(c) Gun cotton
(d) Ice cream

40. Choose the correct set of carbohydrates belonging to the same category:
(a) Ribose, lactose, fructose, maltose
(b) Sucrose, lactose, maltose, trehalose
(c) Maltose, xylose, glyceraldehyde, cellobiose
(d) Starch, fructose, inulin, lactose

41. Polysaccharides are not used for:
(a) Providing energy
(b) Storing energy
(c) Structural support
(d) Storing genetic information

42. Select the incorrect statement:
(a) Chitin is called fungal cellulose
(b) Cellulose is the most abundant organic molecule in nature
(c) Starch, inulin and cellulose are homopolysaccharides
(d) Ribulose and xylulose are aldehyde sugars

43. Select the odd one out with respect to monosaccharides:
(a) Ribose (b) Xylose
(c) Trehalose (d) Mannose

44. Inulin is a low-molecular-weight polysaccharide obtained from the roots of:
(a) *Dahlia* (b) *Raphanus*
(c) *Colchicum* (d) *Brassica*

45. Benedict's reagent test is conducted to confirm the presence of:
(a) Protein (b) Lipids
(c) Reducing sugars (d) Nucleic acid

Proteins

46. Which of the following is an aromatic amino acid?
(a) Alanine (b) Serine
(c) Tryptophan (d) Aspartic acid

47. Which of the following options consists of non-essential amino acids only?
(a) Phenylalanine, methionine, valine
(b) Leucine, isoleucine, tryptophan
(c) Tyrosine, glycine, serine
(d) Threonine, valine, leucine

48. Identify the incorrect statement:
(a) In a polypeptide, amino acids are linked by peptide bonds
(b) Amino acids are substituted methanes
(c) Glycine is a sulphur containing amino acid
(d) Amino acids have an amino group and carboxylic acid group as substituents on the α-carbon

49. Identify the structure given below:

$$\begin{array}{c} NH_2 \\ | \\ H-C-COOH \\ | \\ CH_2OH \end{array}$$

(a) Glycine (b) Alanine
(c) Serine (d) Tyrosine

50. Which structural level enables the protein to function as enzymes?
(a) Primary (b) Secondary
(c) Tertiary (d) Quaternary

51. Identify the incorrect statement with respect to proteins:
(a) The sequence of amino acids in a protein represents the primary structure
(b) Proteins are heteropolymers of amino acids
(c) Adult human haemoglobin consists of two subunits
(d) The helices of proteins are always right handed

52. ________ is an example of an amino acid with a sulphur group in its side chain:
(a) Tryptophan (b) Methionine
(c) Valine (d) Isoleucine

53. The most abundant protein in the animal world is:
(a) Collagen (b) Insulin
(c) Trypsin (d) Myoglobin

54. The alpha helices and beta-pleated sheets represent the ______ structure of proteins:
(a) Primary (b) Secondary
(c) Tertiary (d) Quaternary

55. Which one is an imino acid?
(a) Cysteine (b) Proline
(c) Valine (d) Phenylalanine

56. Upon denaturation, a protein does not lose its:
(a) Primary structure
(b) Secondary structure
(c) Tertiary structure
(d) Quaternary structure

57. Select the incorrect statement:
(a) Glycine is the simplest amino acid
(b) Essential amino acids are not synthesized in our body
(c) Alanine is a mono-aminic, mono-carboxylic amino acid
(d) Secondary structures are stabilised by ester bonds

58. Proteins of cereals and millets are deficient in:
(a) Lysine (b) Cysteine
(c) Tryptophan (d) Both (a) and (c)

59. The first amino acid in protein is present at the:
(a) C-terminal end (b) N-terminal end
(c) Right end (d) Both (a) and (c)

60. ________ structure is absolutely necessary for many biological activities of proteins:
(a) Primary (b) Secondary
(c) Tertiary (d) Straight

61. Which of the following proteins functions as intercellular ground substance?
(a) Insulin (b) Trypsin
(c) Collagen (d) GLUT-4

62. Which of the following amino acids takes part in the synthesis of the thyroxine hormone?
(a) Lysine (b) Glycine
(c) Tyrosine (d) Tryptophan

63. Identify the basic amino acid:
(a) Serine (b) Lysine
(c) Aspartic acid (d) Tryptophan

64. Select the incorrect statement:
(a) Dietary proteins are a source of essential amino acids
(b) GLUT-4 enables glucose transport into cells
(c) Twenty types of amino acids are found in proteins
(d) Secondary structure gives a three-dimensional view of a protein

65. A linear string of spheres or spheres arranged one upon each other in the form of a cube or plate represents which of the following protein structures?

(a) Primary structure
(b) Secondary structure
(c) Tertiary structure
(d) Quaternary structure

66. When a hydroxy methyl group is the side chain attached to the α-carbon in an amino acid, the resultant amino acid is:
(a) Glycine (b) Serine
(c) Cysteine (d) Alanine

67. Tryptophan is not involved in the synthesis of:
(a) Melatonin
(b) Vitamin nicotinamide
(c) Adrenaline
(d) Plant hormone indole-3-acetic acid

68. A protein having a single polypeptide chain cannot have ____ structure:
(a) Primary (b) Secondary
(c) Tertiary (d) Quaternary

69. All of the following are fibrous proteins, except:
(a) Keratin (b) Elastin
(c) Collagen (d) Haemoglobin

70. The bond common to secondary, tertiary and quaternary structures of the protein is:
(a) Ionic bond
(b) Hydrogen bond
(c) Disulphide bond
(d) Electrostatic bond

71. The skin pigment melanin is synthesised from which of the following amino acids?
(a) Alanine (b) Glycine
(c) Tyrosine (d) Tryptophan

72. The long protein chain is folded upon itself like a hollow woollen ball, giving rise to the:
(a) Primary structure
(b) Secondary structure
(c) Tertiary structure
(d) Quaternary structure

73. The single letter code 'E' refers to the amino acid:
(a) Serine (b) Cysteine
(c) Tyrosine (d) Glutamic acid

74. Protein amino acids are asymmetrical and levorotatory, except:
(a) Serine (b) Alanine
(c) Glycine (d) Leucine

75. Select the incorrect statement:
(a) RuBisCO is the most abundant protein in the biosphere
(b) The last amino acid in a polypeptide chain is called the C-terminal amino acid
(c) Lysine and arginine are basic amino acids found in histone proteins
(d) Cysteine and methionine are aromatic amino acids

76. The 'R' group in amino acid alanine is:
(a) Hydrogen
(b) Methyl group
(c) Ethyl group
(d) Hydroxy methyl group

77. Identify the incorrect match with respect to proteins and their corresponding functions:

Protein	Functions
(a) Collagen	Intracellular ground substance
(b) Pepsin	Enzyme
(c) Glucagon	Hormone
(d) Antibody	Fights infectious agents

78. Plant proteins, like those of pulses, are deficient in:
(a) Lysine and tryptophan
(b) Cysteine and methionine
(c) Phenylalanine and tyrosine
(d) Serine and threonine

Lipids

79. Identify the non-essential fatty acid:
(a) Linoleic acid (b) Linolenic acid
(c) Oleic acid (d) Arachidonic acid

80. How many double bonds are present in arachidonic acid?
(a) 1 (b) 2
(c) 3 (d) 4

81. Which of the following fatty acids has 20 carbon atoms including the carboxyl carbon?
(a) Palmitic acid (b) Stearic acid
(c) Oleic acid (d) Arachidonic acid

82. Which of the following is a phosphorylated glyceride found in the cell membrane?
(a) Lecithin
(b) Paraffin wax
(c) Trihydroxypropane
(d) Adenylic acid

83. A typical fat is made up of:
(a) One glycerol and one fatty acid molecule
(b) Three glycerol and three fatty acid molecules

(c) Three glycerol and one fatty acid molecules
(d) One glycerol and three fatty acid molecules

84. Cholesterol is a precursor for each of the following, except:
(a) Testosterone (b) Bile salts
(c) Vitamin D (d) Insulin

85. Identify the saturated fatty acid from the following:
(a) Oleic acid (b) Palmitic acid
(c) Linoleic acid (d) Arachidonic acid

86. Identify the incorrect statement:
(a) Glycerol is trihydroxy propane
(b) Oils have a higher melting point than hard fats
(c) Stearic acid has 18 carbon atoms including the carboxyl carbon
(d) Lipids are insoluble in water and soluble in organic solvents like acetone and chloroform

87. Identify the incorrect match:
(a) Phospholipid – Lecithin
(b) Lanolin – Wax
(c) Glutamic acid – Essential fatty acid
(d) Oleic acid – Unsaturated fatty acid

88. The wax present in the cell wall of bacteria causing leprosy and tuberculosis is:
(a) Paraffin wax (b) Wax-D
(c) Cerumen (d) Lanolin

89. The given chemical structure is of:
$CH_3 - (CH_2)_{14} - COOH$
(a) Oleic acid (b) Stearic acid
(c) Palmitic acid (d) Linolenic acid

90. A large group of lipids that are formed by repeating units called isoprene are known as:
(a) Prostaglandins (b) Terpenoids
(c) Hard fats (d) Lecithin

91. Which of the following is not a part of lecithin?
(a) Fatty acid (b) Glycerol
(c) Cholesterol (d) Choline

92. In animals, cutaneous glands secrete a waxy water insoluble coating on animal fur called:
(a) Lanolin (b) Cerumen
(c) Beeswax (d) Wax-D

93. Which of the following is incorrect with respect to saturated fatty acids?
(a) Lack double bonds
(b) Form hard fats that are solid at room temperature
(c) Undergo hydrogenation reaction
(d) Tend to increase blood pressure

94. Arachidonic acid gives rise to:
(a) Terpenes (b) Prostaglandins
(c) Beeswax (d) Cholesterol

95. Deficiency of essential fatty acids causes:
(a) Phrenoderma
(b) Follicular hyperkeratosis
(c) Both of the above
(d) None of the above

96. A diglyceride differs from a monoglyceride in:
(a) Having two molecules of glycerol
(b) Possessing one additional fatty acid
(c) Having only saturated fatty acid
(d) Having cholesterol instead of glycerol

97. Lipids are insoluble in water because lipid molecules are:
(a) Hydrophilic (b) Hydrophobic
(c) Neutral (d) Zwitter ions

98. Identify the incorrect statement with respect to oils:
(a) They have low melting points
(b) They are liquid at room temperature
(c) Undergo rancidity upon prolonged storage
(d) They are formed by esterification of saturated fatty acids with glycerol

99. Edible oils are converted into hard fats through a process called:
(a) Sublimation (b) Hydrogenation
(c) Drying (d) Suspension

100. Select the incorrect statement:
(a) Camphor, menthol and gibberellins are examples of terpenes
(b) High density lipoproteins (HDL) reduce the risk of atherosclerosis
(c) Beeswax is formed of palmitic acid with meri-cyl alcohol
(d) Lipids are obtained in acid soluble fraction while analysing the chemical composition of the cell

Nucleic Acids

101. The phosphate group is attached to carbon _____ of the sugar residue of its nucleotide:
(a) 1 (b) 2
(c) 4 (d) 5

102. Which of the following is not a nucleoside?
(a) Adenosine (b) Guanosine
(c) Cytosine (d) Uridine

103. The difference between DNA and RNA is because of:
(a) Sugar only
(b) Phosphate and nitrogen base
(c) Sugar and phosphate
(d) Nitrogen base and sugar

104. Which of the following is a nucleotide?
(a) Cytosine (b) Uridine
(c) Adenosine (d) Guanylic acid

105. The nitrogen base not found in DNA is:
(a) Adenine (b) Guanine
(c) Thymine (d) Uracil

106. Select the incorrect statement:
(a) The two strands of DNA are antiparallel
(b) Phosphodiester bond is present between nitrogen base and pentose sugar molecule in a nucleotide
(c) The basic unit of nucleic acid is nucleotide
(d) ATP is the universal energy carrier

107. _____ bond is present between two nucleotides of a polynucleotide:
(a) Covalent (b) Hydrogen
(c) Phosphodiester (d) Disulphide

108. The two strands of DNA are attached to each other by _____ bonds between nitrogen bases:
(a) Covalent (b) Ester
(c) Hydrogen (d) Disulphide

109. Nucleic acids exhibit a wide variety of:
(a) Primary structures
(b) Secondary structures
(c) Tertiary structures
(d) Quaternary structures

110. According to the Watson–Crick model, all of the following are true, except:
(a) DNA exists as a double helix
(b) The two strands of DNA are antiparallel
(c) The backbone of DNA is formed by a sugar-phosphate-sugar chain
(d) The nitrogen bases are projected parallel to the backbone but face outside

111. In a nucleic acid, a phosphate moiety links the:
(a) 3'-carbon of one sugar of one nucleotide to the 5'-carbon of the sugar of the succeeding nucleotide
(b) 5'-carbon of one sugar of one nucleotide to the 3'-carbon of the sugar of the succeeding nucleotide
(c) 3'-carbon of one sugar of one nucleotide to the 3'-carbon of the sugar of the succeeding nucleotide
(d) 5'-carbon of one sugar of one nucleotide to the 5'-carbon of the sugar of the succeeding nucleotide

112. If we draw a line diagram of DNA:
(a) The pitch would be 3.4 nm
(b) The rise per base pair would be 0.34 nm
(c) The pitch would be 0.34 nm
(d) Both (a) and (b) are correct

113. Choose the incorrect statement with respect to B-DNA:
(a) There are 10 base pairs per turn
(b) At each step of the ascent, the DNA strand turns 36°
(c) One turn of the helical structure measures 3.4 nm
(d) The distance between two strands of DNA is 0.34 nm

114. Compared to nucleosides, a nucleotide has:
(a) Acidic nature
(b) Additional sugar
(c) Additional phosphate
(d) Both (a) and (c)

115. Identify the incorrect statement with respect to DNA:
(a) A and G of one strand compulsorily base pairs with C and T, respectively, on the other strand
(b) There are two hydrogen bonds between A and C
(c) There are three hydrogen bonds between G and T
(d) All of the above are incorrect

116. Select the incorrect statement:
(a) Adenine and guanine are substituted purines
(b) Uracil and cytosine are substituted pyrimidines
(c) Nucleic acid containing ribose is ribonucleic acid (RNA)
(d) The nitrogen bases found in DNA are thiamine, cytosine, adenine and guanine

117. Identify the given structure:

HOCH₂ Adenine

OH OH

(a) Adenylic acid
(b) Adenosine
(c) Adenosine mono phosphate
(d) None of the above

118. Identify the labels A and B:

(A) (B)

	A	B
(a)	**Adenine**	**Thymine**
(b)	Guanine	Uracil
(c)	Guanine	Cytosine
(d)	Adenine	Uracil

119. Which of the following is an incorrect statement?
(a) Nitrogen bases are heterocyclic compounds
(b) In a nucleotide, nitrogen bases are attached to carbon atom 1 of the sugars by glycosidic bond
(c) Ribose differs from deoxyribose in having hydroxyl group instead of hydrogen at the third carbon
(d) Pyrimidine has a single ring structure

120. Among the nitrogen bases involved in DNA and RNA formation, the double ring structure is found in:
(a) Uracil (b) Thymine
(c) Guanine (d) Cytosine

Concept of Metabolism

121. Select the incorrect statement:
(a) Metabolites are converted into each other in a series of linked reactions called metabolic pathways
(b) Metabolic pathways are similar to automobile traffic in a city
(c) Metabolic pathways are either linear or circular
(d) Flow of metabolites through a metabolic pathway lacks definite rate and direction

122. In metabolic reactions, every chemical reaction is a ______ reaction:
(a) Uncatalysed (b) Catalysed
(c) Anabolic (d) Both (a) and (c)

123. The proteins with catalytic power are named:
(a) Metabolites (b) Enzymes
(c) Catabolites (d) Bioenergetics

124. The formation of cholesterol from acetic acid is an example of:
(a) Biosynthetic pathway
(b) Anabolic pathway
(c) Catabolic pathway
(d) Both (a) and (b)

125. Identify the incorrect statement:
(a) Anabolic pathways consume energy
(b) Degradation of glucose into lactic acid is an example of catabolic pathway
(c) Living organisms store energy in the form of chemical bonds
(d) Anabolic pathway forms a simple structure from complex structures

126. Read the following statements:
(i) All biomolecules have a turnover
(ii) Breaking and making of biomolecules occurs through chemical reactions called metabolism
(iii) There are some uncatalysed metabolic conversions
(iv) The catalysts which hasten the rate of a given metabolic conversion are called lipids.

Which of the above statements are correct?
(a) (i) and (ii) (b) (ii) and (iii)
(c) (iii) and (iv) (d) (i) and (iv)

127. The most important form of energy currency in a living system is the bond energy in a chemical called:
(a) Adenosine diphosphate
(b) Adenosine triphosphate
(c) Adenosine monophosphate
(d) Adenine diphosphate

128. The living state is a **I** steady-state to be **II** to perform work. Fill in the blanks with correct options:

	I	II
(a)	Equilibrium	Unable
(b)	Non-equilibrium	Able
(c)	Non-equilibrium	Unable
(d)	Equilibrium	Able

129. Which of the following is an incorrect statement?
(a) Living process is a constant effort to prevent falling into equilibrium
(b) Metabolism provides a mechanism for the production of energy
(c) The living state and metabolism are non-synonymous
(d) Without metabolism, there cannot be a living state

130. Select the incorrect statement:
(a) The blood concentration of glucose in a normal healthy individual is 4.2–6.1 mM
(b) Any chemical or physical process moves spontaneously to equilibrium
(c) Systems at equilibrium perform work efficiently
(d) The energy in chemical bonds is utilised for the biosynthetic, osmotic and mechanical work that we perform

Enzymes

131. An active site of an enzyme is a crevice or pocket into which ______ fits:
(a) Inorganic group (b) Coenzyme
(c) Cofactor (d) Substrate

132. Almost all enzymes are:
(a) Nucleic acids (b) Proteins
(c) Carbohydrates (d) Inorganic ions

133. Which of the following is incorrect with respect to the differences between inorganic catalysts and enzymes?
(a) Inorganic catalysts work efficiently at high temperature
(b) All enzymes work efficiently at high temperature
(c) Inorganic catalysts work efficiently at high pressure
(d) Enzymes of thermophilic organisms retain their catalytic power even at high temperature

134. A non-proteinaceous enzyme is:
(a) DNA ligase (b) Trypsin
(c) Ribozyme (d) Hexokinase

135. An allosteric inhibitor of the enzyme acts by binding to the:
(a) Catalytic site (b) Non-catalytic site
(c) Substrate (d) Active site

136. The protein component of a holoenzyme is:
(a) Apoenzyme (b) Cofactor
(c) Coenzyme (d) Holoenzyme

137. The essential chemical components of many coenzymes are:
(a) Vitamins (b) Nucleic acids
(c) Proteins (d) Lipids

138. A competitive inhibitor of succinic dehydrogenase is:
(a) Fumarate (b) Malate
(c) Malonate (d) Isocitrate

139. Which of the following is correct?
(a) Apoenzyme = Holoenzyme + Cofactor
(b) Holoenzyme = Cofactor + Apoenzyme
(c) Cofactor = Holoenzyme + Apoenzyme
(d) Prosthetic group = Cofactor + Prosthetic group

140. Transition state structure of the substrate formed during an enzymatic reaction is:
(a) Transient but stable
(b) Transient but unstable
(c) Permanent but stable
(d) Permanent but unstable

141. Identify the incorrect statement with respect to factors affecting enzyme activity:
(a) Enzymes function in a narrow range of temperature and pH
(b) Each enzyme shows highest activity at optimum temperature and pH
(c) High temperature preserves the enzyme in a temporarily active state
(d) Enzyme activity declines both below and above the optimum value of pH and temperature

142. Enzymes catalysing hydrolysis of ester, ether, peptide and glycosidic bonds are:
(a) Lyases (b) Dehydrogenases
(c) Hydrolases (d) Transferases

143. ______ is a cofactor for the proteolytic enzyme carboxypeptidase:
(a) Magnesium (b) Sodium
(c) Copper (d) Zinc

144. ______ are organic groups whose association with the apoenzyme is only transient, usually occurring during the course of catalysis:
(a) Coenzymes (b) Metal ions
(c) Holoenzymes (d) Prosthetic groups

145. The coenzyme nicotinamide adenine dinucleotide (NAD) and NADP contain the vitamin:
(a) Biotin (b) Niacin
(c) Thiamine (d) Riboflavin

146. Identify the enzyme catalysing the following reaction:

$$\underset{|}{\overset{X}{}}\underset{|}{\overset{Y}{}}\quad C{-}C \rightarrow X{-}Y + C{=}C$$

(a) Oxidoreductase (b) Transferase
(c) Hydrolase (d) Lyase

147. Haem is the prosthetic group for:
(i) Catalase
(ii) Hexokinase
(iii) Peroxidase
(iv) Phosphofructokinase
(a) (i) and (ii) (b) (ii) and (iii)
(c) (i) and (iii) (d) (i) and (iv)

148. Select the incorrect statement:
(a) Enzymes are proteins with three-dimensional structure
(b) Zinc is the cofactor for carbonic anhydrase
(c) All enzymes do not use cofactor for their catalytic activity
(d) Enzymes increase the activation energy of a reaction

149. Competitive inhibitors _______ of the reaction:
(a) Increase the V_{max}
(b) Increase the K_m
(c) Decrease the V_{max}
(d) Decrease the K_m

150. The effect of end product on the rate of enzyme catalysed reaction is called:
(a) Lock and key mechanism
(b) Feedback inhibition
(c) Competitive inhibition
(d) Non-competitive inhibition

151. Which of the following enzymes does not belong to the category hydrolase?
(a) Lactase
(b) Amylase
(c) Nuclease
(d) Succinate dehydrogenase

152. The lock and key hypothesis was given by:
(a) Daniel Koshland (b) Emil Fischer
(c) Louis Pasteur (d) Sumner

153. The substrate concentration at which the chemical reaction catalysed by an enzyme attains half of its maximum velocity is termed:
(a) V_{max} (b) K_m
(c) ½ V_{max} (d) ½ K_m

154. _______ was the first enzyme to be crystallised and purified:
(a) Hexokinase (b) Zymase
(c) Maltase (d) Urease

155. Select the type of enzyme involved in the given reaction: $S\text{-}G + S^1 \rightarrow S + S^1\text{-}G$
(a) Dehydrogenase (b) Ligase
(c) Transferase (d) Hydrolase

156. Which of the following statements is correct?
(a) Cofactor is the thermolabile part of the enzyme
(b) Prosthetic group is loosely attached to the apoenzyme
(c) On increasing the substrate concentration, the rate of reaction initially decreases
(d) K_m value indicates the affinity of the enzyme for its substrate

157. The term 'enzyme' was given by:
(a) Kuhne (b) Sumner
(c) Emil Fischer (d) Pasteur

158. The graph given below represents enzymatic activity in relation to which of the following factors?

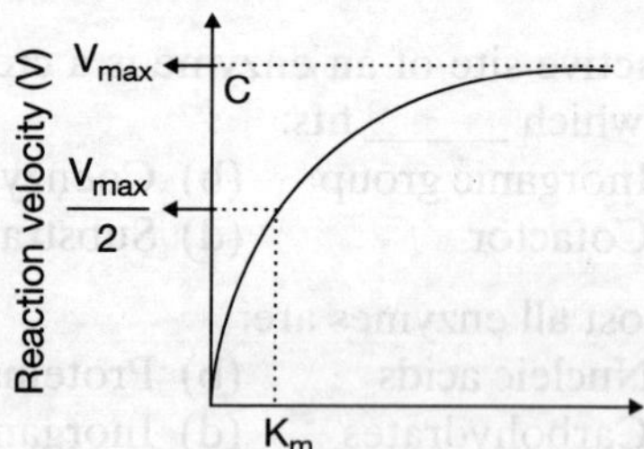

(a) Temperature (b) pH
(c) Substrate (d) Product concentration

159. In the graph given below, the curve b is applicable to which of the following examples?

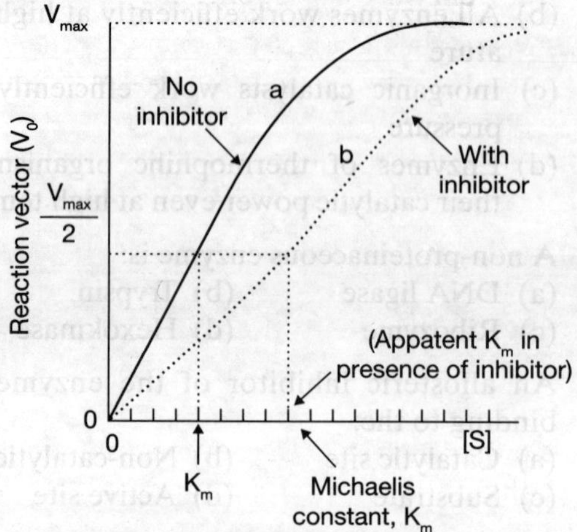

(a) Allosteric inhibition
(b) Competitive inhibition
(c) Non-competitive inhibition
(d) Feedback inhibition

160. The given graph shows the effect of temperature on the rate of reaction of enzyme. What does the graph indicate?

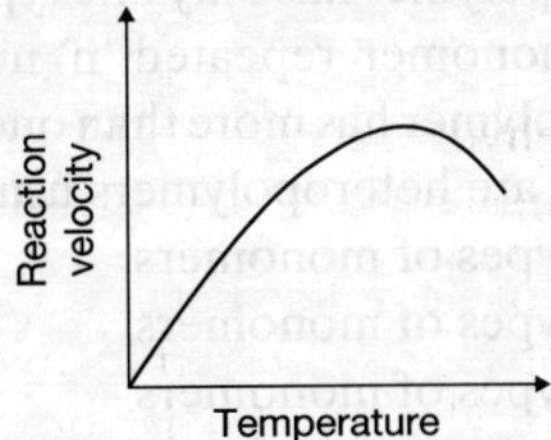

(a) Rate of reaction is highest at minimum temperature
(b) Rate of reaction is highest at maximum temperature
(c) Rate of reaction is not affected by any change in temperature
(d) Rate of reaction is highest at a particular temperature called optimum temperature

161. Which of the following types of enzyme inhibition is used for the control of bacterial pathogens by sulpha drugs?
(a) Allosteric inhibition
(b) Feedback inhibition
(c) Competitive inhibition
(d) Non-competitive inhibition

162. The term feedback refers to:
(a) The effect of substrate on the rate of enzymatic reaction
(b) The effect of end product on the rate of enzymatic reaction
(c) The effect of enzyme concentration on the rate of enzymatic reaction
(d) The effect of external compound on the rate of enzymatic reaction

163. Select the incorrect statement:
(a) Rate of enzyme catalysed reaction doubles or decreases by half for every 10°C change in either direction in the 0–40°C range
(b) Thermal stability is an important quality of enzymes isolated from thermophilic organisms
(c) The fastest enzyme carbonic anhydrase can catalyse 600,000 molecules per second
(d) All the cells of the organism contain the same enzymes

164. Which of the following factor(s) does not effect enzyme activity?
A. Temperature
B. pH
C. Activation energy
D. Enzyme concentration
E. Substrate concentration
F. Product concentration
(a) (C) only (b) (C) and (D)
(c) (E) and (F) (d) (F) only

165. In the graph given below, identify the nature of the reaction with respect to energy kinetics:

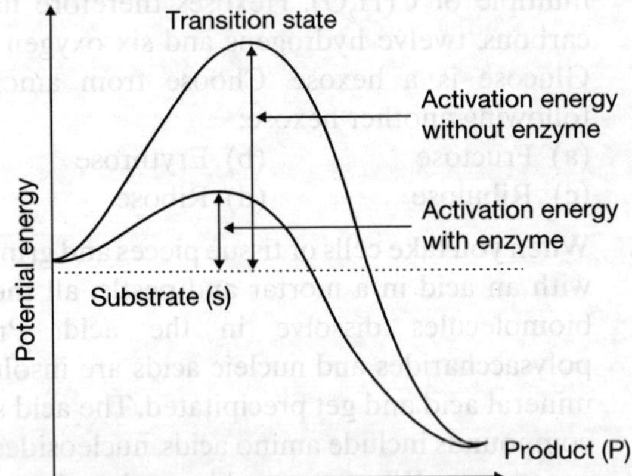

(a) Endothermic
(b) Exothermic
(c) Exothermic first and endothermic later
(d) Varies between the conditions

NCERT EXEMPLAR QUESTIONS

1. It is said that elemental composition of living organisms and that of inanimate objects (like the earth's crust) are similar in the sense that all the major elements are present in both. Then what would be the difference between these two groups? Choose the correct answer from among the following:
(a) Living organisms have more gold in them than inanimate objects
(b) Living organisms have more water in their body than inanimate objects
(c) Living organisms have more carbon, oxygen and hydrogen per unit mass than inanimate objects
(d) Living organisms have more calcium in them than inanimate objects

2. Many elements are found in living organisms either free or in the form of compounds. Which of the following is not found in living organisms?
(a) Silicon (b) Magnesium
(c) Iron (d) Sodium

3. Amino acids have both an amino group and a carboxyl group in their structure. Which one of the following is an amino acid?
(a) Formic acid (b) Glycerol
(c) Glycolic acid (d) Glycine

4. An amino acid under certain conditions can have both positive and negative charges simultaneously in the same molecule. Such a form of amino acid is called:
(a) Acidic form (b) Basic form
(c) Aromatic form (d) Zwitterionic form

5. Sugars are technically called carbohydrates, referring to the fact that their formulae are only multiple of $C(H_2O)$. Hexoses therefore have six carbons, twelve hydrogens and six oxygen atoms. Glucose is a hexose. Choose from among the following another hexose:
(a) Fructose (b) Erythrose
(c) Ribulose (d) Ribose

6. When you take cells or tissue pieces and grind them with an acid in a mortar and pestle, all the small biomolecules dissolve in the acid. Proteins, polysaccharides and nucleic acids are insoluble in mineral acid and get precipitated. The acid soluble compounds include amino acids, nucleosides, small sugars, etc. When one adds a phosphate group to a nucleoside, one gets another acid soluble biomolecule called:
(a) Nitrogen base (b) Adenine
(c) Sugar phosphate (d) Nucleotide

7. When we homogenise any tissue in an acid, the acid-soluble pool represents:
(a) Cytoplasm (b) Cell membrane
(c) Nucleus (d) Mitochondria

8. The most abundant component of living organisms is:
(a) Protein (b) Water
(c) Sugar (d) Nucleic acid

9. A homopolymer has only one type of building block called monomer repeated 'n' number of times. A heteropolymer has more than one type of monomer. Proteins are heteropolymers usually made of:
(a) 20 types of monomers
(b) 40 types of monomers
(c) 30 types of monomers
(d) Only one type of monomer

10. Proteins perform many physiological functions. For example, some function as enzymes. Which of the following represents an additional function that some proteins discharge?
(a) Antibiotics
(b) Pigment conferring colour to skin
(c) Pigments making colours of flowers
(d) Hormones

11. Glycogen is a homopolymer made of:
(a) Glucose units (b) Galactose units
(c) Ribose units (d) Amino acids

12. The number of 'ends' in a glycogen molecule would be:
(a) Equal to the number of branches plus one
(b) Equal to the number of branch points
(c) One
(d) Two, one on the left and the other on the right

13. The primary structure of a protein molecule has:
(a) Two ends (b) One end
(c) Three ends (d) No ends

14. Enzymes are biocatalysts. They catalyse biochemical reactions. In general they reduce activation energy of reactions. Many physico-chemical processes are enzyme mediated. Which of the following reactions is not enzyme-mediated in a biological system?
(a) Dissolving CO_2 in water
(b) Untwining the two strands of DNA
(c) Hydrolysis of sucrose
(d) Formation of peptide bond

PRACTICE TEST PAPER

1. Hydroxymethyl is the side chain of which of the following amino acids?
(a) Serine (b) Glycine
(c) Alanine (d) Valine

2. The activity of the enzyme succinate dehydrogenase can be competitively inhibited by:
(a) Succinate (b) Malonate
(c) Isocitrate (d) Fumarate

3. Monomeric unit is not available for:
(a) Proteins
(b) Lipids
(c) Nucleic acid
(d) Polysaccharide

4. Proteinaceous part of conjugated proteins working as enzymes is called:
(a) Apoenzyme (b) Holoenzyme
(c) Coenzyme (d) Cofactor

5. Sugars such as glycogen are characterised by the presence of which set of functional groups?
(a) Carbonyl and amino
(b) Hydroxyl and carbonyl
(c) Hydroxyl and amino
(d) Amino and sulfhydryl

6. ________ is the monomer unit of the most abundant biomolecule on earth:
(a) Alanine (b) ATP
(c) Glucose (d) N-acetylglucosamine

7. Identify the incorrect match:
(a) Triose – Glyceraldehyde
(b) Tetrose – Xylose
(c) Pentose – Ribose
(d) Hexose – Fructose

8. Dehydration synthesis occurs during:
(a) Peptide bond formation
(b) Glycosidic bond formation
(c) Ester bond formation
(d) All of the above

9. Which of the following is an incorrect statement?
(a) Water is the most abundant constituent of the cell
(b) Collagen is the most abundant protein in the biosphere
(c) Lipids are not strictly biomacromolecules
(d) Amino acids are substituted methanes

10. Identify the incorrect statement with respect to DNA:
(a) It exhibits wide variety of secondary structures
(b) Its backbone is formed by sugar and nitrogen bases
(c) Its structure follows Chargaff's principles
(d) Ribose sugar and nitrogen base uracil are not found in its structure

11. Identify the non-reducing sugar among the following:
(a) Glucose (b) Fructose
(c) Sucrose (d) Glyceraldehyde

12. Which of the following is a non-essential fatty acid?
(a) Oleic acid (b) Linoleic acid
(c) Linolenic acid (d) Arachidonic acid

13. Identify the non-aromatic amino acid:
(a) Tyrosine (b) Phenylalanine
(c) Tryptophan (d) Serine

14. Identify the incorrect match:
(a) Cellulose – Structural homopolysaccharide
(b) Uridylic acid – Nucleoside
(c) Glycogen – Branched storage polysaccharide
(d) Lecithin – Phospholipid

15. Inulin is:
(a) A branched polymer of glucose found in the liver and muscles
(b) Structural polysaccharide found in the exoskeleton of insects
(c) Structural polysaccharide found in the cell wall of plants
(d) Storage polysaccharide of fructose found in *Dahlia*

16. Select the incorrect statement with respect to enzymes:
(a) Almost all enzymes are proteins
(b) The rate of enzyme activity doubles or decreases by half for every 10°C change in either direction
(c) Each enzyme shows its highest activity at a particular temperature and pH called optimum temperature and optimum pH
(d) Magnesium is the cofactor for the enzyme carboxypeptidase

17. Enzymes that catalyse removal of groups from substrates by mechanisms other than hydrolysis leaving double bonds are:
(a) Hydrolases (b) Ligases
(c) Isomerases (d) Lyases

18. The living state is a/an:
(a) Equilibrium steady state
(b) Non-equilibrium steady state to be able to perform work
(c) Equilibrium non-steady state to be able to perform work
(d) Non-equilibrium non-steady state unable to perform work

19. Which of the following is a constituent of DNA?
(a) Thiamine (b) Peptide bond
(c) Uridine (d) Pentose sugar

20. In RNA, glycosidic bond is found between:
(a) Sugar and phosphate
(b) Phosphate and nitrogen base
(c) Sugar and nitrogen base
(d) Two nitrogen bases

21. Identify the mismatch with respect to secondary metabolites:

(a) Alkaloids – Codeine
(b) Drugs – Curcumin
(c) Toxin – Concanavalin A
(d) Pigments – Anthocyanin

22. Choose the homopolymer:
(a) Starch (b) Collagen
(c) DNA (d) Trypsin

23. Select the incorrect match:
(a) Coenzymes – Vitamin derivatives
(b) Heterocyclic ring – Adenine and guanine
(c) Neutral amino and – Lysine
(d) Nucleoside – Adenosine

24. Which of the following is not a polymer of glucose?
(a) Starch (b) Glycogen
(c) Cellulose (d) Chitin

25. Chitin occurs in:
(a) Human hair
(b) Bacterial cell wall
(c) Nails in mammals
(d) Exoskeleton of cockroach

26. The second most abundant component of the cell with respect to percentage of total cellular mass is:
(a) Protein (b) Carbohydrate
(c) Lipid (d) Water

27. Digestive enzymes can be categorised as:
(a) Hydrolases (b) Ligases
(c) Transferases (d) Oxidoreductases

28. Select the incorrect statement with respect to enzymes:
(a) They are substrate specific
(b) They are biocatalysts
(c) They carry out physical changes in the substrate molecule
(d) They have secondary and tertiary structures

29. Difference in amino acids is due to variability in their:
(a) Carboxyl group (b) Amino group
(c) Peptide group (d) R-group

30. An essential amino acid among the following is:
(a) Serine (b) Tyrosine
(c) Phenylalinine (d) Glycine

31. A molecule of ATP bears maximum resemblance to:
(a) Stearic acid (b) Linoleic acid
(c) RNA nucleotide (d) DNA nucleotide

32. Which of the following is true with respect to competitive enzyme inhibition?
(a) Increase in V_{max} and K_m
(b) Decrease in V_{max} and K_m
(c) Unchanged V_{max} and increase in K_m
(d) Unchanged V_{max} and decrease in K_m

33. Which of the following can be found in acid insoluble pool?
(a) Amino acids (b) Simple sugars
(c) DNA (d) Nucleotides

34. Which among the following is the most diverse molecule in animal cells?
(a) Lipids (b) Carbohydrates
(c) Proteins (d) Nucleic acids

35. The biomolecule that cannot be hydrolysed further is:
(a) Insulin (b) Inulin
(c) Lactose (d) Fructose

36. Synthesis of ______ does not require the formation of glycosidic bond. Select the option that fills in the blank correctly:
(a) Cellulose (b) Maltose
(c) Adenosine (d) Deoxyribose

37. Secondary metabolites:
(a) Have identifiable functions and play known roles in normal physiological processs in an organism
(b) Are compounds essential for normal growth and metabolism of an organism
(c) Have only limited or unknown relationship to the synthesis of cell materials and normal growth
(d) More than one option is correct

38. A five-carbon sugar with molecular formula $C_5C_{10}O_4$ is:
(a) Ribose (b) Deoxyribose
(c) Xylose (d) Glucose

39. The biomolecule called the energy currency of a living cell is:
(a) Glucose (b) ATP
(c) Ribose (d) Triglyceride

40. In which of the following biomolecules is the peptide bond found?
(a) Collagen (b) Chitin
(c) Cellulose (d) Lecithin

41. Select the incorrect statement:
(a) Starch can trap iodine in its helices that form its amylose fraction
(b) The right end of the polysaccharide chain is reducing
(c) Cellulose, a branched homopolymer of glucose, is a source of cotton fibres

(d) Both starch and glycogen are homopolymers of glucose

42. Which of the following is an unsaturated fatty acid?
(a) Palmitic acid (b) Stearic acid
(c) Arachidonic acid (d) Linoleic acid

43. Which of the following is not a heteropolymer?
(a) Insulin (b) RNA
(c) Chitin (d) RuBisCO

44. ________ has 16 carbon atoms including the carboxyl carbon. Select the option that fills in the blank correctly:
(a) Palmitic acid (b) Stearic acid
(c) Linoleic acid (d) Arachidonic acid

45. Read the following statements:
(i) Lysine is a diamino monocarboxylic amino acid
(ii) There are three double bonds in arachidonic acid
(iii) Bile pigments are steroidal compounds
(iv) Stearic acid is a saturated fatty acid with 18 carbon atoms
(v) Alanine and serine are non-essential amino acids

Which of the statements given above are incorrect?
(a) (i) and (ii) (b) (ii) and (iii)
(c) (iii) and (iv) (d) (ii) and (v)

ANSWER KEYS

Exercises

1. (c)	2. (c)	3. (b)	4. (d)	5. (d)	6. (c)	7. (c)	8. (d)	9. (b)	10. (d)
11. (a)	12. (d)	13. (b)	14. (c)	15. (d)	16. (c)	17. (a)	18. (c)	19. (c)	20. (b)
21. (b)	22. (c)	23. (b)	24. (d)	25. (d)	26. (b)	27. (b)	28. (c)	29. (b)	30. (d)
31. (a)	32. (a)	33. (a)	34. (c)	35. (d)	36. (d)	37. (b)	38. (d)	39. (c)	40. (b)
41. (d)	42. (d)	43. (c)	44. (a)	45. (c)	46. (c)	47. (c)	48. (c)	49. (c)	50. (c)
51. (b)	52. (b)	53. (a)	54. (b)	55. (b)	56. (a)	57. (d)	58. (d)	59. (b)	60. (c)
61. (c)	62. (c)	63. (b)	64. (d)	65. (d)	66. (b)	67. (c)	68. (d)	69. (d)	70. (b)
71. (c)	72. (c)	73. (d)	74. (c)	75. (d)	76. (b)	77. (a)	78. (b)	79. (c)	80. (d)
81. (d)	82. (a)	83. (d)	84. (d)	85. (b)	86. (b)	87. (c)	88. (b)	89. (c)	90. (b)
91. (c)	92. (a)	93. (c)	94. (b)	95. (c)	96. (b)	97. (b)	98. (d)	99. (b)	100. (d)
101. (d)	102. (c)	103. (d)	104. (d)	105. (d)	106. (b)	107. (c)	108. (c)	109. (b)	110. (d)
111. (a)	112. (d)	113. (d)	114. (d)	115. (d)	116. (d)	117. (b)	118. (a)	119. (c)	120. (c)
121. (d)	122. (b)	123. (b)	124. (d)	125. (d)	126. (a)	127. (b)	128. (b)	129. (c)	130. (c)
131. (d)	132. (b)	133. (b)	134. (c)	135. (b)	136. (a)	137. (a)	138. (c)	139. (b)	140. (b)
141. (c)	142. (c)	143. (d)	144. (a)	145. (b)	146. (d)	147. (c)	148. (d)	149. (b)	150. (b)
151. (d)	152. (b)	153. (b)	154. (d)	155. (c)	156. (d)	157. (a)	158. (c)	159. (b)	160. (d)
161. (c)	162. (b)	163. (d)	164. (a)	165. (b)					

NCERT Exemplar Questions

1. (c)	2. (b)	3. (b)	4. (d)	5. (b)	6. (d)	7. (a)	8. (b)	9. (a)	10. (d)
11. (a)	12. (a)	13. (a)	14. (d)						

Practice Test Paper

1. (a)	2. (b)	3. (b)	4. (a)	5. (b)	6. (c)	7. (b)	8. (d)	9. (b)	10. (b)
11. (c)	12. (a)	13. (d)	14. (b)	15. (d)	16. (d)	17. (d)	18. (b)	19. (d)	20. (c)
21. (c)	22. (a)	23. (c)	24. (d)	25. (d)	26. (a)	27. (a)	28. (c)	29. (d)	30. (c)
31. (c)	32. (c)	33. (c)	34. (c)	35. (d)	36. (d)	37. (c)	38. (b)	39. (b)	40. (a)
41. (c)	42. (d)	43. (c)	44. (a)	45. (b)					

SOLUTIONS

The Cellular Pool, Primary and Secondary Metabolites

1. If we observe the average composition of a cell, the percentage of carbohydrates (3%) is more than lipids (2%), and the percentage of proteins (10–15%) is more than nucleic acid.

2. Biomicromolecules are found in acid soluble fraction and generally have molecular weight less than 1000 daltons.

3. Lipids have molecular weight less than 800 daltons and appear in macromolecular fraction (acid insoluble fraction).

4. In the average composition of a cell, the percentage of lipids is 2% of the total cellular mass.

5. Cellulose is a polymeric substance and an example of secondary metabolite.

6. Lemongrass oil is an example of essential oil.

7. Triglyceride is an example of a lipid, and hence, is found in acid insoluble/retentate fraction.

8. Rubber and cellulose are polymeric secondary metabolites.

9. The molecules in the insoluble fraction (proteins, nucleic acids and polysaccharides), with the exception of lipids, are polymeric substances.

10. Serine is an amino acid found in acid soluble fraction among biomicromolecules.

11. Inorganic compounds like sulphate and phosphate are also seen in acid soluble fraction.

12. The inorganic compounds and elements in living tissue are analysed by treating the tissue with trichloroacetic acid.

13. Ions are least abundant (1%) in the total cellular mass, while studying the average composition of a cell.

14. Option (c) is correct.

15. During chemical analysis of the cell, the acid soluble fraction is called filtrate, while the acid insoluble fraction is called retentate. The biomicromolecules have molecular weight less than 1000 daltons.

16. The percentage of nitrogen in the earth's crust is much lower than in the human body.

17. When we homogenise a tissue, the acid soluble pool represents roughly the cytoplasmic composition.

18. When we grind a tissue, we are disrupting the cell structure. The cell membrane and other membranes are broken into pieces and form vesicles which are not water soluble. Therefore, these membrane fragments in the form of vesicles are separated along with the acid insoluble pool, and hence, in the macromolecular fraction.

19. While primary metabolites have identifiable functions and play known roles in normal physiological processes, we do not, at the moment, understand the roles or functions of all secondary metabolites.

20. Option (b) is correct.

Carbohydrates

21. Glycogen is a homopolysaccharide of glucose.

22. Sucrose is a disaccharide of glucose and fructose.

23. Fructose is an example of ketone sugar.

24. Chitin is a homopolysaccharide of N-acetylgluco samine and found in insects and fungi.

25. Mannose is an example of a monosaccharide, and thus, cannot be further hydrolysed. Maltose, sucrose and lactose are examples of disaccharides.

26. The monosaccharides are linked to each other by glycosidic bonds in a polysaccharide.

27. The monomer unit of cellulose is β – D – glucose.

28. The right end of a polysaccharide chain is reducing.

29. Lactose is a disaccharide of glucose and galactose.

30. Amylose is a straight chain with glucose residues joined by α-1, 4–glycosidic bond. Amylopectin is branched with α-1, 4–glycosidic bonds between glucose molecules in straight chain and α-1, 6–glycosidic bonds at the branch points.

31. Inulin is a homopolymer of fructose.

32. Glycogen is the storage form of glucose in animals and also called animal starch.

33. Cellulose xanthate is commonly called rayon or artificial silk.

34. Ribose is different from deoxyribose in having an extra oxygen at the second carbon.

35. Maltose is commonly called malt sugar. Sucrose is commonly called cane sugar.

36. The husk of *Plantago ovata* is a mucopolysaccharide and used for the treatment of constipation.

37. Sucrose and trehalose are examples of non-reducing disaccharides.

38. Erythrose is an example of tetrose sugar.

39. Cellulose nitrate is found in gun cotton, explosives and rocket propellants.

40. Sucrose, lactose, maltose and trehalose are examples of disaccharides.

41. Polysaccharides are not used for storage of genetic information.

42. Ribulose and xylulose are examples of ketone sugars.

43. Trehalose is a disaccharide of two glucose molecules.

44. Inulin is a low-molecular-weight polysaccharide and obtained from the roots of *Dahlia*.

45. Benedict's reagent test is conducted to confirm the presence of reducing sugars. It is based on the principle that reducing sugar converts cupric to cuprous salts and produces brick-red-coloured precipitates.

Proteins

46. Tryptophan is an example of aromatic amino acid.

47. Option (c) is correct.

48. Glycine is the simplest amino acid and does not contain sulphur. Cysteine and methionine are sulphur containing amino acids.

49. Option (c) is correct.

50. Tertiary structure of the protein enables them to function as enzymes.

51. Adult human haemoglobin has four subunits, 2 alpha and 2 beta chains.

52. Methionine is an example of an amino acid with a sulphur group in its side chain.

53. Collagen is the most abundant protein in the animal world.

54. The alpha helices and beta pleated sheets represent the secondary structure of proteins.

55. Proline is an example of an amino acid.

56. The primary structure of a protein is retained during denaturation: the peptide bonds do not break down.

57. The secondary structures of proteins are stabilised by hydrogen bonds.

58. The proteins of cereals and millets are deficient in lysine and tryptophan.

59. The first protein in the amino acid is present at the N-terminal end.

60. Tertiary structure is absolutely necessary for the biological activities of proteins.

61. Collagen functions as intercellular ground substance.

62. Tyrosine is used for the synthesis of thyroxine.

63. Lysine and arginine are basic amino acids.

64. Tertiary structure gives a three-dimensional view of the protein.

65. Quaternary structure is represented by a linear string of spheres, or spheres arranged one upon each other in the form of a cube or plate.

66. Serine has the hydroxy methyl group as side chain.

67. Adrenaline is synthesised from tyrosine.

68. A protein having single polypeptide chain cannot have quaternary structure.

69. Keratin, collagen and elastin are fibrous proteins, while haemoglobin is an example of

globular protein.

70. Hydrogen bond is found in secondary, tertiary and quaternary structures.

71. Tyrosine is used for the synthesis of melanin.

72. Tertiary structure is formed by folding of long protein chain upon itself, like a hollow woollen ball.

73. The single letter 'E' refers to the amino acid glutamic acid.

74. All protein amino acids are asymmetrical and levorotatory, except glycine.

75. Cysteine and methionine are sulphur containing amino acids. Tyrosine and tryptophan are aromatic amino acids.

76. The 'R' group in alanine is the methyl group.

77. Collagen functions as intercellular ground substance.

78. Plant proteins like pulses are deficient in cysteine and methionine.

Lipids

79. Oleic acid is a non-essential fatty acid.

80. There are four double bonds in arachidonic acid.

81. Arachidonic acid has 20 carbon atoms including carboxyl carbon.

82. Lecithin is a phosphorylated glyceride in the cell membranes.

83. A typical fat molecule is made up of one glycerol and three fatty acids.

84. Cholesterol is a sterol and a precursor of testosterone, bile salts and vitamin D. Insulin is a polypeptide hormone.

85. Palmitic acid is a saturated fatty acid. Oleic acid, linoleic and linolenic acid are unsaturated fatty acids.

86. Oils have low melting points compared to hard fats.

87. Glutamic acid is an example of an acidic amino acid.

88. Wax-D is present in the cell wall of bacteria, causing leprosy and tuberculosis.

89. Option (c) is correct.

90. Terpenoids are a large group of lipids that are formed by repeating units called isoprene.

91. Cholesterol is not a part of lecithin.

92. Lanolin is a waxy water insoluble coating on animal fur secreted by cutaneous glands in animals.

93. Unsaturated fatty acids can undergo hydrogenation reaction, since they have double bonds. Saturated fatty acids cannot undergo hydrogenation reactions.

94. Prostaglandins are derived from arachidonic acid.

95. Deficiency of essential fatty acids causes phrenoderma or follicular hyperkeratosis.

96. A diglyceride has two fatty acids esterified with glycerol, while a monoglyceride has a single fatty acid esterified with glycerol.

97. Lipids are hydrophobic and are thus insoluble in water.

98. Oils are formed by esterification of unsaturated fatty acids with glycerol.

99. Edible oils are converted into hard fats by a process called hydrogenation.

100. Lipids are obtained in acid insoluble fraction during analysis of chemical composition of the cell.

Nucleic Acids

101. The phosphate group is attached to the fifth carbon of the sugar residue of a nucleotide.

102. Cytosine is a nitrogen base and not a nucleoside.

103. DNA differs from RNA in its nitrogen base and sugar. RNA has ribose sugar, while DNA has deoxyribose sugar. The nitrogen bases in DNA are adenine, guanine, cytosine and thymine. RNA has uracil and lacks thymine.

104. Guanylic acid (guanosine mono phosphate) is an example of a nucleotide.

105. Uracil is not found in DNA.

106. Glycosidic bond is present between sugar and nitrogen base.

107. Phosphodiester bond is present between two nucleotides in a polynucleotide chain.

108. Hydrogen bonds are present between the nitrogen bases of the opposite strands of DNA.

109. Nucleic acids exhibit a wide variety of secondary structures.

110. According to the Watson–Crick model, the nitrogen bases are projected perpendicular to the sugar-phosphate backbone, but face inside.

111. In a nucleic acid, a phosphate moiety links the 3'-carbon of one sugar of one nucleotide to the 5'-carbon of the sugar of the succeeding nucleotide.

112. If we draw a line diagram of DNA, the pitch would be 3.4 nm and the rise per base pair would be 0.34 nm.

113. The distance between the two strands of DNA is approximately 20 angstroms.

114. Compared to a nucleoside, nucleotide has an additional phosphate and is acidic in nature.

115. In DNA, A and G of one strand compulsorily base pairs with T and C, respectively, on the other strand. Thus, the base pairing of opposite strands is stereochemically selective, and adenine always pairs with thymine, while guanine base pairs with cytosine. Two and three hydrogen bonds are formed in the A-T and G-C base pairs, respectively.

116. The nitrogen bases found in DNA are adenine, guanine, thymine and cytosine.

117. Option (b) is correct.

118. Option (a) is correct.

119. Ribose differs from deoxyribose in having a hydroxyl group instead of hydrogen at the second carbon.

120. Adenine and guanine are double ring structures found in nucleic acids.

Concept of Metabolism

121. The flow of metabolites in a metabolic pathway has definite rate and direction.

122. In metabolic pathways, every chemical reaction is a catalysed reaction.

123. The proteins with catalytic power are called enzymes.

124. The formation of cholesterol from acetic acid is an example of anabolic or biosynthetic pathway.

125. Anabolic pathways form complex structures from simple structures.

126. There are no uncatalysed metabolic conversions in a living system. The catalysts which hasten the rate of a given metabolic conversion are called enzymes.

127. The most important form of energy currency in a living system is a bond energy in a chemical called adenosine triphosphate.

128. The living state needs to be in a non-equilibrium steady-state to be able to perform work.

129. The living state and metabolism are synonymous terms.

130. Systems at equilibrium do not perform work. Physical or chemical processes move spontaneously to equilibrium.

Enzymes

131. An active site of an enzyme is a crevice or pocket in which the substrate fits.

132. Almost all enzymes are proteins. There are some nucleic acids that behave like enzymes.

133. Inorganic catalysts work efficiently at high temperatures (say, above 40°C). However, enzymes isolated from organisms who normally live under extremely high temperatures (hot vents and sulphur springs) are stable and retain their catalytic power even at high temperature (up to 80°–90°C).

134. Ribozymes are RNA molecules that act as enzymes.

135. An allosteric inhibitor of the enzyme acts by binding to an allosteric site which is other than the active site or the catalytic site.

136. The protein portions of the conjugated enzymes are called apoenzyme.

137. The essential chemical components of many coenzymes are vitamins. For example, coenzyme nicotinamide adenine dinucleotide (NAD) contains the vitamin niacin.

138. When the inhibitor closely resembles the substrate in its molecular structure and inhibits the activity of the enzyme, it is known as competitive inhibitor. For example, malonate is the competitive inhibitor of succinic dehydrogenase.

139. Option (b) is correct.

140. Transition state structures of the substrate formed during an enzymatic reaction are transient but unstable.

141. Each enzyme shows its highest activity at a particular temperature called optimum temperature. Low temperature preserves the enzyme in a temporarily inactive state whereas high temperature destroys enzymatic activity because proteins are denatured by heat.

142. Enzymes catalysing hydrolysis of ester, ether, peptide and glycosidic bonds are hydrolases.

143. Zinc is a cofactor for the proteolytic enzyme carboxypeptidase.

144. Coenzymes are organic groups whose association with the apoenzyme is only transient, usually occurring during the course of catalysis.

145. The coenzyme nicotinamide adenine dinucleotide (NAD) and NADP contain the vitamin niacin.

146. Option (d) is correct.

147. Haem is the prosthetic group and part of the active part of enzymes peroxidase and catalase, which catalyse the breakdown of hydrogen peroxide to water and oxygen.

148. Enzymes lower the activation energy of a reaction.

149. Competitive inhibitors increase the K_m of a reaction.

150. The effect of end product on the rate of enzyme catalysed reaction is called feedback inhibition.

151. Succinate dehydrogenase belongs to the class of enzymes called oxidoreductases or dehydrogenses.

152. The lock and key hypothesis was given by Emil Fischer.

153. The substrate concentration at which the chemical reaction catalysed by an enzyme attains half its maximum velocity is called K_m or Michaelis constant.

154. Urease was the first enzyme to be crystallised and purified.

155. Option (c) is correct.

156. Cofactor is the thermostable part of the enzyme. Prosthetic groups are tightly bound to the apoenzyme. On increasing the substrate concentration, the rate of reaction initially increases.

157. The term enzyme was given by Willy Kuhne.

158. Option (c) is correct.

159. Option (b) is correct.

160. Option (d) is correct.

161. Sulpha drugs competitively inhibit the enzyme bacterial folate synthase since they resemble p-aminobenzoic acid.

162. The term feedback refers to the effect of end product on the rate of enzymatic reaction.

163. All the cells of an organism do not contain the same enzymes.

164. Option (a) is correct.

165. Option (b) is correct.

10 Cell Cycle and Cell Division

EXERCISES

Cell Cycle

1. A cell cycle consists of:
(a) G_1, S and G_2 phases
(b) Mitosis and meiosis
(c) Prophase, metaphase, anaphase, telophase
(d) Interphase and mitosis

2. Interphase lasts more than what percentage of the duration of the cell cycle?
(a) 95% (b) 85%
(c) 70% (d) 60%

3. The sequence of the cell cycle is:
(a) S, M, G_1, G_2
(b) G_1, S, G_2, M
(c) M, G_1, G_2, S, M
(d) G_1, G_2, S, M

4. Choose the correct option with respect to the duration of the cell cycle in a typical human cell and a yeast cell:

	Human cell	**Yeast cell**
(a)	24 hours	90 hours
(b)	90 minutes	24 minutes
(c)	24 hours	90 minutes
(d)	90 minutes	20 minutes

5. What will be the DNA content of the cells at the G_1, S and G_2 phases if the DNA contents after the M phase is 2C?
(a) 2C, 4C, 4C (b) 4C, 4C, 2C
(c) 2C, C, 2C (d) 4C, 4C, 2C

6. Read the following statements and choose the correct option regarding the sequence of events given below:
A. Duplication of DNA
B. Synthesis of tubulin protein
C. Synthesis of deoxyribonucleotides, amino acids and ATP
D. Karyokinesis
(a) C→B→A→D (b) C→A→B→D
(c) A→C→D→B (d) A→B→D→C

7. The interval between mitosis and initiation of DNA replication in actively dividing cells is:
(a) G_1 phase (b) G_0 phase
(c) G_2 phase (d) M phase

8. Some cells in the adult animal do not divide; they exit the G_1 phase and enter an inactive stage which is called:
(a) G_2 phase (b) G_0 phase
(c) S phase (d) M phase

9. What is incorrect about the G_0 phase?
(a) Cell metabolism continues in G_0
(b) Cell growth occurs in G_0
(c) Biocatalyst helps exit G_0
(d) Mitosis occurs after G_0

10. DNA replication occurs in:
(a) G_1 phase (b) S phase
(c) M phase (d) G_2 phase

11. A plant cell has 20 pg DNA in the phase where centriole duplication takes place. Calculate the amount of DNA in the same cell at the post-mitotic gap phase:
(a) 40 pg (b) 10 pg
(c) 20 pg (d) 90 pg

12. Read the following statements with respect to the quiescent stage of the cell cycle and select the correct option:
A. Cells proliferate rapidly in this stage
B. Inactive stage, but cells remain metabolically inactive
C. It is observed in those cells which occasionally divide
(a) Only (A) (b) Only (B)
(c) Only (C) (d) Both (B) and (C)

13. The stage of the cell cycle where the cell decides to undergo differentiation is:
(a) G_0 (b) G_1
(c) G_2 (d) M

14. Number of DNA strands present in chromosomes during the G_2 phase is:
(a) One (b) Two
(c) Four (d) Eight

15. Progression of the cell cycle is regulated by concentration of which type of molecule?
(a) Centrosomes
(b) Cyclin-dependent kinases
(c) Cyclins
(d) Microtubules

16. Fibroblast cells in our body are those that are arrested in:
(a) G_0 phase (b) G_1 phase
(c) G_2 phase (d) Yet to start division

17. Which of the following processes occur in both G_1 and G_2 phases of the cell cycle?
A. Synthesis of deoxyribonucleotides
B. Duplication of cell organelle
C. RNA synthesis
D. Protein synthesis
(a) Only (C) and (D)
(b) Only (D)
(c) Only (B), (C) and (D)
(d) (A), (B), (C) and (D)

18. Read the following statements and select the correct option:
A. Cell growth (cyloplasmic increase) is a continuous process of the cell cycle
B. DNA synthesis occurs during one specific stage of the cell cycle
C. During cell division, cell growth and DNA replication take place in a random way
D. All events of the cell cycle are themselves under genetic control
(a) Only (A), (B) and (D)
(b) Only (A), (B) and (C)
(c) Only (B) and (D)
(d) (A), (B), (C) and (D)

19. Interphase:
(a) Is also called resting phase of the cell cycle
(b) Covers approximately 95% of the duration of the cell cycle
(c) Is the phase between two successive M phases
(d) All of the above

20. Consider the following statements and select the correct option for A, B and C:
(i) The phase of the cell cycle where DNA polymerase is active is ___**A**___
(ii) During the ___**B**___ phase, most of the cell organelles increase in the cell
(iii) The ___**C**___ phase is the most dramatic period of the cell cycle

	A	B	C
(a)	S	G_1	M
(b)	G_1	G_2	M
(c)	S	G_2	S
(d)	G_1	S	G_2

21. The phase of the cell cycle in which histone protein is synthesized is also related with:
(a) Duplication of mitochondria and chloroplast
(b) Tubulin protein synthesis
(c) Centriole duplication
(d) Synthesis of cyclin-dependent protein kinase

22. In the G_0 phase of the cell cycle, the cell:
(a) Is metabolically active
(b) Shows differentiation
(c) Does not divide further
(d) All of the above

23. State true (T) or false (F) for the following statements and select the correct option:
A. The interval between mitosis and the synthetic phase is characterised by the synthesis of tubulin protein
B. Interphase is metabolically a very active phase because of biosynthesis of a number of molecules
C. Decision of a cell to divide occurs in the M phase of the cell cycle
D. Cell growth ensures no disturbance in the nucleocytoplasmic ratio

	A	B	C	D
(a)	F	F	T	T
(b)	T	T	F	F
(c)	T	F	F	T
(d)	F	T	T	F

24. Before the S phase, if the amount of DNA is 2C, then by the end of the G_2 phase, the amount of DNA would be:
(a) C (b) 2C
(c) 4C (d) 8C

25. Which of the following cells in an adult animal do not divide?
(a) Nerve cell
(b) Cells lining the stomach
(c) Upper layer of the epidermis
(d) (a) and (c)

26. In which phase of the cell cycle are sister chromatids available as a template for repair?
(a) G_1 (b) G_2
(c) S (d) M

27. Study the given diagrammatic view of the cell cycle indicating the formation of two cells from one cell and select the most appropriate option as answer:

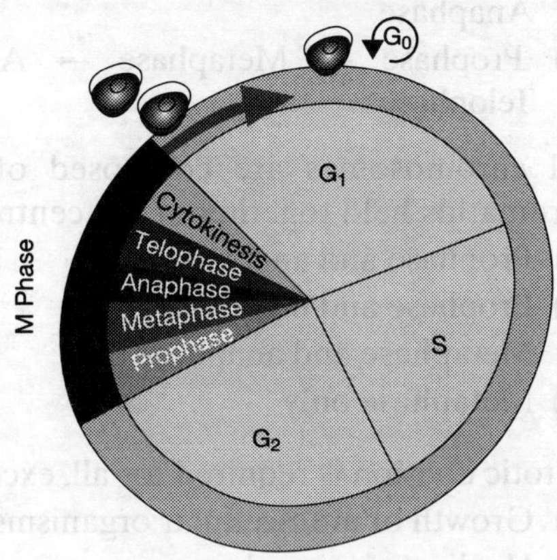

A. The G_1 phase represents the cell as metabolically active and growing continuously
B. The S phase marks the period during which DNA duplication takes place
C. The amount of DNA and the number of chromosomes doubles during the G_2 phase
D. The M phase represents the phase when actual cell division occurs.
E. In the G_0 phase, the cell become metabolically inactive and cannot proliferate in any circumstances

The correct statements are:
(a) (A), (B) and (D)
(b) (A), (B), (C) and (D)
(c) (B), (C) and (D)
(d) (B), (D) and (E)

28. Antiphase is the phase in which ATP is synthesized during cell division. It refers to:
(a) G_0 phase (b) G_1 phase
(c) S phase (d) G_2 phase

29. What happens in the S phase during the cell cycle?
(a) DNA replication occurs in the nucleus
(b) Chromosome number becomes double
(c) Synthesis of tubulin protein and RNA
(d) Double membrane–bound cell organelles duplicate

30. Cells in the G_0 phase of the cell cycle:
(a) Exit the cell cycle
(b) Enter the cell cycle
(c) Terminate the cell cycle
(d) Suspend the cell cycle

31. The checkpoint in the cell cycle plays an important role in:
(a) Repair of DNA damage
(b) Assessment of DNA damage
(c) Apoptosis
(d) Inhibition of cell division

32. The centriole duplicates in the cytoplasm during the ______ phase of the cell cycle:
(a) G_1 phase (b) M phase
(c) S phase (d) G_2 phase

33. If there are 14 chromosomes in each root cell of an onion, what will be the number of chromosomes in the G_2 phase and G_1 phase, respectively?
(a) 28, 14 (b) 14, 7
(c) 7, 14 (d) 14, 14

34. In the 24 hours average duration of the cell cycle of a human cell, cell division proper lasts for only about:
(a) Four hours (b) 90 minutes
(c) An hour (d) 10 hours

35. In the cell cycle, two DNA molecules are present in a chromosome during:
(a) G_1 phase (b) Beginning of S phase
(c) G_2 phase (d) End of M phase

36. Consider the following statements and select the correct option for A, B and C:
(i) Cells in the ___**A**___ phase of the cell cycle are least susceptible to radiation damage
(ii) ___**B**___ is the shortest duration of the cell cycle
(iii) Maximum cytoplasmic growth occurs in ___**C**___

	A	B	C
(a)	S	M	G_1
(b)	G_2	M	G_1
(c)	G_1	S	G_2
(d)	G_2	S	G_1

37. In the resting phase of the cell cycle:
(a) There are no metabolic processes in the cell
(b) Cell growth and DNA replication occur in an orderly manner
(c) Only RNA and protein synthesis occurs in the cell
(d) The cell is metabolically active but no DNA, RNA and protein synthesis occurs

38. Nucleocytoplasmic ratio is restored through:
(a) M phase (b) G_1 phase
(c) G_2 phase (d) S phase

39. Radioactive thymidine is fed to cells about to enter the S phase. It will make radioactive:
(a) Euchromatin only
(b) Heterochromatin only
(c) Neither euchromatin nor heterochromatin
(d) Both euchromatin and heterochromatin

40. Which does not occur in interphase of the eukaryotic cell cycle?
(a) Increase in ATP synthesis
(b) Increase in DNA synthesis
(c) Increase in RNA synthesis
(d) Reduction in cell size

41. The decision of a cell to divide occurs in the ___ phase of the cell cycle:
(a) G_1 (b) S
(c) G_2 (d) M

42. Organelle lacking DNA, but capable of duplication, is:
(a) Ribosome (b) Centriole
(c) Nucleus (d) Semiautonomous organelle

43. Antiphase is:
(a) Telophase of mitosis
(b) Failure of meiosis II
(c) The stage in the cell cycle when energy source is stored for the completion of cell division; usually the G_1 phase
(d) Bouquet stage

Mitosis

44. Mitosis is responsible for:
A. Growth of multicellular organisms
B. Cell repair
C. Formation of four haploid cells
D. Healing and regeneration of cells
Choose the correct option:
(a) (A) and (B) only
(b) Only (C)
(c) (A), (B), (C) and (D)
(d) (A), (B) and (D)

45. Plant material suited for studying mitosis is:
(a) Shoot apex (b) Root apex
(c) Leaf tip (d) Anther

46. Choose the correct sequence with respect to the stages of karyokinesis:
(a) Prophase → Anaphase → Metaphase → Telophase
(b) Prophase → Telophase → Anaphase → Metaphase
(c) Prophase → Metaphase → Telophase → Anaphase
(d) Prophase → Metaphase → Anaphase → Telophase

47. All chromosomes are composed of two sister chromatids held together at the centromere in:
(a) Prophase and anaphase
(b) Prophase and metaphase
(c) Metaphase and anaphase
(d) Metaphase only

48. Mitotic division is required for all, except:
(a) Growth of multicellular organisms
(b) Healing of injured tissue
(c) Regeneration of organisms
(d) Formation of gametes in animals

49. The bead-like thickened portions of early prophase chromosomes are:
(a) Puffs (b) Chromomeres
(c) Centromeres (d) Genes

50. During prophase, chromosomes are:
(a) Large and coiled (b) Large and straight
(c) Thick and uncoiled (d) Thick and straight

51. Mark the option which truly represents the events of the cell division stage in the given diagram:

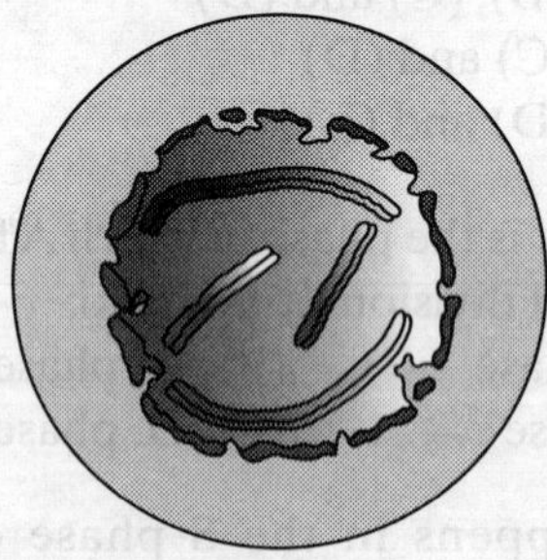

(a) Nuclear membrane, nucleolus and ER re-form
(b) Chromosomes decondense and lose their individuality
(c) Number of chromosomes becomes double in the cell
(d) Cell does not show Golgi complex, ER and nuclear envelope

52. The shape of chromosomes can be best studied during:
(a) Metaphase (b) Telophase
(c) Anaphase (d) Prophase

53. Find the incorrect match:
(a) Chromosomes move to the equator – Anaphase
(b) Identity of chromosomes is lost – Telophase
(c) Centrosome begins to move towards opposite poles of the cell – Prophase
(d) Centromeres of all the chromosomes lie over the equator – Metaphase

54. Phragmoplast is a precursor of:
(a) Cell plate (b) Equatorial plate
(c) Cell furrow (d) Nuclear membrane

55. Which of the following phases of mitosis is the longest?
(a) Prophase (b) Metaphase
(c) Anaphase (d) Telophase

56. A duplicated chromosome has how many chromatids?
(a) One (b) Two
(c) Three (d) Four

57. Select the correct statement with respect to telophase:
(a) Chromosomes are thickest
(b) Chromosome number reduces to half
(c) Nucleolus disappears
(d) Nucleolus, ER and Golgi complex re-form

58. Match the following columns and select the correct option:

	Column I		Column II
A.	Prophase	(i)	Congression
B.	Metaphase	(ii)	Disintegration of spindle fibres
C.	Anaphase	(iii)	Condensation of chromatin
D.	Telophase	(iv)	Best stage to study shape of chromosomes

(a) A – (iii), B – (i), C – (iv), D – (ii)
(b) A – (iii), B – (iv), C – (i), D – (ii)
(c) A – (iv), B – (i), C – (ii), D – (iv)
(d) A – (ii), B – (iii), C – (iv), D – (i)

59. Consider the following statements and select the option that correctly fills in the blank:
(i) During mitosis, mitotic spindle formation is completed in ___A___
(ii) During mitosis, ER and nucleolus begin to disappear at ___B___
(iii) Nuclear envelope assembles around the chromosome cluster at ___C___

	A	B	C
(a)	Metaphase	Early prophase	Telophase
(b)	Prophase	Metaphase	Telophase
(c)	Telophase	Anaphase	Late prophase
(d)	Late prophase	Metaphase	Telophase

60. In which stage of mitosis are the chromosomes composed of two chromatids?
(a) Metaphase and anaphase
(b) Prophase and metaphase
(c) Metaphase and telophase
(d) Prophase and telophase

61. Nuclear membrane is formed around the group of daughter chromosomes during telophase by:
(a) ER (b) Lysosomes
(c) Golgi body (d) Microbodies

62. A stage in mitosis that begins in mid-anaphase and is completed with telophase is:
(a) Crossing over (b) Karyokinesis
(c) Cytokinesis (d) Interkinesis

63. Cytokinesis is:
(a) Division of the nucleus
(b) Division of the cytoplasm
(c) Duplication of the centriole in the cytoplasm
(d) Synthesis of the cytoplasm during cell growth

64. Cytokinesis in plant cells occurs by:
(a) The formation of cell furrow
(b) Formation of cell plate
(c) Centrifugal growth of cell plate
(d) Both (b) and (c)

65. Cytokinesis in animal cells:
(a) Is centrifugal
(b) Involves furrow formation
(c) Occurs by cell plate formation
(d) Both (a) and (b)

66. Syncytium formation is due to:
(a) Failure of cytokinesis after karyokinesis
(b) Failure of both cytokinesis and karyokinesis
(c) Occurrence of cytokinesis and karyokinesis
(d) Occurrence of karyokinesis after cytokinesis

67. Read the following statements and select the correct option:
A. A large cell has a higher surface area to volume ratio than a smaller cell
B. Cell growth results in disturbing the ratio between the nucleus and cytoplasm
(a) Only (A) is correct
(b) Only (B) is correct
(c) Both (A) and (B) are correct
(d) Both (A) and (B) are incorrect

68. In the spireme stage:
(a) Congression of chromosomes occurs
(b) Arm of chromosomes is attached to nuclear membrane

(c) Chromosomal ends appear like a ball of wool
(d) Spindle fibre

69. Mitotic apparatus is formed by which amongst the following?
(a) Centrosome, spindles and chromosomes
(b) Centriole, asters, spindles, chromosomes
(c) Asters and spindles
(d) Asters, spindles and chromosomes

70. Each daughter chromosome in anaphase moves towards opposite poles at the speed of 1 μm per minute with the expenditure of energy:
(a) 5 ATP (b) 30 ATP
(c) 38 ATP (d) 60 ATP

71. Mitotic cell division is seen in the diploid somatic cell of:
(a) Only plants (b) Only animals
(c) Bacteria (d) Both plant and animals

72. The two chromatids of a metaphase chromosome represent:
(a) Replicated chromosomes to be separated at anaphase
(b) Homologous chromosomes of a diploid set
(c) Non-homologous chromosomes joined at the centromere
(d) Maternal and paternal chromosomes formed at the centromere

73. All statements are correct with respect to mitosis, except:
(a) Chromosomes are shortest and thickest at metaphase
(b) Shape of chromosome can be studied at anaphase
(c) Mitotic divisions in meristematic tissues result in continuous growth of plants throughout their life
(d) It increases genetic variability in a population

74. Mitosis is inhibited by the drug colchicine. How does this drug control mitosis?
(a) It inhibits cytokinesis
(b) It inhibits chromosome movement
(c) It inhibits aster formation
(d) It inhibits the centriole

75. Chromosomes move towards different poles at anaphase. This is because of:
(a) Sliding filament of actin
(b) Polymerisation and depolymerisation of tubulin
(c) Ability of kinetochore to move by using ATP
(d) Genetically determined process

76. A stage of cell division is depicted in the given figure.

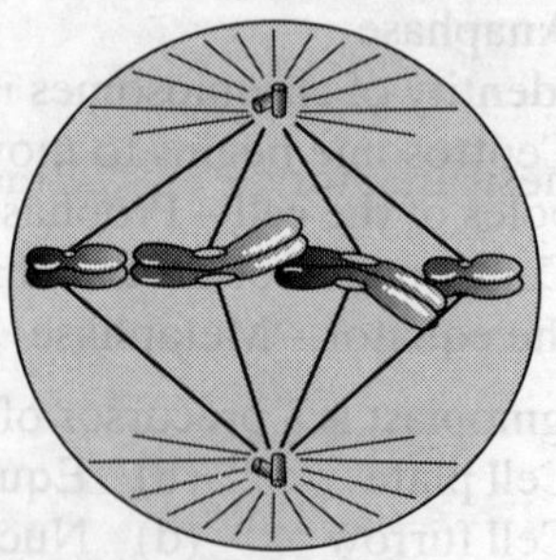

Select the most appropriate option which gives the correct identification of the stage:
(a) Centromere of each chromosome splits length-wise
(b) It is the stage at which the morphology of chromosomes is most easily studied
(c) Chromosome material becomes untangled
(d) Chromosome identity is lost as discrete elements

77. A loose and distinct network of nucleoprotein fibres can be observed in:
(a) Late prophase (b) Metaphase
(c) Anaphase (d) Interphase nucleus

78. In mitosis, the spindle is:
(a) Bipolar (b) Multipolar
(c) Unipolar (d) Random

79. Many cells function properly and divide mitotically even though they do not have:
(a) Plasma membrane (b) Cytoskeleton
(c) Mitochondria (d) Plastids

80. Mark the option which truly represents the event of the cell division stage in the given figure:

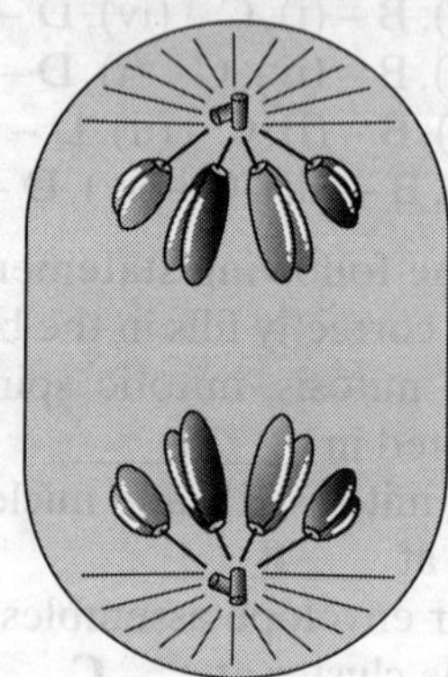

(a) Its absence leads to the syncytium condition in liquid endosperm of coconut
(b) Chromosome remains partially condensed at this stage

(c) Centromere is the leading edge and chromated arms trail behind during the movement of chromosomes
(d) Mitotic spindle formation is completed

81. Synthesis and storage of ATP molecules required for cell division takes place in:
(a) G_1 phase (b) G_2 phase
(c) M phase (d) S phase

82. An anaphase chromosome contains:
(a) 1 DNA molecule
(b) 3 DNA molecules
(c) 2 DNA molecules
(d) 4 DNA molecules

83. Which chromosome may be lost during cell division?
(a) Giant chromosome
(b) Acentric chromosome
(c) Polycentric chromosome
(d) Telocentric chromosome

84. Read the following statements and select the correct option:
A. At the time of cytokinesis, organelles like mitochondria and plastids are distributed between the daughter cells
B. Cell plate represents the middle lamella between the walls of two adjacent cells
(a) Only (A) is correct
(b) Only (B) is correct
(c) Both (A) and (B) are incorrect
(d) Both (A) and (B) are correct

85. What is true for mitosis?
(a) It has two divisions
(b) It maintains the number of chromosomes
(c) It occurs in somatic cells only
(d) It occurs in somatic cells as well as gonads

86. Azide and cyanide inhibit:
(a) Metaphase (b) Prophase
(c) Anaphase (d) Telophase

87. Select the correct sequence of events taking place during mitotic anaphase:
A. Chromatids separate
B. Centromeres split
C. Chromatids move to opposite poles
(a) B→A→C (b) C→B→A
(c) A→B→C (d) A→C→B

88. Read the following statements:
A. Condensation of chromosome is completed
B. Mitotic spindle formation is completed
C. Centromere lies at equator and arms remain directed towards poles
The above characteristics are attributed to:
(a) Prophase (b) Metaphase
(c) Anaphase (d) Telophase

89. Spindle fibres which extend from pole to kinetochores are:
(a) Chromosomal fibres (b) Tractile fibres
(c) Interzonal fibres (d) Both (a) and (b)

90. Reason for chromosomal movement in anaphase:
(a) Astral rays
(b) Centrioles
(c) Kinetochore only
(d) Kinetochore and spindle fibre

91. Chromosomes exhibit high level of coiling at which phase of karyokinesis?
(a) Interphase (b) Metaphase
(c) Telophase (d) Prophase

92. The phragmoplast is organised:
(a) At the beginning of anaphase
(b) At the end of anaphase
(c) At the beginning of telophase
(d) At the end of telophase

93. Movement of chromosomes during anaphase:
(a) Is independent of spindle fibre
(b) Is due to cytoplasmic streaming
(c) Is due to spindle fibre and kinetochore
(d) Is dependent upon the shape of the chromosome

94. Observe the given figure and select the correct option with respect to the phase of mitosis it shows:

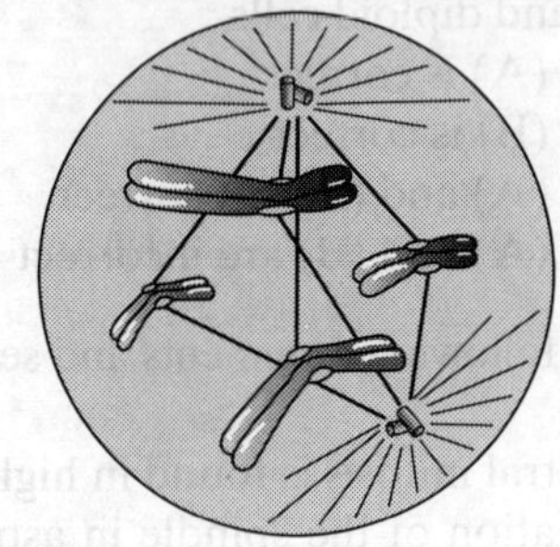

(a) Late prophase
(b) Early prophase
(c) Transition to metaphase
(d) Anaphase

95. Match the columns and select the correct answer:

A. Mitosis	(i) Occurs in diploid cells only
B. Meiosis	(ii) Occurs in both haploid and diploid cells
	(iii) Daughter and parent cells have the same chromosome number
	(iv) Synapsis of homologous chromosomes

(a) (i) (ii) (b) (ii) (iii)
(c) (iii) (iv) (d) (iv) (i)

96. How many generations are required by a cell of meristem to produce 256 cells?
(a) 8 (b) 7
(c) 255 (d) 9

97. Read the following statements carefully:
(i) Mitosis is usually restricted to the diploid cell only
(ii) In some lower plants and in some social insects, haploid cells divide by mitosis
(iii) Mitosis results in the production of genetically identical cells
(iv) A very significant contribution of mitosis is cell repair
(v) Growth of multicellular organisms is due to mitosis
How many statement(s) is/are correct?
(a) Three (b) Four
(c) Five (d) One

98. Arms of chromosomes are trailing behind in:
(a) Telophase (b) Metaphase
(c) Anaphase (d) Prophase

99. Read the following statements and select the correct option:
A. In animals, mitotic cell division is only seen in diploid somatic cells
B. Plants can show mitotic divisions in both haploid and diploid cells
(a) Only (A) is correct
(b) Only (B) is correct
(c) Both (A) and (B) are correct
(d) Both (A) and (B) are incorrect

100. Read the following statements and select the correct option:
A. Anastral mitosis is found in higher plants
B. Formation of the spindle in astral mitosis is a mechanism that leads to the distribution of the centrioles between two daughter cells
(a) Only (A) is correct
(b) Only (B) is correct
(c) Both (A) and (B) are correct
(d) Both (A) and (B) are incorrect

Meiosis and Amitosis

101. Anaphase II differs from anaphase I, as the former:
(a) Occurs in haploid cells
(b) Involves division of centromere
(c) Is the substage of homotypic division
(d) All of the above

102. Choose the incorrect statement with respect to meiosis:
(a) Involves pairing of homologous chromosomes
(b) Produces four haploid cells
(c) Involves two sequential cycles of nuclear and cell division
(d) Has a single cycle of nuclear and cell division but two cycles of DNA replication

103. Meiosis differs from mitosis in having:
(a) Duplication of DNA
(b) Interphase
(c) Synapsis and crossing over
(d) Short interphase

104. Choose the correct sequence of events in prophase I:
(a) Leptotene → Zygotene → Pachytene → Diplotene → Diakinesis
(b) Leptotene → Pachytene → Zygotene → Diplotene → Diakinesis
(c) Leptotene → Diplotene → Zygotene → Pachytene → Diakinesis
(d) Leptotene → Diakinesis → Diplotene → Zygotene → Pachytene

105. Meiosis does not involves two sequential cycles of:
(a) Karyokinesis (b) Cytokinesis
(c) DNA replication (d) Centriole duplication

106. In which stage do chromosomes appear as thin, long thread-like structures?
(a) Leptotene (b) Zygotene
(c) Pachytene (d) Metaphase

107. Recombination nodule is formed:
A. Between non-sister chromatids of homologous chromosomes
B. At the site where crossing over occurs
C. In the pachytene stage
(a) Only (A) is correct
(b) Only (B) is correct
(c) Only (A) and (C) are correct
(d) (A), (B) and (C) are correct

108. Consider the following statements and select the correct options for A, B and C:
(i) Bivalent chromosomes are clearly visible as tetrad in ____A____
(ii) In oocytes of some vertebrates, ____B____ lasts for months or years
(iii) Recombinase enzyme is activated in ____C____ stage of prophase I

	A	B	C
(a)	Zygotene	Diplotene	Pachytene
(b)	Pachytene	Diplotene	Pachytene

(c) Zygotene Pachytene Diplotene
(d) Pachytene Diplotene Zygotene

109. Which of the following events occurs once in meiosis?
(a) Karyokinesis
(b) Crossing over
(c) Centriole duplication
(d) Equitorial plate formation

110. Reduction in chromosome number occurs at:
(a) Anaphase I (b) Anaphase II
(c) Prophase II (d) Metaphase I

111. What will be the amount of DNA (A) and the number of chromosomes (B), respectively, in a pollen grain, if its pollen mother cell has 32 pg of DNA and 16 chromosomes in the G_1 phase?

	A	B
(a)	16 pg	16
(b)	32 pg	8
(c)	16 pg	8
(d)	8 pg	8

112. A student observed 24 chromosomes in every cell of a diploid plant. If this plant forms haploid gametes during the zygotene stage of meiosis I, how many bivalents will be seen by this student?
(a) 6 (b) 48
(c) 24 (d) 12

113. Synaptonemal complex is:
(a) Formed in zygotene stage
(b) Formed between homologous chromosomes
(c) Required for recombination process
(d) All of the above

114. Which event is not associated with prophase I?
(a) Synepsis (b) Congression
(c) Terminalisation (d) Crossing over

115. Meiosis does not help in:
(a) Bringing out variations
(b) Maintaining the same chromosome number generation after generation
(c) Formation of gametes
(d) Repair process

116. Mitotic prophase and prophase I of meiosis are similar in:
A Condensation of chromatin material
B. Disappearance of nuclear membrane and nucleolus
C. Disappearance of ER and Golgi bodies
D. Crossing over between homologous chromosomes
(a) A only (b) A, B and C
(c) A and B (d) A, B and D

117. In metaphase I, chromosomes are in:
(a) Tetrad stage (b) Dyad stage
(c) Monovalent stage (d) Monad stage

118. Give the name of the phase of meiosis in which:
A. Monovalent chromosomes align on the equatorial plate
B. Centromere of each chromosome splits and monads move towards opposite poles

	A	B
(a)	Metaphase I	Anaphase I
(b)	Metaphase II	Anaphase I
(c)	Metaphase I	Anaphase II
(d)	Metaphase II	Anaphase II

119. A pollen mother cell has 20 chromosomes and 80 pg of DNA at the S phase. Calculate the total number of chromosomes and chromatids, and the amount of DNA present in each daughter cell after telophase II:

	Number of chromosomes	Number of chromatids	Amount of DNA
(a)	10	10	20
(b)	10	5	20
(c)	20	20	40
(d)	20	20	80

120. Segregation of Mendelian factors (Aa) occurs during:
(a) Diplotene (b) Anaphase I
(c) Anaphase II (d) Zygotene

121. In synapsis, two homologous chromosomes are connected at the:
(a) Centromeres (b) Chromosomes
(c) Telomeres (d) None of the above

122. Endomitosis is:
(a) Mitosis without the nucleus
(b) Mitosis within the nucleus
(c) Frequent nucleus
(d) Mitosis in the uterine wall

123. During the cell cycle, two molecules of DNA are present in the chromosome during:
(a) G_1 phase (b) Beginning of S phase
(c) G_2 phase (d) End of M phase

124. Read the following statements and select the correct option:
A. Meiosis involves two successive divisions without any DNA replication occurring between them
B. The number of chromosomes in daughter cells after meiosis II is reduced to half but the amount of DNA remains the same
(a) Only (A) is correct
(b) Only (B) is correct
(c) Both (A) and (B) are correct
(d) Both (A) and (B) are incorrect

125. If the meiospore has 20 kilograms of DNA, then what was the DNA content in its meiocyte at the G_1 and G_2 phases, respectively:
(a) 20, 40 (b) 40, 80
(c) 40, 40 (d) 80, 40

126. From the given features, how many occur at the pachytene stage of meiosis?
(i) Synapsis
(ii) Chromosomes clearly visible at tetrad
(iii) Formation of chiasmata
(iv) Recombination nodule
(v) Terminalisation
(vi) Exchange of genetic material
(vii) Disjunction
(a) Four (b) Five
(c) Two (d) Three

127. Read the following statements:
A. Chromosomes clearly visible as tetrad
B. Exchange of non-sister chromatids between homologous chromosomes
The above characteristics are attributed to:
(a) Zygotene (b) Diplotene
(c) Pachytene (d) Leptotene

128. Chromosome appears beaded during:
(a) Pachytene (b) Leptotene
(c) Diplotene (d) Diakinesis

129. The events listed below occur during different phases of meiosis. Identify them and select the correct option:
A. All chromosomes of a cell are directed towards one side and are attached to the nuclear membrane
B. Pairing of homologous chromosomes
C. Chiasmata slip from their original position and pass outwards
D. Exchange of chromatid segments between the homologous chromosomes

	A	B	C	D
(a)	Leptotene	Zygotene	Diplotene	Pachytene
(b)	Leptotene	Diplotene	Diakinesis	Pachytene
(c)	Diakinesis	Diplotene	Zygotene	Leptotene
(d)	Pachytene	Zygotene	Diakinesis	Diplotene

130. At anaphase II of meiosis, each chromosome contains:
(a) 4 DNA (b) 3 DNA
(c) 2 DNA (d) 1 DNA

131. Significance of meiosis lies in:
(a) Production of genetic variability
(b) Maintaining constancy of chromosome number during sexual reproduction
(c) Reduction of chromosome number to half
(d) All of the above

132. Number of chromosome groups at the equatorial plate of metaphase I of a plant body having 2N = 50 chromosomes shall be:
(a) 25 (b) 50
(c) 100 (d) 75

133. Intra-meiotic interphase is a resting phase between meiosis I and meiosis II, in which:
A. DNA replicates
B. Centriole pairs replicate in animal cells
C. Chromosomes decondense to form chromatin
D. RNA and proteins are synthesized
(a) (B), (C) and (D)
(b) (B) and (D) only
(c) (A), (B) and (C) only
(d) (A), (C) and (D) only

134. Final stage of meiotic prophase I is marked by:
(a) Formation of chiasmata
(b) Terminalisation of chiasmata
(c) Disjunction of chromosomes
(d) Dispersion of chromatin material

135. At the end of meiosis II, how many cells are formed from a single diploid parent cell?
(a) Two haploid cells
(b) Two diploid cells
(c) Four haploid cells
(d) Four diploid cells

136. Read the following statements with respect to meiosis and select the correct option stating which ones are true and which ones are false:
A. Meiosis increases genetic variability in the population which is essential for evolution
B. Synaptonemal complex develops between two synapsed homologous chromosomes
C. Each chromosome of a bivalent attaches with two spindles in metaphase I
D. At anaphase I, the centromere splits but sister chromatids do not separate

	A	B	C	D
(a)	T	T	F	F
(b)	F	T	T	F
(c)	F	F	T	T
(d)	T	F	F	T

137. Which stage of meiosis do the following figures (A) and (B) represent:

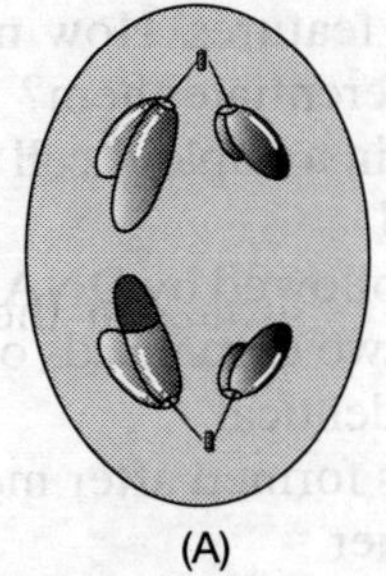
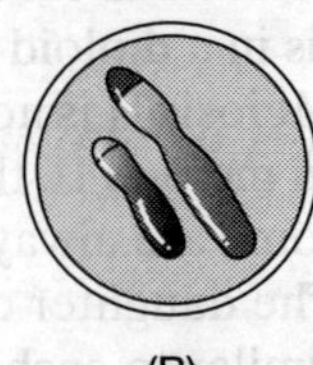
(A) (B)

A	B
(a) Anaphase II	Metaphase I
(b) Anaphase I	Metaphase II
(c) Anaphase I	Telophase I
(d) Anaphase II	Telophase II

138. Amitosis is:
(a) Division involving formation of chromosome bridges
(b) Division involving spindle formation
(c) Division in which chromosomes are unequally distributed
(d) Cleavage of nucleus without recognisable chromosome distribution

139. Amitosis is seen in:
(a) Bacteria
(b) Protozoans
(c) Mammalian cartilage cells
(d) All of the above

140. The RNA and proteins required during meiosis II are synthesized during which of the following stages?
(a) M phase
(b) Interphase
(c) Synthetic phase
(d) Intra-meiotic interphase

141. How many meiotic divisions are required to produce 100 pollen grains in an angiospermic plant?
(a) 25 (b) 4
(c) 125 (d) 100

142. Arrange the following events of meiosis in their correct sequence of occurrence and choose the correct option:
A. Chromosomal synapsis
B. Dissolution of synaptonemal complex
C. Chromosomes clearly visible as tetrad
D. Alignment of monovalent at equator
E. Terminalisation of chiasmata
(a) A→C→B→E→D
(b) A→C→E→B→D
(c) A→D→C→B→E
(d) B→A→E→D→C

143. If the number of bivalents is 16 in metaphase I, what will be the number of chromosomes in daughter cells after meiosis I and meiosis II, respectively?
(a) 32, 16 (b) 16, 8
(c) 16, 16 (d) 9, 8

144. Brachymeiosis consists of:
(a) Two reductional divisions and one equational division
(b) One reductional division and one equational divisions
(c) One reductional division and two equational divisions
(d) Two reductional divisions and two equational divisions

145. Meiosis ensures the production of the ___A___ phase in the life cycle of a sexually reproducing organism whereas fertilisation restores the ___B___ phase:

A	B
(a) Diploid	Haploid
(b) Haploid	Diploid
(c) Haploid	Triploid
(d) Diploid	Triploid

146. The male gamete of a plant has 20 pg DNA. What would be the amount of DNA in the meiocyte of the same plant at its G_2 phase of the cell cycle?
(a) 20 pg (b) 40 pg
(c) 80 pg (d) 160 pg

147. "A cell has 15 bivalents in its zygotene stage". Select the correct option regarding the above statement:
A. The cell has a total of 30 chromosomes
B. Each daughter cell produced from the above cell has 15 chromosomes
C. The cell has a total of 60 chromatids
(a) Only (A) (b) Only (C)
(c) Only (A) and (B) (d) (A), (B) and (C)

148. If the number of chromosomes and the amount of DNA in a meiocyte is 20 and 20 pg, respectively, in the G_1 phase, what will be the number of chromosomes and amount of DNA in the daughter cell after meiosis II?
(a) 10, 10 pg (b) 10, 5 pg
(c) 10, 20 pg (d) 5, 5 pg

149. The X-shaped structure that appears during the ________ stage of prophase I is:
(a) Diplotene (b) Zygotene
(c) Pachytene (d) Diakinesis

150. Activity of recombinase enzyme is seen at:
(a) Leptotene (b) Pachytene
(c) Diplotene (d) Diakinesis

151. Which of the following is not true for meiosis?
(a) Production of genetic variability
(b) Maintaining constancy of chromosome number during sexual reproduction
(c) Reduction of chromosome number to half
(d) Production of diploid cells

152. Meiosis II is also called equational division like mitosis, but meiosis II and mitosis differ from each other in several features. How many of the following features differentiate them?
A. Meiosis II occurs in a haploid cell while mitosis in a diploid cell
B. Meiosis II is not followed by DNA replication
C. In meiosis II, the two chromatids of a chromosome are always identical
D. The daughter cells formed after meiosis II are similar to each other
(a) Two (b) Three
(c) One (d) Four

NCERT EXEMPLAR QUESTIONS

1. Meiosis in diploid organisms results in:
(a) Production of gametes
(b) Reduction in the number of chromosomes
(c) Introduction of variation
(d) All of the above

2. At which stage of meiosis is the genetic constitution of gametes finally decided?
(a) Metaphase I (b) Anaphase II
(c) Metaphase II (d) Anaphase I

3. Meiosis occurs in organisms during:
(a) Sexual reproduction
(b) Vegetative reproduction
(c) Both sexual and vegetative reproduction
(d) None of the above

4. During anaphase I of meiosis:
(a) Homologous chromosomes separate
(b) Non-homologous chromosomes separate
(c) Sister chromatids of chromosomes separate
(d) Non-sister chromatids of chromosomes separate

5. Mitosis is characterised by:
(a) Reduction division
(b) Equal division
(c) Both reduction and equal division
(d) Pairing of homologous chromosomes

6. A bivalent of meiosis I consists of:
(a) Two chromatids and one centromere
(b) Two chromatids and two centromeres
(c) Four chromatids and two centromeres
(d) Four chromatids and four centromeres

7. Cells which are not dividing are likely to be at:
(a) G_1 (b) G_2
(c) G_0 (d) S phase

8. Which of the events listed below is not observed during mitosis?
(a) Chromatin condensation
(b) Movement of centrioles to opposite poles
(c) Appearance of chromosomes with two chromatids joined together to the centromere
(d) Crossing over

9. Identify the incorrect statement about meiosis:
(a) Pairing of homologous chromosomes
(b) Four haploid cells are formed
(c) At the end of meiosis, the number of chromosomes is reduced to half
(d) Two cycles of DNA replication occur

10. Select the correct statement about the G_1 phase:
(a) Cell is metabolically inactive
(b) DNA in the cell does not replicate
(c) It is not a phase of synthesis of macromolecules
(d) Cell stops growing

PRACTICE TEST PAPER

1. During which phase (S) of the cell cycle does the amount of DNA in a cell remain at the 4C level if the initial amount of DNA is denoted as 2C?
(a) G_0 and G_1 (b) G_1 and G_2
(c) Only S (d) S and G_2

2. The complex formed by a pair of synapsed homologous chromosomes is called:
(a) Bivalent (b) Tetrad
(c) Recombination nodule (d) Both (a) and (b)

3. During mitosis, the endoplasmic reticulum and nucleolus begin to disappear at:
(a) Early prophase (b) Late prophase
(c) Early metaphase (d) Interphase

4. The stage of the cell cycle where the cell has undergone differentiation is:
(a) G_0 (b) G_1
(c) G_2 (d) S

5. Antiphase is the phase in which ATP is synthesized during cell division. It refers to:
(a) G_0 phase (b) G_1 phase
(c) S phase (d) G_2 phase

6. Histone protein synthesis occurs during:
(a) G_1 phase (b) G_2 phase
(c) G_0 phase (d) S phase

7. The best material for the study of meiosis in the laboratory is:
(a) Anther (b) Root tip
(c) Leaf tip (d) Ovary

8. Colchicine is a mitotic poison because it:
(a) Causes splitting of chromosomes
(b) Inhibits the formation of mitotic spindle
(c) Causes non-pairing of chromosomes
(d) Agglutinates the chromosomes

9. The figure shown below depicts which stage of cell division?

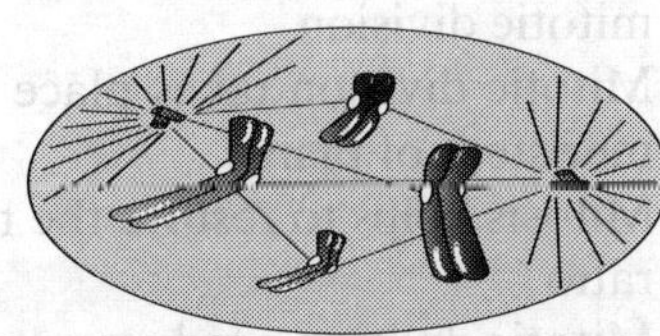

(a) Metaphase (b) Prophase
(c) Transition to metaphase (d) Anaphase

10. What would be the DNA content (A) and number of chromosomes (B) in the meiocyte at the telophase II stage, if at the G_1 phase, it possesses 10 pg of DNA and 20 chromosomes?

	A	B
(a)	5	10
(b)	10	20
(c)	5	20
(d)	10	10

11. The microtubules from the opposite poles of the spindle attach to a pair of homologous chromosomes in:
(a) Metaphase I (b) Metaphase II
(c) Metaphase (d) Both (a) and (c)

12. Synthesis of the following substances takes place in interkinesis, except:
(a) RNA (b) Protein
(c) Enzyme (d) DNA

13. Choose the odd one out with respect to the significance of mitosis:
(a) Growth (b) Healing of wound
(c) Genetic variations (d) Asexual reproduction

14. In metaphase, each chromosome is attached with the help of ___ spindle fibres:
(a) 1 (b) 2
(c) 4 (d) Many

15. Match the following columns and select the correct option:

	Column I		Column II
A.	Synaptonemal complex formation	(i)	Diplotene
B.	Terminalisation start phase	(ii)	Pachytene
C.	Recombinase enzyme activity	(iii)	Diakinesis
D.	Disappearance of nuclear membrane	(iv)	Zygotene

(a) A – (iv), B – (i), C – (ii), D – (iii)
(b) A – (iv), B – (ii), C – (i), D – (iii)
(c) A – (iii), B – (iv), C – (i), D (ii)
(d) A – (ii), B – (iii), C – (iv), D – (i)

16. Which of the following is not related to the S phase of the cell cycle?
(a) Centriole duplication occurs in the cytoplasm
(b) DNA replication occurs in the nucleus

(c) Amino acids of histone proteins are synthesized
(d) Ploidy level remains the same

17. Which of the following events occurs twice in the meiotic cell cycle?
(a) Karyokinesis, cytokinesis and DNA replication
(b) DNA replication and karyokinesis
(c) Cytokinesis and DNA duplication
(d) Karyokinesis and centriole duplication

18. In which of the following stages of meiosis does the centromere not divide but the number of chromosomes becomes half at both ends?
(a) Metaphase I (b) Anaphase I
(c) Metaphase II (d) Anaphase II

19. For which of the following stages is the maximum time taken during meiosis?
(a) Leptotene (b) Zygotene
(c) Pachytene (d) Diplotene

20. If the DNA content in the sperm is 30 pg, what was the DNA content in its mother cell at the end of meiosis I?
(a) 60 pg (b) 120 pg
(c) 30 pg (d) 240 pg

21. In which of the following stages of cell division is each chromosome made up of two DNA molecules?
(a) Metaphase, Anaphase
(b) Metaphase, Metaphase I and Anaphase I
(c) Anaphase, Metaphase I and Anaphase II
(d) Anaphase, Anaphase I and Anaphase II

22. A cell entering the M phase of the cell cycle is:
(a) Always diploid and with duplicated chromosomes
(b) Either haploid or diploid but always with duplicated chromosomes
(c) Always haploid and with duplicated chromosomes
(d) Always haploid or diploid but always with unduplicated chromosomes

23. Pole-ward movement of dyad chromosomes is a feature of:
(a) Anaphase I (b) Anaphase II
(c) Metaphase I (d) Metaphase II

24. Read the following statements and select the correct option:
A. Heart cells in animals do not appear to exhibit cell division
B. M phase is the most dramatic period of the cell cycle
C. The stage between meiosis I and meiosis II is generally long
(a) (A) and (C) are correct
(b) (B) and (C) are correct
(c) (A) and (B) are correct
(d) (A), (B) and (C) are correct

25. Terminalisation start phase of meiosis is also concerned with:
(a) Formation of bivalent stage
(b) Dissolution of synaptonemal complex
(c) Crossing over
(d) Reformation of synaptonemal complex

26. If a meiocyte contains 20 chromosomes, what will be the chromosome number at each pole during anaphase I?
(a) 40 (b) 20
(c) 5 (d) 10

27. In a complete cell cycle of a human cell, the time taken by the mitotic phase is approximately:
(a) 25% of the total time
(b) 5% of the total time
(c) 95% of the total time
(d) 50% of the total time

28. Mark the similar feature between meiosis I and meiosis II:
(a) Both are equational divisions
(b) Both divisions are heterotypic divisions
(c) Both the divisions involve pairing of homologous chromosomes and recombination between them
(d) Both occur at the time of gamete formation in diploid organisms

29. In which stage of meiosis is the lamp brush chromosome observed?
(a) Anaphase II (b) Diplotene
(c) Diakinesis (d) Zygotene

30. Identity the following statements as true (T) or false (F) and choose the option accordingly:
A. In some social insects, haploid cells divide by mitotic division
B. Mitotic division takes place only in the apical meristem of plants
C. Mitosis helps to restore the nucleocytoplasmic ratio
D. Mitotic division produces daughter cells which are identical to each other as well as to the parent cell

	A	B	C	D
(a)	T	F	T	T
(b)	T	F	F	T
(c)	F	F	F	T
(d)	F	T	T	F

31. Separation of chromosomes with associated chromatids at the centromere occurs in:
(a) Anaphase
(b) Anaphase I
(c) Anaphase II
(d) Anaphase I and Anaphase

32. Mention the DNA content of the cell at G_2, prophase I and metaphase II:
(a) 4C, 4C, 2C (b) 2C, 2C, 1C
(c) 4C, 2C, 1C (d) 4C, 2C, 2C

33. How many bivalents are formed in the meiocyte of humans?
(a) 46 (b) 21
(c) 44 (d) 23

34. A duplicated chromosome has how many chromatids?
(a) One (b) Two
(c) Three (d) Four

35. Disjunction process takes place in:
(a) Prophase (b) Metaphase
(c) Anaphase (d) Telophase

36. Which stage is regarded as the reverse of prophase?
(a) Interphase (b) Telophase
(c) Anaphase (d) Metaphase

37. In mitosis, chromosomal condensation is completed at:
(a) Late prophase (b) Early prophase
(c) Anaphase (d) Metaphase

38. If a cell has 20 chromosomes with 20 picogram DNA at its G_1 phase, what will be the number of chromatids and the amount of DNA in prophase?
(a) 40, 20 (b) 20, 40
(c) 40, 40 (d) 20, 20

39. The phenomenon of bringing the chromosomes to the equator of the spindle is called:
(a) Decondensation (b) Disjunction
(c) Congression (d) Synapsis

40. The longest phase of meiosis I is:
(a) The stage phase of meiosis I
(b) Divided into five sub-stages
(c) Characterised by division of centromere
(d) Responsible for separation of homologous chromosomes

41. Which of the following is not associated with prophase I?
(a) Synapsis (b) Terminalisation
(c) Disjunction (d) Condensation

42. The enzyme recombinase is required at which stage of meiosis?
(a) Leptotene (b) Pachytene
(c) Diplotene (d) Diakinesis

43. Which of the following phases represent the phase between two successive M phases?
(a) Synthetic phase (b) Interkinesis
(c) Interphase (d) All except (a)

44. How many of the following events occur once in meiosis?
Synapsis, DNA duplication, Centriole duplication, Crossing over, Karyokinesis, Cytokinesis, Chiasmata formation, Splitting of centromere
(a) Four (b) Five
(c) Three (d) Six

45. The cell cycle G_1 phase is followed by _____ and metaphase follows _____:
(a) S, Anaphase (b) S, Prophase
(c) G_2, Anaphase (d) G_0, Anaphase

ANSWER KEYS

Exercises

1. (d) **2.** (a) **3.** (b) **4.** (c) **5.** (a) **6.** (b) **7.** (a) **8.** (b) **9.** (d) **10.** (b)
11. (b) **12.** (d) **13.** (a) **14.** (b) **15.** (c) **16.** (b) **17.** (c) **18.** (a) **19.** (d) **20.** (a)
21. (c) **22.** (d) **23.** (b) **24.** (c) **25.** (a) **26.** (c) **27.** (a) **28.** (b) **29.** (b) **30.** (d)
31. (b) **32.** (c) **33.** (d) **34.** (c) **35.** (c) **36.** (a) **37.** (b) **38.** (a) **39.** (d) **40.** (d)
41. (a) **42.** (b) **43.** (c) **44.** (d) **45.** (b) **46.** (d) **47.** (b) **48.** (d) **49.** (b) **50.** (a)
51. (d) **52.** (c) **53.** (a) **54.** (a) **55.** (a) **56.** (b) **57.** (d) **58.** (a) **59.** (a) **60.** (b)
61. (a) **62.** (c) **63.** (b) **64.** (d) **65.** (b) **66.** (a) **67.** (b) **68.** (c) **69.** (c) **70.** (b)
71. (d) **72.** (a) **73.** (d) **74.** (c) **75.** (b) **76.** (b) **77.** (d) **78.** (a) **79.** (d) **80.** (c)
81. (b) **82.** (a) **83.** (b) **84.** (d) **85.** (d) **86.** (b) **87.** (a) **88.** (b) **89.** (d) **90.** (d)
91. (b) **92.** (b) **93.** (c) **94.** (c) **95.** (c) **96.** (a) **97.** (c) **98.** (c) **99.** (c) **100.** (a)
101. (d) **102.** (d) **103.** (c) **104.** (a) **105.** (c) **106.** (a) **107.** (d) **108.** (b) **109.** (b) **110.** (a)
111. (c) **112.** (d) **113.** (d) **114.** (b) **115.** (d) **116.** (b) **117.** (a) **118.** (d) **119.** (d) **120.** (b)
121. (b) **122.** (b) **123.** (c) **124.** (a) **125.** (b) **126.** (d) **127.** (c) **128.** (b) **129.** (b) **130.** (d)
131. (d) **132.** (a) **133.** (b) **134.** (b) **135.** (c) **136.** (a) **137.** (d) **138.** (d) **139.** (d) **140.** (d)
141. (a) **142.** (a) **143.** (c) **144.** (a) **145.** (b) **146.** (c) **147.** (d) **148.** (b) **149.** (a) **150.** (b)
151. (d) **152.** (a)

NCERT Exemplar Questions

1. (d) **2.** (b) **3.** (a) **4.** (a) **5.** (b) **6.** (c) **7.** (c) **8.** (d) **9.** (d) **10.** (b)

Practice Test Paper

1. (d) **2.** (d) **3.** (a) **4.** (a) **5.** (b) **6.** (d) **7.** (a) **8.** (b) **9.** (c) **10.** (a)
11. (a) **12.** (d) **13.** (c) **14.** (b) **15.** (a) **16.** (c) **17.** (d) **18.** (b) **19.** (d) **20.** (a)
21. (b) **22.** (b) **23.** (a) **24.** (c) **25.** (b) **26.** (d) **27.** (b) **28.** (d) **29.** (b) **30.** (a)
31. (b) **32.** (a) **33.** (d) **34.** (b) **35.** (c) **36.** (b) **37.** (d) **38.** (b) **39.** (c) **40.** (b)
41. (c) **42.** (b) **43.** (c) **44.** (b) **45.** (b)

SOLUTIONS

Cell Cycle

1. The sequence of events by which a cell duplicates its genome, synthesizes the outer constituent of the cell and eventually divides into two daughter cells is called cell cycle. Cell cycle consists of interphase and mitosis.
2. Interphase constitutes more than 95% of the duration of the cell cycle.
3. Option (b) is correct.
4. Yeast cell divides once every 90 minutes, while a typical human cell divides once approximately every 24 hours.
5. 2C, 4C, 4C
6. Option (b) is correct.
7. G_1 phase is the interval between mitosis and initiation of DNA replication. It is also called post-mitotic gap phase.
8. G_0 phase or quiescent stage
9. The non-dividing cells enter the quiescent stage.
10. DNA replication occurs in the S phase of the cell cycle.
11. In the S phase, DNA content = 20 pg. Amount of DNA in G_1 (post-mitotic gap phase) = 10 pg. In the S phase, the amount of DNA doubles but the number of chromosomes remains the same.
12. Both (B) and (C) are correct. In the G_0 phase, the cells remain metabolically active but do not proliferate, unless they are called on to do so.
13. G_0 phase
14. Each chromosome has two chromatids called sister chromatids.
15. Cyclins are proteins that activate protein kinases to regulate the eukaryotic cell cycle.
16. Fibroblast cells in our body are those that are arrested in the G_1 phase.
17. (B), (C) and (D) are correct. Synthesis of deoxyribonucleotides occurs in the G_1 phase.
18. (A), (B) and (D) are correct. The cell synthesizes its constituents in a sequential pattern to ensure the proper division and distribution of components.
19. All the statements are correct.
20. A – S, B – G_1, C – M
21. In the S phase, the following events occur: DNA replication, centriole duplication, histone protein synthesis.
22. All the statements are correct.
23. (C) and (D) are false. Decision of a cell to divide occurs in the G_1 phase of the cell cycle. Cell growth disturbs the nucleocytoplasmic ratio.
24. 4C
25. In adult animals, the nerve cells do not divide.
26. In the S phase, each chromosome has two chromatin threads.
27. (A), (B) and (D) are correct.
28. G_1 phase
29. In the S phase, DNA replication occurs in the nucleus and centriole duplication occurs in the cytoplasm.
30. In the G_0 phase of the cell cycle, cells suspend the cell cycle.
31. Assess the DNA damage
32. S phase
33. Chromosome number remains the same.
34. One hour is the average time taken by the M phase, which is 4% of the total time taken by a cell cycle.
35. G_2 phase
36. A – S, B – M, C – G_1 phase. Cells are the least sensitive in the S phase, then the G_1 phase, then the G_2 phase and the most sensitive in the M phase of the cell cycle.
37. In the resting phase (interphase), cell growth and DNA replication occur in an orderly manner.

38. M phase restores the nucleocytoplasmic ratio.

39. Both euchromatin and heterochromatin

40. Cell size increases.

41. G_1 phase

42. Centriole duplication occurs in the S phase of the cell cycle.

43. The stage in the cell cycle when an energy source is stored for the completion of cell division is called antiphase. It is usually in the G_1 phase.

Mitosis

44. (A), (B) and (D) are correct. Plants can show mitotic divisions in both haploid and diploid cells.

45. Root apex

46. Prophase – Metaphase – Anaphase – Telophase

47. In prophase and metaphase, the chromosomes are composed of two sister chromatids held together at the centromere. In anaphase, the sister chromatids of each chromosome separate.

48. Meiosis forms gametes, which is important for sexual reproduction.

49. Chromosomes are small, bead-shaped and heavily staining masses of coiled chromatin that are linearly arranged along the chromosome during early prophase.

50. During prophase, chromosomes are large and straight.

51. The given figure represents late prophase. Cells at the end of prophase, when viewed under the microscope, do not show Golgi complexes, ER, nucleolus and nuclear envelope.

52. At anaphase, the chromosomes appear to be V, L, J and I shaped.

53. At anaphase, the chromosomes move towards opposite poles.

54. Phragmoplast is the precursor of the cell plate. They are vesicles bounded out from the Golgi apparatus, rich in pectin, and will form the middle lamella.

55. Prophase is the longest phase of the M phase.

56. Two chromatids

57. During telophase, a nuclear membrane is formed around each set of chromosomes at opposite poles. Nucleolus, Golgi apparatus and ER reform.

58. a(iii), b(i), c(iv), d(ii)

59. A – Metaphase, B – Early Prophase, C – Telophase

60. Prophase and Metaphase

61. A new nuclear membrane is formed around each set of chromosomes at opposite poles by ER.

62. Cytokinesis

63. Cytokinesis is the division of the cytoplasm.

64. Cell wall formation is achieved by the appearance of the cell plate in the centre of the cell. Furrowing cannot occur in plant cells as it is surrounded by a cell wall. Cell plate formation is centrifugal.

65. Cytokinesis in animal cells is achieved by the appearance of cell furrow in the plasma membrane; furrowing occurs in a centripetal manner.

66. In some organisms, division of the nucleus occurs but division of the cytoplasm does not occur. As a result, a large number of nuclei are present in the cytoplasm. This leads to the formation of syncytium; that is, a cell containing a large number of nuclei.

67. Only (B) is correct. A smaller cell has a larger surface area volume ratio than a large cell.

68. During early prophase, chromosomes look like a ball of wool. This stage is called the spireme stage.

69. Mitotic apparatus is formed by asters and spindles.

70. 30 ATP

71. In animals, mitosis is seen only in the diploid somatic cells, except in social insects (where haploid cells are also divided by mitosis). Against this, plants can show mitotic division in both haploid and diploid cells.

72. Replicated chromosomes are separated at anaphase.

73. In meiosis, genetic variations are created due to crossing over.

74. Colchicine cause disassembly of spindle fibres. Colchicine causes metaphasic arrest and doubling of chromosomes.

75. Anaphasic movement of chromosomes occurs by shortening (dipolymerisation) of chromosomal fibres and extension (polymerisation) of interzonal fibres.

76. Metaphase is characterised by all the chromosomes that come to lie at the equator. At this stage, chromosome morphology is most easily studied.

77. Interphase nucleus

78. In mitosis, the spindle is bipolar. Both the poles of the mitotic spindle are made up of centrosomes and the chromosomes are lined up at the spindle equator to ensure their correct bi-orientation and segregation.

79. Plastids have no role in cell division.

80. Centromeres of chromosomes are found towards the poles, thus leading the path, while the arms of the chromosomes trail behind.

81. In the G_2 phase, actual preparation of the M phase occurs.

82. Anaphasic chromosomes contain one chromatid, two arms and one molecule of ds-DNA.

83. Acentric chromosomes

84. Both (A) and (B) are correct.

85. Option (d) is correct.

86. Azide and cyanide inhibit prophase.

87. B – A – C

88. Metaphase

89. Chromosomal fibres or tractile fibres extend from the pole to the chromosome.

90. Kinetochore and spindle fibre

91. At metaphase, condensation of chromosomes is completed.

92. The phragmoplasts are organised at the end of anaphase.

93. Spindle fibre and kinetochore

94. Transition to metaphase

95. A – (iii), B – (iv)

96. Number of generations of mitosis for producing X cells = 2n.

97. All five statements are correct.

98. Anaphase stage

99. Both (A) and (B) are correct.

100. Only (A) is correct. Formation of the spindle in asteral mitosis is a mechanism that leads to the distribution of the sister chromatids between two daughter cells.

Meiosis and Amitosis

101. Option (d) is correct.

102. Meiosis involves two sequential cycles of nuclear and cell division with one cycle of DNA replication.

103. Meiosis involves pairing of homologous chromosomes (synapsis) and recombination between them (crossing over).

104. Leptotene – Zygotene – Pachytene – Diplotene – Diakinesis

105. Meiosis involves a single cycle of DNA replication.

106. Leptotene stage

107. (A), (B) and (C) are correct.

108. A – Pachytene, B – Diplotene, C – Pachytene

109. Crossing over occurs at the pachytene stage of meiosis I.

110. There occurs true reduction in the number of chromosomes at anaphase I because homologous chromosomes separate at this stage.

111. A – 16 pg, B – 8.

At the S phase, number of chromosomes = 16, NA amount = 64 pg.

At the end of meiosis I, a cell has 8 chromosomes and 32 pg DNA.

At the end of meiosis II, a cell has 8 chromosomes and 16 pg DNA.

112. The number of bivalents in a cell = Number of haploid chromosomes = 12.

113. All are correct.

114. The phenomenon of bringing chromosomes to the equator of the spindle is called congression. It occurs at metaphase.

115. Mitosis is the process of repairing or replacing dead/old cells by new cells.

116. (A), (B) and (C) are correct.

117. At metaphase I, bivalent (tetrad) chromosomes arrange themselves on the equator of the bipolar spindle to produce a double equatorial plate.

118. A – Metaphase II, B – Anaphase II

119. Number of chromosomes = 10.

Number of chromatids = 10.

Amount of DNA = 20 pg.

120. Segregation of Mendelian factors (Aa) occurs at anaphase I because the homologous chromosomes are separate.

121. In synapsis, two homologous chromosomes are separated at the chromomeres.

122. Endomitosis is mitosis within the nucleus.

123. In the G_2 phase, one chromosome has two chromatids, four arms and two molecules of ds-DNA.

124. Only (A) is correct. Reduction in DNA from 2C to 1C occurs only in meiosis II.

125. At the G_1 stage = 40 kg; G_2 stage = 80 kg.

126. Three features (2, 4 and 6) are correct for pachytene.

127. Pachytene

128. Leptotene. Chromosomes appear beaded during the leptotene stage.

129. A – Leptotene, B – Diplotene, C – Diakinesis, D – Pachytene

130. At anaphase II, each chromosome has one chromatid, two arms and one DNA.

131. All the statements are correct.

132. At metaphase I, bivalent chromosomes arrange themselves on the equator to produce a double equatorial plate. Each plate will have half the number of diploid chromosomes.

133. Interkinesis or intrameiotic interphase involves RNA and protein synthesis and centriole pair duplication. DNA replication does not occur.

134. The last stage of prophase I is diakinesis and it is characterised by terminalisation of chiasmata.

135. Four haploid cells are formed.

136. (C) and (D) are false. The microtubules of the spindle fibres from opposite poles attach to the centromere of the chromosome facing towards it. Univalent (dyad) chromosomes containing two chromatids held together at the centromere separate and move towards opposite poles at the anaphase I.

137. A – Anaphase II, B – Telophase II

138. During amitosis, the nucleus of the cell elongates, then a constriction appears in the nucleus, which gradually deepens and divides the nucleus into two daughter nuclei.

139. All are correct.

140. Intrameiotic interphase is the gap between meiosis I and meiosis II. At this phase, a cell prepares itself for meiosis II.

141. 25

142. a – c – b – e – d

143. Number of bivalents = 16 because number of chromosomes = 32.

Chromosome number after meiosis I = 16.

Chromosome number after meiosis II = 16.

144. Brachy meiosis is a different type of meiosis observed in some fungi. It involves three types of divisions: the first and second are reductional and the third is equational.

145. A – Haploid, B – Diploid

146. 80 pg DNA

147. Each divalent has two chromosomes, 4 chromatids, 8 arms and 4 molecules of linear ds-DNA.

148. Number of chromosomes = 10.

Number of DNA = 5 pg.

149. Diplotene stage

150. Pachytene stage

151. Four haploid cells are produced at the end of meiosis II.

152. Two

11 Transport in Plants

EXERCISES

Introduction, Means of Transport

1. Transport of substances over short distances occurs by:
(a) Diffusion
(b) Cytoplasmic streaming movement
(c) Xylem and phloem
(d) Both (a) and (b)

2. Transport over longer distances proceeds through ________:
(a) Cytoplasmic streaming movement
(b) Osmosis
(c) Xylem and phloem
(d) Diffusion

3. Multidirectional flow of a variety of organic and inorganic solutes occurs through:
(a) Xylem (b) Phloem
(c) Symplast (d) Active transport

4. Read the following statements and state whether true (T) or false (F):
A. Short-distance transport proceeds through vascular tissue
B. Transport in xylem is essentially unidirectional
C. Organic and mineral nutrients undergo multi-directional transport
D. Hormones or plant growth regulators and other chemical stimuli are transported in small amounts in a polarised or unidirectional manner always

	A	B	C	D
(a)	F	T	T	F
(b)	F	T	T	T
(c)	T	F	F	T
(d)	T	F	F	T

5. Which of the following is not a characteristic of diffusion?
(a) It is a slow process
(b) Dependent on living systems
(c) More common in liquids and gases
(d) Only means of movement of gases within the plant

6. Select the correct feature for facilitated diffusion:
(a) It is uphill transport
(b) Does not require membrane proteins
(c) It is a highly specific selection process
(d) Rate of transport does not saturate

7. Select the incorrect statement:
(a) Long-distance transport proceeds through vascular tissue
(b) When any plant part undergoes senescence, nutrients may be withdrawn from such regions and moved to the growing parts
(c) Diffusion rate is not affected by the gradient of concentration
(d) Hydrophilic substances cannot pass through the cell membrane due to the presence of lipids in the cell membrane

8. Gaseous movement within the plant body is due to:
(a) Simple diffusion only
(b) Actual transport only
(c) Diffusion and sometimes by active transport
(d) Facilitated diffusion and active transport

9. Consider the given properties:
A. Highly selective in nature
B. Energy expenditure
C. Uphill transport
D. Transport saturation
E. Requirement of special membrane proteins

How many of the above proteins are related to facilitated diffusion and active transport?
(a) (C), (E) (b) (E), (C)
(c) (C), (D) (d) (D), (E)

10. Porins are found in all, except:
(a) Outer membrane of plastids
(b) Outer membrane of mitochondria
(c) Outer membrane of some bacteria
(d) Inner membrane of both mitochondria and plastids

11. The process of active transport is different from facilitated diffusion due to:
(a) Involvement of membrane proteins
(b) Its response to protein inhibitors
(c) Being an uphill process
(d) Transport saturation

12. Aquaporins:
(a) Form small pores for the entry of large proteins
(b) Form water channels
(c) Allow the uphill transport of water molecules
(d) Help in the movement of hydrophobic substances

13. Water channels are made up of:
(a) Eight different types of aquaporins
(b) Four different types of aquaporins
(c) Two different types of aquaporins to facilitate diffusion
(d) Three different types of aquaporins

14. Match the means of transport given in Column I with its characteristic feature in Column II:

Column I	Column II
A. Simple diffusion	(i) Semipermeable membrane
B. Facilitated diffusion	(ii) Carrier protein and energy
C. Active transport	(iii) Downhill movement
D. Osmosis	(iv) Porins

(a) A – (ii), B – (iii), C – (iv), D – (i)
(b) A – (ii), B – (iii), C – (i), D – (iv)
(c) A – (iii), B – (iv), C – (ii), D – (i)
(d) A – (i), B – (iv), C – (ii), D – (iv)

15. Read the following statements and select the correct option:
A. Diffusion is a slow process and is not dependent on living systems
B. Diffusion of solids rather than in solids is more likely
(a) Only (A) is correct
(b) Only (B) is correct
(c) Both (A) and (B) are correct
(d) Both (A) and (B) are incorrect

16. Two diffusion molecules transported across the membrane in opposite directions is called:
(a) Uniport (b) Symport
(c) Antiport (d) Active transport

17. When a molecule moves across a membrane independent of other molecules, the process is called:
(a) Uniport (b) Symport
(c) Antiport (d) Diffusion

18. Identify the parts labelled A , B , C and D:

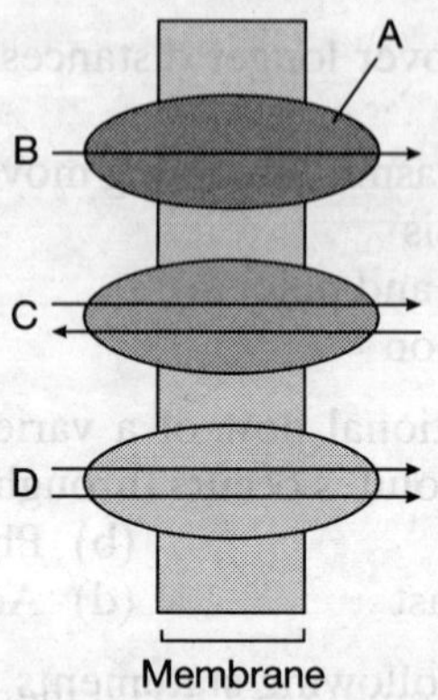

	A	B	C	D
(a)	Carrier protein	Uniport	Symport	Antiport
(b)	Carrier protein	Uniport	Antiport	Symport
(c)	Pumps	Uniport	Antiport	Symport
(d)	Pump	Co-transport	Symport	Antiport

19. Substances that have a ____ moiety find it difficult to pass through the membrane:
(a) Hydrophilic (b) Hydrophobic
(c) Neutral (d) Lipophilic

20. Pumps are related to:
(a) Simple diffusion
(b) Facilitated diffusion

(c) Active transport
(d) Both (b) and (c)

21. Simple diffusion differs from facilitated diffusion in:
(a) Showing downhill transport
(b) Requiring energy
(c) Showing transport saturation
(d) No requirement of special membrane proteins

22. In active transport, transporter proteins are sensitive to inhibitors that react with:
(a) Protein backbone
(b) Lipid bilayer
(c) Polar tail of lipids
(d) Protein side chains

23. Proteins in the membrane responsible for facilitated diffusion and active transport:
(a) Are highly selective
(b) Are liable to saturate
(c) Do not respond to inhibitors
(d) Are under hormonal regulation

24. Select the incorrect statement for diffusion:
(a) A gradient must already be present for diffusion to occur
(b) The diffusion rate depends on the size of substances
(c) Small substances diffuse faster
(d) Diffusion of any substance across a membrane never depends on its solubility in lipid

25. Which does not pertain to facilitated transport?
(a) Uphill transport
(b) Highly selective
(c) Transport saturation
(d) Requirement of special membrane proteins

26. Select the incorrect statement with respect to the means of transport in the given diagram:

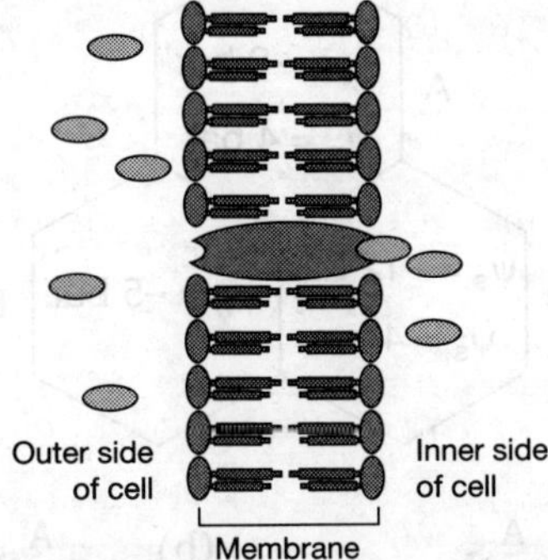

(a) The process is non-sensitive to inhibitors
(b) Transport saturation takes place
(c) It is very specific and allows the cell to select substances for uptake
(d) It occurs along the concentration gradient

Plant–Water Relations (Water Potential, Osmosis, Plasmolysis, Imbibition)

27. What determines the diffusion of water from one cell to another?
(a) WP (b) TP
(c) Ψw (d) OP

28. Osmosis is driven by:
(a) Concentration of dissolved solutes in a solution
(b) Pressure difference
(c) Membrane composition
(d) Both (a) and (b)

29. Water flows from:
(a) High water potential to low water potential
(b) High OP to low OP
(c) Hypertonic solution to hypotonic solution
(d) All of the above

30. Physiological demonstration of osmosis is carried out by:
(a) Porometer
(b) Bell jar experiment
(c) Thistle funnel whose mouth is tied with egg membrane
(d) Thistle funnel whose mouth is tied with paper

31. Numerically, osmotic pressure is equivalent to ________ but the sign is opposite:
(a) Osmotic potential
(b) Solute potential
(c) Water potential
(d) All except (c)

32. The osmotic expansion of a cell kept in water is chiefly regulated by:
(a) Vacuoles (b) Plastids
(c) Mitochondria (d) Cytoskeleton

33. If the external solution balances the osmotic pressure of the cytoplasm, it is said to be:
(a) Hypertonic (b) Hypotonic
(c) Isotonic (d) Plasmolysed

34. Osmosis refers to the diffusion of water across:
A. Impermeable membrane
B. Permeable membrane
C. Semipermeable membrane
D. Selectively permeable membrane
(a) (B) and (C) (b) (C) and (D)
(c) (B) and (C) (d) (C) only

35. The more the solute molecules, the ________ is the solute potential:
(a) Higher (b) More positive
(c) More negative (d) Neutral

36. Difference between osmotic pressure and turgor pressure is:
(a) Osmotic potential
(b) DPD
(c) Turgor potential
(d) Solute potential

37. Water potential is equal to:
(a) $\Psi s + \Psi p$ (b) $\Psi s - \Psi p$
(c) $\Psi p + \Psi w$ (d) $\Psi s + \mho p$

38. For a solution, at atmospheric pressure, Ψw is:
(a) Zero
(b) Negative
(c) Equal to Ψs
(d) Both (b) and (c)

39. Which one is not true for a fully turgid cell?
(a) Ψw = maximum (b) DPD = OP
(c) DPD = 0 (d) OP = TP

40. Water potential:
(a) Of a cell is always positive
(b) Of a solution increases by adding solutes
(c) Of pure water is zero at atmospheric pressure
(d) Is independent of any external pressure

41. Water potential of a cell is not affected by:
(a) Solute potential
(b) Pressure potential
(c) Free energy
(d) Placing it in isotonic solution

42. Water potential and osmotic potential of pure water are:
(a) 100 and 0 (b) 0 and 0
(c) 0 and 100 (d) 100 and 100

43. The ratio of osmotic pressure exerted by 1 M of sucrose and 1 M of NaCl solution will be:
(a) 1 (b) 2
(c) 0.1 (d) 0.5

44. DPD of a fully turgid cell is:
(a) Zero (b) Equal to TP
(c) Equal to OP (d) Negative

45. Phenomenon that is not related to plasmolysis:
(a) Usually reversible
(b) Withdrawal of protoplast from the cell wall
(c) Exosmosis from the cytoplasm
(d) TP of the cell becomes positive

46. Study the following figure and the statements:

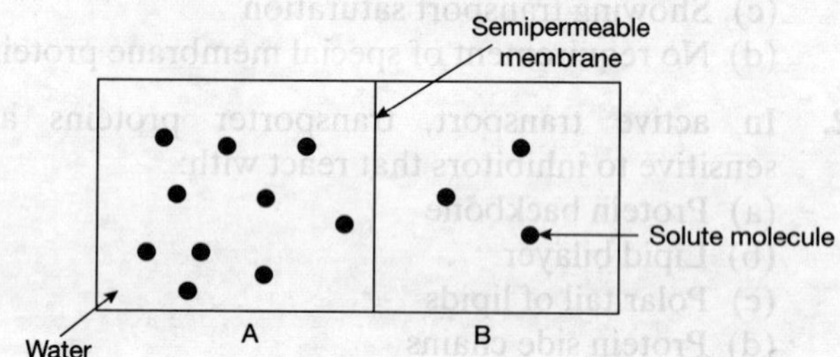

(a) Solution in chamber B has higher water potential
(b) Solution in chamber A has lower solute potential
(c) Movement of solute molecules takes place from A to B
(d) At equilibrium, both the chambers have equal water potential

Which of the above statements are correct?
(a) Only (a) and (b)
(b) Only (d)
(c) (a), (b) and (d)
(d) (a), (b), (c) and (d)

47. Cell wall shows:
(a) Semipermeability
(b) Differential permeability
(c) Impermeability
(d) Complete permeability

48. Find the correct pathway for the movement of water in the given representation of a cell:

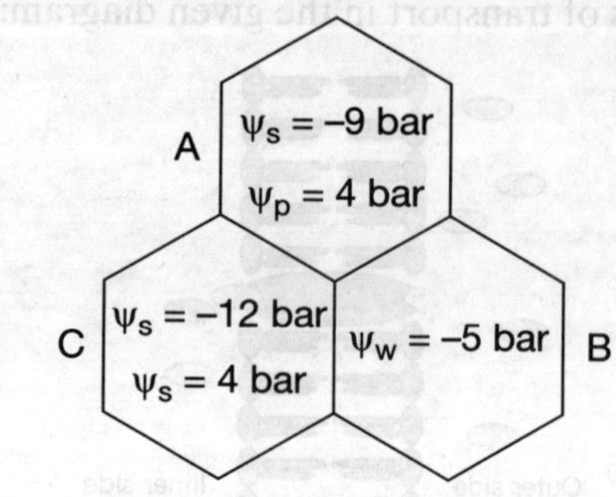

(a) 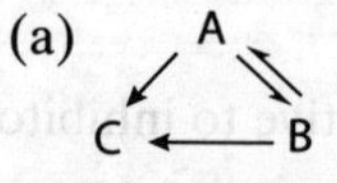

(b) 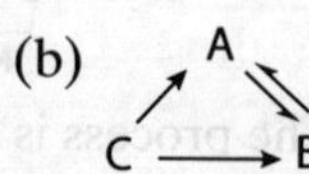

(c) 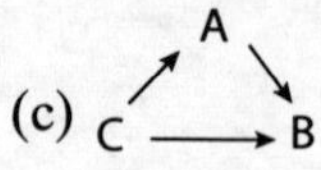

(d) 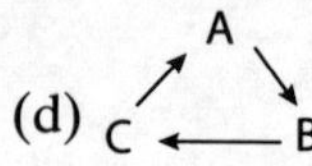

49. In a turgid cell, DPD is equal to:
(a) TP (b) OP + TP
(c) Zero (d) OP

50. The absorption of water by dry seeds and dry wood takes place by a special type of diffusion:
(a) Osmosis
(b) Facilitated diffusion
(c) Imbibition
(d) Active transport

51. Imbibition is not shown by:
(a) Hydrocelluloids (b) Suberin
(c) Starch (d) Protein

52. The best imbibant among the following is:
(a) Agar (b) Cellulose
(c) Starch (d) Protein

53. What occupies the space between the cell wall and the shrunken protoplast in a plasmolysed cell?
(a) Air
(b) Pure water
(c) Hypertonic solution
(d) Hypotonic solution

54. Fill in the blanks and select the correct option:
(i) Diffusion pressure of pure water is ____A____ and addition of solute to water leads to ____B____ in DPD
(ii) A solution is said to be hypotonic if its osmotic pressure is ____C____ than that of cell sap
(iii) Solutes present in any solution ____D____ the free energy of water

	A	B	C	D
(a)	Maximum	Increase	Lower	Reduces
(b)	Minimum	Increase	Higher	Increases
(c)	Maximum	Decrease	Lower	Reduces
(d)	Minimum	Decrease	Higher	Has no effect on

55. Pressure potential in a plasmolysis cell is:
(a) Positive (b) Zero
(c) Negative (d) Remains the same

56. A plasmolysis cell can be de-plasmolysed by placing it in:
(a) Isotonic solution
(b) Saturated solution
(c) Pure water and hypotonic solution
(d) Hypertonic solution

57. Plasmolysis occurs when water moves out of the cell and the cell membrane of a plant cell shrinks away from its cell wall. Identify the different stages of plasmolysis:
A. First stage of plasmolysis, where pressure potential become zero and the osmotic concentration of the cell is almost the same as the external solution
B. The protoplast withdraws itself from corners of the cell wall
C. Protoplast shrinks and withdraws from the cell wall, except at one or a few points

	A	B	C
(a)	Evident plasmolysis	Limiting plasmolysis	Incipient plasmolysis
(b)	Limiting plasmolysis	Incipient plasmolysis	Evident plasmolysis
(c)	Limiting plasmolysis	Evident plasmolysis	Incipient plasmolysis
(d)	Evident plasmolysis	Incipient plasmolysis	Limiting plasmolysis

58. Which phenomenon is shown in the given figure?

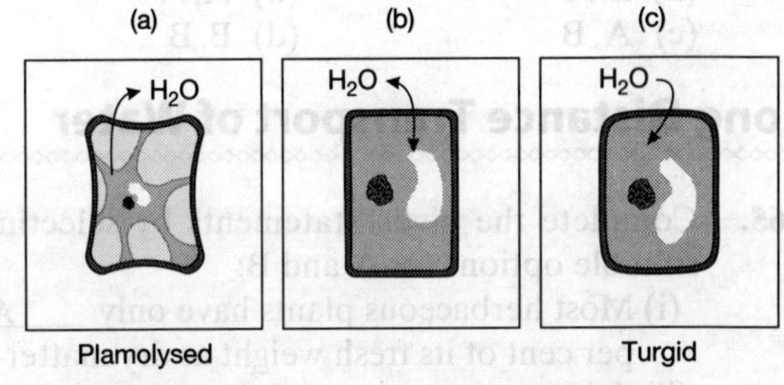

(a) Imbibition (b) Plasmolysis
(c) Deplasmolysis (d) Melting

59. Without which pressure would the seedling not have been able to emerge out of the soil into the open?
(a) Osmotic pressure
(b) Suction pressure
(c) Turgor pressure
(d) Imbibition pressure

60. Which of the following is ultimately responsible for enlargement and extension growth of cells?
(a) Osmotic pressure
(b) Turgor pressure
(c) Wall pressure
(d) Osmotic potential

61. Osmotic pressure depends upon:
(a) Concentration of solutes
(b) Temperature
(c) Ionisation
(d) All of the above

62. As a result of exosmosis, Ψw of the cell:
(a) Increases
(b) Decreases
(c) Remains the same
(d) Becomes zero

63. Osmotic pressure (OP) of a flaccid cell is:
(a) Equal to TP
(b) Equal to DPD
(c) Zero
(d) All except (b)

64. Given below are the values of Ψw and Ψs of two chambers (A and B):

A	B
Ψs = –20	Ψs = –24
Ψw = –10	Ψs = –12

The solution of which chamber has lower solute potential and higher Ψ_s, respectively?
(a) B, A (b) A, A
(c) A, B (d) B, B

Long Distance Transport of Water

65. Complete the given statements by selecting the suitable options for A and B:
(i) Most herbaceous plants have only ___A___ per cent of its fresh weight as dry matter
(ii) A mature corn plant absorbs almost ___B___ of the water in a day

	A	B
(a)	10–15	Five litres
(b)	10–15	Three litres
(c)	90–95	Three litres
(d)	90–95	Five litres

66. Xylem is associated with translocation of:
(a) Water only
(b) Water and non-reducing sugar
(c) Water and minerals
(d) Hormones and sucrose

67. The bulk movement of substances through the vascular tissues of plants is called:
(a) Transpiration (b) Translocation
(c) Active transport (d) Transfaction

68. Most of the nitrogen in plants that is transported is:
(a) Organic via the phloem
(b) Organic via the xylem
(c) Inorganic via the phloem
(d) Inorganic via the xylem

69. Select the incorrect statement with respect to transpiration pull:
(a) It acts as the main driving force for translocation of water in tall trees
(b) Its magnitude is maximum during the day
(c) It generates a strong negative pressure in the xylem
(d) It is a manifestation of active water absorption

70. Which is not true regarding passive water absorption?
(a) Occurs only when transpiration is high
(b) Force develops in root
(c) Does not require energy
(d) Living cells are not essential

71. Most of the water absorbed by root hairs occurs by/through:
(a) Passive absorption
(b) Active absorption
(c) Transpiration pull
(d) All except (b)

72. The movement of a molecule across a typical plant cell having a diameter of 50 μm takes approximately:
(a) 2.5 seconds (b) 25 seconds
(c) 25 minutes (d) 5 minutes

73. Cohesion/Transpiration pull theory of ascent of sap was proposed by:
(a) Munch (b) Dixon and Joly
(c) Bose (d) Stephen Hales

74. Magnitude of root pressure ranges:
(a) 2–5 atm (b) 1–5 atm
(c) 0.1–0.2 atm (d) 4–6 atm

75. Root pressure is a:
(a) Negative hydrostatic pressure in the xylem of the root
(b) Positive hydrostatic pressure in the xylem due to metabolic activities of the root
(c) Positive hydrostatic pressure in the xylem due to passive absorption of water by the root
(d) Negative hydrostatic pressure in the xylem that develops in the root due to transpiration pull

76. Guttation is an effect of:
(a) Transpiration pull
(b) Positive hydrostatic pressure
(c) Root pressure
(d) Both (b) and (c)

77. Root pressure is responsible for:
A. Providing modest push in overall process of water transport
B. Re-establishing continuous chain of water molecules in the xylem
C. Water movement up tall trees
(a) Only (A) and (B) (b) Only (A) and (C)
(c) (A), (B) and (C) (d) Only (A)

78. When root pressure is high, some herbaceous plants lose a small quantity of excess water in the form of drops from the margin of the tip of leaves. This phenomenon is called:
(a) Diffusion (b) Guttation
(c) Imbibition (d) Transpiration

79. State true (T) or false (F) for the given statements and choose the correct option:
A. Seeds of pines cannot germinate properly without mycorrhizal associations
B. During passive absorption of water by the root, a positive pressure is developed in the xylem
C. Root pressure plays a major role in water transport up tall trees
D. Most plants meet their water needs by transpiration pull

	A	B	C	D
(a)	T	T	T	T
(b)	F	F	F	T
(c)	T	F	F	T
(d)	T	T	F	F

80. Mark the incorrect statement for apoplastic movement of water:
(a) It is a system of interconnected protoplasts
(b) Movement occurs exclusively through the intercellular spaces and wall of the cells
(c) It does not involve crossing of cell membrane by water molecules
(d) It does not provide any barrier to water movement

81. Apoplastic pathway becomes symplastic at the:
(a) Epidermis (b) Endodermis
(c) Xylem (d) Cortex

82. A plant can adjust the quantity and type of solute reaching the xylem at the:
(a) Cortex (b) Endodermis
(c) Pericycle (d) Hypodermis

83. Apoplast pathway differs from symplast pathway in:
(a) Offering more resistance to movement of water
(b) Being slightly slower
(c) Moving the water through the cell wall and intercellular space exclusively
(d) Being affected by the metabolic status of the root cell

84. Transport protein as control point determines the type and quantity of solutes in the xylem. It is present on/in:
(a) Endodermal cell membrane
(b) Endodermal cell wall
(c) Cortical cell membrane
(d) Hypodermal cell wall

85. Transpiration and root pressure cause water to rise in plants by:
(a) Pulling it upward
(b) Pushing it upward
(c) Pulling and pushing it upward, respectively
(d) Pushing and pulling it upward, respectively

86. Water is pushed by ___A___ and pulled by ___B___:

	A	B
(a)	Transpiration	Root pressure
(b)	Root pressure	Transpiration
(c)	Cytoplasmic streaming movement	Active transport
(d)	Passive transport	Active transport

87. According to Dixon and Jolly, the ascent of xylem sap depends mainly on the following properties of water, except:
(a) Root pressure (b) Adhesion
(c) Cohesion (d) Surface tension

88. The fungal partner in mycorrhiza:
(a) Provides sugar to the plant
(b) Fixes atmospheric nitrogen
(c) Absorbs water and minerals for the plant
(d) Provide nitrogen containing compounds and sugar to the plants

89. The apoplast system is continuous throughout the plant, except which of the following tissues in the roots?
(a) Epidermis (b) Endodermis
(c) Cortex (d) Xylem

90. Rate of flow of water upward through the xylem in a plant is approximately:
(a) Up to 1 metre per hour
(b) Up to 5 metres per hour
(c) Up to 15 metres per hour
(d) Up to 50 metres per hour

91. In the root, the apoplast path of water transport is through:
(a) Cell wall only
(b) Cell wall and intercellular spaces
(c) Intercellular spaces
(d) Intercellular spaces, cell wall and endodermis

92. Passive absorption occurs due to:
(a) Tension in root
(b) Tension in xylem sap
(c) ATP
(d) Root pressure

93. Guttation takes place through specialised pores called:
(a) Lenticel (b) Hydathodes
(c) Stomata (d) Pneumatothodes

94. Study the given diagram carefully:

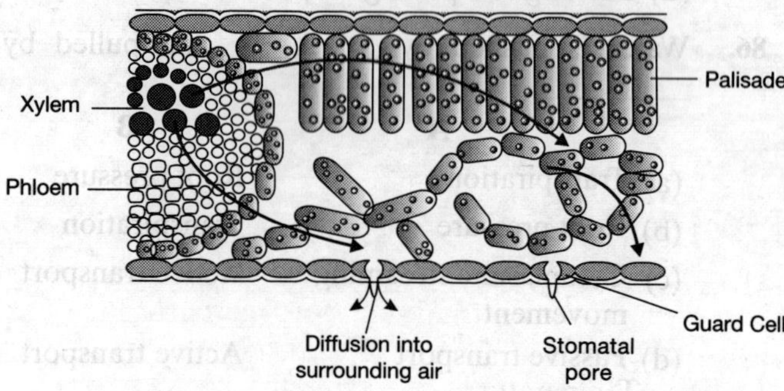

In the above figure, pull is generated due to:
(a) Lower capillarity and high tensile strength
(b) Lower concentration of water vapour in the atmosphere
(c) Diffusion of water vapour from the substomatal cavity to the atmosphere
(d) More than one option is correct

95. Transpiration can create pressure sufficient to lift a xylem-sized column of water approximately ________ high:
(a) 13 metres (b) 130 metres
(c) 30 metres (d) 3 metres

96. Cohesive force of water is due to:
(a) O bonds (b) OH bonds
(c) S bonds (d) H bonds

Transpiration

97. Transpiration is:
(a) Loss of water in the form of vapour through lenticels
(b) Loss of water in the form of vapour through stomata
(c) Loss of water in the form of liquid droplets through stomata
(d) Both (b) and (c)

98. Maximum transpiration occurs in:
(a) Mesophytes (b) Xerophytes
(c) Thallophytes (d) Pteredophytes

99. Major part of total transpiration occurs through:
(a) Cuticle (b) Bark
(c) Stomata (d) Lenticels

100. The orientation of cellulose microfibres in the cell wall of guard cells is:
(a) Radial (b) Longitudinal
(c) Oblique (d) Tangential

101. Opening and closing of stomata is due to the:
(a) Hormonal changes in guard cells
(b) Changes in turgor pressure of guard cells
(c) Gaseous exchanges
(d) Respiration

102. Read the following statements with respect to stomata:
A. Stomata are turgor-operated
B. The inner wall of guard cells is thin and elastic
C. Cellulose microfibres are oriented radially in guard cells
D. In grass, stomata are sunken
E. Isobilateral leaves are amphistomatic
How many of the above statements are incorrect?
(a) Two (b) Three
(c) One (d) Four

103. At the time of opening of stomata, the pH of the guard cell in comparison to that of the subsidiary cell is:
(a) Equal (b) Lower
(c) Uncertain (d) Higher

104. The incorrect feature with respect to guard cells is:
(a) Presence of chloroplast
(b) The inner wall towards the aperture is thick and elastic
(c) Longitudinally oriented cellulose microfibrils
(d) Bordered by one or more subsidiary cells

105. The major solute taken in by the guard cells is:
(a) Na^+ (b) K^+
(c) H^+ (d) Ca^{++}

106. What is the action spectrum of transpiration?
(a) Blue and red
(b) Red and far red
(c) Blue-violet
(d) Orange and red

107. Increased level of K^+ and malate ion in the vacuole of guard cells leads to:
(a) Increase in osmotic concentration of guard cells
(b) Increase in turgor pressure of guard cells
(c) Endosmosis into guard cells
(d) All of the above

108. Transpiration is useful for plants in all, except:
(a) Cooling of leaf surface
(b) Generating root pressure
(c) Creating transpiration pull
(d) Maintaining shape and structure of plants

109. Stomatal movement is not affected by:
(a) O_2 concentration
(b) Light
(c) Temperature
(d) CO_2 concentration

110. Rate of transpiration is measured by:
(a) Ganong's potometer
(b) Porometer
(c) Auxanometer
(d) Photometer

111. At constant temperature, the rate of transpiration will be higher at:
(a) Sea level
(b) 1 km below sea level
(c) 1.5 km above sea level
(d) 1 km above sea level

112. Rate of transpiration is high in:
(a) C_3 plants
(b) C_4 plants
(c) CAM plants
(d) Both C_3 and C_4 plants

113. When dry cobalt chloride paper strip comes in contact with the water vapour of transpiring leaf surface, it turns:
(a) Blue (b) Red
(c) Yellow (d) Pink

114. Cobalt chloride paper test is used for checking:
(a) Root pressure (b) Transpiration
(c) Phloem (d) Ascent of sap

115. Transpiration can cool the leaf surface by evaporative cooling up to:
(a) 20–25°C (b) 10–15°C
(c) 15–20°C (d) 2°C

116. Identify the parts labelled A and B with respect to guard cells:

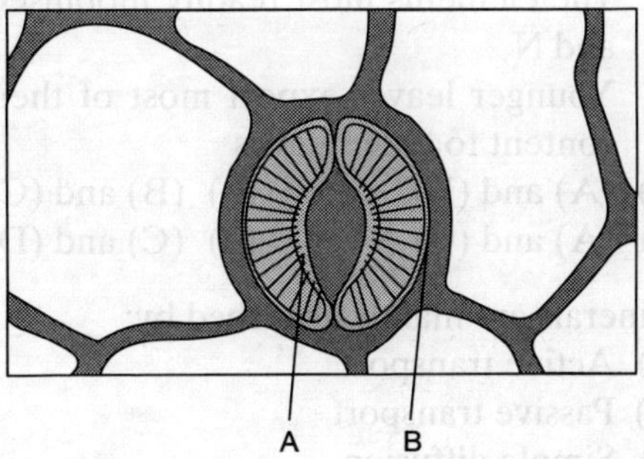

	A	B
(a)	Thick and elastic inner wall	Thin and elastic outer wall
(b)	Thick and inelastic inner wall	Thin and elastic outer wall
(c)	Thin and elastic inner wall	Thick and elastic outer wall
(d)	Thick and elastic inner wall	Thick and inelastic outer wall

117. Transpiration rate decreases with increase in:
(a) Relative humidity
(b) Available soil water
(c) Temperature
(c) Light intensity

118. Which of the following will not directly affect transpiration?
(a) Temperature
(b) Light
(c) Wind speed
(d) Chlorophyll content of leaves

Uptake and Transport of Mineral Nutrients

119. Choose the odd one out for chief sink for mineral elements:
(a) Apical meristem
(b) Developing flower and fruit
(c) Senesced leaves
(d) Lateral meristem

120. The element that cannot remobilised is:
(a) Nitrogen (b) Phosphorus
(c) Potassium (d) Calcium

121. Read the following statements carefully and select the correct option having two correct statements:
A. Some elements that are structural components like calcium are not remobilised
B. Mineral ions are slowly remobilised
C. The elements most readily mobilised are P, K and N
D. Younger leaves export most of their mineral content to older leaves
(a) (A) and (B) (b) (B) and (C)
(c) (A) and (C) (d) (C) and (D)

122. Minerals are mainly absorbed by:
(a) Active transport
(b) Passive transport
(c) Simple diffusion
(d) Facilitated diffusion

123. Which of the following elements is more readily mobilised?
(a) P (b) N
(c) K (d) All of the above

124. Transport protein, as a control point to determine the type and quantity of solutes in the xylem, is present on/in:
(a) Endodermal cell membrane
(b) Endodermal cell wall
(c) Cortical cell membrane
(d) Pericycle cell membrane

125. Read the following statements with respect to the transport of mineral ions and select the correct option:
A. Minerals are removed to other parts before leaf fall in deciduous plants
B. Exchange of minerals does not take place between xylem and phloem
C. After ions reach the xylem, their further transport up the stem is through the transpiration stream
(a) (B) and (C) are correct
(b) (B) and (C) are correct
(c) (B) is incorrect
(d) (A) (B), (C) are correct

Phloem Transport

126. To explain the translocation of food, pressure flow hypothesis was proposed by:
(a) Graft (b) E Munch
(c) Levitt (d) Joseph Priestley

127. Sugar translocation in flowering plants occurs by ____A____ in the form of ____B____ and is a ____C____ process:

	A	B	C
(a)	Sieve tubes	Sucrose	Bidirectional
(b)	Sieve tubes	Glucose	Bidirectional
(c)	Companion cell	Sucrose	Bidirectional
(d)	Companion cell	Starch	Multidirectional

128. Find the odd one out with respect to translocation of food:
(a) Phloem sap is mainly water and sucrose
(b) It is transported by the phloem from a source to the sink
(c) Loading is an active and unloading is a passive process
(d) Water is required for translocation of food

129. According to the pressure flow hypothesis in phloem transport, organic solutes flow from the regions of:
(a) Low diffusion pressure
(b) Low osmotic pressure
(c) High turgor pressure
(d) Low turgor pressure

130. Phloem sap is mainly:
(a) Water and starch
(b) Fructose and glucose
(c) Water and sucrose
(d) Water and fructose

131. Select the incorrect statement with respect to phloem transport:
(a) Translocation in phloem is bidirectional
(b) The source–sink relationship is invariable
(c) At the sink, osmotic pressure must be reduced
(d) Active transport is necessary to move sucrose out of the phloem sap at the sink

132. From the mesophyll cell, sucrose moves into the companion cell and then into sieve tube cells by:
(a) Active transport (b) Simple diffusion
(c) Passive transport (d) Facilitated diffusion

133. Read the following statements with respect to the girdling experiment and state true (T) or false (F):
A. It is an experimental demonstration of translocation of food by the phloem in two directions
B. Swelling of bark above the girdled part is due to accumulation of nutrients
C. In this experiment, upward movement of water is not effected initially
D. In a girdled plant, the shoot dies first, then the root

	A	B	C	D
(a)	T	F	F	T
(b)	F	T	F	F
(c)	F	T	T	F
(d)	T	F	T	T

134. A girdled plant will eventually die because of:
(a) Absence of downward movement of water
(b) Absence of upward movement of water
(c) Absence of upward movement of organic food
(d) Absence of downward movement of organic food

135. Girdling experiment was used to prove that:
(a) Xylem is the water conducting channel
(b) Food translocation is by the phloem
(c) Translocation through xylem is essentially unidirectional
(d) All except (b)

136. Meaningful ringing experiments cannot be carried out on sugarcane because:
(a) Phloem is present inside the xylem
(b) It cannot tolerate the injury
(c) Vascular bundles are scattered
(d) It has closed vascular bundles

137. Translocation of photosynthesis end products in sieve tubes is:
(a) 3–5 mm/h (b) 3–5 cm/h
(c) 1–15 cm/h (d) 60–100 cm/h

NCERT EXEMPLAR QUESTIONS

1. Which of the following statements does not apply to reverse osmosis?
(a) It is used for water purification
(b) In this technique, pressure greater than osmotic pressure is applied to the system
(c) It is a passive process
(d) It is an active process

2. Which one of the following will not directly affect transpiration?
(a) Temperature
(b) Light
(c) Wind speed
(d) Chlorophyll content of leaves

3. The lower surface of the leaf will have more stomata in a:
(a) Dorsiventral leaf
(b) Isobilateral leaf
(c) Both (a) and (b)
(d) None of the above

4. The form of sugar transported through the phloem is:
(a) Glucose (b) Fructose
(c) Sucrose (d) Ribose

5. The process of guttation takes place:
(a) When the root pressure is high and the rate of transpiration is low
(b) When the root pressure is high and the rate of transpiration is high
(c) When the root pressure equals the rate of transpiration
(d) When the root pressure as well as rate of transpiration are high

6. Which of the following is an example of imbibition?
(a) Uptake of water by root hair
(b) Exchange of gases in stomata
(c) Swelling of seed when put in soil
(d) Opening of stomata

7. When a plant undergoes senescence, the nutrients may be:
(a) Accumulated
(b) Bound to the cell wall
(c) Translocated
(d) None of the above

8. Water potential of pure water at standard temperature is equal to:
(a) 10 (b) 20
(c) Zero (d) None of the above

9. Choose the correct option. Mycorrhiza is a symbiotic association of a fungus with a root system which helps in:
A. Absorption of water
B. Mineral nutrition
C. Translocation
D. Gaseous exchange

(a) Only (A)
(b) Only (B)
(c) Both (A) and (B)
(d) Both (B) and (C)

10. Based on the figure given below, which of the following statements is not correct?

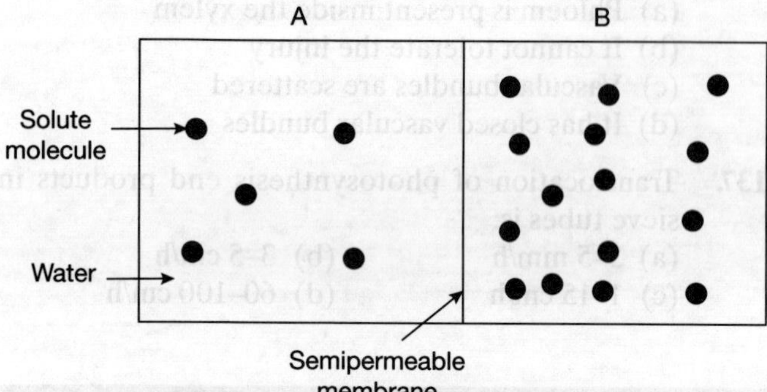

(a) Movement of solvent molecules will take place from chambers A and B
(b) Movement of solute will take place from A to B
(c) Presence of a semipermeable membrane is a pre-requisite for this process to occur
(d) The direction and rate of osmosis depends on the pressure gradient and concentration gradient

11. Match the following and choose the correct option:

Column I	Column II
A. Leaves	(i) Anti-transpirant
B. Seed	(ii) Transpiration
C. Root	(iii) Negative
D. Aspirin	(iv) Imbibition
E. Plasmolysed cell	(v) Absorption

(a) A – (ii), B – (iv), C – (v), D – (i), E – (iii)
(b) A – (iii), B – (ii), C – (iv), D – (i), E – (v)
(c) A – (i), B – (ii), C – (iii), D – (iv), E – (v)
(d) A – (v), B – (iv), C – (iii), D – (ii), E – (i)

12. Mark the mismatched pair:
(a) Amyloplast – Stores protein granule
(b) Elaioplast – Stores oils or fats
(c) Chloroplasts – Contains chlorophyll
(d) Chromoplasts – Contains coloured pigment other than chlorophyll

PRACTICE TEST PAPER

1. Select the incorrect statement with respect to active transport:
(a) It uses energy to carry substances along a concentration gradient
(b) Also called uphill transport
(c) Transport saturation takes place
(d) Requires special membrane proteins

2. Porins are proteins that form huge pores in the outer membrane of:
(a) Plastids
(b) Mitochondria
(c) Gram-negative bacteria
(d) All of the above

3. Which is not a result of diffusion?
(a) Distribution of water and gases
(b) Supply of CO_2 from atmosphere to leaves
(c) Transport of photosynthates
(d) Loss of water vapour from leaves to the atmosphere

4. Read the following statements and select the correct option:
A: Root pressure is observed at night and early morning
B: Root pressure is a positive hydrostatic pressure
(a) Only (A) is correct
(b) Both (A) and (B) are correct
(c) Only (B) is correct
(d) Both (A) and (B) are incorrect

5. A waterlogged soil is called physiological dry soil because of:
(a) Absence of O_2
(b) Increased salt concentration
(c) Increased viscosity of water
(d) Abundance of salt

6. Stomated opening decreases in the presence of:
(a) High CO_2 concentration
(b) Low pH
(c) High ABA
(d) All of the above

7. Most plants meet their water needs by:
(a) Transpiration pull
(b) Root pressure
(c) Osmotic pressure
(d) Both (a) and (b)

8. The phloem translocates a variety of:
(a) Organic solutes (b) Inorganic solutes
(c) Hormones (d) All of the above

9. Select the incorrect statement for water potential:
(a) If some solute is dissolved in pure water, its water potential decreases.
(b) All solutions have a lower water potential than pure water
(c) The higher the solute molecules, the less negative the water potential
(d) Water molecules will move from high water potential to low water potential

10. What will be the direction of movement of water in two cells?

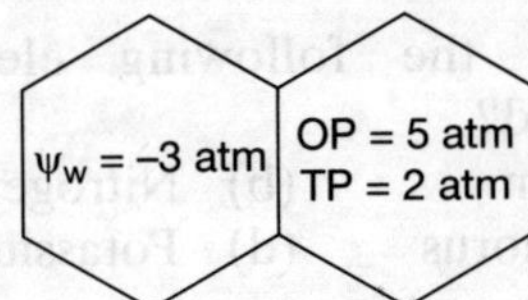

(a) A to B (b) B to A
(c) No net movement (d) Data deficient

11. Wooden doors swell up and get stuck during the rainy season due to:
(a) Capillarity (b) Imbibition
(c) Deplasmolysis (d) Exosmosis

12. Photosynthetics are translocated from the source to sink organs mainly in the form of:
(a) Sucrose (b) Glucose
(c) Fructose (d) Starch

13. The apoplast is located:
(a) Outside the plasma membrane
(b) In the entire cytosol
(c) On both sides of the plasma membrane
(d) In the plastidial content

14. Which of the following values of OP, DPD and TP appear to be correct for a fully turgid cell?

	OP (atm)	DPD (atm)	TP (atm)
(i)	9	5	4
(ii)	10	10	0
(iii)	15	7	8
(iv)	20	0	20

15. An analysis of xylem exudates shows that much of nitrogen is carried as:
(a) Inorganic ions
(b) Organic form as amino acids
(c) Organic form as organic acids
(d) Ammonium salts

16. __________ are the sites through which hydrophilic molecules cross the membrane:
(a) Cholesterol
(b) Glycolipids
(c) Membrane proteins
(d) Phospholipids

17. Which feature of guard cells helps in the opening of stomata?
(a) Thick outer wall
(b) Longitudinally oriented microfibrils
(c) Radially oriented microfibrils
(d) Thin outer and inner walls

18. Which does not affect the rate of diffusion of a substance?
(a) Temperature and concentration gradient
(b) Electrical charges of the diffusing materials
(c) Size or mass of particles
(d) Presence of other substances in a solution

19. For a solution/cell:
(a) Ψw is negative but Ψs is positive
(b) Ψw is positive but Ψs is negative
(c) Both Ψw and Ψs are negative
(d) Both Ψw and Ψs are positive

20. Process of water exudation from hydathodes is known as:
(a) Guttation (b) Transpiration
(c) Evaporation (d) Bleeding

21. Which of the following external factors affects the rate of transpiration?
A. Light
B. Number of stomata
C. Temperature
D. Humidity
E. Canopy structure
F. Plant water status
(a) (A), (C), (E) and (F)
(b) (A), (C) and (D)
(c) (A), (D) and (E)
(d) (B), (E) and (F)

22. Select the mismatched pair:
(a) Phloem sap – Mainly water and reducing sugar
(b) Dorsiventral – Hypostomatic
(c) Transpiration – Cool leaf surface 10–15°C
(d) Tensile strength – An ability to resist a pulling force

23. The most accepted mechanism for the translocation of sugar from source to sink is:
(a) Active K^+ transport theory
(b) Cohesion–tension theory
(c) Pressure flow hypothesis
(d) Active transport theory

24. Deplasmolysis occurs due to:
(a) Endosmosis of water
(b) Exosmosis of water
(c) Hypertonic solution
(d) Isotonic solution

25. Water potential is determined by:
(a) Solute potential and pressure potential
(b) Osmotic pressure and DPD
(c) DP and TP
(d) DP and DPD

26. Read the following statements with respect to root pressure:
A. It provide a modest push in the overall process of water transport
B. Its magnitude is maximum during early morning and at night
C. It is a negative hydrostatic pressure developed in the xylem
(a) Only (A) is correct
(b) Only (B) and (C) are correct
(c) Only (C) is incorrect
(d) (A) (B) and (C) are correct

27. The opening of stoma is aided by the orientation of the microfibrils in the cell wall of guard cells. These microfibrils are:
(a) Arranged longitudinally
(b) Lipid derived
(c) Cellulosic
(d) Proteinaceous

28. State true (T) or false (F) for the following statements and select the correct option:
A. Diffusion in solids rather than of solids is more likely
B. The positive pressure that develops in the xylem is called root pressure
C. The greater the solute molecules, the greater the solute potential
D. Water, minerals and food are generally translocated by a mass flow or bulk flow system

	A	B	C	D
(a)	T	T	F	T
(b)	T	T	T	F
(c)	F	T	F	T
(d)	F	F	F	T

29. Facilitated diffusion and active transport:
(a) Energy dependent
(b) Uphill transport
(c) Not highly selective
(d) Dependent on special membrane proteins

30. Inner wall of guard cell is:
(a) Thin and elastic
(b) Thin and non-elastic
(c) Thick and elastic
(d) Thick and non-elastic

31. Which of the following is not a similarity between facilitated diffusion and active transport?
A. Transport rate reaches maximum when all the proteins transporters are saturated
B. Protein transporters are sensitive to inhibitors that react with protein side chains
C. Protein transporters use energy to carry substances across the cell membrane
D. Substances are transported from low concentration to high concentration
(a) (A) and (C) (b) (B) and (C)
(c) (B) and (C) (d) (C) and (D)

32. Which of the following elements is not remobilised?
(a) Calcium (b) Nitrogen
(c) Phosphorus (d) Potassium

33. Select the incorrectly matched pair:
(a) Guttation – Root pressure
(b) Transpiration – Evaporative loss of water
(c) Imbibition – Endothermic
(d) Plasmolysis – Due to exosmosis

34. The greater the concentration of water in a system, the greater is its:
(a) Kinetic energy (b) Solute potential
(c) Water potential (d) Both (a) and (c)

35. Select the correct statement with respect to stomatal opening:
A. Immediate cause is change in turgidity of the guard cell
B. Radial orientation of the cellulosic microfibrils in the cell wall of guard cells makes the process easier
C. It is a passive process
D. Explained by potassium pump theory
(a) (A) and (B) only
(b) (A), (B) and (D) only
(c) (A), (B) and (C) only
(d) (A), (B), (C), (D)

36. Girdling experiment was used to prove that:
(a) Xylem is the water conducting channel
(b) Food translocation is by pressure flow

(c) Pholem translocates food in one direction
(d) Phloem is not responsible for food translocation

37. Loading of sucrose in phloem:
(a) Is dependent on energy
(b) Sets up a water potential gradient between source and sink
(c) Produces hypertonic condition in phloem
(d) All of the above

38. A column of water within the xylem does not break because of some physical properties of water such as:
A. Cohesion
B. Adhesion
C. Surface tension
(a) (A) and (B) only (b) (A) only
(c) (B) and (C) only (d) (A), (B) and (C)

39. The apoplastic movement of water occurs exclusively through:
(a) Cytoplasmic streaming movement
(b) Plasmodesmatal connection
(c) Intercellular spaces and wall of cell
(d) Cytoplasm and vacuole

40. A cell increases in volume if the external medium is/has:
(a) Hypotonic (b) Hypotonic
(c) Isotonic (d) More solute concentration

41. Which of the following plant factors affect the rate of transpiration?
(a) Cuticle and wax coating on leaves
(b) Sunken stomata
(c) Percentage of open stomata in plant
(d) All of the above

42. Effect of root pressure is observable during/when:
(a) Middle of the day
(b) Early morning and late night when transpiration is low
(c) Transpiration is high and absorption is low
(d) Salt concentration in soil is high as compared to plants

43. Movement in phloem is _____ due to _____ in plants:
(a) Unidirectional, presence of sucrose and water
(b) Bidirectional, presence of non-reducing sugar
(c) Bidirectional, variability in source and sink
(d) Unidirectional, invariability in source and sink

44. On a hot summer day, plant cooling occurs due to:
(a) Transport of water from root to aerial parts of the plant
(b) Loss of liquid water
(c) Water loss from entire plant
(d) Loss of water from aerial surface

45. The following percentage of water absorbed by herbaceous plants is lost in transpiration:
(a) 80 (b) 99
(c) 60 (d) 40

ANSWER KEYS

Exercises

1. (d)	**2.** (c)	**3.** (b)	**4.** (a)	**5.** (b)	**6.** (c)	**7.** (c)	**8.** (a)	**9.** (a)	**10.** (d)
11. (c)	**12.** (b)	**13.** (a)	**14.** (c)	**15.** (a)	**16.** (c)	**17.** (a)	**18.** (b)	**19.** (a)	**20.** (c)
21. (d)	**22.** (d)	**23.** (a)	**24.** (d)	**25.** (a)	**26.** (a)	**27.** (c)	**28.** (d)	**29.** (a)	**30.** (c)
31. (d)	**32.** (a)	**33.** (c)	**34.** (b)	**35.** (c)	**36.** (b)	**37.** (a)	**38.** (d)	**39.** (b)	**40.** (c)
41. (d)	**42.** (b)	**43.** (d)	**44.** (a)	**45.** (d)	**46.** (c)	**47.** (d)	**48.** (a)	**49.** (c)	**50.** (c)
51. (b)	**52.** (a)	**53.** (c)	**54.** (a)	**55.** (c)	**56.** (c)	**57.** (b)	**58.** (b)	**59.** (c)	**60.** (b)
61. (d)	**62.** (b)	**63.** (b)	**64.** (a)	**65.** (a)	**66.** (c)	**67.** (b)	**68.** (b)	**69.** (d)	**70.** (b)
71. (d)	**72.** (a)	**73.** (b)	**74.** (a)	**75.** (b)	**76.** (d)	**77.** (a)	**78.** (b)	**79.** (c)	**80.** (a)
81. (b)	**82.** (b)	**83.** (c)	**84.** (a)	**85.** (c)	**86.** (b)	**87.** (a)	**88.** (c)	**89.** (b)	**90.** (c)
91. (b)	**92.** (b)	**93.** (d)	**94.** (d)	**95.** (b)	**96.** (d)	**97.** (b)	**98.** (a)	**99.** (c)	**100.** (a)
101. (b)	**102.** (a)	**103.** (b)	**104.** (c)	**105.** (b)	**106.** (a)	**107.** (d)	**108.** (b)	**109.** (a)	**110.** (a)
111. (c)	**112.** (a)	**113.** (d)	**114.** (b)	**115.** (b)	**116.** (a)	**117.** (a)	**118.** (d)	**119.** (c)	**120.** (d)
121. (c)	**122.** (a)	**123.** (d)	**124.** (a)	**125.** (c)	**126.** (b)	**127.** (a)	**128.** (c)	**129.** (c)	**130.** (c)
131. (b)	**132.** (a)	**133.** (c)	**134.** (d)	**135.** (b)	**136.** (c)	**137.** (d)			

NCERT Exemplar Questions

1. (d)	**2.** (d)	**3.** (a)	**4.** (c)	**5.** (a)	**6.** (c)	**7.** (b)	**8.** (c)	**9.** (c)	**10.** (b)
11. (a)	**12.** (a)								

Practice Test Paper

1. (a)	**2.** (d)	**3.** (c)	**4.** (b)	**5.** (a)	**6.** (d)	**7.** (a)	**8.** (d)	**9.** (c)	**10.** (c)
11. (b)	**12.** (a)	**13.** (a)	**14.** (d)	**15.** (b)	**16.** (c)	**17.** (c)	**18.** (d)	**19.** (c)	**20.** (a)
21. (b)	**22.** (a)	**23.** (c)	**24.** (a)	**25.** (a)	**26.** (c)	**27.** (c)	**28.** (a)	**29.** (d)	**30.** (c)
31. (d)	**32.** (a)	**33.** (c)	**34.** (d)	**35.** (b)	**36.** (c)	**37.** (d)	**38.** (d)	**39.** (c)	**40.** (b)
41. (d)	**42.** (b)	**43.** (c)	**44.** (d)	**45.** (b)					

SOLUTIONS

Introduction, Means of Transport

1. Over small distances, substances move by diffusion and by cytoplasmic streaming movement supplemented by active transport.
2. Transport over long distances occurs through the vascular system and is called translocation.
3. Phloem
4. (A) and (D) are false. (A) Long distance transport proceeds through vascular tissue. (D) Hormones and other growth regulators are not always transported in a unidirectional manner.
5. Diffusion is not dependent on a living system.
6. Facilitated diffusion occurs along the concentration gradient (downhill movement) through special membrane proteins and transport reaches maximum when all the proteins transported are being used (saturation).
7. Diffusion rates are affected by gradient of concentration.
8. Gaseous movement within the plant body is due to simple diffusion only.
9. Facilitated diffusion (A, D, E) Active transport (A, B, C, D, E)
10. Porins are proteins that form huge pores in the outer membrane of the plastids, mitochondria and some bacteria.
11. Uphill transport involves movement of particles against the concentration gradient.
12. Aquaporins are water channels.
13. Water channels are made up of eight different types of aquaporins.
14. Simple diffusion – Downhill movement, Facilitated diffusion – Porins, Active transport – Carrier protein and energy, Osmosis – Semipermeable membrane
15. Only (A) is correct. Diffusion in solids rather than of solids is more likely.
16. Antiport transport
17. Uniport
18. A – Carrier protein, B – Uniport, C – Antiport, D – Symport
19. Substances that have a hydrophilic moiety find it difficult to pass through the membrane, so their movement has to be facilitated.
20. Pumps are proteins that use energy to carry substances across the cell membrane.
21. No requirement for special membrane proteins.
22. In active transport, transport proteins are sensitive to inhibitors that react with protein side chains.
23. Carrier proteins are highly selective in nature.
24. Diffusion of any substance across a membrane also depends on its solubility in lipids, the major constituent of the membrane.
25. Uphill transport is related to active transport.
26. Carrier proteins are sensitive to inhibitors, which react with protein side chains.

Plant–Water Relations (Water Potential, Osmosis, Plasmolysis, Imbibition)

27. Water potential determines the diffusion of water from one cell to another.
28. Concentration of dissolved solutes in a solution and pressure difference
29. Water flows from high water potential to low water potential.
30. Thistle funnel, whose mouth is tied with egg membrane
31. Osmotic potential is equal to osmotic pressure but has negative value.
32. Vacuoles maintain turgidity of the cells.
33. If the external solution balances the osmotic pressure of the cytoplasm, it is said to be isotonic.
34. Osmosis refers to the diffusion of water across a semipermeable or selectively permeable membrane.

35. The more the solute molecules, the higher the negative solute potential.

36. DPD = OP – TP

37. $\Psi w = \Psi s + \Psi p$

Ψs = Solute potential, Ψp = Pressure potential

38. For a solution at atmosphere pressure, Ψw is either negative or equal to Ψs.

39. For a fully turgid cell, OP = TP, DPD = 0.

40. The greater the concentration of water in a system, the greater is its kinetic energy or water potential. Hence, it is obvious that pure water will have the greatest water potential.

41. If a plant cell is placed in an isotonic solution, its water potential will not be affected.

42. Water potential and osmotic potential of pure water are zero.

43. Osmotic pressure of an electrolyte is two times greater than that of a non-electrolyte.

44. DPD = OP – TP. For a fully turgid cell, OP = TP, DPD = 0.

45. Plasmolysis is the shrinkage of the protoplast of a cell under the influence of a hypertonic solution; hence, turgor pressure of the cell becomes negative.

46. (A), (B) and (D) are correct. Movement of solvent molecules takes place from solution B to solution A.

47. Cell wall shows complete permeability.

48. Option (a) is correct.

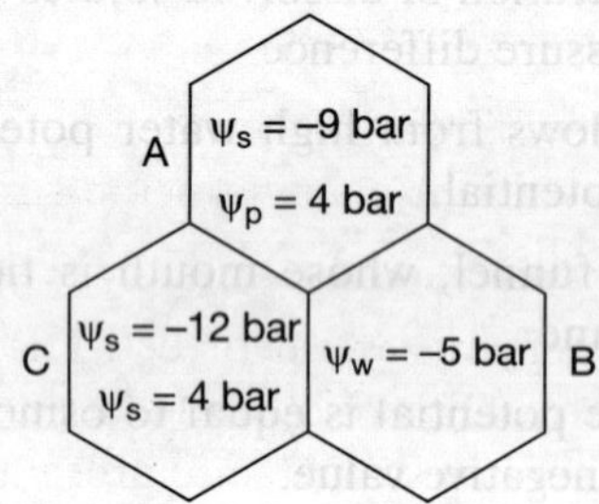

Water molecules will move from higher Ψw (less -) to lower Ψw (more -)

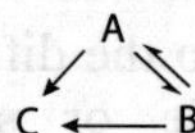

49. In a turgid cell, DPD is equal to zero.

50. Inhibition is a special type of diffusion where a liquid (water) is absorbed by solid colloids (dry seeds and wood). The solid substance is called imbibant and the liquid is called imbibate.

51. Imbibition is not shown by suberin because it is a hydrophobic substance.

52. Agar is the best imbibant because it is hydrophilic in nature.

53. In the final stages of plasmolysis, the hypertonic solution enters the cell between the protoplast and the cell wall.

54. A – Maximum, B – Increase, C – Lower, D – Reduces

55. Pressure potential in a plasmolysed cell is negative.

56. The swelling up of a plasmolysed cell under the influence of a hypotonic solution is called deplasmolysis. It is due to endosmosis.

57. A – Limiting plasmolysis, B – Incipient plasmolysis, C – Evident plasmolysis

58. Plasmolysis

59. Seedlings emerge out of the soil into the open due to imbibition.

60. Turgor pressure is ultimately responsible for enlargement and extension growth of cells.

61. OP = miRT

m = molar concentration, i = ionisation constant, R = universal gas constant and T = absolute temperature

OP of a solution depends upon the solute concentration, temperature and ionisation.

62. Exosmosis results in decrease in Ψw of the cell.

63. In a flaccid cell, TP = 0 and DPD = OP – TP. Hence, DPD = OP.

64. Solution B has lower solute potential (Ψs) and Solution A has higher water potential (Ψw).

Long Distance Transport of Water

65. A – 10–15%, B – 5 litres

66. Xylem is associated with translocation of water and mineral salts.

67. The bulk movement of substances through conducting or vascular tissues (xylem and phloem) of plants is called translocation.

68. Most of the nitrogen in plants is transported in organic form through the xylem.

69. Transpiration pull plays a major role in passive water absorption.

70. For passive water absorption, force develops in aerial parts.

71. Passive water absorption is responsible for 96% of total water uptake.

72. 2.5 seconds

73. Cohesion/Transpiration pull theory of ascent of sap was proposed by Dixon and Joly.

74. Magnitude of root pressure ranges 2–5 atm.

75. Root pressure is a positive hydrostatic pressure in the xylem due to metabolic activation of the root, and it is responsible for pushing up water to small heights in the stem.

76. Guttation is an effect of root pressure (positive hydrostatic pressure).

77. (A) and (B) are correct.

78. Guttation

79. (B) and (C) are false. During active water absorption by roots, a positive pressure is developed in the xylem, called root pressure. Transpiration pull plays a major role in water transport up tall trees.

80. Apoplastic pathway is composed of non-living structures like intercellular space and cell wall.

81. Endodermis

82. Endodermis

83. In the apoplastic pathway, water moves through the cell wall and intercellular space.

84. Transport proteins as control points determine the type and quantity of solutes in xylem and is present in the cell membrane of endodermal cells.

85. Pulling and pushing pressure, respectively

86. A – Root pressure, B – Transpiration

87. Root pressure

88. The fungal hyphae has a very large surface area that absorbs mineral ions and water from the soil from a much larger volume of soil, which perhaps a root cannot do.

89. Endodermis is impervious to water because of a band of suberised matrix called the Casparian strip.

90. The flow of water upwards through the xylem in plants can achieve fairly high rates of up to 15 metres/hour.

91. Cell wall and intercellular spaces

92. Passive water absorption occurs due to tension in xylem sap.

93. Guttation takes place through special pores called hydathodes.

94. (B) and (C) are correct.

95. 130 metres

96. Cohesive force is the mutual attraction between water molecules (H bonds).

Transpiration

97. Transpiration is loss of water in the form of vapour through the stomata.

98. Maximum transpiration occurs in Mesophytes.

99. Major part of total transpiration occurs through the stomata. Stomatal transpiration constitutes 50–97% of total transpiration.

100. Cellulosic microfibrils are oriented radially in the cell wall of guard cells.

101. The immediate cause of the opening or closing of stomata is a change in the turgidity of the guard cells.

102. Two statements are incorrect. The inner wall of the guard cell is thick and elastic. In xerophyte plants, sunken stomata are present.

103. Due to efflux of H^+, the pH of guard cells is higher as compared to subsidiary cells.

104. Guard cells have radially oriented cellulosic microfibrils.

105. K^+ ion

106. Red and blue light constitute the action spectra of transpiration.

107. Increased K^+ and malate ions are stored in vacuoles of the guard cells, increasing their osmotic concentration. Hence, water enters the guard cells by endosomes.

108. Option (b) is correct.

109. Stomatal movement is not affected by O_2 concentration.

110. Rate of transpiration is measured by Ganong's potometer.

111. 1.5 km above sea level

112. Rate of transpiration is higher in C_3 plants.

113. Dry cobalt chloride paper of blue colour becomes pink in contact with water vapour of a transpiring leaf surface.

114. Transpiration

115. Transpiration can coat the leaf surface by evaporative cooling up to 10–15°C.

116. A – Thick and elastic inner cell wall, B – Thin and elastic outer wall

117. In humid atmosphere (when the relative humidity is high), the rate of transpiration decreases.

118. Chlorophyll content of leaves

Uptake and Transport of Mineral Nutrients

119. The chief sinks for the mineral elements are the growing regions of plants such as the apical and lateral meristems, young leaves, developing flowers, fruits and seeds and the storage organs.

120. Only those elements incorporated in the structural components are not remobilised. For example, calcium.

121. (A) and (C) are correct.

122. Minerals are mainly absorbed by active transport.

123. Elements most readily mobilised are P, N and K.

124. Endodermal cell membrane

125. Only (B) is incorrect.

Phloem Transport

126. Pressure flow hypothesis was proposed by E Munch.

127. A – Sieve tubes, B – Sucrose, C – Bidirectional

128. Loading and unloading are active processes.

129. Phloem transports organic solutes from the region of high turgor pressure.

130. Phloem sap is mainly water and non-reducing sugar sucrose.

131. Transport through the phloem is bidirectional because the source and sink are reversible.

132. Active transport

133. (A) and (D) are false. In a girdled plant, the roots die first.

134. A girdled plant will eventually die because of absence of downward movement of water.

135. Food translocation is through phloem.

136. Meaningful ringing experiments cannot be carried out on sugarcane because vascular bundles are scattered.

137. 60–100 cm/h

12 Mineral Nutrition

EXERCISES

Mineral Requirements of Plants, Essential Mineral Nutrients, Criteria for Essentiality, Micro and Macronutrients

1. The technique of growing plants in an artificial nutrient medium was first developed by:
(a) Jan Ingenhousz
(b) Julius von Sachs
(c) Van Niel
(d) Joseph Priestley

2. Hydroponics is a technique in which plants are grown in:
(a) Greenhouse
(b) Water-saturated sand
(c) Balanced nutrient solution
(d) Purified distilled water

3. Which of the following statements is correct with reference to hydroponics:
(a) It determines the mineral nutrients that are essential for plants
(b) It involves the culture of plants in a soil with defined mineral solution
(c) It requires purified water with non-defined mineral nutrient salt
(d) It is not employed as a technique for the commercial production of vegetables

4. Which of the following is the method by which essential elements were identified in plants?
(a) Plant ash analysis
(b) Hydroponics
(c) Plant tissue culture
(d) Chromatography

5. What does the given experiment set-up depict?

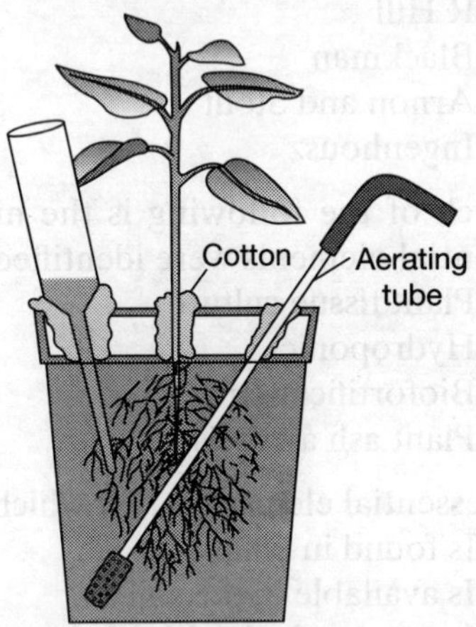

Choose the correct option accordingly:
(a) Measurement of the growth of a plant
(b) Nutrient solution culture
(c) CO_2 is essential for photosynthesis
(d) O_2 is required for aerobic respiration

6. Which of the following is not a benefit of hydroponics?
(a) Can regulate pH for a particular crop
(b) Controls soil-borne pathogens
(c) Cannot control the problem of weeding
(d) Can be used in areas having thin, infertile and dry soils

7. Hydroponics has been employed as a technique for the commercial production of:
(a) Tomato (b) Lettuce
(c) Seedless cucumber (d) All of the above

8. How many types of mineral elements have been reported in plants?
(a) 60 (b) 105
(c) 17 (d) 10

9. Read the following statements:
A. The element must be absolutely necessary for supporting normal growth and reproduction
B. The requirement of the element must be specific and not replaceable by another element
The above statements are applicable for:
(a) Criteria for toxicity
(b) Criteria for essentiality
(c) Criteria for hydroponics
(d) Criteria for critical elements

10. Who stated the criteria of essentiality?
(a) R Hill
(b) Blackman
(c) Arnon and Stout
(d) Ingenhousz

11. Which of the following is the method by which essential elements were identified in plants?
(a) Plant tissue culture
(b) Hydroponics
(c) Biofortification
(d) Plant ash analysis

12. An essential element is that which:
(a) Is found in plant ash
(b) Is available in the soil
(c) Improves the health of plants
(d) Is irreplaceable and indispensable for the growth of plants

13. What is the significance of ash in the study of mineral nutrients of plants?
(a) It reveals what minerals are present in the soil
(b) It informs us about which element is essential and in which amount it is necessary for a particular plant
(c) It is of no practical significance
(d) It indicates how much irrigation is needed by the plant

14. The micronutrient concentration present in the dry matter of plants is ______ in mole/kg:
(a) > 10 (b) < 0.1
(c) < 10 (d) <1

15. The essential elements that are present in plant tissues in excess of 10 m mole per kg of dry matter are:
(a) Micronutrients (b) Macronutrients
(c) Toxic elements (d) Beneficial elements

16. The lowest concentration of minerals that can be detected in plants by the present technique is:
(a) 10^{-8} g/ml (b) 10^{-8} mg/ml
(c) 10^{-10} mg/ml (d) 0.1 mg/ml

17. Non-essential but beneficial elements are:
(a) Silicon, Sodium, Molybdenum
(b) Cobalt, Sodium, Selenium
(c) Gold, Nickel, Chlorine
(d) Strontium, Selenium, Iron

18. Essential elements that are components of energy-related chemical compounds in plants:
(a) P, Mg (b) N, S
(c) K, P, Mg (d) N, P, K

19. Plants growing near nuclear test sites take up radioactive ______ from the soil:
(a) Gold (b) Selenium
(c) Strontium (d) Iron

20. Essential elements are mainly divided into how many categories on the basis of:
A. Quantitative presence
B. Diverse function

	A	B
(a)	4	2
(b)	2	4
(c)	2	3
(d)	2	6

21. Fe, Mn, Cu, Ca, Mg, K, Cl, Zn, Mo, Mg. How many elements from this list are macronutrients?
(a) Five (b) Seven
(c) Six (d) Eight

22. Of the following:
A. RuBisCO
B. PEPCo
C. Alcohol dehydrogenase
D. Netrogenase
(a) (A), (B) and (C) are activated by Mg
(b) (C) and (D) are activated by Zn and Mo, respectively
(c) (B) and (D) are activated by Mn
(d) All are activated by Mg

23. All are trace elements, except:
(a) Ni (b) Mg
(c) Mo (d) Fe

24. Read the following statements and state whether they are true (T) or false (F):
A. Carbon, hydrogen and oxygen are structural elements of the cell as they are components of biomolecules
B. Magnesium and phosphorus are components of energy-related compounds
C. Many essential elements are responsible for activating or inhibiting enzymes
D. Essential elements have no role in maintaining the osmotic potential of the cell

	A	B	C	D
(a)	T	T	T	F
(b)	T	F	T	F
(c)	T	F	T	T
(d)	F	T	F	F

25. _______ plays an important role in the opening and closing of stomata:
(a) Mg (b) P
(c) K (d) Ca

26. An essential element which is an activator of RuBisCO and PEPCo is:
(a) Zn^{++} (b) Mg^{++}
(c) Mo (d) Mn^{++}

27. Fertiliser elements are:
(a) Critical elements
(b) Non-mineral elements
(c) Beneficial elements
(d) Micronutrients

28. Criterion of essentiality is applicable to:
(a) Only micronutrients
(b) Only macronutrients
(c) Both micro and macronutrients
(d) All elements found in the soil

29. Find the incorrectly matched pair:
(a) Critical elements – N, P, K
(b) Framework elements – C, H, O
(c) Toxic elements – Al, As, Mg, Pb
(d) Beneficial elements – Mg, Mo, Mn

30. Which of the following is not an essential element for plants?
(a) Potassium (b) Iron
(c) Iodine (d) Zinc

Role of Micro and Macronutrients, Deficiency Symptoms of Essential Elements, Toxicity of Micronutrients, Mechanism of Absorption of Elements, Translocation of Solutes, Soil as a Reservoir of Essential Elements

31. Essential elements participate in various metabolic processes in plant cells, such as:
A. Permeability of cell membrane
B. Maintenance of osmotic concentration of cell sap
C. Electron transport systems
D. Buffering action
(a) Only (A) and (B)
(b) Only (A) and (C)
(c) Only (b) and (C)
(d) (A), (B), (C) and (D)

32. Which of the following minerals activates the enzymes involved in respiration?
(a) Nitrogen and Phosphorus
(b) Magnesium and Manganese
(c) Potassium and Calcium
(d) Sulphur and Iron

33. Necrosis means:
(a) Death of the tissue
(b) Yellow spot on leaves
(c) Excessive enlargement of leaf
(d) Darkening of green colour in leaves

34. The following functions are related to:
A. Required by plants in the greatest amount
B. Required by all parts of the plant, particularly meristematic tissues
C. One of the major constituents of protein, nucleic acids, vitamins and hormones
(a) Ca (b) Fe
(c) N (d) Mg

35. Identify the element on the basis of the given statements:
A. It activates various enzymes, especially carboxylases
B. It is required in the biosynthesis of auxin
(a) Zn (b) B
(c) Mn (d) Fe

36. Match the following:

Column I	Column II
A. Copper	(i) Carbohydrate translocation
B. Sulphur	(ii) Nitrogen metabolism
C. Boron	(iii) Biotin
D. Molybdenum	(iv) Tyrosinase

(a) A – (iv), B – (i), C – (iii), D – (ii)
(b) A – (iii), B – (iv), C – (i), D – (ii)
(c) A – (iv), B – (iii), C – (i), D – (ii)
(d) A – (iii), B – (iv), C – (ii), D – (i)

37. Which one is a component of CoA?
(a) Zn (b) Cu
(c) S (d) Fe

38. Select the mismatched pair:
(a) Zn – Tryptophan synthesis
(b) S – Methionine
(c) Cu – Cytochrome oxidase
(d) Mo – Nitrite reductase

39. Which of the following elements plays an important role in photosynthetic oxygen evolution?
(a) Manganese (b) Magnesium
(c) Molybdenum (d) Boron

40. If chlorophyll is burnt, what element will be left?
(a) Ca (b) Mg
(c) Na (d) Mn

41. For which of the following minerals is the best defined function the splitting of water to liberate oxygen during photosynthesis?
(a) Mn (b) Mg
(c) Mo (d) Fe

42. Minerals that maintain cation–anion balance in cells are:
(a) Cl^- and K^+ (b) K^+ and Fe^{++}
(c) Cl^- and Mg^{++} (d) Ca^{++} and Mg^{++}

43. Inhibition of cell division occurs due to lack or low level of:
(a) Mo, S, Ni (b) K, S, Mo, N
(c) Mg, K, Fe (d) Ca, Cu, K, Mg

44. All of the following aspects are related to which of the following minerals?
A. Involved in protein synthesis
B. Opening and closing of stomata
C. Activates many enzymes
D. Helps in the maintenance of the turgidity of cells
(a) Mg (b) Fe
(c) P (d) K

45. Uptake of nitrogen by plants occurs in which of the given forms:
(a) NO_3^- (b) NO_2^-
(c) N_2O (d) NH_4^+

46. The element necessary for middle lamella is:
(a) Mg (b) Ca
(c) Fe (d) Both (a) and (b)

47. Select the incorrectly matched pair:
(a) Magnesium – Involved in the synthesis of DNA and RNA
(b) Potassium – Required in more abundant quantities in the meristematic tissue buds, leaves and root tips
(c) Phosphorus – Required for all phosphorylation reactions
(d) Zinc – Needed for the biosynthesis of PGR that helps overcome apical dominance

48. Sulphur is not a component of:
(a) Methionine (b) Thiamine
(c) Ferredoxin (d) Adenosine

49. Which of the following is the mismatched pair?

Mineral element	Form that is absorbed by plant
(a) Iron	Fe^{++}
(b) Copper	Cu^{++}
(c) Molybdenum	MoO_2^{++}
(d) Boron	$B_4O_7^{-}$

50. Which of the following is not involved in the photolysis of water?
(a) Mn (b) Cl
(c) Mg (d) Ca

51. Iron:
A. Required in larger amounts in comparison to other micronutrients
B. It is an important constituent of ferredoxin and cytochromes
(a) Only (A) is correct
(b) Only (B) is correct
(c) Both (A) and (B) are correct
(d) Both (A) and (B) are incorrect

52. Identify the mineral nutrients on the basis of the given features:
A. It is required in the formation of mitotic spindle
B. It is required for uptake and utilisation of calcium ions
C. It helps to maintain ribosome structure
D. It activates the catalase enzyme

	A	B	C	D
(a)	Cu	Mn	K	S
(b)	Ca	B	Mg	Zn
(c)	Ca	B	Mg	Fe
(d)	Cl	S	Mn	Cu

53. Which of the following is not true for phosphorus?
(a) It is absorbed in the form of $H_2PO_4^-$
(b) It is one of the major constituents of all proteins
(c) It is a constituent of the cell membrane
(d) It is required for all phosphorylation reactions

54. The most abundant intracellular cation is:
(a) K^+ (b) Na^+
(c) Ca^{++} (d) H^+

55. Select the pair from the following in which both the elements share common functions in the splitting of water molecules during photosynthesis in plants:
(a) Cl, Mg (b) Mn, Cl
(c) B, Mo (d) K, P

56. The functions given below are performed by which of the following mineral elements?
(i) Required by meristematic and differentiating tissue
(ii) Accumulates in older leaves
(iii) Essential for normal function of the cell membrane
(a) Ca (b) Fe
(c) Mg (d) Cu

57. Select the mismatched pair:
(a) Mn – Splitting of water to liberate oxygen
(b) Mo – Nitrate reductase
(c) Mg – Activator of RuBisCO
(d) Fe – Component of plastocyanin

58. Sulphur is absorbed by plants as:
(a) SO_3 from soil (b) SO_2 from air
(c) SO_4 from soil (d) Both (b) and (c)

59. The most abundant element found in plants is:
(a) Carbon (b) Nitrogen
(c) Iron (d) Manganese

60. Which of the following elements is essential for root growth?
(a) Zn (b) Mo
(c) Ca (d) S

61. Read the following statements regarding various functions of essential elements:
A. Constituent of ring structure of chlorophyll
B. Involved in the synthesis of DNA and RNA
C. Constituent of thiamine, biotin and co-enzyme A
D. Required for uptake and utilisation of Ca^{++}
E. Essential for pollen germination
F. Essential for carbohydrate translocation

These functions of essential elements indicate different elements:
(a) Two (b) Four
(c) Three (d) Five

62. An element is said to be deficient, when present:
(a) Below critical concentration
(b) above critical concentration
(c) At critical concentration
(d) All except (a)

63. Phosphorus is a structural element in:
(a) Cell membrane (b) Curtain proteins
(c) All nucleic acids (d) All of the above

64. Exanthema, a disease of citrus plants, is caused by the deficiency of:
(a) Zinc (b) Magnesium
(c) Copper (d) Manganese

65. Deficiency of which element is not responsible for the death of leaf tissues?
(a) K (b) Cu
(c) Ca (d) Mn

66. Find the correct match:
(i) ___A___ is a constituent of the ring structure of chlorophyll
(ii) ___B___ is a constituent of coenzyme A
(iii) ___C___ is responsible for maintaining turgidity of the cell

	A	B	C
(a)	Fe	Cu	P
(b)	Fe	S	Cl
(c)	Mg	Fe	K
(d)	Mg	S	K

67. Which of the following is not caused by deficiency of mineral nutrition?
(a) Inhibition of cell division
(b) Chlorosis
(c) Etiolation
(d) Delayed flowering

68. Grey spots of oat is caused by deficiency of:
(a) Cu (b) Zn
(c) Mn (d) Fe

69. Elements found in ferredoxin are:
(a) Fe and S (b) Fe and Ca
(c) Mg and Zn (d) Fe and Cu

70. Chlorophyll synthesis will be largely affected due to deficiency of which of the following sets of elements?
(a) K, Fe, Mg (b) Fe Mg Ca
(c) Fe, Mg, Zn (d) K, Fe, Zn

71. Read the following statements with respect to toxicity of nutrients:
A. The toxicity symptoms are difficult to identify
B. Toxicity levels for any element vary for different plants
C. Excess of an element may inhibit the uptake of another element

Select the correct statements:
(a) Only (C)
(b) Only (B) and (C)
(c) Only (A) and (C)
(d) (A), (B) and (C)

72. Deficiency of symptoms of nitrogen and potassium are visible first in:
(a) Young leaves (b) Roots
(c) Older leaves (d) Buds

73. Excess of manganese inhibits translocation of ____ to the shoot apex:
(a) Calcium (b) Potassium
(c) Iron (d) Magnesium

74. The elements most readily mobilised are all, except:
(a) S (b) P
(c) K (d) N

75. Manganese competes with ____ and magnesium for uptake and with ____ for binding with enzymes:
(a) Fe, Ca (b) Fe, Mo
(c) Fe, Mg (d) Ca, Fe

76. Deficiency of how many of the following elements can show symptoms like chlorosis?

Mo	K	N	Cu	Zn	Mg
Ca	Mn	S	Zn		

(a) 5 (b) 6
(c) 7 (d) 4

77. Read the following statements carefully:
A. The concentration of the element below which plant growth is retarded
B. Any mineral ion concentration in tissue that gives a 10% reduction in dry matter

Select the correct option according to the statements given above:

	A	B
(a)	Critical Concentration	Toxic Concentration
(b)	Toxic Concentration	Critical Concentration
(c)	Micro Concentration	Macro Concentration
(d)	Critical Concentration	Micro Concentration

78. Deficiency symptoms first appear in the younger leaf if the soil is deficient in:
(a) S (b) Ca
(c) K (d) (a) and (b)

79. Select the incorrect statement:
(a) Plants absorb copper as cupric ion
(b) Necrosis occurs due to deficiency of Ca, Cu, K and Mg
(c) Potassium is required by plants in the greatest amount
(d) Manganese activates many enzymes involved in photosynthesis, respiration and nitrogen metabolism

80. Mn in higher concentration produces toxicity and it can cause:
A. Decrease in fresh weight of tissues by about 10%
B. The appearance of yellow spots surrounded by green veins and veinlets
C. Inhibition of calcium translocation to root apex
D. Manganese competes with magnesium for binding with enzymes

The correct statements are:
(a) (A) and (B) (b) (A), (B) and (D)
(c) (D) Only (d) (C) and (D)

81. All of the given are mobile elements, except:
(a) Nitrogen (b) Calcium
(c) Phosphorus (d) Chlorine

82. The symptoms of Mn toxicity may actually be a deficiency symptom of:
(a) Fe (b) Ca
(c) Mg (d) All of the above

83. The primary symptom of manganese toxicity is:
(a) Appearance of brown spots surrounded by chlorotic veins
(b) Loss of apical dominance
(c) Presence of white bud
(d) Senescent leaves

84. Ni is an essential part of which of the following enzymes?
(a) Urease
(b) Nitrogenase

(c) Transaminase
(d) Glutamate synthetase

85. Deficiency of mobile elements first appears on:
(a) Older leaves (b) Apical meristem
(c) Younger leaves (d) Floral parts

86. Sulphur-containing amino acid is/are:
(a) Methionine (b) Cysteine
(c) Cystine (d) All of the above

87. Deficiency symptoms of essential elements:
A. Are called hunger signs
B. Vary from element to element
C. Depend upon motility of elements also
D. Disappear when the deficient mineral nutrient is provided to the plant
The correct statement(s) is/are:
(a) Only (A)
(b) Only (A) and (D)
(c) Only (A), (C) and (C)
(d) (A), (B), (C) and (D)

88. The symptom related to Mo deficiency is/are:
(a) Chlorosis
(b) Inhibition of cell division
(c) Delayed flowering
(d) All of the above

89. Which of the following statements is not correct?
(a) Nitrogen deficiency symptoms first appear in younger leaves
(b) Phosphorus is a structural component of cellular energy currency
(c) The ionic concentration of potassium in guard cells regulates stomatal movement
(d) Cytochrome is an iron-containing protein

90. All minerals cannot be absorbed passively because:
A. Concentration of minerals in soil is usually lower than the concentration of minerals in root cells
B. Minerals are present in soil as charged particles
(a) Only (A) is correct
(b) Only (B) is correct
(c) Both (A) and (B) are correct
(d) Both (A) and (B) are incorrect

91. Find the incorrect statement about chlorine with respect to plants:
(a) Absorbed as a chloride anion from the soil
(b) Along with Na^+ and K^+ helps in determining the solute concentration
(c) Essential for the water splitting reaction in photosynthesis
(d) Associated with certain enzymes associated with redox reactions

92. Which element is required in the germination of pollen grain?
(a) Chlorine (b) Potassium
(c) Boron (d) Calcium

93. The deficiency disease "dieback of shoots" of plants is caused by:
(a) Boron (b) Zinc
(c) Manganese (d) Copper

94. Which mineral element is essential for root hair growth:
(a) Zn (b) Ca
(c) Mo (d) S

95. Minerals absorbed by the roots move to the leaf through:
(a) Xylem (b) Phloem
(c) Pressure gradient (d) Both (b) and (c)

96. Mineral transport in plants is:
(a) Controlled by transpiration pull
(b) Dependent on sugar translocation
(c) Dependent on energy expenditure
(d) An osmotic phenomenon

97. Which of the following are not true about the mechanism of element absorption in plants?
A. In the initial phase, there is rapid uptake of ions into the outer space of the cell
B. The first phase of mineral absorption is active
C. The symplast comprises the cytoplasm and vacuole
D. The entry or exit of ions to and from the inner space does not require metabolic energy
(a) (A) and (B) (b) (B) and (C)
(c) (B) and (D) (d) (C) and (D)

98. Initial phase of ion uptake:
(a) Is passive and slow
(b) Is not dependent on metabolic energy
(c) Occurs through apoplast
(d) All of the above

99. Active absorption of minerals is dependent on:
(a) Temperature (b) Apoplast
(c) Light (d) Availability of oxygen

100. The element involved in redox reactions is:
(a) Cu (b) Fe
(c) Zn (d) All except (c)

Metabolism of Nitrogen (Nitrogen Cycle, Biological Nitrogen Fixation, Nodule Formation and Fate of Ammonia)

101. Refer to the figure given below and select the correct option which indicates the name of the microorganism that carried out reactions A and B:

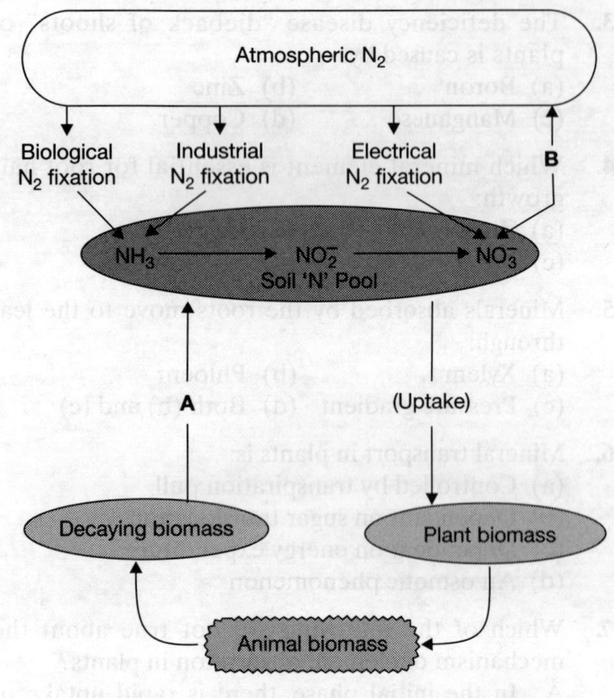

	A	**B**
(a)	*Bacillus ramosus*	*Pseudomonas*
(b)	*Nitrosomonas*	*Thiobacillus*
(c)	*Nitrobacter*	*Nitrosomonas*
(d)	*Pseudomonas*	*Bacillus vulgaris*

102. During non-biological nitrogen fixation, the nitrogen is converted to nitrogen oxide with the help of:
(a) Lightning (b) UV radiation
(c) *Rhizobium* (d) Both (A) and (B)

103. Nitrifying bacteria are:
(a) Autotrophs (b) Chemoautotrophs
(c) Heterotrophs (d) Decomposers

104. The enzyme nitrogenase is a:
(a) Cu–Fe protein (b) Ni–Fe protein
(c) Mo–Fe protein (d) Ni–Cu protein

105. An anaerobic bacterium capable of nitrogen fixation is:
(a) *Clorobium* (b) *Rhodospirillum*
(c) *Azotobacter* (d) *Clostridium*

106. $N_2 + 8e^- + 8H^+ + 16ATP \rightarrow 2NH_3 + H_2 + 16ADP + 16Pi$

The above equation refers to:
(a) Ammonification
(b) Nitrogen fixation
(c) Nitrification
(d) Denitrification

107. Nitrogen fixation in the root nodule of *Alnus* is brought about by:
(a) *Azorhizobium* (b) *Glomus*
(c) *Frankia* (d) *Rhizobium*

108. Which of the following is incorrectly matched?

	Microorganisms	Nitrogenase activity	Photosynthetic activity
(a)	*Rhizobium*	+	–
(b)	*Frankia*	+	+
(c)	*Anabaena*	+	+
(d)	*Azotobacter*	+	–

109. Nitrogenase enzyme:
A. Requires ATP for reduction of nitrogen
B. Is an Mn–Fe protein
C. Is present in prokaryotes only
D. Is oxygen insensitive
(a) (A) and (B) (b) (B) and (D)
(c) (B) and (C) (d) (A) and (C)

110. The region of the nodule where *Rhizobium* is active, lacks:
(a) ATP
(b) Nitrogenase enzyme
(c) Molecular oxygen
(d) Leghemoglobin

111. Mark the incorrect option regarding the given step of the nitrogen cycle: $2NO_2^- + O_2 \rightarrow 2NO_3^-$
(a) The bacteria involved is *Nitrobacter*
(b) It shows nitrate assimilation
(c) It is a part of nitrification
(d) The involved bacteria is a chemoautotroph

112. Denitrification is carried out by:
(a) *Pseudomonas* and *Thiobacillus*
(b) *Bacillus remosus* and *Bacillus vulgaris*
(c) *Nitrosomonas* and *Nitrococcus*
(d) *Pseudomonas* and *Nitrococcus*

113. Yield of rice is enhanced by:
(a) *Azolla* (b) *Anabaena*
(c) *Nostoc* (d) *Clostridium*

114. Select the correct option to fill in the blanks in the following statements:

(i) The enzyme ____A____ is capable of nitrogen reduction, which is present exclusively in ____B____

(ii) If a plant is transferred from water to soil solution, its rate respiration will be ____C____

(iii) *Clostridium* is anaerobic and a ____D____ nitrogen fixing bacterium

	A	B	C	D
(a)	Nitrogenase	Cyanobacteria	Decrease	Symbiotic
(b)	Nitrogenase	Prokaryotes	Increase	Free living
(c)	Transaminase	Prokaryotes	Decrease	Symbiotic
(d)	Nitrate reductase	Cyanobacteria	Increase	Free living

115. *Frankia*:
(a) Is a chemoautotroph
(b) Is capable of nitrogen fixation in free-living conditions
(c) May establish symbiotic relations with the roots of non-leguminous plants
(d) Is an ammonifying bacteria

116. The part of the plant root where nitrogen fixing bacteria live and form nodules is:
(a) Epidermis (b) Cortex
(c) Pith (d) Hypodermis

117. Plants cannot accumulate NH_4^+ ion because it is:
(a) Oxidising agent (b) Reducing agent
(c) Toxic to plants (d) Oxidatively aminated

118. The process of conversion of proteins → amino acids → NH_3 is called:
(a) Nitrification (b) Nitrogen fixation
(c) Nitrate assimilation (d) Ammonification

119. Amino acid, which plays a central role in nitrogen metabolism, is:
(a) Glutamic acid (b) Methionine
(c) Phenylalaine (d) Aspartic acid

120. The process that is the opposite of nitrogen fixation is:
(a) Nitrification (b) Denitrification
(c) Ammonification (d) Nitrate reduction

121. Read the following statements carefully with respect to conversion of atmospheric nitrogen to ammonia by the nitrogenase enzyme found in nitrogen fixing bacteria:
A. The enzyme is not operational in microbes under free-living conditions
B. Nodules have adaptations that ensure that the enzyme is protected from oxygen
C. Ammonia synthesis by the enzyme requires a very high input of energy
D. The energy required is obtained from the respiration of microbes
Select the correct statements:
(a) (B) and (C) (b) (B), (C) and (D)
(c) (A), (B) and (C) (d) Only (B)

122. Nitrifying bacteria help in:
(a) Reducing nitrates to free nitrogen
(b) Oxidising ammonia to nitrates
(c) Converting free nitrogen to nitrogen compound
(d) Converting proteins into ammonia

123. $A + NH_4^+ + B \xrightarrow{C} \text{Glutamate} + H_2O + D$
Consider the above reaction:

	A	B	C	D
(a)	Glutamic acid	$FADH_2$	Glutamate dehydrogenase	FAD
(b)	OAA	$NADPH_2$	Glutamate oxidase	NADP
(c)	Ketoglutaric acid	NADPH	Glutamate dehydrogenase	NADP
(d)	α-ketoglutaric acid	NADH	Transaminase	NAD

124. The two most important amides which are found in legume plants are a structural part of proteins:
(a) Glutamine and aspartic acid
(b) Glutamine and asparagine
(c) Alanine and asparagine
(d) Alanine and glutamine

125. *Azotobacter* and *Beijerinckia* are examples of:
(a) Symbiotic nitrogen fixer
(b) Ammonifying bacteria
(c) Free-living nitrogen fixer
(d) Nitrifying bacteria

126. Mark the incorrect statement for ureides:
(a) Transported via the phloem
(b) Have high nitrogen to carbon ratio
(c) Formed in soya bean plant
(d) Produced inside nodules

127. The nitrogen fixed in the root nodules of *Glycine max* is exported to other parts of the plant as:
(a) Ureids (b) Asparagine
(c) Nitrates (d) Ammonia

128. Reaction of α-ketoglutaric acid with ammonia to form glutamic acid is:
(a) Oxidative amination
(b) Reductive amination

(c) Transamination
(d) Ammonification

129. Read the following steps during root nodule formation and arrange them sequentially:
A. Root nodule formation
B. Rhizobia colonise the surrounding roots
C. Curling of root hair
D. Rhizobia released into cortical cells
E. Infection thread is produced
(a) (B), (C), (E), (D), (A)
(b) (B), (C), (D), (E), (A)
(c) (A), (B), (C), (D), (E)
(d) (B), (C), (A), (D), (E)

130. Section of root nodule in sweetpea plant shows pinkish colour due to the presence of:
(a) Anthocyanin (b) Leghemoglobin
(c) Xanthophyll (d) Carotene

131. Select the correct option to fill in the blanks in the following statements:
(i) The process of transfer of amino group from one amino acid to the keto group of keto acid is called ___A___
(ii) ___B___ is the main amino acid from which transfer of amino group takes place and other amino acids are formed
(iii) Amides are different from amino acids as they contain more ___C___

	A	B	C
(a)	Transamination	Aspartic acid	Nitrogen
(b)	Reductive amination	Glutamic acid	Oxygen
(c)	Transamination	Glutamic acid	Nitrogen
(d)	Reductive amination	Pyruvic acid	Oxygen

132. How many electrons, protons and ATP are required for the synthesis of one ammonia molecule in N_2 fixation metabolism?
(a) $4e^-$, $4H^+$ and 8 ATP
(b) $4e^-$, $4H^+$ and 16 ATP
(c) $8e^-$, $8H^+$ and 16 ATP
(d) $8e^-$, $8H^+$ and 8 ATP

133. The function of leghemoglobin in the root nodule of legumes is:
(a) Expression of nif gene
(b) Inhibition of nitrogenase activity
(c) Oxygen removal
(d) Nodule differentiation

134. The enzyme which helps in the formation of a number of amino acids from glutamic acid is:
(a) Glutamate synthetase
(b) Glutamate dehydrogenase
(c) α-ketoglutarate dehydrogenase
(d) Transaminase

135. Consider the following reactions :
A. $NH_3 \rightarrow NO_3^-$
B. $NO_3^- \rightarrow N_2$
A and B reactions are:

	A	B
(a)	Oxidation	Reduction
(b)	Reduction	Reduction
(c)	Oxidation	Oxidation
(d)	Reduction	Oxidation

136. In the following reaction, which important step of amino acid synthesis is shown:

$$R_1-\underset{NH_3}{\overset{H}{C}}-CO\bar{O} + R_2-\underset{O}{\overset{H}{C}}-CO\bar{O} \rightarrow R_1-\underset{O}{\overset{H}{C}}-CO\bar{O} + R_2-\underset{NH_3^+}{\overset{H}{C}}-CO\bar{O}$$

(a) Reductive amination
(b) Deamination
(c) Transamination
(d) Catalytic amydation

137. Read the following statements carefully:
(i) Every mineral element that is present in soil is required by plants
(ii) Nitrogen is a nutrient element which is highly immobile in plants
(iii) Denitrification can be considered as the opposite of nitrogen fixation
(iv) During nitrogen fixation, reduction of one molecule of nitrogen into two ammonia molecules consumes 8 ATP
(v) Insectivorous plants grow in nitrogen-deficient soil
How many of these statement(s) is/are correct?
(a) Three (b) Four
(c) Two (d) One

138. Which of the following reactions is performed by *Bacillus ramosus*?
(a) $N_2 \rightarrow NH_3$
(b) $NH_3 \rightarrow NO_2^- \rightarrow NO_3^-$
(c) Protein $\rightarrow$ Amino acid $\rightarrow NH_3$
(d) $NO_3^- \rightarrow N_2$

139. Consider the following statements and select the correct option stating which ones are true (T) and which ones are false (F):
A. Both *Rhizobium* and *Frankia* are free living in soil and can fix atmospheric nitrogen in free-living conditions

B. *Rhizobium* is a rod-shaped bacteria which has a symbiotic relationship with the roots of several leguminous plants
C. Root nodules establish a direct vascular connection with the host for exchange of nutrients
D. The inner portion of the nodule is red or pink due to the presence of nitrogenase enzyme

	A	B	C	D
(a)	F	T	T	F
(b)	F	T	T	T
(c)	T	F	F	F
(d)	F	F	T	T

140. Nitrate absorbed by the plant root is transported to the leaves, where it is reduced to form ammonia. This conversion is called:
(a) Nitrogen fixation
(b) Transamination
(c) Nitrate assimilation
(d) Ammonification

141. Read the following statements and select the correct option:
A. The enzyme nitrogenase is highly sensitive to molecular oxygen; it requires anaerobic conditions
B. To protect the nitrogenase enzyme, the nodule contains leghemoglobin, which acts as an oxygen scavenger
C. Nitrogen fixation by the nitrogenase enzyme requires a very high input of energy
D. The energy required for N_2 fixation is obtained from the respiration of the host cell
(a) Only (A) and (B)
(b) Only (A), (B) and (C)
(c) Only (A), (B) and (D)
(d) (A), (B), (C) and (D)

142. Which of the following is not a requirement of transamination?
(a) Amino donor (b) Amino acceptor
(c) High energy input (d) Transaminase

143. Plants cannot accumulate NH_4^- ion because:
(a) It is an oxidising agent
(b) It is a reducing agent
(c) It is toxic to plants
(d) It is oxidatively aminated

144. Read the following statements and select the correct option:
A. Glutamine and asparagine are a structural part of plant proteins, along with amino acids
B. They are transported to other parts of the plant via the sieve tube element
(a) Only (A) is correct
(b) Only (B) is correct
(c) Both (A) and (B) are correct
(d) Both (A) and (B) are incorrect

145. Both nitrogenase and RuBisCO activity are exemplified by:
(a) *Rhodospirillum* and *Azotobacter*
(b) *Beijerinckia* and *Rhisobium*
(c) *Oscillatoria* and *Nostoc*
(d) *Anabaena* and *Frankia*

NCERT EXEMPLAR QUESTIONS

1. Which of the two elements have the closest common deficiency symptoms?
(a) N and S (b) N and B
(c) Cu and B (d) P and Ni

2. During hydroponics, the mineral solution is aerated. Why?
(a) A solution that is not aerated becomes depleted of oxygen, a condition called anoxia
(b) The dissolved oxygen of the mineral solution is continuously utilised by the root cells for their respiration
(c) To mix minerals of the nutrient solution
(d) Both (a) and (b)

3. Match the columns with respect to the functions of beneficial elements:

Beneficial elements	Function
A. Sodium	(i) Prevents lodging of grasses
B. Silicon	(ii) Generally toxic to most plants
C. Cobalt	(iii) Essential for C_4 plants
D. Selenium	(iv) Essential for growth of legumes

(a) A – (i), B – (ii), C – (iii), D – (iv)
(b) A – (iv), B – (iii), C – (i), D – (ii)
(c) A – (iii), B – (i), C – (iv), D – (ii)
(d) A – (i), B – (iv), C – (ii), D – (iii)

4. Which of the following plant hormones have nitrogen in their structure?
(a) Auxin and cytokinin
(b) GA and ABA
(c) ABA and auxin
(d) GA and cytokinin

5. 'Critical toxicity levels':
(a) Is the concentration of a toxic element that gives a 10% reduction in dry weight
(b) Vary widely between micronutrients as well as between plant species
(c) Is for macronutrients only
(d) More than one option is correct

6. Which of the following is incorrect?
(a) With deficiency of nitrogen and sulphur, impaired protein synthesis occurs
(b) Sulphur is readily mobilised in most plant species
(c) Chlorosis is the indirect effect of sulphur and nitrogen deficiency
(d) Sulphur is taken by plants from the soil in the form of SO_4^{2-}

7. The most abundant cellular cation (with respect to essential elements) is ______ and so is required in large amounts by most plants:
(a) Mn^{2+} (b) Mg^{2+}
(c) Ca^{2+} (d) K^{+}

8. The pungent flavour of mustard and turnip of the family Brassicaceae is due to:
(a) Sulphur-containing compounds
(b) Thiocyanates and isothiocyanates
(c) Phosphorous-containing compounds
(d) More than one option is correct

9. Which is true for iron as an essential element in plants?
(a) Absorbed by plants in the form of Fe^{3+}
(b) Iron is not a constituent of chlorophyll but is required for its synthesis
(c) Iron is a constituent of several oxidase enzymes
(d) All are correct

10. Siderophores:
(a) Are high-affinity, iron-chelating compounds secreted by microbes and plant roots
(b) Serve to transport all essential elements into the root cells
(c) Are Ca^{2+} binding agents
(d) Do not occur naturally and are synthetic

PRACTICE TEST PAPER

1. Criteria for essentiality of elements in plant growth is all, except:
(a) Absolutely necessary for supporting growth
(b) Directly involved in plant metabolism
(c) Specific role in plant growth and replaceable by other elements
(d) Directly involved in plant metabolism

2. The deficiency symptoms of the immobile element tend to appear first in young tissues. Select that deficient element:
(a) Mg (b) Ca
(c) K (d) Zn

3. Pollen germination and carbohydrate translocation are affected by the essential element:
(a) Boron (b) Molybdenum
(c) Chlorine (d) Manganese

4. Deficiency of which of the following elements leads to delay in flowering?
(a) Mo, S, N (b) Fe, Mn, Zn
(c) Ca, K, Mg, Cu (d) S, K, Mo, N

5. Sodium, silicon, cobalt and selenium are:
(a) Essential elements
(b) Toxic elements
(c) Beneficial elements
(d) Critical elements

6. Essential element present in plant tissue in an amount of more than 10 m mole kg^{-1} of dry matter is/are:
(a) Mn (b) Zn
(c) Cu (d) All of the above

7. Which of the following is a limiting nutrient for both natural and agricultural ecosystems?
(a) Carbon (b) Nitrogen
(c) Boron (d) Potassium

8. Select the incorrect match with respect to the activator element:
(a) Mg^{++} – PEPCo
(b) Fe – Catalase
(c) Ni^{++} – Urease
(d) Mo – Nitrite reductase

9. Which of the following minerals aids in the biosynthesis of tryptophan?
(a) Manganese (b) Zinc
(c) Magnesium (d) Iron

10. The conversion of nitrate into ammonium takes place in:
(a) Mitochondria (b) Chloroplast
(c) Golgi body (d) Peroxysome

11. Which of the following enzymes is required in reductive amination?
(a) Transaminase
(b) Glutamate dehydrogenase
(c) Nitrogenase
(d) Nitrate reductase

12. First phase of mineral uptake is:
(a) Rapid and active
(b) Slow and passive
(c) Rapid and apoplastic
(d) Through specific pumps in the plasma membrane

13. Mn toxicity in plants:
A. Inhibits absorption of Mg and Fe
B. Competes with Mg for binding with enzymes
C. Inhibits calcium translocation
D. Decreases the dry weight of tissue by about 10%
(a) Only (A) and (D) are correct
(b) Only (B), (C) and (D) are correct
(c) Only (D) is correct
(d) (A), (B), (C) and (D) are correct

14. Nitrococcus is:
(a) Ammonifying bacteria
(b) Nutrifying bacteria
(b) Denitrifying bacteria
(c) Nitrogen fixing bacteria

15. Which of the following statements is incorrect about leghemoglobin?
(a) It is a reddish or pink-coloured pigment
(b) It prevents oxygen poisoning during N_2 fixation
(c) It induces root nodule formation
(d) It occurs in root nodules of leguminous plants only

16. Along with the transpiration stream, the nodule of soya bean exports fixed nitrogen as:
(a) Aspartate and glutamate
(b) Asparagine and glutamine
(c) Ureids
(d) Amides

17. Which of the following reactions of nitrogen metabolism is performed by chemoautotrophic bacteria?
(a) $N_2 \rightarrow NH_3$
(b) $NH_3 \rightarrow NO_2^- \rightarrow NO_3^-$
(c) $NO_3^- \rightarrow N_2$
(d) Proteins $\rightarrow$ Amino acids $\rightarrow NH_3$

18. Identify the following statements as true (T) or false (F) and select the option accordingly:
A: Sulphur is the main constituent of biotin, thiamine and ferredoxin
B: The technique of growing plants in a nutrient solution is known as hydroponics
C: Toxicity symptoms of mineral ions are easy to identify
D: Mo is required for splitting of water molecule to release oxygen

	A	B	C	D
(a)	T	T	F	F
(b)	T	F	T	T
(c)	T	T	T	F
(d)	F	T	F	F

19. In symbiotic N_2 fixation by *Rhizobium leguminoserum*, energy is provided by:
(a) Bacteria
(b) Host cell
(c) Bacteria and host cell
(d) Not an energy-requiring process

20. Find out the odd one out with respect to the role of potassium in plants:
(a) Synthesis of middle lamella
(b) Opening of stomata
(c) Involved in protein synthesis
(d) Maintains anion–cation balance

21. During biological nitrogen fixation, the energy input is:
(a) 16 ATP for each NH_3
(b) 8 ATP for each NH_3
(c) 32 ATP for each NH_3
(d) 16 ATP for two NH_3

22. Deficiency symptoms of elements can be visualised by what kind of changes?
(a) Physiological changes
(b) Morphological changes
(c) Chemical changes
(d) Anatomical changes

23. How many electrons and protons are required for reduction of one molecule of nitrogen?
(a) $6e^-$ and $6H^+$ (b) $16e^-$ and $16H^+$
(c) $12e^-$ and $12H^+$ (d) $8e^-$ and $8H^+$

24. Match the columns and choose the correct option:

Column I (Microbe)	Column II (Process)
A. *Nitrobacter*	(i) Ammonification
B. *Anabaena*	(ii) Denitrification
C. *Pseudomonas*	(iii) Nitrogen fixation
D. *Bacillus*	(iv) Nitrification

(a) A – (iii), B – (iv), C – (ii), D – (iii)
(b) A – (iv), B – (iii), C – (i), D – (ii)
(c) A – (iv), B – (iii), C – (ii), D – (i)
(d) A – (ii), B – (iii), C – (i), D – (iv)

25. Heterotrophic filamentous prokaryote which fixes nitrogen in root nodules is:
(a) Alfa-alfa (b) *Alnus*
(c) Lentils (d) *Pisum*

26. Potassium mineral deficiency causes:
A. Chlorosis
B. Necrosis
C. Inhibition of cell division
D. Delayed flowering
(a) (A), (B) and (D)
(b) (B), (C) and (D)
(c) (A), (B) and (C)
(d) (A), (C) and (D)

27. Mark the odd one out with respect to the function of calcium:
(a) Constitutent of middle lamella
(b) Helps in the formation of mitotic spindle
(c) Maintains ribosomal structure
(d) Involved in normal functioning of cell membrane

28. Plants require sulphur for:
(a) ATP synthesis (b) Protein synthesis
(c) Glucose synthesis (d) DNA replication

29. During biological nitrogen fixation, division and growth of which root cells leads to formation of root nodules?
(a) Epidermis cells only
(b) Cortical cells and pith cells
(c) Endodermal cells and pith cells
(d) Cortical cells and pericycle cells

30. Glutamic acid is a key amino acid formed in plants mainly by the process of:
(a) Nitrogen fixation
(b) Reductive amination
(c) Nitrate assimilation
(d) Transamination

31. Salt respiration is:
(a) Linking of ion movement with respiratory chain
(b) Active increase in respiration during mineral absorption
(c) Decrease in respiration during salt absorption
(d) Secretion of salt through respiratory channel

32. Conversion of ammonia to nitrate by microbes is:
(a) Nitrification (b) N_2 fixation
(c) Ammonification (d) Denitrification

33. Leghemoglobin functions as:
(a) Oxygen scavenger
(b) Nitrogen scavenger
(c) CO_2 scavenger
(d) Hydrogen carrier

34. Element related to nitrogen metabolism is:
(a) Mn (b) Mg
(c) Mo (d) Mi

35. Denitrification is actually a process of:
(a) Redox (b) Oxidation
(c) Reduction (d) Diamination

36. Micronutrients:
(a) Are less important than macronutrients
(b) Are as important as macronutrients
(c) Have a minor role in plant nutrition
(d) Have no role in plant nutrition

37. Plastoquinone as hydrogen carrier contains:
(a) Mg (b) Fe
(c) S (d) Cu

38. Which plays an important role in energy metabolism?
(a) Calcium (b) Sodium
(c) Phosphorous (d) Sulphur

39. If deficiency symptoms of any element appear first in the senescent leaves, this element cannot be:
(a) Calcium (b) Nitrogen
(c) Potassium (d) Magnesium

40. Plants mostly absorb nitrogen from the soil in the form of:
(a) Nitrate (b) Nitrite
(c) Di nitrogen (d) Ammonium ion

41. Nitrogen fixation by organisms requires conditions that are:
(a) Highly alkaline (b) Anaerobic
(c) Saturated sunlight (d) Free of water

42. Which of the following is not a micronutrient for plants?
(a) Cu (b) B
(c) Zn (d) Cr

43. Which element is essential for the growth of the root tip?
(a) Fe (b) Ca
(c) Mn (d) Zn

44. Plants require two metallic compounds for chlorophyll synthesis. They are:
(a) Fe and Ca (b) Fe and Mg
(c) Cu and Ca (d) Ca and K

45. An essential element is that which:
(a) Is found in plant ash
(b) Is available in the soil
(c) Improves the health of plants
(d) Is irreplaceable and indispensable for the growth of plants

ANSWER KEYS

Exercises

1. (b) **2.** (c) **3.** (a) **4.** (b) **5.** (b) **6.** (c) **7.** (d) **8.** (a) **9.** (b) **10.** (c)
11. (b) **12.** (d) **13.** (b) **14.** (c) **15.** (b) **16.** (a) **17.** (b) **18.** (a) **19.** (c) **20.** (b)
21. (a) **22.** (b) **23.** (b) **24.** (a) **25.** (c) **26.** (b) **27.** (a) **28.** (c) **29.** (d) **30.** (c)
31. (d) **32.** (b) **33.** (a) **34.** (c) **35.** (a) **36.** (c) **37.** (c) **38.** (d) **39.** (a) **40.** (b)
41. (a) **42.** (a) **43.** (b) **44.** (d) **45.** (c) **46.** (d) **47.** (d) **48.** (d) **49.** (a) **50.** (c)
51. (c) **52.** (c) **53.** (b) **54.** (a) **55.** (b) **56.** (a) **57.** (d) **58.** (c) **59.** (a) **60.** (c)
61. (c) **62.** (a) **63.** (d) **64.** (c) **65.** (d) **66.** (d) **67.** (c) **68.** (c) **69.** (a) **70.** (a)
71. (d) **72.** (c) **73.** (a) **74.** (a) **75.** (c) **76.** (c) **77.** (a) **78.** (d) **79.** (c) **80.** (c)
81. (b) **82.** (d) **83.** (a) **84.** (a) **85.** (a) **86.** (d) **87.** (d) **88.** (d) **89.** (a) **90.** (c)
91. (d) **92.** (c) **93.** (d) **94.** (b) **95.** (a) **96.** (a) **97.** (c) **98.** (d) **99.** (d) **100.** (d)
101. (a) **102.** (d) **103.** (b) **104.** (c) **105.** (d) **106.** (b) **107.** (c) **108.** (b) **109.** (d) **110.** (c)
111. (b) **112.** (a) **113.** (a) **114.** (b) **115.** (c) **116.** (b) **117.** (c) **118.** (d) **119.** (a) **120.** (b)
121. (a) **122.** (b) **123.** (c) **124.** (b) **125.** (c) **126.** (a) **127.** (a) **128.** (b) **129.** (a) **130.** (b)
131. (c) **132.** (a) **133.** (c) **134.** (d) **135.** (a) **136.** (c) **137.** (a) **138.** (c) **139.** (a) **140.** (c)
141. (d) **142.** (c) **143.** (c) **144.** (a) **145.** (c)

NCERT Exemplar Questions

1. (a) **2.** (d) **3.** (c) **4.** (a) **5.** (d) **6.** (b) **7.** (d) **8.** (d) **9.** (d) **10.** (a)

Practice Test Paper

1. (c) **2.** (b) **3.** (a) **4.** (a) **5.** (c) **6.** (d) **7.** (b) **8.** (d) **9.** (b) **10.** (b)
11. (b) **12.** (c) **13.** (d) **14.** (b) **15.** (c) **16.** (c) **17.** (b) **18.** (a) **19.** (b) **20.** (a)
21. (b) **22.** (b) **23.** (d) **24.** (c) **25.** (b) **26.** (c) **27.** (c) **28.** (b) **29.** (d) **30.** (b)
31. (b) **32.** (a) **33.** (a) **34.** (c) **35.** (c) **36.** (b) **37.** (d) **38.** (c) **39.** (a) **40.** (a)
41. (b) **42.** (d) **43.** (b) **44.** (b) **45.** (d)

SOLUTIONS

Mineral Requirement of Plants, Essential Mineral Nutrients, Criteria of Essentiality, Macro and Micronutrients

1. Julius von Sachs
2. Commercial technique of soil-less culture is called hydroponics, in which plants are grown in a balanced nutrient solution.
3. Hydroponics determines the mineral nutrients that are essential for plants.
4. Hydroponics
5. Nutrient solution culture
6. Hydroponics controls the problem of weeding.
7. Tomato, Seedless Cucumber and Lettuce
8. 60
9. Criteria of essentiality
10. Arnon and Stout gave the criteria of essentiality.
11. Option (b) is correct.
12. The requirement of the element must be specific and not replaceable by another element.
13. Which element is essential and in which amount it is necessary for a particular plant
14. Micronutrients or trace elements are required in very small amounts (less than 10 m mole/kg of dry matter).
15. Macronutrients
16. There are techniques that can detect minerals, even at a very low concentration of 10^{-8} g/ml.
17. There are some beneficial elements such as sodium, silicon, cobalt and selenium, which are required by higher plants.
18. P and Mg (P in ATP and Mg in chlorophyll)
19. Strontium
20. A – 2, B – 4
21. Five elements are macronutrients (Ca, Mg, K, P and S).
22. Mg^{++} is an activator of both RuBisCO and PEPCo. Zn^{++} is an activator of alcohol dehydrogenase and Mo is an activator of nitrogenase.
23. Mg is a macronutrient.
24. Only (D) is false. Some essential elements can alter the osmotic potential of a cell.
25. Potassium plays an important role in the opening and closing of stomata.
26. Mg^{++}
27. Critical elements (N, P, K) are called fertiliser elements.
28. Criterion of essentiality is applicable to both micro and macroelements.
29. Sodium, selenium, silicon and cobalt are beneficial elements.
30. Iodine is not an essential element for plants.

Role of Micro and Macronutrients, Deficiency Symptoms of Essential Elements, Toxicity of Micronutrients, Mechanism of Absorption of Elements, Translocation of Solutes, Soil as a Reservoir of Essential Elements

31. (A), (B), (C) and (D) are correct.
32. Magnesium and Manganese
33. Necrosis is the death of plant tissue.
34. Nitrogen
35. Zinc
36. Copper – Activator of tyrosinase, Sulphur – Constituent of biotin, Boron – Helps in carbohydrate translocation, Molybdenum – Related to nitrogen metabolism
37. Sulphur is a constituent of coenzyme A.
38. Mo is an activator of nitrate reductase.
39. Manganese is essential for photolysis of water to release oxygen during photosynthesis.

40. Mg is a part of the ring structure of chlorophyll.

41. Mn

42. Cl^- and K^+ maintain cation–anion balance in the cells.

43. K, S, Mo, N

44. All the functions are related to the K^+ ion.

45. Nitrogen is mainly absorbed as NO^-_3 though some are also taken up as NO^{-2} or NH^+_4.

46. Middle lamella is made up of calcium and magnesium pectate.

47. Zinc is essential for the biosynthesis of auxin that induces apical dominance.

48. Sulphur is not a component of adenosine.

49. Plants obtain iron in the form of ferric ions (Fe^{+++}).

50. For photolysis of water, Mn, Cl and Ca are required.

51. Both (A) and (B) are correct.

52. A – Ca, B – B, C – Mg, D – Fe

53. Phosphorous is a constituent of certain proteins.

54. K^+ is the most free ion.

55. Mn and Cl

56. All the functions are related to calcium.

57. Copper is a component of plastocyanin.

58. SO_4 from soil

59. Carbon is the most abundant element found in plants.

60. Calcium is required by meristematic and differentiating tissues.

61. The given functions are related to three elements: Mg – a, b; S – c; B – d, e, f

62. An element is said to be deficient when present below the critical concentration.

63. Phosphorous is a structural element in the cell membrane, certain proteins and all nucleic acids.

64. Deficiency of copper may lead to the disease exanthema in citrus plants. This disease is characterised by progressive death of twigs, branches, shoots and roots.

65. Necrosis or tissue death is due to deficiency of Ca, Cu, K, Mg.

66. A – Mg, B – S, C – K

67. Etiolation is characterised by long weak stems, smaller leaves due to longer internodes and a pale yellow colour. It occurs due to deficiency of light.

68. Mn

69. Ferredoxin contain Fe and S.

70. K, Fe, Mg

71. (A), (B) and (C) are correct.

72. Nitrogen and phosphorous are mobile elements, their deficiency symptoms first appear in older leaves.

73. Excess of manganese inhibits translocation of calcium to the shoot apex.

74. S is an immobile element.

75. Fe, Mg

76. Seven elements causes chlorosis. These are Mo, K, Mn, S, Zn, N, Mg.

77. A – Critical elements, B – Toxic elements

78. S and Ca

79. Nitrogen is required by plants in the greatest amount.

80. Only D is correct. Mn toxicity decreases the dry weight of tissue by about 10% and the prominent symptom of Mn toxicity is brown spots surrounded by chlorotic veins.

81. Calcium is an immobile element.

82. The symptoms of Mn toxicity may actually be the deficiency symptom of Fe, Ca and Mg.

83. Appearance of brown spots surrounded by chlorotic veins

84. Ni is an activator of the enzyme uriase.

85. Deficiency of symptoms of mobile elements first appear in older leaves.

86. Cysteine, cystine and methionine are sulphur-containing amino acids.

87. (A) (B) (C) and (D) are correct.

88. All are correct.

89. The deficiency symptoms of nitrogen are visible first in the senescent leaves.

90. Both (A) and (B) are correct.

91. Copper is associated with certain enzymes associated with redox reactions.

92. Boron is required for pollen germination.

93. Dieback of shoots of plants is caused by copper.

94. Calcium is required for root hair formation.

95. Mineral salts are translocated through the xylem, along with ascending stream of water.

96. Mineral transport in water is controlled by transpiration pull.

97. (B) and (D) are not true. The first phase of mineral absorption is passive; in this phase, uptake of ions occurs into the free space or outer space of the cell. In the second phase, the ions are taken into the inner space, that is, the symplast, with requirement of metabolic energy.

98. All statements are correct.

99. Active absorption of minerals requires metabolic energy and so it is dependent on the availability of oxygen.

100. Cu and Fe are associated with certain enzymes involved in redox reactions.

Metabolism of Nitrogen (Nitrogen Cycle, Biological Nitrogen Fixation, Nodule Formation, Fate of Ammonia)

101. A – *Bacillus ramosus* (Ammonifying bacteria), B – *Pseudomonas* (Denitrifying bacteria)

102. Lightning and UV radiation

103. Nitrifying bacteria are chemoautotrophs.

104. The enzyme nitrogenase is an Mo–Fe protein.

105. *Clostridium* is an anaerobic bacteria.

106. Nitrogen fixation

107. *Frankia* fixes nitrogen in the root nodule of non-leguminous plants like *Alnus* and *Casuarina.*

108. *Frankia* shows only nitrogenase activity, not carboxylase activity.

109. (A) and (C) are correct. Nitrogenase enzyme is sensitive to oxygen.

110. Presence of molecular oxygen inactivates the nitrogenase enzyme, which is essential for nitrogen fixation.

111. The given reaction shows nitrification.

112. Nitrate present in the soil is also reduced to nitrogen by the process of denitrification. Denitrification is carried out by bacteria *Pseudomonas* and *Thiobacillus.*

113. Azolla is used as biofertiliser.

114. A – Nitrogenase, B – Prokaryotes, C – Increase, D – Free living

115. *Frankia* fixes nitrogen in the root nodules of non-leguminous plants.

116. Cortical cells of roots divide and form root nodules.

117. NH^+_4 is toxic to plants.

118. Ammonification

119. Glutamic acid

120. Denitrification is the opposite of N_2 fixation. Denitrification involves reduction of nitrate into free nitrogen.

121. (B) and (C) are correct. The energy required for nitrogen fixation is obtained from the respiration of the host cells.

122. Nitrifying bacteria oxidise ammonia into nitrates.

123. A – Glutamic acid, B – NADPH, C – Glutamate dehydrogenase, D – NADP

124. Glutamine and Asparagine

125. *Azotobacter* and *Beijerinekia* are free-living nitrogen fixing bacteria.

126. Ureids are transported by the xylem vessels.

127. Ureids

128. Reductive amination

129. B – C – E – D – A

130. Nodules are pink in colour due to leghaemoglobin.

131. A – Transamination, B – Glutamic acid, C – Nitrogen

132. $N_2 + 8e^- + 8H^+ + 16ATP \rightarrow 2NH_3 + H_2 + 16\,ADP + 16\,Pi$

For synthesis of one NH_3, $4e^-$, $4H^+$ and 8 ATP are required.

133. Leghaemoglobin acts as oxygen scavenger.

134. Transaminase

135. A – Oxidation, B – Reduction

136. Transaminase

137. Three statements (III, IV and V) are correct.

138. Protein – Amino acid – NH_3 (Ammonification)

139. (A) and (D) are false. Both *Rhizobium* and *Frankia* cannot fix nitrogen in free-living conditions.

140. Nitrate assimilation

141. (A), (B), (C) and (D) are correct.

142. High-energy input is required for nitrogen fixation.

143. NH^+_4 is highly toxic to plants.

144. Only (A) is correct. Amides are transported to other parts of the plant by the xylem vessels.

145. Cyanobacteria (*Oscildatoria* and *Nostoc*) show both nitrogenase and carboxylase activity.

13 Photosynthesis in Higher Plants

EXERCISES

Introduction, Early Experiments, Site of Photosynthesis, Photosynthetic Pigments

1. Which scientist first found out the essential role of sunlight in the process of photosynthesis?
(a) Joseph Priestley (b) Jan Ingenhouz
(c) Julius von Sachs (d) van Niel

2. The following set-up is the first experiment on photosynthesis carried out by Priestley. He hypothesised that:

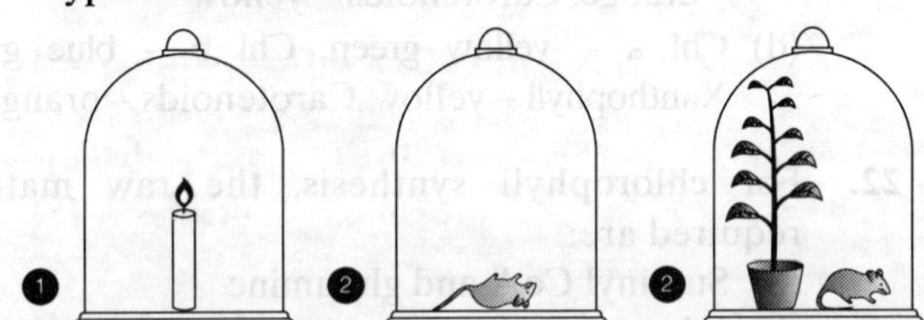

(a) Plants restore to the air whatever breathing animals and burning candles remove
(b) CO_2 is essential for photosynthesis
(c) Sunlight is essential for plant processes that purify the air
(d) All of the above

3. Who provided evidence for the production of glucose when plants grow and that glucose is usually stored as starch?
(a) Julius von Sachs (b) van Helmont
(c) van Niel (d) Melvin Calvin

4. The first action spectrum of photosynthesis was described from the experiment results of:
(a) Joseph Priestley (b) TW Engelmann
(c) Calvin (d) Ingenhousz

5. Match the following:

	Column I		Column II
A.	TW Engelmann	(i)	Mint
B.	Cornelius van Niel	(ii)	*Cladophora*
C.	Joseph Priestley	(iii)	An aquatic plant
D.	Jan Ingenhousz	(iv)	Bacteria

(a) A – (iv), B – (i), C – (ii), D – (iii)
(b) A – (ii), B – (iv), C – (i), D – (iii)
(c) A – (iii), B – (iv), C – (i), D – (ii)
(d) A – (ii), B – (i), C – (iv), D – (iii)

6. $2H_2A + CO_2 \rightarrow 2A + CH_2O + H_2O$
In higher plants, what is A?
(a) CO_2 (b) O_2
(c) S (d) $C_6H_{12}O_6$

7. A simplified equation of photosynthesis is given by: $CO_2 + 2H_2O \rightarrow (CH_2O)_n + H_2O + O_2$
(a) van Niel (b) van Helmont
(c) Ingenhousz (d) Melvin Calvin

8. In photosynthesis:
(i) Water is chemically oxidised
(ii) CO_2 is chemically reduced
(iii) Water molecule is split in the presence of light to release oxygen
The correct statement(s) is/are:
(a) Only (i)
(b) Only (i) and (iii)
(c) Only (ii) and (iii)
(d) (i), (ii) and (iii)

9. Who proved experimentally that oxygen during photosynthesis comes from water and not from CO_2?
(a) van Niel (b) Ruben and Kamen
(c) Ingenhousz (d) Aarmon

10. $6\,CO_2 + 12\,H_2O \xrightarrow{\text{Light}} C_2H_{12}O_{16} + 6\,H_2O + 6\,O_2$
In the above equation, what is the source of released oxygen?
(a) Only CO_2 (b) Only H_2O
(c) Both CO_2 and H_2O (d) Glucose

11. In the overall process of photosynthesis, the number of CO_2, water, sugar and oxygen molecules utilised and produced is:
(a) 12 (b) 19
(c) 36 (d) 31

12. Priestley performed the experiment which hypothesised that plants restore to air whatever breathing animals and burning candles remove. Which plant was utilised by him?
(a) Tomato (b) Mint
(c) *Hydrilla* (d) Potato

13. Who proved that oxygen comes from water and not from CO_2 in photosynthesis? It was based on experiments of photosynthesis in purple and green bacteria.
(a) van Niel (b) Engelmann
(c) Arnold (d) Arnon

14. Who showed that sunlight is essential to the plant process that purifies the air fouled by burning candles or breeding animals?
(a) Engelman (b) Priestley
(c) Ingenhousz (d) van Niel

15. In the half-leaf experiment, a compound is used to absorb CO_2. This compound is:
(a) NaOH (b) KOH
(c) NaCl (d) HCl

16. Moll's half-leaf experiment shows:
(a) CO_2 is essential for photosynthesis
(b) Light is essential for photosynthesis
(c) Unequal transpiration from two surfaces of a leaflet
(d) Relation between transpiration and photosynthesis

17. Who found that only the green parts of plants release oxygen?
(a) Joseph Priestly (b) Jan Ingenhousz
(c) Engelmann (d) van Niel

18. A photosynthetic cell is provided with $C^{14}O_2$ and H_2O^{18}. They form:
(a) $C_6H_{12}O_6 + O_2$ (b) $C^{14}{}_6H_{12}O^{18}{}_6 + O_2$
(c) $C^{14}{}_6H_{12}O_6 + O^{18}{}_2$ (d) $C^{14}{}_6H_{12}O^{18}{}_6 + O^{18}{}_2$

19. van Niel showed that during photosynthesis, hydrogen released from a suitable oxidisable compound reduces CO_2 to carbohydrates. To demonstrate this, he used:
(a) Purple sulphur bacteria
(b) Green sulphur bacteria
(c) *Cladophora* and aerobic bacteria
(d) Both (a) and (b)

20. Leaf pigments, Chl a, Chl b, xanthophylls and carotene can be separated by:
(a) PAGE
(b) Autoradiography
(c) Paper chromatography
(d) Column chromatography

21. A chromatographic separation of the leaf pigments shows that colour in leaves is due to four pigments, which are:
(a) Chlorophyll a – blue green, Chl b – yellow green, Xanthophyll – yellow, Carotenoids – yellow orange
(b) Chl a – yellow green, Chl b – blue green, Xanthophyll – yellow, Carotenoids – yellow orange
(c) Chl a – blue green, Chl b – red, Xanthophyll – orange, Carotenoids – yellow
(d) Chl a – yellow green, Chl b – blue green, Xanthophyll – yellow, Carotenoids – orange

22. For chlorophyll synthesis, the raw material required are:
(a) Succinyl CoA and glutamine
(b) α-ketoglutaric acid and glutamic acid
(c) Succinyl CoA and glycine
(d) Acetyl CoA and glycine

23. Chlorophyll a is:
(i) Bright and blue green in chromatograph
(ii) Chief pigment of light harvesting complex (LHC)
(iii) Called universal photosynthetic pigment
(a) All are correct
(b) Only (i) is correct
(c) (i) and (ii) are correct
(d) (i) and (iii) are correct

24. First action spectrum of photosynthesis was described by:
(a) TW Engelmann (b) JV Sachs
(c) Arnon (d) Ingelhousz

25. Which of the following is not an accessory photosynthetic pigment?
(a) Chlorophyll a (b) Chlorophyll b
(c) Xanthophylls (d) Carotenoids

26. Read the following statements and select the correct option:

A. Chl a and Chl b absorb equal amounts of blue light
B. Chl b absorbs more blue wavelength than Chl a
C. Chl a and Chl b absorb equal amounts of red light
D. Carotenoids absorb light of the blue-violet region of the split spectrum

(a) (A) and (B) (b) (A) and (D)
(c) (B) and (C) (d) (B) and (D)

27. Carotenoids protect plants from:
(a) Photo oxidation
(b) Desiccation
(c) Photorespiration
(d) Photosynthesis

28. Maximum photosynthesis takes place in:
(a) Red light (b) Green light
(c) Blue light (d) Orange light

29. The amount of carbon dioxide (in grams) assimilated by one gram of chlorophyll in an hour is:
(a) Photosynthetic number
(b) Quantum number
(c) Redox potential
(d) Action spectrum

30. The experiment performed by Jan Ingenhousz concluded that:
(a) Plants restore to the air whatever breathing animals and burning candles remove
(b) Sunlight is essential for the plant process that purifies the air fouled by burning candles
(c) Glucose is produced when plants grow and glucose is usually stored as starch
(d) Oxygen evolved by green plants comes from H_2O, not from carbon dioxide

31. Chloroplasts are located at the ____ with their broad surfaces _______ to the cell wall of the mesophyll cells:
(a) Outer margins, at right angles
(b) Inner margins parallel
(c) Outer margins, parallel
(d) Inner margins, at right angles

32. Chlorophyll a and b are very similar in their molecular structure, except that the ______ group is present in chlorophyll a and ____ group in chlorophyll b:
(a) $-CH_3, -CHO$ (b) $-CHO, -CH_3$
(c) $-C_2H_5, -CHO$ (d) $-CH_3, =C=O$

33. TW Engelmann determined the action spectrum of photosynthesis as the:
(a) Amount of CO_2 released
(b) Amount of CO_2 reduced
(c) Amount of O_2 released
(d) Both (b) and (c)

34. In the given graph, a, b, c represent the absorption spectrum of which of the following pigments, respectively?

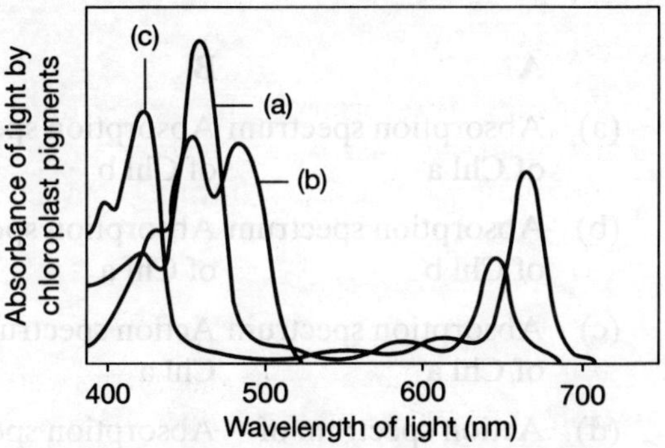

	A	B	C
(a)	Chlorophyll b	Carotenoid	Chlorophyll a
(b)	Chlorophyll a	Chlorophyll b	Chlorophyll b
(c)	Chlorophyll a	Carotenoid	Chlorophyll b
(d)	Carotenoid	Chlorophyll a	Chlorophyll b

35. Which plays a protection role by absorbing excess light and diverting it away from chlorophyll, preventing photooxidation?
(a) Phycolilins (b) Carotenoids
(c) Anthocyanin (d) NADP reductase

36. Chlorophyll is composed of four pyrrole rings and a central core of:
(a) Iron (b) Manganese
(c) Magnesium (d) Sulphur

37. Photosynthetic pigments found in chloroplasts occur in:
(a) Thylakoid membranes
(b) Lumen of thylakoid
(c) Stroma
(d) Chloroplast envelope

38. Choose the correct option with respect to the given graph:

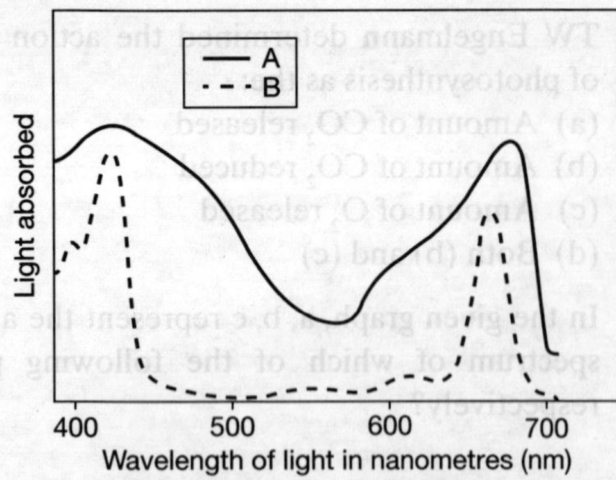

	A	B
(a)	Absorption spectrum of Chl a	Absorption spectrum of Chl b
(b)	Absorption spectrum of Chl b	Absorption spectrum of Chl a
(c)	Absorption spectrum of Chl a	Action spectrum of Chl a
(d)	Action spectrum of Chl a	Absorption spectrum of Chl a

39. The wavelength of light to carry out photosynthesis in bacteria is:
(a) Ultraviolet (b) Blue
(c) Red (d) Far red

40. The Nobel prize was awarded to this scientist for discovering the pathway of carbon assimilation:
(a) Calvin (b) Parnus
(c) Kreb (d) Watson

Light Reaction, Electron Transport, Splitting of Water, Cyclic and Non-cyclic Photo Phosphorylation, Chemiosmotic Hypothesis

41. The photochemical phase of photosynthesis occurs in:
(a) Periplastidial space
(b) Lumen of thylakoid
(c) Stroma
(d) Grana

42. Light in the wavelength between 400 nm and 700 nm is called:
(a) Photosynthetic active radiation (PAR)
(b) Solar constant
(c) Coefficient of utilisation
(d) Heat units

43. Photochemical phase of photosynthesis involves
(a) Photolysis of water molecule
(b) Photo phosphorylation
(c) Reduction of CO_2
(d) All except (c)

44. First step in light reaction of photosynthesis is:
(a) Photoexitation of cholophyll molecule
(b) Photolysis of water
(c) Photo phosphorylation
(d) Fluoresence

45. Red drop in photosynthesis in green algae refers to decrease in the rate of photosynthesis in:
(a) Blue light
(b) Green light
(c) Light greater than wavelength of 680 nm
(d) Light at wavelength of 680 nm

46. Which ions play a prominent role in the photolysis of water?
(a) Magnesium, Calcium and Chloride
(b) Manganese, Calcium, Iron
(c) Manganese, Iron, Chloride
(d) Manganese, Calcium and Chloride

47. $2H_2O \rightarrow 2H^+ + O_2 + 4e^-$
Based on the above equation, answer the following questions and select the correct option:
A. Where does this reaction take place in plants?
B. What is the by-product of this reaction?

	A	B
(a)	Stroma	Dark reaction
(b)	Lumen of chloroplast	Light reaction
(c)	Inner side of thylakoid membrane	Oxygen release
(d)	Outer side of thylakoid membrane	Oxygen release

48. Examine the figure given below and name the component marked A:

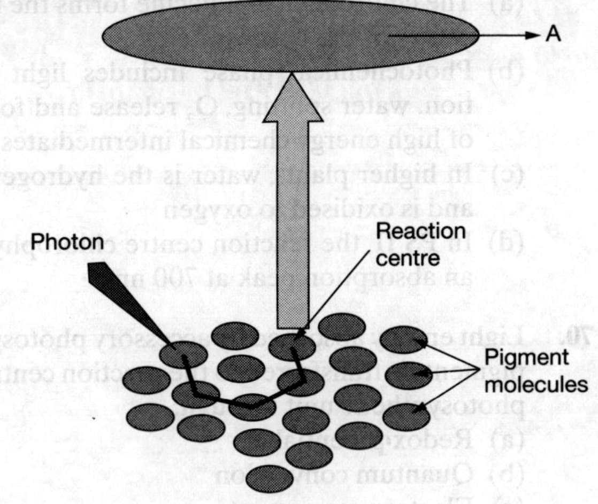

(a) Primary electron acceptor
(b) LHC
(c) Antenna molecule
(d) NADP reductase

49. Stroma lamellae lack:
(a) PS I only
(b) PS II only
(c) Both PS I and PS II
(d) PS II and NADP reductase

50. For the evolution of one molecule of oxygen, how many photons are required?
(a) 4 (b) 8
(c) 12 (d) 10

51. Which of the following is not a product of light reaction of photosynthesis?
(a) NADPH (b) NADH
(c) ATP (d) Oxygen

52. Cyclic photo phosphorylation operates when:
A. Poor availability of CO_2
B. Low light intensity
C. Anaerobic condition
(a) Only (A) (b) Only (A) and (B)
(c) Only (B) and (C) (d) (A), (B) and (C)

53. Cyclic photo phosphorylation normally occurs in:
(a) Stroma (b) Matrix
(c) Grana lamellae (d) Stroma lamellae

54. Emerson's enhancement effect and red drop were instrumental in the discovery of:
(a) Photo phosphorylation and non-cyclic electron transport
(b) Two photosystems operating simultaneously
(c) Photo phosphorylation and cyclic electron transport
(d) Oxidative phosphorylation

55. NADPH:
(a) Is the only product of light reaction
(b) Is generated through non-cyclic ETS only
(c) Is generated through cyclic and non-cyclic transport
(d) Both (a) and (c)

56. Complete the given analogy with respect to absorption peak of reaction centre in a photosystem:
PS I : 700 nm : PS II: ______
(a) 680 nm (b) 720 nm
(c) 580 nm (d) 490 nm

57. Photosystems (PS I and PS II) are named:
(a) In the sequence in which they function during light reaction
(b) In the sequence of their discovery
(c) In the order of their pigments
(d) In the sequence in which they receive the electrons

58. The LHC are made up of hundreds of pigment molecules bound to:
(a) Starch (b) Proteins
(c) Lipid (d) Fat droplets

59. Which of the following is not concerned with non-cyclic photo phosphorylation?
(a) Splitting of water molecule
(b) Oxygen evolution
(c) Synthesis of assimilatory power
(d) Reduction of NAD^+

60. A reduction in O_2 evolution occurs when light wavelength is:
(a) More than 680 nm (b) Less than 680 nm
(c) Less than 660 nm (d) 680 nm

61. Non-cyclic photo phosphorylation is the result of an interaction of PS I and PS II. There is a continuous flow of electrons from:
(a) Water $\rightarrow$ PS I $\rightarrow$ PS II $\rightarrow$ NADP
(b) Water $\rightarrow$ PS II $\rightarrow$ PS I $\rightarrow$ NADP
(c) Water $\rightarrow$ PS I $\rightarrow$ NADP $\rightarrow$ PS II
(d) PS II $\rightarrow$ Water $\rightarrow$ PS I $\rightarrow$ NADP

62. Water splitting complex:
A. Is associated with reaction centre of PS II
B. Maintains a continuous electron supply to the reaction centre of PS II

C. Is associated with non-cyclic photo phosporylation
D. Is situated on the inner side of the thylakoid membrane

The correct statements are:
(a) Only (A) and (B)
(b) Only (C) and (D)
(c) Only (A), (B) and (C)
(d) (A), (B), (C) and (D)

63. A single ____ molecule forms the reaction centre in the photosystem:
(a) Chlorophyll a (b) Chlorophyll b
(c) Carotenoid (d) Xanthophyll

64. Reaction centre in pigment system is:
(a) Site of photolysis
(b) Site of photo phosphorylation
(c) Site of electron
(d) Site of ATP synthesis

65. Chemiosmotic hypothesis for generation of ATP during light reaction was first explained by:
(a) Hill (b) P Mitchell
(c) van Niel (d) Calvin

66. If a plant is growing in a condition where only light of wavelength beyond 680 nm is available, then find the correct option:
A. Only PS I is active
B. Non-cyclic photo phosphorylation occurs
C. Only $NADPH_2$ formation occurs
D. Splitting of water molecule occurs
E. Electrons return to reaction centre
(a) (A), (B) and (C) (b) (A), (C) and (E)
(c) (A) and (E) only (d) (B) and (D) only

67. Chemiosmotic hypothesis of ATP synthesis in chloroplast is based on:
(a) Accumulation of Na^+
(b) Proton gradient
(c) Increase in pH of lumen
(d) All except (a)

68. What is/are the function(s) of accessory pigments?
A. These pigments make photosynthesis more efficient by absorbing a wider range of wavelength
B. They absorb light and transfer the energy to the reaction centre
C. They protect the reaction centre from photooxidation
(a) Only (A) is correct
(b) Only (A) and (B) are correct
(c) Only (A) and (C) are correct
(d) (A) (B) and (C) are correct

69. Which of the following is incorrect with respect to photosynthesis in plants?
(a) The chlorophyll a molecule forms the reaction centre of the photosystems
(b) Photochemical phase includes light absorption, water splitting, O_2 release and formation of high energy chemical intermediates
(c) In higher plants, water is the hydrogen donor and is oxidised to oxygen
(d) In PS II, the reaction centre chlorophyll a has an absorption peak at 700 nm

70. Light energy absorbed by accessory photosynthetic pigments is transferred to the reaction centre in the photosynthetic unit through:
(a) Redox potential
(b) Quantum conversion
(c) Electron movement
(d) Electron spin transfer

71. The first electron acceptor molecule in the transport chain from the reaction centre of PS II (A) and PS I (B) is:

	A	B
(a)	Pheophytin	FeS
(b)	FeS	Pheophytin
(c)	Plastoquinone	Plastocyanin
(d)	Pheophytin	Plastocyanin

72. Which of the following is common between cyclic and non-cyclic photo phosphorylation?
(a) Synthesis of $NADPH_2$
(b) Splitting of H_2O molecules
(c) Involvement of PS II
(d) Synthesis of ATP

73. In non-cyclic photo phosphorylation, the last electron acceptor is:
(a) H_2O (b) ADP
(c) $NADP^+$ (d) NAD^+

74. During photosynthesis, photolysis of water takes place:
(a) Towards the outer side of the thylakoid membrane
(b) In the stroma
(c) Towards the inner side of the thylakoid membrane
(d) With the help of ATP synthetase enzyme

75. Z-scheme is:
(i) Cyclic photo phosphoylation
(ii) Non-cyclic photo phosphorylation
(iii) Circulation of electron within the photosystem

(a) Only (i) (b) Only (ii)
(c) (i) and (ii) (d) (ii) and (iii)

76. Examine the figure given below and select the incorrect option:

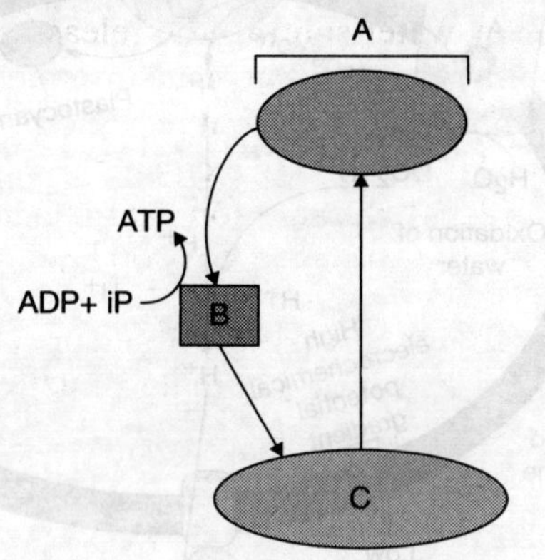

(a) A – Primary electron acceptor pheophytin
(b) B – Electron transport system
(c) C – Absorbs 680 nm light
(d) Both (a) and (c)

77. Fill in the blanks by choosing the correct option regarding chemiosmosis:
(i) NADP reductase enzyme is located on the ___A___ side of the membrane
(ii) Protons in the stroma ___B___ in number while in the ___C___, there is accumulation of protons
(iii) ___D___ part of the ATPase enzyme forms a trans-membrane channel for facilitated diffusion of protons across the membrane

	A	B	C	D
(a)	Stroma	Decrease	Lumen	F_o
(b)	Lumen	Decrease	Lumen	F_o
(c)	Stroma	Decrease	Lumen	F_1
(d)	Lumen	Decrease	Lumen	F_1

78. Which of the following is not a requirement of chemiosmosis?
(a) Proton pump
(b) Proton gradient
(c) Thylakoid membrane
(d) More proton accumulation in the stroma than in the lumen of the thylakoid

79. During chemiosmosis, the proton gradient breaks:
(a) Proton moves across membrane through plastoquinone
(b) Movement of proton occurs from stroma to lumen
(c) pH of lumen increases
(d) Movement of proton occurs from lumen to stroma

80. What is the role of the antenna molecule in photosynthesis?
(a) Protects the chlorophyll from photooxidation in high light intensity
(b) Forms reaction centre
(c) Constitutes LHC in both the photosystems
(d) Both (a) and (c)

81. Assimilatory power photosynthetic bacteria are:
(a) NADH and ATP (b) NADPH and ATP
(c) Only ATP (d) Only NADPH

82. In the following diagram, label A, B and C:

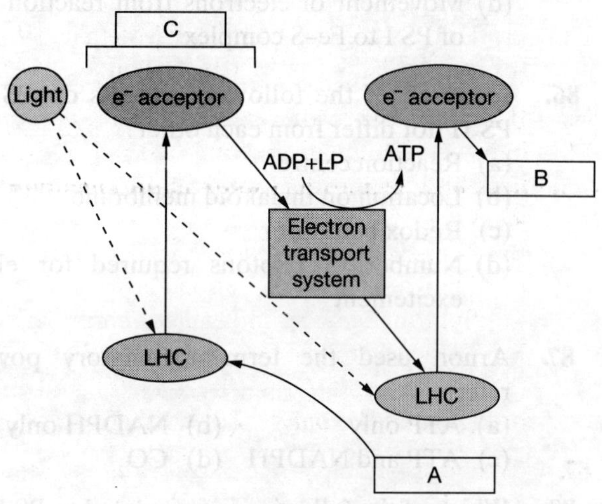

	A	B	C
(a)	Water	NADP	PS II
(b)	NADP	Water	PS II
(c)	Water	NADP	PS I
(d)	NADP	Water	PS I

83. In photosynthesis of purple and green sulphur bacteria, ____ is the hydrogen donor:
(a) H_2O (b) H_2S
(c) Glucose (d) Both (a) and (b)

84. Development of proton gradient on two sides of the thylakoid membrane for ATP synthesis is due to all, except:
(a) Movement of protons across the membrane to the stroma through the transmembrane channel of the F_o–F_1 complex
(b) Protons in the stroma decrease in number, while in the lumen, there is accumulation of protons
(c) Photolysis of water in the lumen side of the thylakoid membrane
(d) Continuous withdrawal of protons from the stroma for reduction of $NADP^+$

85. Which of the following represents uphill movement of electrons in the Z-scheme?
(a) Movement of electrons from pheophytin to plastocyanin
(b) Movement of electrons within cytochromes
(c) Movement of electrons from ferredoxin to $NADP^+$
(d) Movement of electrons from reaction centre of PS I to Fe–S complex

86. In which of the following aspects do PS I and PS II not differ from each other?
(a) Reaction centre
(b) Location on thylakoid membrane
(c) Redox behaviour
(d) Number of photons required for electron excitement

87. Arnon used the term assimilatory power to refer:
(a) ATP only (b) NADPH only
(c) ATP and NADPH (d) CO_2

88. Which of the following is not related to PS I?
(a) Reduction of NADP+
(b) Involved in both cyclic and non-cyclic photo phosphorylation
(c) The reaction centre is P_{700}
(d) Present in appressed region of granal thylakoid

89. In light reaction, the electron-deficient P_{680} is able to restore electrons from:
(a) Water molecule (b) $NADPH_2$
(c) P_{700} (d) Both (a) and (b)

90. In the following diagram, label A, B and C:

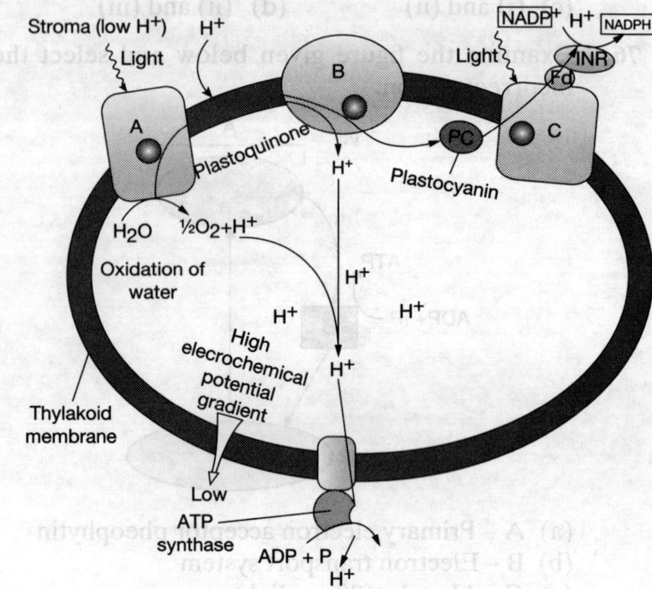

	A	B	C
(a)	PS I	Cyt b6-f	PS II
(b)	PS II	Cyt b6-f	PS I
(c)	PS I	Cyt a-a3	PS II
(d)	PS II	Cyt a-a3	PS I

91. During cyclic photo phosphorylation:
(a) Electrons from photosystem I are not passed to NADP
(b) NADP is no longer available in oxidised state
(c) Electrons are transferred back to the reaction centre of PS I
(d) All of the above

92. Select the incorrect match:
(a) Cyclic photo phosphorylation – Produces ATP only
(b) Non-cyclic photo phosphorylation – Z-scheme
(c) Red drop – Inactivates PS I
(d) Etiolation – Absence of light

93. In the electron transport chain of photosynthesis with the breakdown of water, electrons are generated which are then passed to the:
(a) Oxidised P_{700} (b) Oxidised P_{680}
(c) Reduced P_{680} (d) Reduced P_{700}

94. ATP and $NADPH_2$ from the light reaction are used in:
(a) Biochemical phase to reduce CO_2
(b) To produce assimilatory power
(c) Photolysis of water
(d) All of the above

95. $NADPH^+$ produced in light reaction is a:
(a) Powerful reducing agent
(b) Weak reducing agent
(c) Weak oxidising agent
(d) Powerful oxidising agent

Dark Reaction (Calvin Cycle, C_4 Pathway, Photorespiration)

96. Read the following statements with respect to dark reaction and select the correct option:
A. It can occur in the dark only
B. It depends on products of light reaction
C. It cannot occur during the day
D. It produces sugars
(a) (A) and (B) (b) (B) and (C)
(c) (A) and (D) (d) (B) and (D)

97. Dark reaction of photosynthesis occurs in:
(a) Granal thylakoid membrane
(b) Lumen of thylakoid
(c) Stroma of chloroplast
(d) Periplastidial space

98. Algae employed by Calvin et al. to experiment on photosynthesis belong to:
(a) *Euglena* and *Scenedesmus*
(b) *Chara*
(c) *Chlorella* and *Chlamydomonas*
(d) *Chlorella* and *Scenedesmus*

99. Calvin cycle involves:
(a) Oxidative carboxylation
(b) Reductive carboxylation
(c) Phosphorylation
(d) Oxidative phosphorylation

100. Dark reaction is dependent on the production of '___A___' and '___B___' of light reaction. What are 'A' and 'B'?

	A	B
(a)	NADPH	FAD
(b)	ATP	NADPH
(c)	ATP	NAD
(d)	CO_2	ATP

101. Fill in the blanks by choosing the correct option:
(i) ___X___ is the biosynthetic phase of photosynthesis
(ii) Calvin cycle was described using radioactive ___Y___
(iii) The formation of PGA in dark reactions is called ___Z___

	X	Y	Z
(a)	Dark reaction	C^{14}	Carboxylation
(b)	Hill reaction	C^{14}	Reduction
(c)	Blackman reaction	O^{18}	Reduction
(d)	Dark reaction	O^{18}	Carboxylation

102. During the Calvin cycle, formation of phosphoglyceraldehyde from phosphoglyceric acid is:
(a) Hydrolysis (b) Oxidation
(c) Reduction (d) Electrolysis

103. Which phase of the Calvin cycle does/does not use any energy-rich molecule?
(a) Carboxylation and regeneration
(b) Carboxylation only
(c) Reduction only
(d) Reduction and regeneration

104. Which of the following is not a step in the Calvin cycle?
(a) Carboxylation
(b) Reduction
(c) Regeneration
(d) Photo phosphorylation

105. RuBisCO occurs in high quantity as it is:
(a) An oxygenase
(b) Catalysing reversible reaction
(c) Degraded rapidly
(d) Very slow acting enzyme

106. How many ATP molecules are required in the Calvin cycle for fixation of CO_2 molecule?
(a) 9 (b) 15
(c) 54 (d) 6

107. The form of carbon used for carboxylation of phosphenol pyruvate in C_4 plants is:
(a) H_2CO_3 (b) C_2H_4
(c) HCO_3^- (d) CH_4

108. The first CO_2 acceptor molecules in the C_3 and C_4 pathways are, respectively:
(a) RuBP, PGA (b) RuBP, PEP
(c) PEP, RuBP (d) PEP, Malic acid

109. For each molecule of glucose formed in the plant, the number of molecules of ATP and $NADPH_2$ required are:
(a) 12 and 18 (b) 18 and 12
(c) 30 and 18 (d) 30 and 18

110. The carboxylase enzymes RuBisCO and PEPCo are activated by:
(a) Mn (b) Mo
(c) Zn (d) Mg

111. The process which makes a major difference between C_3 and C_4 plants is:
(a) Calvin cycle (b) Photorespiration
(c) C-2 cycle (d) All except (a)

112. For synthesis of a molecule of glucose, the requirement of ATP and NADPH is:
(a) 15 and 10 (b) 33 and 20
(c) 12 and 8 (d) 18 and 12

113. C_3 plant differs from C_4 plant in that:
(a) They lack Kranz anatomy
(b) They have greater productivity of biomass
(c) They tolerate higher temperature
(d) They lack photorespiration

114. (i) Primary CO_2 acceptor in hatch and slack pathway
(ii) Present in mesophyll cell
(iii) Three-carbon compound

Which of the following is defined by the above statements?
(a) Malic acid (b) OAA
(c) PEP (d) RuBP

115. During the Calvin cycle, the total number of CO_2, ATP, NADPH molecules utilised and glucose, ADP and NADP molecules generated is:
(a) 31 (b) 36
(c) 61 (d) 67

116. Which of the following is not associated with Kranz anatomy of C_4 plants?
(a) Kranz means wreath
(b) Several layers of bundle sheath cells
(c) Thick cell wall of bundle sheath cell impervious to gaseous exchange
(d) Presence of granal chloroplast in bundle sheath cells

117. Which of the following is a similarity between the mesophyll cell and bundle sheath cell of C_4 plants?
(a) Calvin cycle enzyme
(b) Carboxylase activity
(c) Agranal chloroplast
(d) Similar enzymes of light reaction

118. To fix six CO_2 molecules, how many ATP molecules are required via the C_4 pathway?
(a) Two (b) Twelve
(c) Five (d) Three

119. C_4 plants have better productivity because:
(a) C_4 plants absorb more light
(b) C_4 plants absorb more CO_2
(c) C_4 plants lack photorespiration
(d) All of the above

120. The C_4 plants are photosynthetically more efficient than C_3 plants because:
(a) They have more chloroplasts
(b) The CO_2 compensation point is higher
(c) CO_2 generated during photorespiration is trapped and recycled through PEP carboxylase
(d) The CO_2 efflux is not presented

121. For the cyclic pathway given below, which of the following options is incorrect?

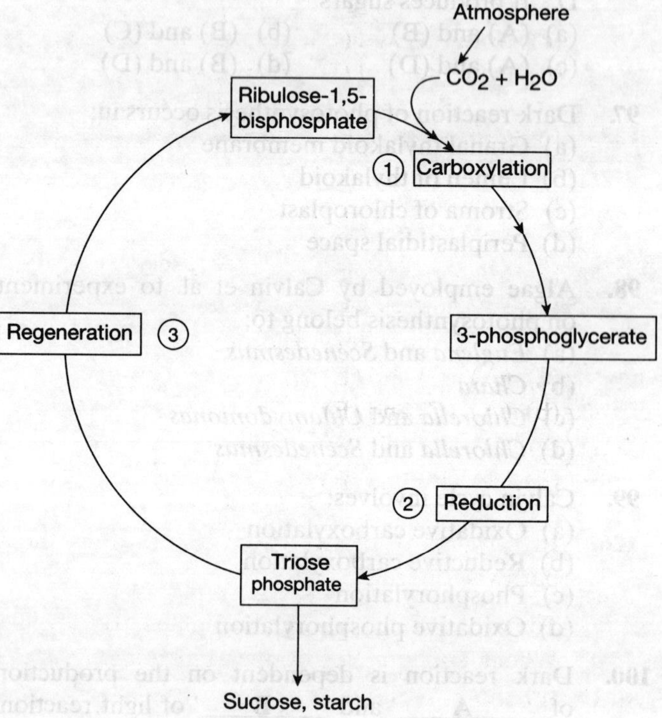

(a) The first stage is the most crucial step
(b) During the second stage, ATP and $NADPH_2$ molecules are utilised
(c) Activity of RuBP carboxylase results in the formation of the first stable product
(d) ATP as well as RuBP are formed during regeneration

122. The site of carboxylase activity of RuBisCO in bundle sheath cells is:
(a) Tomato (b) Maize
(c) Bell pepper (d) Potato

123. Select the incorrectly matched pair:
(a) Regeneration of PEP – ATP-dependent process
(b) RuBP oxygenase – C_2 cycle
(c) Photorespiratory loss – Occurs in C_3 plants
(d) Decarboxylation in bundle sheath cell – C_4 cycle

124. Read the following statements and select the correct option:
A. Immediately after light becomes unavailable, the biosynthetic phase of photosynthesis continues for some time and then stops
B. The biosynthetic phase of photosynthesis is dependent on the product of light reaction
(a) Only (A) is correct
(b) Only (B) is correct
(c) Both (A) and (B) are correct
(d) Both (A) and (B) are incorrect

125. In Kranz anatomy, the bundle sheath cells have:
A. Thick walls
B. Thin walls
C. No intercellular spaces
D. Many intercellular spaces
E. Few chloroplasts
F. Large number of chloroplasts
(a) (A), (C) and (E)
(b) (A), (C) and (F)
(c) (B), (E) and (F)
(d) (B), (D) and (E)

126. In how many of the listed plants does the C_3 pathway operate?
Tomato, Potato, Maize, Sugarcane, Barley, Pea, Wheat, *Opuntia*, Bell pepper, Pineapple
(a) Ten (b) Five
(c) Four (d) Eight

127. In the following diagram, identify labels A, B and C:

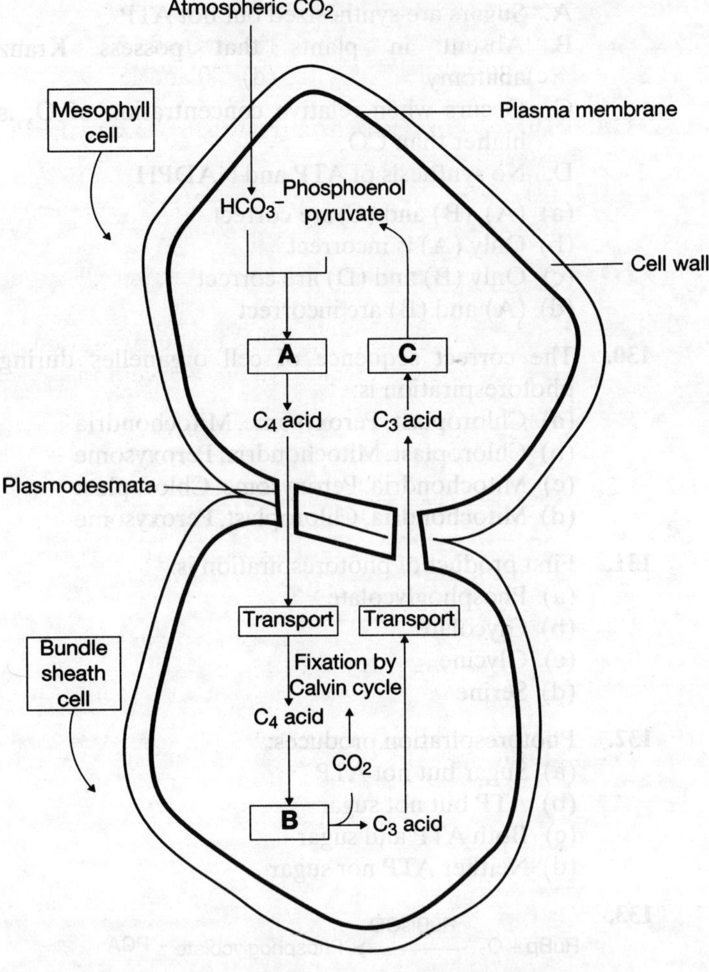

	A	B	C
(a)	Fixation	Decarboxylation	Recarboxylation
(b)	Fixation	Decarboxylation	Regeneration
(c)	Decarboxylation	Carboxylation	Regeneration
(d)	Decarboxylation	Regeneration	Carboxylation

128. Photorespiration is favoured by:
(a) High oxygen and low carbon dioxide
(b) High carbon dioxide and low oxygen
(c) High temperature and low oxygen
(d) High humidity and temperature

129. Mark the statements that are correct for photorespiratory pathway:
A. Sugars are synthesized but not ATP
B. Absent in plants that possess Kranz anatomy
C. Occurs when relative concentration of O_2 is higher than CO_2
D. No synthesis of ATP and NADPH
(a) (A), (B) and (C) are correct
(b) Only (A) is incorrect
(c) Only (B) and (D) are correct
(d) (A) and (B) are incorrect

130. The correct sequence of cell organelles during photorespiration is:
(a) Chloroplast, Peroxysome, Mitochondria
(b) Chloroplast, Mitochondria, Peroxysome
(c) Mitochondria, Peroxysome, Chloroplast
(d) Mitochondria, Chloroplast, Peroxysome

131. First product of photorespiration is:
(a) Phosphoglycolate
(b) Glycolate
(c) Glycine
(d) Serine

132. Photorespiration produces:
(a) Sugar but not ATP
(b) ATP but not sugar
(c) Both ATP and sugar
(d) Neither ATP nor sugar

133.

$$\text{RuBp} + O_2 \xrightarrow{\text{RuBisCO}} \text{Phosphoglycolate} + \text{PCA}$$

This reaction occurs in:
(a) Mitochondria
(b) Cytoplasm of mesophyll cell
(c) Chloroplast
(d) Peroxysome

134. Read the statements given below and select the correct option:

A: Photorespiration decreases photosynthetic output
B: In the photorespiratory pathway, neither ATP nor NADPH is produced
(a) Only A is correct
(b) Only B is correct
(c) Both A and B are correct
(d) Both A and B are incorrect

135. Read the following statements and state true (T) or false (F):
A. RuBisCO has much greater affinity for oxygen than CO_2
B. C_4 pathway is observed only in monocots
C. C_4 plants respond to higher temperature and show higher rate of photosynthesis
D. C_4 pathway is more efficient than C_3 pathway due to absence of photorespiration in C_4 plants

	A	B	C	D
(a)	T	F	T	T
(b)	T	F	F	T
(c)	F	T	F	F
(d)	F	F	T	T

136. All are characteristic features of C_4 plants, except:
(a) Responsiveness to high light intensity
(b) Lack of photorespiration
(c) Scotoactive stomata
(d) Greater productivity

137. A process that makes an important difference between C_3 and C_4 plants is:
(a) Photosynthesis (b) Photorespiration
(c) Transpiration (d) Glycolysis

138. C_4 and C_3 plants differ from each other in which aspect?
A. Type of pigments involved in photosynthesis
B. The primary acceptor of CO_2 during carbon fixation
C. Type of end product of photosynthesis
D. Total number of ATP consumed during the biosynthetic phase of photosynthesis
(a) (A) and (B) (b) (A) and (D)
(c) (B) and (D) (d) (B) and (C)

139. Which is correct for C_4 plants?

	Mesophyll	Bundle sheath
(a)	PEPCO–C_4 pathway	RuBisCO–C_3 pathway
(b)	PEPCO–C_3 pathway	RuBisCO–C_4 pathway
(c)	RuBisCO–C_4 pathway	PEPCo–C_3 pathway
(d)	RuBisCO–C_3 pathway	PEPCo–C_4 pathway

140. There is no photorespiratory loss in C_4 plants, because:
(a) They have a large number of chloroplasts
(b) Greater affinity of RuBisCO for CO_2
(c) Concentric arrangement of mesophyll
(d) Increased CO_2 concentration at enzyme site

141. Consider the following four statements regarding C_4 plants and select the correct option which includes all incorrect ones only:
A. Calvin cycle occurs in mesophyll cells
B. RuBisCO enzyme is present in bundle sheath cells

C. First CO_2 acceptor is PEP
D. Mesophyll cell chloroplasts are larger and agranal
(a) (A) and (B) (b) (B) and (C)
(c) (B) and (D) (d) (A) and (D)

142. During photorespiration, the chemical which passes into the peroxisome from the mitochondria is:
(a) Glycolate (b) Serine
(c) Glycine (d) Hydroxy pyruvate

143. During photorespiration, CO_2 is released in:
(a) Mitochondria
(b) Choloroplast
(c) Peroxisome
(d) Both peroxisome and mitochondria

144. Find the incorrect statement for CAM plants:
(a) Scotoactive stomata
(b) Acidification occurs at night
(c) RuBisCO activity only during the day
(d) Absence of hatch and slack pathway

145. How many additional ATPs are required by C_4 plants for the synthesis of one molecule of glucose?
(a) 30 (b) 12
(c) 18 (d) 15

146. Read the following statements with respect to C_4 plants:
(i) They have greater productivity of biomass
(ii) They have CO_2 concentration phase at the RuBisCO site
(iii) They have double carboxylation steps separated by space only
(iv) Primary CO_2 fixation is in the mesophyll cell, which is used as a source of CO_2 for the second carboxylation step
How many of the statements are correct?
(a) Two (b) Three
(c) Four (d) Five

147. CAM helps plants in:
(a) Disease resistance (b) Reproduction
(c) Conserving water (d) Photorespiration

148. Select the odd one out with respect to CAM plants:
(a) Pineapple (b) *Opuntia*
(c) Cactus (d) Sorghum

149. Number of CO_2 acceptor and carboxylating enzymes in a CAM plant are:
(a) 2, 2, respectively (b) 1, 1, respectively
(c) 2, 1, respectively (d) 1, 2, respectively

150. CAM plants differ from C_4 plants in:
(a) Performing primary CO_2 fixation using RuBisCO
(b) Having dimorphic chloroplast
(c) Forming starch in bundle sheath cells
(d) Having double carboxylation step separated by time

151. One reason for poorer adaptability of C_4 plants to low temperature is:
(a) Low light saturation point
(b) Sensitivity to higher CO_2 levels
(c) Temperature sensitivity of RuBisCO
(d) Temperature sensitivity of pyruvate dikinase

152. Read the following statements and select the correct option:
A. Photorespiration protects plants from photooxidation
B. Light-induced reactions may damage photosynthetic machinery
(a) Only (A) is correct
(b) Only (B) is correct
(c) Both (A) and (B) are correct
(d) Both (A) and (B) are incorrect

153. The first stable product identified in the Calvin cycle of C_3 (A) and C_4 (B) plants is:

	A	B
(a)	3-PGA	3-PGA
(b)	RuBP	PEP
(c)	PGA	OAA
(d)	PGA	Malic acid

154. C_4 plants can tolerate saline conditions because of:
(a) Spatial separation of two CO_2 fixation steps
(b) Presence of organic acids in them
(c) Presence of Kranz anatomy in them
(d) Presence of low temperature sensitive enzyme PEP synthetase

155. How many carbon atoms are present in?
A. Primary CO_2 acceptor in C_3 plant
B. Primary CO_2 acceptor in C_4 plant
C. Primary CO_2 acceptor in CAM plant

	A	B	C
(a)	5	3	5
(b)	5	3	3
(c)	3	4	4
(d)	3	4	3

156. Primary CO_2 fixation in C_4 plants occurs in:
(a) Bundle sheath cell chloroplast
(b) Mesophyll cell chloroplast
(c) Bundle sheath cell cytoplasm
(d) Mesophyll cell cytoplasm

157. How many ATP molecules are required for regeneration of 6 molecules of RuBP in the Calvin cycle?

(a) 18 (b) 3
(c) 6 (d) 1

158. Read the following statements with respect to the CAM cycle and select the correct option:

A. It is also called diurnal acid cycle
B. Related to chloroplast dimorphism
C. Found in plants with scotoactive stomata

(a) Only (A) is correct
(b) Only (C) is correct
(c) (A), (B) and (C) are correct
(d) Only (B) is incorrect

159. Bundle sheath cells are similar to mesophyll cells in:

(a) Having grana in chloroplast
(b) Having large number of chloroplasts
(c) Being involved in CO_2 fixation
(d) Their primary CO_2 acceptor

160. C_4 plants are better than C_3 plants in:

(a) Utilisation of water
(b) Having photorespiration
(c) Producing more yield at very high CO_2 concentration
(d) Being sensitive to high temperature

161. Dual activity of RuBisCO is seen in:

(a) C_3 plants (b) C_4 plants
(c) CAM plants (d) Both (b) and (c)

162. Which is the correct ratio of ATP utilisation in the reduction and regeneration steps of the Calvin cycle?

(a) 1:1 (b) 2:1
(c) 1:2 (d) 3:1

163. To fix 1 CO_2, how many ATP (A) and $NADPH_2$ (B) molecules are required via the C_4 pathway in C_4 plants?

	A	B
(a)	3	2
(b)	2	0
(c)	5	2
(d)	3	0

164. The sensitivity of C_4 photosynthesis to low temperature as compared to C_3 plants is due to a particular enzyme called:

(a) Pyruvate phosphate dikinase
(b) PEPCase
(c) Transaminase
(d) Malate dehydrogenase

165. In this figure, X and Y are:

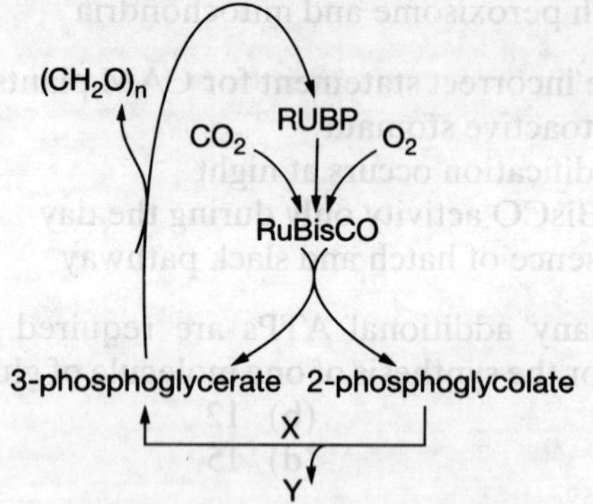

	X	Y
(a)	C_4 pathway	CO_2
(b)	CAM pathway	O_2
(c)	Photorespiration	CO_2
(d)	C_3 pathway	O_2

Factors Affecting Photosynthesis (Light, CO_2 Concentration, Temperature and Water)

166. The law of limiting factors was given by:

(a) Hatch and Slack (b) Calvin
(c) Lindeman (d) Blackman

167. According to Blackman, photosynthesis is affected by several factors, but the role of photosynthesis is limited by the:

(a) Fastest step in the pathway
(b) Slowest step in the pathway

(c) Factor present in least amount
(d) Factor present in large amount

168. In the given graph, what is the value of D?

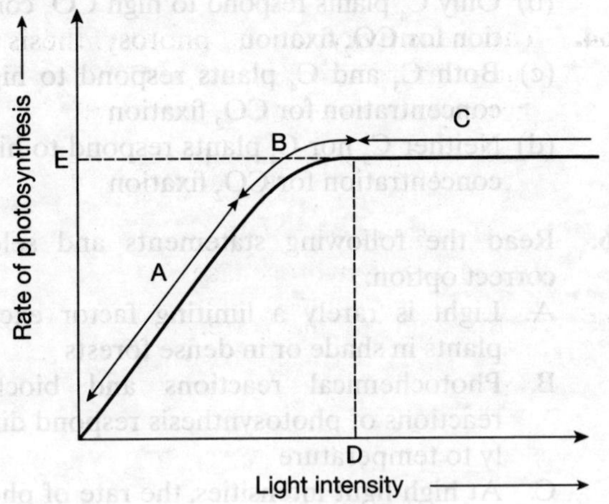

(a) 10% (b) 5%
(c) 2% (d) 30%

169. The present concentration of CO_2 is very low in the atmosphere:
(a) 500 ppm (b) 360–400 ppm
(c) 1000 ppm (d) < 300 ppm

170. Relationship between incident light and CO_2 fixation rate is:
(a) Sigmoid at higher light intensity
(b) Linear at low intensity of light
(c) Not affected if the available intensity is below saturation
(d) Linear at high light intensity

171. Read the following statements and state true (T) or false (F):
A. C_4 plants respond to higher CO_2 concentration by showing increased rates of photosynthesis, leading to higher productivity
B. The optimum temperature for photosynthesis depends on the habitat that they are adapted to
C. The effect of water as a factor is more, through its effect on the plant, rather than directly on photosynthesis

	A	B	C
(a)	F	T	T
(b)	T	T	T
(c)	T	F	F
(d)	F	T	F

172. In low light conditions, which of the following plants respond to high CO_2 conditions?
(a) Only C_3 plants
(b) Only C_4 plants
(c) Neither C_3 nor C_4 plants
(d) Both C_3 and C_4 plants

173. When CO_2 concentration in atmosphere is increased, the rate of photosynthesis in C_3 and C_4 plants will change as follows:
(a) Increased in C_3 and C_4 plants
(b) Increased in C_3 and no effect in C_4 plants
(c) Increased in C_3 and decreased in C_4 plants
(d) No effect on both C_3 and C_4 plants

174. Decreased rate of photosynthesis at high concentration of oxygen is referred to as:
(a) Pasteur effect
(b) Emerson effect
(c) Warburg effect
(d) Richmond–Lang effect

175. Select the incorrect statement with respect to factors affecting photosynthesis:
(a) Water is rarely a limiting factor in photosynthesis
(b) Dark reactions are not temperature controlled
(c) Maximum photosynthesis occurs in red and blue light of visible spectrum
(d) C_3 and C_4 plants respond differently to CO_2 concentrations

176. With reference to factors affecting the rate of photosynthesis, which of the following statements is not correct?
(a) Increasing atmospheric CO_2 concentration up to 0.5 % can enhance CO_2 fixation rate
(b) C_3 plants respond to higher temperature with enhanced photosynthesis while C_4 plants have much lower optimum temperature
(c) Tomato is a greenhouse crop which can be grown in a CO_2-enriched atmosphere for higher yield.
(d) Light saturation for CO_2 fixation occurs at 10% of full sunlight

177. At CO_2 compensation point:
(a) The rate of photosynthesis is equal to the rate of respiration
(b) Photosynthesis stops and respiration becomes slow
(c) Rate of photosynthesis and respiration is high
(d) Rate of photosynthesis is decreased but respiration is not affected

178. Select the option that correctly fills in the blanks:
(i) ___A___ is the major limiting factor influencing the rate of photosynthesis
(ii) C_4 plants show CO_2 saturation at about ___B___ µl / L
(iii) In the majority of plants, light saturation occurs at ___C___ of sunlight

	A	B	C
(a)	CO_2	360	10%
(b)	CO_2	450	1%
(c)	Light	450	20%
(d)	Light	360	10%

179. The most efficient light for photosynthesis is:
(a) Red (b) Blue
(c) Green (d) White

180. Optimum temperature for photosynthesis in C_3 plants is:
(a) 20–25ºC (b) 15–20ºC
(c) 10–15ºC (d) 35–40ºC

181. Increase in CO_2 concentration up to ________ per cent can cause an increase in CO_2 fixation rate; beyond this level, it can become damaging over long periods:
(a) 0.03 (b) 0.04
(c) 0.05 (d) 0.045

182. Lower CO_2 compensation point in C_4 plants can be attributed to:
(a) Low photorespiratory activity
(b) CO_2 concentrating mechanism
(c) Shows CO_2 fertilisation
(d) Greater affinity of CO_2 fixation through PEPCo

183. Which of the following does not affect the rate of photosynthesis?
(a) Light quality
(b) Duration of light
(c) Concentration of CO_2
(d) Wavelength of light

184. Plants which respond to higher temperature and show greater rate of photosynthesis are:
(a) C_3 plants (b) C_2 plants
(c) C_4 plants (d) Both C_2 and C_4 plants

185. At low light intensities:
(a) Only C_3 plants respond to high CO_2 concentration for synthesis of glucose
(b) Only C_4 plants respond to high CO_2 concentration for CO_2 fixation
(c) Both C_3 and C_4 plants respond to high CO_2 concentration for CO_2 fixation
(d) Neither C_3 nor C_4 plants respond to high CO_2 concentration for CO_2 fixation

186. Read the following statements and select the correct option:
A. Light is rarely a limiting factor except for plants in shade or in dense forests
B. Photochemical reactions and biochemical reactions of photosynthesis respond differently to temperature
C. At high light intensities, the rate of photosynthesis decreases
(a) Only (A) is correct
(b) Only (B) and (C) are correct
(c) (A), (B) and (C) are correct
(d) All are incorrect

187. C_3 plants attain photosynthesis saturation at CO_2 concentration of:
(a) 360 µl L^{-1} (b) 500 µl L^{-1}
(c) 450 µl L^{-1} (d) 120 µl L^{-1}

188. Both ________ of light influence photosynthesis:
(a) Quality and intensity
(b) Quality and duration
(c) Duration and intensity
(d) Quality and quantity

189. How many plants in the given box can be allowed to grow in a CO_2-enriched atmosphere for higher yield?

Sugar cane, Sorghum, Maize, Amaranthus, Tomato, Bell pepper

(a) Four (b) Three
(c) Two (d) Six

190. Which of the following seldom acts as a limiting factor in photosynthesis?
(a) Temperature (b) Light
(c) CO_2 (d) Water

NCERT EXEMPLAR QUESTIONS

1. Which metal ion is a constituent of chlorophyll?
(a) Iron (b) Copper
(c) Magnesium (d) Zinc

2. Which pigment acts directly to convert light energy to chemical energy?
(a) Chlorophyll a (b) Chlorophyll b
(c) Xanthophyll (d) Carotenoid

3. Which range of wavelength (in nm) is called photosynthetically active radiation (PAR)?
(a) 100–390 (b) 390–430
(c) 400–700 (d) 760–100

4. Which light range is least effective in photosynthesis?
(a) Blue (b) Green
(c) Red (d) Violet

5. Chemosynthetic bacteria obtain energy from:
(a) Sun
(b) Infrared rays
(c) Organic substances
(d) Inorganic chemicals

6. Energy required for ATP synthesis in PS II comes from:
(a) Proton gradient
(b) Electron gradient
(c) Reduction of glucose
(d) Oxidation of glucose

7. During light reaction in photosynthesis, the following are formed:
(a) ATP and sugar
(b) Hydrogen, O_2 and sugar
(c) ATP, hydrogen donor and O_2
(d) ATP, hydrogen and O_2 donor

8. Dark reaction in photosynthesis is so called because:
(a) It can occur in the dark also
(b) It does not directly depend on light energy
(c) It cannot occur during the day
(d) It occurs more rapidly at night

9. PEP is a primary CO_2 acceptor in:
(a) C_4 plants (b) C_3 plants
(c) C_2 plants (d) Both C_3 and C_4 plants

10. Splitting of water is associated with:
(a) Photosystem I
(b) Lumen of thylakoid
(c) Both photosystem I and II
(d) Inner surface of thylakoid membrane

11. The correct sequence of flow of electrons in the light reaction is:
(a) PS II, plastoquinone, cytochromes, PS I, ferredoxin
(b) PS I, plastoquinone, cytochromes, PS II, ferredoxin
(c) PS I, ferredoxin, PS II
(d) PS I, plastoquinone, cytochromes, PS II, ferredoxin

12. The enzyme that is not found in a C_3 plant is:
(a) RuBP carboxylase
(b) PEP carboxylase
(c) NADP reductase
(d) ATP synthase

13. The reaction that is responsible for the primary fixation CO_2 is catalysed by:
(a) RuBP carboxylase
(b) PEP carboxylase
(c) RuBP carboxylase and PEP carboxylase
(d) PGA synthase

14. When CO_2 is added to PEP, the first stable product synthesized is:
(a) Pyruvate
(b) Glyceraldehyde-3-phosphate
(c) Phosphoglycerate
(d) Oxaloacetate

PRACTICE TEST PAPER

1. The following hypothesis "Plants restore to the air whatever breathing animals and burning candles remove" was given by:
(a) Joseph Priestley (b) Ingenhousz
(c) Julius van Sachs (d) van Niel

2. Identify the following statements as true (T) or false (F) and choose the option accordingly:
A. Chloroplasts align themselves along the walls of mesophyll cells
B. There is clear-cut division of labour within the chloroplast
C. Sugar is synthesized in the lumen of chloroplasts during dark reaction
D. Chlorophyll a is protected from photo oxidation by chlorophyll b only

	A	B	C	D
(a)	T	F	T	T
(b)	T	F	T	F
(c)	F	T	F	F
(d)	F	T	T	F

3. Photosystem concerned with splitting of water molecule:
(a) PS II (b) PS I
(c) Both PS II and PS I (d) P_{700}

4. The protons formed by splitting of water are released in the:
(a) Lumen of thylakoid
(b) Outer side of membrane
(c) Stroma of thylakoid
(d) Membrane of thylakoid

5. Radioactive isotope of oxygen O^{18} was used to find out the source of oxygen released through photosynthesis by:
(a) Hill
(b) van Niel
(c) Ruben Hassid and Kamen
(d) Hatch and Slack

6. Which fraction of the visible spectrum of solar radiation is primarily absorbed by carotenoids of higher plants?
(a) Blue and green (b) Green and red
(c) Red and violet (d) Violet and blue

7. Which of the following is a characteristic of PS I?
(a) It is active only up to 680 nm of light
(b) The reaction centre of PS I is P_{680}
(c) It is reduced by electrons released in photolysis of water
(d) PS I is involved in both cyclic and non-cyclic photo phosphorylation

8. In C_4 plants, the bundle sheath cells:
(a) Have thin walls to facilitate gaseous exchange
(b) Have large intercellular spaces
(c) Are rich in PEP carboxylase
(d) Have a high density of chloroplast

9. How many Calvin cycles yield one molecule of glucose?
(a) Four (b) Six
(c) Eight (d) Two

10. Which of the following is not a product of light reaction of photosynthesis?
(a) Oxygen (b) ATP
(c) NADPH (d) NADH

11. Thomas Engelmann illuminated a filament of algae with light that passed through a prism, thus exposing different segments of the algal filament to different wavelengths of light. He added aerobic bacteria and found that these bacteria accumulated in the areas illuminated by red and blue light. If you ran the same experiment without passing light through a prism, what would you predict?
(a) There would be no difference in results
(b) The number of bacteria would decrease along the entire length of the filament
(c) The bacteria would be relatively distributed along the length of the filament
(d) The number of bacteria would increase along the entire length of the filament

12. Accessory pigments:
(a) Play no role in photosynthesis
(b) Release electrons and are oxidised
(c) Transfer electrons to NADP
(d) Allow plants to harvest visible light of wider range wavelength

13. The van Niel experiment was on:
(a) A green algae, *Cladophora*
(b) *Chlorella* and *Scendesmus*
(c) Purple and green sulphur bacteria
(d) Mint plant

14. Photorespiration involves oxidation of:
(a) Chlorophyll (b) PGA
(c) PEP (d) RuBP

15. Calvin cycle represents which of the following phenomena?
(a) Oxidative carboxylation
(b) Substrate level phosphorylation
(c) Photosynthetic carbon oxidative cycle
(d) Reductive carboxylation

16. Conversion of pyruvate into PEP by use of ATP in the Cu pathway occurs in:
(a) Mesophyll cell's cytoplasm
(b) Mesophyll cell's chloroplast
(c) Bundle sheath cell's cytoplasm
(d) Bundle sheath cell's chloroplast

17. Why is the Calvin cycle called the C_3 cycle?
(a) Primary CO_2 acceptor is C_3 compound
(b) Many intermediate compounds are C_3 compounds
(c) First stable product is 3PGA which is a C_3 compound
(d) All except (a)

18. Net energy outcome of cyclic phosphorylation is:
(a) ATP and $NADPH_2$
(b) ATP
(c) Sugar
(d) $NADPH + H^+$

19. Biosynthetic phase of photosynthesis needs:
(a) Only ATP and CO_2
(b) ATP, $NADPH_2$, CO_2 and water
(c) Only ATP and CO_2
(d) Only CO_2

20. Photosystem is composed of:
(a) LHC
(b) Reaction centre
(c) Accessory pigments
(d) All of the above

21. Dark reaction of photosynthesis is so called because it:
(a) Can also occur in the dark
(b) Does not require light energy
(c) Cannot occur during the day
(d) Occurs more rapidly at night

22. In the Calvin cycle, which stage requires energy?
A. Carboxylation
B. Reduction
C. Regeneration
(a) Only (A) (b) (A) and (B)
(c) (B) and (C) (d) (A) and (C)

23. How many molecules of water are needed by a green plant to produce one molecule of hexose/reduce molecules of CO_2?
(a) 6 (b) 12
(c) 24 (d) 10

24. The Z-scheme refers to:
(a) The type of photosynthesis used in plants found in areas with low light intensity
(b) Arrangement of granal thylakoids within the stroma of chloroplast
(c) Double carboxylation process in C4 plants
(d) An energy diagram for the transfer of electrons in light reactions or photosynthesis in plants

25. Consider the following statements with respect to C_4 plants and select the correct option:
A. They have higher optimum temperature than C_3 plants
B. They possess dimorphic chloroplasts
C. They are adapted to tropical regions
D. They do not show response to low CO_2 concentrations
(a) (A) and (D) are correct
(b) (B) and (D) are incorrect
(c) (A), (B) and (C) are correct
(d) (A), (B), (C), and (D) are correct

26. Leaves of sugarcane have:
(a) Agranal chloroplast in all the cells
(b) Oxigenase activity of RuBisCO in mesophyll cells
(c) Double carboxylation catalysed by RuBisCO only
(d) Agranal chloroplasts in bundle sheath cells

27. CAM plants differ from C_4 plants in:
(a) Absence of dimorphic chloroplast
(b) Presence of Kranz anatomy
(c) Having double carboxylation steps separated by space
(d) Formation of OAA as first stable product

28. In photorespiration, use of O_2 and evolution of CO_2 occur in:
(a) Chloroplast, Peroxisome
(b) Chloroplast, Mitochondria
(c) Mitochondria, Chloroplast
(d) Peroxisome, Chloroplast

29. NADP reductase enzyme is located:
(a) Inside the lumen
(b) On the stroma side of the thylakoid membrane
(c) On the outer side of the inner membrane of the chloroplast
(d) On the lumen side of the thylakoid membrane

30. Read the following statements and select the correct option:

A: PS I is located on the non-oppressed region of the granal thylakoid as well as stroma lamellae
B: PS I participates in non-cyclic photo phosphorylation only
(a) Only (A) is correct
(b) Only (B) is correct
(c) Both (A) and (B) are correct
(d) Both (A) and (B) are incorrect

31. Select the incorrect statement for LHC:
(a) Made up of hundreds of pigment molecules
(b) Pigment molecules are bound to lipids only
(c) Harvests different wavelengths of light
(d) Helps to make photosynthesis more efficient

32. The energetically costly but most efficient cycle of dark reaction in plants adapted to dry tropical regions:
(a) C_3 cycle (b) C_2 cycle
(c) CAM cycle (d) Hatch–Slack cycle

33. TW Engelmann experimented on:
(a) Chemosynthetic bacteria
(b) *Cladophora*
(c) *Hydrilla*
(d) Photosynthetic bacteria

34. Competitive binding of O_2 and CO_2 is shown by which of the following enzymes in the chloroplast?
(a) PEP Carboxylase
(b) RuBisCO
(c) PEP Synthetase
(d) ATP Synthetase

35. PS I does not have:
(a) Chl a in the reaction centre
(b) Antenna molecule
(c) Pigment molecules bound to protein
(d) Pheophytin as primary acceptor

36. Read the following statements carefully:
A. Dark reaction is affected by temperature but is not directly dependent on light
B. CO_2 is required in all major steps of the Calvin cycle
C. Sugarcane is more efficient in CO_2 fixation
D. Stroma lamellae do not play an important role in trapping of light and synthesis of NADPH

How many of these statements are correct?
(a) Two (b) One
(c) Four (d) Three

37. Which of the following is not associated with Z-scheme?
(a) Splitting of water molecule
(b) Oxygen release
(c) NAD^+ reduction
(d) ATP synthesis

38. Which is the evidence to show that O_2 released in photosynthesis comes from water molecules?
(a) Isotopic O_2 supplied in H_2O, appears in released oxygen
(b) Isolated chloroplast in water releases O_2, if supplied potassium ferrocyanide or some other electron acceptors
(c) Photosynthetic bacteria use H_2S, and CO_2 to make carbohydrates and release sulphur
(d) All of the above

39. Choose the correct fate of the first carboxylation product in C_4 plants during photosynthesis:
(a) C_3 acid $\rightarrow$ C_4 acid $\rightarrow$ PEP
(b) C_4 acid $\rightarrow$ C_3 acid $\rightarrow$ PEP
(c) C_3 acid $\rightarrow$ PEP $\rightarrow$ C_4 acid
(d) C_4 acid $\rightarrow$ PEP $\rightarrow$ C_3 acid

40. All the statements are correct for C_4 plants, except:
(a) More chloroplast number in bundle sheath cells
(b) Thin-walled cells around the vascular bundles
(c) PEPCo in cytosol of mesophyll cells
(d) First stable product is OAA in C_4 pathway

41. How many plants from the given box may represent the C_4 cycle?

Sugarcane, Maize, Rice, Tomato, Potato, Wheat, Pineapple, Cactus, *Opuntia*, Bell pepper, *Eleusine*

(a) Ten (b) Five
(c) Seven (d) Four

42. Read the following statements and select the correct option:
A. C_4 plants can concentrate CO_2 at RuBisCO site in bundle sheath cells
B. CAM plants show double carboxylation separated by time
(a) Only (A) is correct
(b) Only (B) is correct
(c) Both (A) and (B) are correct
(d) Both (A) and (B) are incorrect

43. Dimorphic chloroplasts are seen in:
(a) C_4 plants (b) C_3 plants
(c) CAM plants (d) Both (a) and (c)

44. Select the incorrect statement with respect to carotenoids:

(a) Shows yellow-orange colour in chromatogram
(b) Absorbs the light of the blue-violet region of the visible spectrum
(c) Main component of reaction centre of PS II
(d) Protects the chlorophyll molecule from photo oxidation and high light intensity

45. Which of the following is correct for photorespiration?
(a) PEP + $CO_2 \rightarrow$ OAA
(b) RuBP + $CO_2 \rightarrow$ 2PGAP
(c) RuPB + $O_2 \rightarrow$ Phosphoglycolate + PGA
(d) PEP + $O_2 \rightarrow$ Phosphoglycolate + PGA

ANSWER KEYS

Exercises

1. (b) **2.** (a) **3.** (a) **4.** (b) **5.** (b) **6.** (b) **7.** (a) **8.** (d) **9.** (b) **10.** (b)
11. (d) **12.** (b) **13.** (a) **14.** (c) **15.** (b) **16.** (a) **17.** (b) **18.** (c) **19.** (d) **20.** (c)
21. (a) **22.** (c) **23.** (d) **24.** (a) **25.** (a) **26.** (d) **27.** (a) **28.** (a) **29.** (a) **30.** (b)
31. (c) **32.** (a) **33.** (c) **34.** (a) **35.** (b) **36.** (c) **37.** (a) **38.** (d) **39.** (d) **40.** (a)
41. (d) **42.** (a) **43.** (d) **44.** (a) **45.** (c) **46.** (d) **47.** (c) **48.** (d) **49.** (d) **50.** (b)
51. (b) **52.** (d) **53.** (d) **54.** (b) **55.** (b) **56.** (a) **57.** (b) **58.** (b) **59.** (d) **60.** (a)
61. (b) **62.** (d) **63.** (a) **64.** (c) **65.** (b) **66.** (c) **67.** (b) **68.** (d) **69.** (d) **70.** (d)
71. (a) **72.** (d) **73.** (c) **74.** (c) **75.** (b) **76.** (d) **77.** (a) **78.** (d) **79.** (d) **80.** (d)
81. (a) **82.** (a) **83.** (b) **84.** (a) **85.** (d) **86.** (c) **87.** (d) **88.** (d) **89.** (a) **90.** (b)
91. (d) **92.** (c) **93.** (b) **94.** (a) **95.** (a) **96.** (d) **97.** (c) **98.** (d) **99.** (b) **100.** (b)
101. (a) **102.** (c) **103.** (b) **104.** (d) **105.** (d) **106.** (a) **107.** (c) **108.** (b) **109.** (b) **110.** (d)
111. (c) **112.** (d) **113.** (a) **114.** (c) **115.** (c) **116.** (d) **117.** (b) **118.** (b) **119.** (d) **120.** (a)
121. (d) **122.** (b) **123.** (d) **124.** (c) **125.** (b) **126.** (a) **127.** (b) **128.** (a) **129.** (b) **130.** (a)
131. (a) **132.** (d) **133.** (c) **134.** (c) **135.** (a) **136.** (c) **137.** (b) **138.** (c) **139.** (a) **140.** (d)
141. (d) **142.** (c) **143.** (a) **144.** (d) **145.** (b) **146.** (c) **147.** (c) **148.** (d) **149.** (a) **150.** (d)
151. (d) **152.** (c) **153.** (a) **154.** (b) **155.** (b) **156.** (d) **157.** (c) **158.** (d) **159.** (c) **160.** (c)
161. (a) **162.** (b) **163.** (b) **164.** (a) **165.** (c) **166.** (d) **167.** (c) **168.** (a) **169.** (b) **170.** (b)
171. (a) **172.** (c) **173.** (b) **174.** (c) **175.** (b) **176.** (b) **177.** (a) **178.** (a) **179.** (d) **180.** (a)
181. (c) **182.** (d) **183.** (b) **184.** (c) **185.** (d) **186.** (c) **187.** (c) **188.** (a) **189.** (c) **190.** (d)

NCERT Exemplar Questions

1. (c) **2.** (a) **3.** (c) **4.** (b) **5.** (d) **6.** (a) **7.** (c) **8.** (b) **9.** (a) **10.** (d)
11. (a) **12.** (b) **13.** (c) **14.** (d)

Practice Test Paper

1. (a) **2.** (c) **3.** (a) **4.** (a) **5.** (c) **6.** (d) **7.** (d) **8.** (d) **9.** (b) **10.** (d)
11. (c) **12.** (d) **13.** (b) **14.** (d) **15.** (d) **16.** (a) **17.** (c) **18.** (b) **19.** (b) **20.** (d)
21. (b) **22.** (c) **23.** (b) **24.** (d) **25.** (c) **26.** (d) **27.** (a) **28.** (b) **29.** (b) **30.** (a)
31. (b) **32.** (d) **33.** (b) **34.** (b) **35.** (d) **36.** (a) **37.** (c) **38.** (d) **39.** (b) **40.** (b)
41. (a) **42.** (c) **43.** (a) **44.** (c) **45.** (c)

SOLUTIONS

Introduction, Early Experiments, Site of Photosynthesis and Photosynthetic Pigments

1. Jan Ingenhousz stated that sunlight is essential for the plant processes that purify the air fouled by burning candles or breathing animals.
2. Plants restore to the air whatever breathing animals and burning candles remove.
3. Julius von Sachs provided evidence for the production of glucose when plants grow.
4. TW Engelmann stated the first action spectrum of photosynthesis.
5. A – (ii), B – (iv), C – (i), D – (iii)
6. O_2
7. van Niel
8. (i) (ii) and (iii) are correct.
9. Ruben and Kamen stated that the oxygen obtained during photosynthesis evolved from water and not from CO_2.
10. Only H_2O
11. $6CO_2 + 12H_2O \rightarrow C_6H_{12}O_6 + 6H_2O + 6O_2$ $(6 + 12 + 1 + 6 + 6 = 31)$
12. Mint plant
13. van Niel
14. Ingenhousz
15. In the half-leaf experiment, the compound KOH is used which absorbs CO_2.
16. Moll's half-leaf experiment is performed to show that CO_2 is essential for photosynthesis.
17. Jan Ingenhousz found that it is only the green part of plants that could release oxygen.
18. $C^{14}_6H_{12}O_6 + O^{18}_2$
19. van Niel studied purple and green bacteria and reported that photosynthesis is essentially a light-dependent process in which hydrogen from a suitable oxidisable compound reduces CO_2 to carbohydrate.

    $2H_2A + CO_2$ in the presence of sunlight gives $2A + CH_2O + H_2O$

    When H_2S is a hydrogen donor for purple and green bacteria, the oxidation product is sulphur.
20. Leaf pigments can be separated by paper chromatography.
21. Chlorophyll a – Blue green, Chlorophyll b – Yellow green, Xanthophyll – Yellow, Carotenoids – Yellow orange
22. Chlorophyll synthesis starts with the condensation of succinyl CoA and glycine.
23. (i) and (iii) are correct. Chlorophyll a acts as reaction centre.
24. TW Engelmann
25. Pigments other than chlorophyll a are called accessory photosynthetic pigments.
26. (B) and (D) are correct. Chlorophyll b absorbs more blue light than chlorophyll a, while chlorophyll a absorbs more red light than chlorophyll b.
27. Carotenoids are called shield pigments as they protect chlorophyll from photooxidation.
28. Maximum photosynthesis takes place in red light.
29. Photosynthetic number
30. Plants restore to the air whatever breathing animals and burning candles remove.
31. Outer margins, parallel
32. –CH3, –CHO
33. TW Engelmann determined the action spectrum of photosynthesis as the amount of O_2 released.
34. A – Chlorophyll b, B – Carotenoid, C – Chlorophyll a
35. Carotenoids
36. Chlorophyll is composed of four pyrrole rings and a central core of magnesium.

37. Photosynthetic pigments found in chloroplasts occur in the thylakoid membrane.

38. A – Action spectrum of chl a, B – Absorption spectrum of chl a

39. Far red

40. Calvin

Light Reaction, Electron Transport, Splitting of Water, Cyclic and Non-cyclic Photo Phosphorylation, Chemiosmotic Hypothesis

41. Photochemical phase (light reaction) occurs in grana.

42. Photosynthetic active radiation

43. Photochemical phase involves photolysis of water and photo phosphorylation. Reduction of CO_2 occurs in dark reaction.

44. Photo excitation of the chlorophyll molecule is the first step of photosynthesis.

45. Emerson found that the rate of photosynthesis suddenly dropped in the far red region of the spectrum (above 680 nm).

46. Manganese, Calcium and Chloride

47. A – Lumen of thylakoid, B – Oxygen release

48. Primary electron acceptor

49. Stroma lamellae lack PS II and NADP reductase enzyme.

50. The number of light quanta needed for the production for one oxygen molecule is eight.

51. NADH

52. Cyclic photo phosphorylation occurs mostly in stroma lamella and unstacked thylakoid. It is functional under low light intensity, anaerobic conditions and when CO_2 availability is less.

53. Stroma lamella

54. Two photosystems operating simultaneously

55. NADPH, ATP and oxygen are products of the light reaction. NADPH is generated through non-cyclic ETS only.

56. Reaction centre of PS II is P_{680}. It absorbs light of 680 nm.

57. Photosystems PS I and PS II are named in the sequence of their discovery.

58. Proteins

59. In non-cyclic photo phosphorylation, reduction of NADP+ occurs.

60. Photolysis of water occurs at PS II. Its reaction centre absorbs light of 680 nm wavelength.

61. Water – PS II – PS I – NADP+

62. (A), (B), (C) and (D) are correct.

63. A single chlorophyll a molecule forms the reaction centre in the photosystem. The reaction centre is P_{700} in PS I and P_{680} in PS II.

64. Release of electrons occurs only from chlorophyll a (reaction centre).

65. P Mitchell

66. (A) and (E) are correct. When only light of wavelength beyond 680 nm is available, only PS I is active and only cyclic photo phosphorylation occurs. No external electron donor is required. No photolysis of water takes place and it produces ATP only.

67. Proton gradient is responsible for ATP synthesis on the head of CF_0 – CF_1 particles plugged on the thylakoid membrane.

68. (A), (B) and (C) are correct.

69. In PS II, the reaction centre is P_{580} which absorbs light of 680 nm.

70. Antenna molecules collect the light energy and transfer this energy by inductive resonance process to the reaction centre.

71. A – Pheophytin, B – FeS

72. Synthesis of ATP

73. NADP+ is the last electron acceptor in non-cyclic photo phosphorylation.

74. Photolysis of water occurs towards the inner side of the thylakoid membrane.

75. The whole scheme of transfer of electrons from PS II to PS I and finally to $NADP^+$ to form NADPH + H^+ during non-cyclic photo phosphorylation called the Z-scheme, due to its characteristic shape.

76. A – Primary electron acceptor FeS, C – Absorbs 700 nm light

77. A – Stroma, B – Decrease, C – Lumen, D – F_0
78. Chromosomes require more H^+ accumulation in the lumen of the thylakoid than in stroma.
79. Movement of protons occurs across the membrane from the thylakoid lumen to the stroma.
80. Both (A) and (C) are correct. One molecule of chlorophyll a functions as the reaction centre.
81. NADH and ATP
82. A – Water, B – NADP, C – PS II
83. Photosynthetic bacteria use H_2S as hydrogen donor.
84. Movement of protons across the membrane to the stroma is through the transmembrane channel of the CF_0–CF_1 complex.
85. Movement of electrons from the reaction centre of PS I to the FeS complex is uphill movement.
86. Number of photons required for electron excitement
87. NADPH +, H+ and ATP produced in light reaction constitute assimilatory power.
88. PS I is present in non-oppressed region of the granal thylakoid.
89. Water molecule acts as electron donor during non-cyclic electron transport
90. A – PS II, B – Cyt b_{6-f}, C – PS I
91. All the statements are correct.
92. Red drop – inactivates PS II
93. Oxidised P_{680}
94. Assimilatory power (NADPH +, H^+ and ATP) used in reduction of CO_2 during dark reaction
95. Powerful reducing agent

Dark Reaction (Calvin Cycle, C_4 Pathway, Photorespiration)

96. (B) and (D) are correct. Dark reaction is a light-independent process.
97. Stroma of chloroplast
98. Melvin Calvin discovered the C_3 cycle in *Chlorella* and *Scendesmus.*
99. Calvin cycle involves reductive carboxylation.
100. A – ATP, B – NADPH
101. X – Dark reaction, Y – C^{14}, Z – Carboxylation
102. Reduction step involves conversion of PGA into PGAL.
103. During the Calvin cycle, energy is used in two steps: reduction and regeneration.
104. Photo phosphorylation occurs during light reaction of photosynthesis.
105. Option (d) is correct.
106. For synthesis of one glucose molecule from $6CO_2$ molecules, 18 ATP and 12 $NADPH^+H^+$ are required because for fixation of $3CO_2$ molecule, 9 ATP are required.
107. HCO^-_3
108. RuBP, PEP
109. 18 ATP and 12 $NADPH_2$
110. Mg acts as activator of both RuBisCO and PEPG.
111. Photorespiration or the C_2 cycle occurs in C_3 plants.
112. Option (d) is correct.
113. C_4 plants are characterised by the presence of Kranz anatomy in the leaves. Kranz means 'wreath' and it is a reflection of the arrangement of cells.
114. PEP
115. 67 ($6CO_2$ + 18ATP + 12NADPH + 1Glucose + 18ATP + $12NADP^+$)
116. Bundle sheath chloroplasts are larger and agranal.
117. Both mesophyll and bundle sheath cells show carboxylase activity.
118. In C_4 plants for fixation of 6 CO_2 molecules, 30 ATP are required. In the C4 pathway, 12 ATP. In the C_3 pathway, 18 ATP.
119. All the statements are correct.
120. C_4 plants have more chloroplast and show CO_2 concentration effect.
121. The regulation step requires one ATP for phosphorylation to form RuBP.

122. C_4 plants (maize) show RuBisCO activity in the bundle sheath cells.

123. In C_4 plants, decarboxylation occurs in bundle sheath cells to release CO_2 for carboxylase activity of RuBisCO.

124. Both (A) and (B) are correct.

125. Kranz anatomy is characterised by the presence of large number of chloroplasts, thick walls impervious to gaseous exchange and no intercellular spaces.

126. All photosynthetic plants synthesize glucose through the C_3 pathway.

127. A – Fixation, B – Decarboxylation, C – Regeneration

128. Photorespiration is favoured by high oxygen and low CO_2 concentration.

129. (B), (C) and (D) are correct for photorespiration. Neither sugar nor ATP is synthesized.

130. Chloroplast, Peroxysome, Mitochondria

131. O_2 + RuBP in the presence of RuBiSCO gives phosphoslycolate (2C).

132. Neither ATP nor sugar

133. Chloroplast

134. Both (A) and (B) are produced.

135. Only (B) is false. C_4 plants are mostly monocots but some are dicots.

136. CAM plants possess scotoactive stomata.

137. Photorespiration occurs only in C_3 plants.

138. (B) and (D) are correct.

139. Mesophyll cell – PEPCo – C_4 pathway; Bundle sheath cell – RuBisCO – C_3 pathway

140. In C_4 plants, there is high concentration of CO_2 at the RuBisCO site to minimise photorespiratory loss.

141. (A) and (D) are incorrect. The Calvin cycle occurs in bundle sheath chloroplast and bundle sheath cells have larger and agranal chloroplast.

142. Glycine

143. During photorespiration, CO_2 and NH_3 are reduced in mitochondria.

144. CAM plants follow the C_4 pathway at night when the stomata are open.

145. 12 ATP

146. Four statements (I, II, III, IV) are correct. C_4 plants have double decarboxylation steps separated by spaces.

147. CAM helps plants in conserving water.

148. Sorghum is a C_4 plant.

149. CAM plants have two CO_2 acceptors (PEP and RuBP) and two carboxylase enzymes (PEPCo and RuBiSCO).

150. In CAM plants, the C_4 and C_3 pathways are separated by time.

151. Pyruvate dikanase enzyme is sensitive to low temperature and is essential for the synthesis of PEP, which is a primary CO_2 acceptor.

152. Both (A) and (B) are correct.

153. A – PGA, B – PGA

154. Due to the presence of organic acids, C_4 plants can tolerate saline conditions.

155. (a) RuBP (5C), (b) PEP (3C), (c) PEP (3C)

156. In C_4 plants, the C_4 pathways occurs in the cytoplasm of the mesophyll cell because the PEPCo enzyme is present in the cytoplasm of the mesophyll cell.

157. 6 ATP

158. Only (B) is incorrect. C_4 plants show chloroplast dimorphism.

159. Both bundle sheath and mesophyll cells are involved in carboxylation activity.

160. Option (a) is correct.

161. Dual activity of RuBiSCO (carboxylase and oxygenase) is seen in C_3 plants.

162. Reduction step – 12 ATP; Regeneration step – 6 ATP; Ratio = 2:1

163. A – 2, B – 0

164. Enzyme pyruvate phosphate dikinase is sensitive to low temperature.

165. X – Photorespiration, Y – CO_2

Factors Affecting Photosynthesis (Light, CO_2 Concentration, Temperature and Water)

166. Blackman proposed law of limiting factors.

167. Law of limiting factors states that when a biological process is controlled as to its rapidity by a number of separate factors, the rate of the process is limited by the pace of the slowest factor.

168. 10%

169. Normal concentration of CO_2 in the atmosphere is 360 ppm.

170. Linear at low intensity of light

171. Only (A) is false. C_3 plants show saturation at 450 ppm and thus the current CO_2 concentration is limiting for C_3 plants.

172. At low light intensity, neither C_3 nor C_4 plants respond to high CO_2 concentration.

173. Increased in C_3 plants (CO_2 fertilisation effect) and no effect in C_4 plants

174. Warburg effect

175. The dark reaction being enzymatic is temperature controlled.

176. C_4 plants (tropical plants) have a higher temperature optimum than the plants adapted to temperate climates (C_3 plants).

177. At CO_2 compensation point, the rate of photosynthesis is equal to the rate of respiration.

178. A – CO_2, B – 360, C – 10%

179. Maximum photosynthesis occurs in white light.

180. Optimum temperature for C_3 plant is 20–25°C.

181. 0.05%

182. Greater affinity for CO_2 fixation through PEPCo

183. The duration of light does not affect the rate of photosynthesis, but it affects the total photosynthesis.

184. C_4 plants respond to higher temperature.

185. Neither C_3 nor C_4 plants respond to high CO_2 concentration for CO_2 fixation.

186. Both (B) and (C) are correct.

187. 450 microlitre/litre

188. Both quality and intensity of light influence photosynthesis

189. Tomato, Bell pepper (two plants)

190. Water is rarely a limiting factor for photosynthesis.

14 Respiration in Plants

EXERCISES

Introduction – Do Plants Breathe?

1. The breaking of the C–C bonds of compounds through oxidation within the cells, leading to the release of a considerable amount of energy, is called:
(a) Metabolism (b) Cellular respiration
(c) Breathing (d) Anabolism

2. Energy released by oxidation of respiratory substrates:
(a) Comes out in a single step to increase the possibility of maximum ATP production
(b) Is not used directly
(c) Is trapped as chemical energy in the energy currency of the cell
(d) Both (b) and (c)

3. Respiration is regarded as a/an:
(a) Anabolic process
(b) Catabolic process
(c) Endergonic process
(d) Both (b) and (c)

4. Which substrate provides twice as much energy as carbohydrate?
(a) Oil and fats (b) Protein
(c) Vitamins (d) Minerals

5. During floating respiration, the respiratory substrate comprises:
(a) Carbohydrates only
(b) Proteins
(c) Carbohydrates and fats
(d) Protein and fats

6. Common respiratory substrate is:
(a) Fructose (b) Glucose
(c) Lactose (d) Sucrose

7. Read the following statements and select the correct option:
A. Energy required for life processes is obtained by oxidation of some macromolecules
B. Carbohydrates are the preferred respiratory substrate
C. Plants have no specialised organs for gaseous exchange but they have stomata and lenticels
(a) Only (A) is correct
(b) Only (B) and (C) are correct
(c) Only (A) and (C) are correct
(d) (A), (B) and (C) are correct

8. In aerobic respiration, the total number of ATP molecules obtained from 1 glucose molecule is:
(a) 28 ATP (b) 38 ATP
(c) 30 ATP (d) 15 ATP

9. In ATP molecules, energy is stored in:
(a) Chemical bonds (b) Hydrogen bonds
(c) Carbon bonds (d) Pyrophosphate bonds

10. ________ act(s) as the energy currency of the cell:
(a) Glucose (b) Sunlight
(c) ATP (d) Organic compounds

11. Read the statements given below and select the correct option:
A: Energy is released in a series of slow, stepwise reactions during respiration
B: In prokaryotes, breakdown of complex molecules to yield energy takes place in the cytoplasm only
(a) Only (A) is correct
(b) Only (B) is correct
(c) Both (A) and (B) are correct
(d) Both (A) and (B) are incorrect

12. Plants do not breathe like animals because:
(a) They have specialised organs for gaseous exchange
(b) They have stomata and hydathodes for gaseous exchange
(c) Each plant part takes care of its own gas exchange
(d) Root, stem and leaves respire at higher rates than animals

13. What is the substrate for protoplasmic respiration?
(a) Organic acids (b) Carbohydrates
(c) Proteins (d) Fats

14. An anaerobically respiring yeast cell consumes 30 glucose molecules per hour. If a little amount of oxygen is provided, what will be the consequence?
(a) Respiration will stop abruptly
(b) The rate of glucose consumption and ATP production will decrease
(c) The rate of glucose consumption will increase but the rate of ATP production will decrease
(d) The rate of glucose consumption will decrease but ATP production will increase

15. What type of food would you suggest to an athlete to consume before participating in an event?
(a) Protein (b) Fat
(c) Carbohydrate (d) Both (a) and (b)

16. Usable energy of respiration is:
(a) Used in charging biomolecules into activity
(b) Stored as heat
(c) Immediately consumed in cellular activities
(d) Trapped in ATP molecules

17. Nearly every living organism uses glucose as a nutrient source of energy. Why?
(a) Glucose is the only molecule capable of producing the energy required to produce ATP
(b) The structure of glucose is very similar to ATP
(c) Glucose has more potential energy than any other respiratory substrate
(d) The ability to harvest energy from glucose appeared very early in biological evolution

18. The reason(s) why plants can get oxygen without respiratory organs is/are:
(a) Almost all living cells in a plant have their surfaces exposed to air
(b) Plants have little demand for gaseous exchange
(c) Every part takes care of its own gaseous energy, so very little transport of gases is required from one part to another
(d) All of the above

19. The amount of energy given by one mole to ATP is:
(a) 7.3 kcal (b) 721 kcal
(c) 7600 kcal (d) 700 kcal

20. Respiration differs from combustion in:
(a) Liberating more energy as compared to combustion
(b) Liberating of all the energy at once in contrast to combustion
(c) Liberating energy gradually in steps in contrast to combustion
(d) Both (a) and (b)

21. The main purpose of cellular respiration is to:
(a) Convert potential to kinetic energy
(b) Convert kinetic to potential energy
(c) Create energy in the cell
(d) Convert energy stored in the chemical bonds of glucose to energy that the cell can use

Glycolysis

22. The term glycolysis originated from the Greek word which means:
(a) Glucose + lysis (b) Gluco + lysis
(c) Glycose + lysis (d) Glycos + lysis

23. Glucose $\xrightarrow{\text{EMP pathway}}$ 2 molecules of ____X____.
X is:
(a) Citric acid (b) Pyruvic acid
(c) Ethanol (d) Acetyl-CoA

24. Glycolytic process is characterised by all of these, except:
(a) Occurs in cytoplasm
(b) Partial oxidation of glucose
(c) Oxidation
(d) Net production of ATP is four

25. Sucrose is converted into glucose and fructose by the enzyme:
(a) Zymase (b) Invertase
(c) PFK (d) Kinase

26. Glycolysis:
(a) Occurs in the mitochondria of the eukaryotic cell
(b) Is the complete oxidation of glucose
(c) Gives acetyl-CoA as the key product
(d) Occurs in all living organisms

27. Phosphorylation of glucose during glycolysis is catalysed by:
(a) Phosphofructokinase
(b) Hexokinase
(c) Aldolase
(d) Pyruvate kinase

28. Mark the correct option with respect to the following statements:
A. Glycolysis involves both decarboxylation and dehydrogenation
B. Glycolysis occurs in the cytoplasm of cells of prokaryotes and eukaryotes
(a) Only (A) is correct
(b) Only (B) is correct
(c) Both (A) and (B) are correct
(d) Both (A) and (B) are incorrect

29. Two redox equivalents are removed during EMP pathway in:
(a) Formation of PGAL
(b) Conversion of PEP
(c) Breakdown of fructose 1–6 bisphosphate
(d) Formation of BPGA

30. How many molecules of CO_2 are released during the glycolysis of one molecule of glucose?
(a) Two (b) Zero
(c) Four (d) One

31. Pacemaker enzyme of EMP pathway is:
(a) PFK (b) Enolase
(c) Pyruvate kinase (d) Hexokinase

32. Which is the connecting link between respiration and photosynthesis?
(a) Acetyl-CoA (b) α-ketoglutaric acid
(c) PGAL (d) OAA

33. Glycolysis was discovered by:
(a) Embden (b) Mayerhof
(c) Parnas (d) All of the above

34. Choose the odd one out with respect to the number of carbon atoms:
(a) 3-PGA
(b) Glyceraldehyde-3-phosphate
(c) Pyruvic acid
(d) Acetyl-CoA

35. To maximise ATP yield via glycolysis:
(a) No O_2 is utilised
(b) Substrate-level phosphorylation is adequate
(c) Electron needs to be donated to oxygen atoms
(d) Both (a) and (b)

36. What is the net gain to ATP molecules in the glycolytic pathway?
(a) 4 (b) 2
(c) 8 (d) 1

37. Only source of energy in RBCs is:
(a) Krebs cycle
(b) Glycolysis
(c) Hexose monophosphate shunt
(d) ETS

38. Enzyme involved in dehydration reaction of glycolysis is:
(a) Phosphofructokinase
(b) Hexokinase
(c) Pyruvate kinase
(d) Enolase

39. Which is false for glycolysis?
(a) First step is phosphorylation of glucose molecule
(b) ATP is produced
(c) The end products are pyruvate, water and CO_2
(d) ATP is consumed

40. For two molecules of glucose, glycolysis uses and produces ATP molecules:
(a) 4 and 8 (b) 2 and 4
(c) 2 and 8 (d) 2 and 2

41. Glycolysis yields approximately ____________ of the total energy of glucose:
(a) 5% (b) 10%
(c) 15% (d) 12%

42. Read the following statements with respect to glycolysis:
(i) Glycolysis is the ten-step process of non-enzymatic reactions
(ii) Glucose undergoes partial oxidation to form one molecule of pyruvic acid
(iii) Four ATP are produced through substrate-level phosphorylation
(iv) No decarboxylation and oxidation occur during glycolysis
(v) It is only the source of energy in human RBCs
How many of these statements are correct?
(a) Two (b) Four
(c) Three (d) One

43. Excess of glucose 6-phosphate inhibits:
(a) Phosphofructokinase
(b) Hexokinase
(c) Pyruvate kinase
(d) PEP synthase

44. Excess of ATP inhibits respiration by inhibiting one of the following enzymes:
(a) Phosphofructokinase
(b) Hexokinase
(c) Pyruvic decarboxylase
(d) Aldolase

45. Enzyme pair common to both C_3 cycle and EMP pathway is:

(a) Aldolase and enolase
(b) Aldolase and triosephosphate isomerase
(c) Phosphoglyceromutase and triosephosphase isomerase
(d) Cytochrome oxidase and enolase

46. Read the following statements with respect to glycolysis in cellular respiration and state whether true (T) or false (F):
A. Subsequent steps of metabolism of glucose and fructose are the same
B. ATP molecules are synthesized by fructose 6-phosphate and fructose 1, 6-biphosphate
C. NAD^+ is reduced to $NADH_2$ between PGAL and BPGA
D. Water molecule is released between 2-phosphoglycerate and PEP

	A	B	C	D
(a)	T	F	T	T
(b)	T	F	T	F
(c)	F	F	T	T
(d)	F	T	F	F

47. In glycolysis, the redox equivalents are removed from:
(a) Triose phosphate
(b) Triose bisphosphate
(c) PEP
(d) Fructose 1, 5-bisphosphate

48. Glucose and fructose are phosphorylated to give rise to ___A___ by the activity of enzyme ___B___:

	A	B
(a)	Fructose 6-phosphate	PFK
(b)	Glucose 6-phosphate	Hexokinase
(c)	Glucose 6-phosphate	Invertase
(d)	Fructose 6-phosphate	Triose phosphate isomerase

49. Why are different enzymes involved in each step of glycolysis?
(a) Each step occurs in a different subcellular location
(b) Each step occurs in different cells
(c) Each step involves a different change in potential energy
(d) Each step involves a different chemical reaction

50. NAD^+ is:
(a) A protein
(b) An oxidising agent
(c) A reducing agent
(d) Formed under only aerobic conditions

51. Which of the following is a triose sugar?
(a) Fructose 1, 6-bisphosphate
(b) Fructose 6-phosphate
(c) Glucose 6-phosphate
(d) 3-phosphoglyceraldehyde

Fermentation

52. Find the odd one out with respect to the fate of pyruvic acid produced by glycolysis:
(a) Production of lactic acid
(b) Production of CO_2 and H_2O
(c) Production of starch
(d) Production of alcohol

53. Fermentation was discovered by:
(a) Gay Lussac (b) Buchner
(c) Lipman (d) Pasteur

54. Common pathway of aerobic respiration and fermentation is:
(a) Formation of pyruvic acid in cytoplasm
(b) Oxidative decarboxylation of pyruvic acid
(c) Oxidation of $NADH_2$
(d) All except (b)

55.

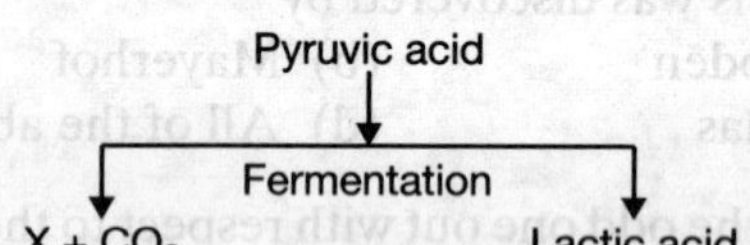

Identify X:
(a) Acetyl-CoA (b) Succinic acid
(c) PEP (d) Ethanol

56. Yeast poison themselves to death when concentration of alcohol is about:
(a) 9% (b) 13%
(c) 30% (d) 3%

57. Ethanol is formed from acetaldehyde by an enzyme called:
(a) Pyruvate kinase
(b) Pyruvate dehydrogenase
(c) Alcohol dehydrogenase
(d) Alcohol decarboxylase

58. How much energy present in glucose is released during lactic acid and alcohol fermentation?

(a) 7% (b) Less than 7%
(c) More than 7% (d) Only 2%

59. Which is never a product of anaerobic respiration?
(a) CO_2 (b) Alcohol
(c) H_2O (d) ATP

60. In the figure given below, the correct match is:

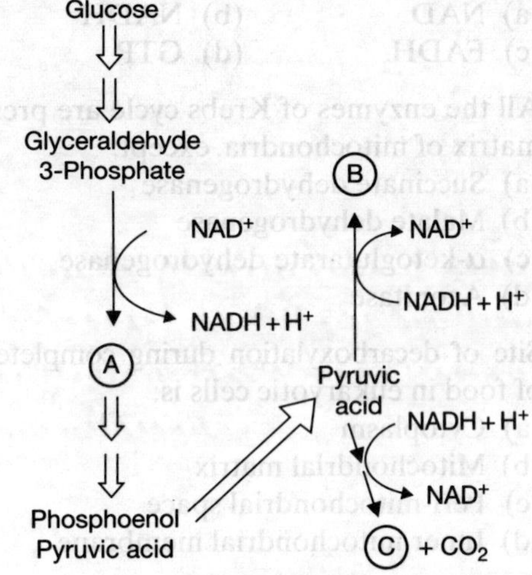

	A	B	C
(a)	Phosphoglyceric acid	Lactic acid	Ethanol
(b)	Phosphoglyceric acid	Ethanol	Lactic acid
(c)	α-ketoglutarate	Lactic acid	Ethanol
(d)	OAA	Lactic acid	Butyric acid

61. Find the incorrect statement with respect to anaerobic respiration:
(a) Pyruvic acid is reduced to lactic acid in animal cells
(b) Pyruvic acid is first decarboxylated then reduced to form ethanol in plant cells
(c) Net gain is 2ATP per glucose molecule
(d) $NADH_2$ is oxidised to NAD^+ very vigorously as compare to aerobic respiration

62. How many of the following are not associated with alcoholic fermentation?
(i) Oxidation of NADH to NAD^+
(ii) Pyruvate dehydrogenase activity
(iii) Net gain is 2ATP
(iv) Less than 7% energy in glucose is released and not all of it is trapped as high energy bonds of ATP
(a) Two (b) One
(c) Four (d) Three

63. In yeast cell, if the TCA cycle is shut down due to lack of oxygen, glycolysis will probably:
(a) Shut down
(b) Increase
(c) Produce more ATP/glucose molecules
(d) Produce more $NADH_2$/glucose molecules

64. First step of alcohol fermentation from pyruvate:
(a) Dehydrogenation (b) Decarboxylation
(c) Condensation (d) Oxidation

65. Fermentation is different from aerobic respiration:
(a) In having partial breakdown of glucose
(b) In producing less ATP per glucose
(c) In having slow oxidation of $NADH_2$ to NAD^+
(d) All of the above

66. Which enzyme converts glucose into alcohol?
(a) Invertase (b) Lipase
(c) Zymase (d) Diastase

67. For formation of ethanol, pyruvic acid is first changed to acetaldehyde by the enzyme:
(a) Pyruvate dehydrogenase
(b) Pyruvate decarboxylase
(c) Alcohol dehydrogenase
(d) Alcohol oxidase

68. In muscles during exercise, when O_2 is inadequate for cellular respiration, pyruvic acid is reduced to:
(a) Acetyl-CoA (b) C_2H_5OH
(c) Lactic acid (d) Methanol

69. What is the total gain of energy during anaerobic respiration?
(a) One molecule of ATP
(b) Two molecules of ATP
(c) Four molecules of ATP
(d) Eight molecules of ATP

70. Inhibition of sugar breakdown due to the presence of oxygen under anaerobic conditions is called:
(a) Pasteur effect (b) Warburg effect
(c) Crabtree effect (d) Kutusky effect

71. Which product of glycolysis is consumed in alcohol fermentation?
(a) $NADH_2$ (b) ATP
(c) $NADH_2$ and ATP (d) $FADH_2$

72. Fermentation takes place under anaerobic conditions in:
(a) All prokaryotes
(b) Unicellular eukaryotes only

(c) Many prokaryotes and unicellular eukaryotes
(d) Few prokaryotes and all eukaryotes

73. Select the incorrect statement with respect to lactic acid fermentation:
(a) Pyruvic acid is reduced to lactic acid in the presence of lactate dehydrogenase in muscle cells during exercise
(b) Glycolytic $NADH_2$ is used to reduce pyruvic acid
(c) Lactobacillus converts lactose sugar into lactic acid without glycolysis
(d) CO_2 is not released as a by-product like alcoholic fermentation

Aerobic Respiration (TCA Cycle, ETS, Oxidative Phosphorylation)

74. Under aerobic conditions, pyruvic acid, the key product of glycolysis, forms:
(a) Lactic acid
(b) Ethanol + CO_2
(c) Acetyl-CoA + CO_2
(d) Phosphoenolpyruvate and CO_2

75. Which of the following represents the correct sequence of the events that occur in the catabolism of glucose in the presence of oxygen?
(a) Glycolysis, Krebs cycle, oxidative decarboxylation
(b) EMP pathway, electron transport chain, Krebs cycle
(c) Glycolysis, Krebs cycle, electron transport chain
(d) Citric acid cycle, glycolysis, electron transport chain

76. The following equation represents:
Pyruvic acid + CoA + NAD^+ $\xrightarrow{Mg^{++}}$ Acetyl-CoA + CO_2 + NADH + H^+
(a) TCA cycle
(b) EMP pathway
(c) Breakdown of pyruvate in cytosol
(d) Oxidative decarboxylation

77. A $\xrightarrow{\text{LINK REACTION}}$ B (CoA, NAD+ in; CO_2, NADH + H+ out)

How many statements are correct for the above reaction?
A. Mg^{++} is involved in the process
B. A is a three-carbon compound
C. B is acetyl-CoA
D. Catalyst is pyruvate decarboxylase
E. B involved in both aerobic and anaerobic respiration
(a) Two (b) Three
(c) Four (d) Five

78. All of the following are formed in the citric acid cycle, except:
(a) NAD (b) NADH
(c) $FADH_2$ (d) GTP

79. All the enzymes of Krebs cycle are present in the matrix of mitochondria, except:
(a) Succinate dehydrogenase
(b) Malate dehydrogenase
(c) α-ketoglutarate dehydrogenase
(d) Aconitase

80. Site of decarboxylation during complete oxidation of food in eukaryotic cells is:
(a) Cytoplasm
(b) Mitochondrial matrix
(c) Peri-mitochondrial space
(d) Inner mitochondrial membrane

81. Site for Krebs cycle is:
(a) Cytoplasm
(b) Peri-mitochondrial space
(c) Inner membrane of mitochondria
(d) Matrix of mitochondria

82. Site for citric acid cycle in bacteria is:
(a) Cytoplasm
(b) Mitochondrial matrix
(c) Nucleoid
(d) Mesosome

83. Name of enzyme that brings about condensation of the acetyl group with OAA and water:
(a) Pyruvate dehydrogenase
(b) Cis-aconitase
(c) Citrate synthetase
(d) Succinate synthetase

84. Aerobic respiration completes in:
(a) Cell cytoplasm
(b) Two processes, one in the cytoplasm and the other in the mitochondrial matrix
(c) Mitochondrial matrix and inner mitochondrial membrane
(d) Mitochondrial matrix only

85. Identify the given reaction:

Pyruvic acid + CoA + NAD^+ $\xrightarrow[Mg^{++}]{\text{Pyruvate dehydrogenase}}$ Acetyl-CoA + CO_2 + NADH + H^+

(a) Oxidative decarboxylation

(b) Oxidative phosphorylation
(c) Reductive amination
(d) Dehydrogenation and carboxylation

86. Krebs cycle starts with condensation of acetyl-CoA with:
(a) OAA (b) Water
(c) NAD (d) Both (a) and (b)

87. GTP is formed during conversion of:
(a) Malic acid ⟶ Oxalo-acetic acid
(b) Fumaric acid ⟶ Malic acid
(c) Succinic acid ⟶ Fumaric acid
(d) Succinyl-CoA ⟶ Succinic acid

88. Enzyme α-ketoglurarate dehydrogenase brings about:
(a) Oxidation only
(b) Decarboxylation only
(c) Oxidation and decarboxylation
(d) Oxidation and reduction

89. Single turn of citric acid cycle yields:
(a) 3NADH$_2$, 1FADH$_2$, 1GTP
(b) 3NADH$_2$, 1FADH$_2$, 2GTP
(c) 4NADH$_2$, 1FADH$_2$, 1GTP
(d) 6NADH$_2$, 2FADH$_2$, 1GTP

90. Which of the following is the only 5-carbon compound formed during Krebs cycle?
(a) Malic acid (b) Succinic acid
(c) ISO citrate (d) α-ketoglutaric acid

91. Net gain of ATP through substrate level phosphorylation when only one molecule of glucose completely oxidises in aerobic respiration is:
(a) 2 (b) 4
(c) 10 (d) 6

92. Number of ATP molecules produced after complete breakdown of one molecule of acetyl-CoA is:
(a) 11 (b) 12
(c) 15 (d) 30

93. Read the following statements carefully:
A. Starts with the condensation of the acetyl group with four carbons containing organic acid
B. Involves the two steps of decarboxylation
C. A single step of substrate level phosphorylation
These statements are related to:
(a) EMP pathway (b) Link reaction
(c) HMP shunt (d) TCA cycle

94. Electron transport system (ETS) is located in the mitochondrial:
(a) Outer membrane
(b) Matrix
(c) Inner membrane
(d) Inter membrane space

95. Mitochondrial matrix electron transport chain is:
(a) Cyclic phosphorylation
(b) Non-cyclic phosphorylation
(c) Oxidative phosphorylation
(d) Substrate level phosphorylation

96. Oxidation of NADH occurs in:
(a) Anaerobic respiration
(b) Aerobic respiration
(c) Fermentation
(d) More than one option is correct

97. In mitochondrial ETS, Cyt C acts as a mobile electron carrier between complexes:
(a) III and IV (b) II and III
(c) I and II (d) IV and V

98. Terminal electron donor in mitochondrial ETS is:
(a) Cyt a3 (b) Cytochrome A
(c) Oxygen (d) Cytochrome C

99. In the citric acid cycle, which of the following intermediates is a result of two successive decarboxylations?
(a) α-ketoglutaric acid (b) Malic acid
(c) Succinyl-CoA (d) Succinic acid

100. Which of the following membrane-bound complexes in mitochondria is not a proton pump?
(a) NADH dehydrogenase
(b) Succinate dehydrogenase
(c) Cytochrome b, c
(d) Cytochrome c oxide

101. Which of the following is a mobile electron carrier present in the inner membrane of mitochondria?
(a) Cytochrome a (b) Cytochrome a3
(c) Cytochrome c (d) Cytochrome b

102. Final acceptor of electron in ETS is:
(a) Oxygen (b) Complex IV
(c) Cytochrome c (d) Complex V

103. How many ATP are produced by ETS when only one molecule of glucose is oxidised completely in a bacterial cell?
(a) 6 (b) 4
(c) 15 (d) 34

104. During ETS in mitochondria, UQ receives electrons from:
(a) Complex I and II (b) Complex I only
(c) Complex III and V (d) Complex III only

105. In ETS in the mitochondrial membrane, how many H^+ are pumped into the peri-mitochondrial space for oxidation of 1$FADH_2$?
(a) 2 (b) 4
(c) 6 (d) 8

106. The cytochrome that contains copper besides iron is:
(a) Cyt b (b) Cyt c
(c) Cyt c1 (d) Cyt a3

107. During chemiosmosis, the accumulator of H^+ ions occurs in/on:
(a) Peri-mitochondrial space
(b) Outer membrane
(c) Inner membrane
(d) Matrix

108. The phenomenon of synthesis of ATP is:
(a) Phosphorylation and exergonic
(b) Phosphorylation and endergonic
(c) Dephosphatisation and exergonic
(d) Dephosphatisation and endergonic

109. Choose the incorrect option for the figure given below:

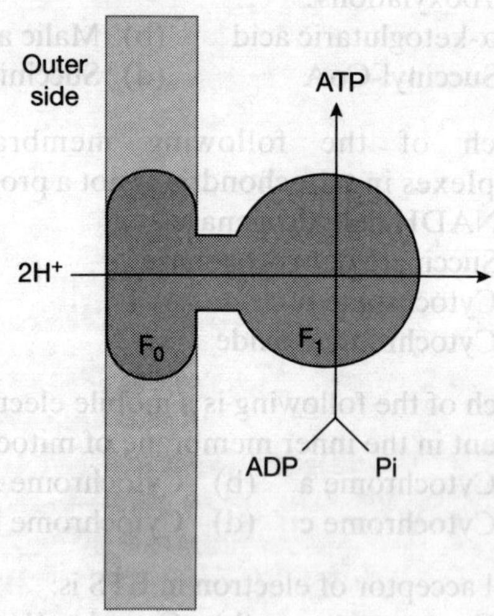

(a) A represents the site for oxidative phosphorylation
(b) Showing movement of protons down the electrochemical proton gradient
(c) B is an integral protein complex
(d) For production of ATP, protons pass through the F_0 complex, from the matrix to the inter membrane space

110. How many water molecules are generated in ETS when reduced coenzymes from one glucose molecule are oxidised?
(a) 0 (b) 1
(c) 12 (d) 24

111. Which multiprotein complex on the inner mitochondrial membrane can easily be removed on giving the temperature?
(a) Complex I
(b) Complex III
(c) Complex IV
(d) Headpiece of Complex V

112. How many ATP molecules will be produced from 6 pairs of H^+ in ETS by oxidative phosphorylation?
(a) 6 ATP (b) 5 ATP
(c) 4 ATP (d) 2 ATP

113. Redox potential is maximum of:
(a) NADH dehydrogenase
(b) Molecular oxygen
(c) Cytochrome C
(d) Complex V

114. Mobile electron carriers in ETS of mitochondria are:
(a) Co Q and Cyt a
(b) Co Q and Cyt c
(c) UQ and Cyt a
(d) Co Q and Cyt c

115. Cytochrome c acts as a mobile electron carrier for the transfer of electrons between Complex III and Complex IV of ETS. It is located:
(a) In the peri-mitochondrial space
(b) In the matrix of mitochondria
(c) On the outer side of the inner membrane
(d) On the inner side of the inner membrane

116. Two molecules of $NADH_2$ require ______ molecules of oxygen for oxidation through ETS:
(a) 1 (b) 1/2
(c) 2 (d) 6

117. What is the function of molecular oxygen in cellular respiration?
A. It causes breakdown as pyruvic acid
B. It combines with glucose to produce CO_2
C. It combines with carbon from organic molecules to produce CO_2
D. It combines with hydrogen from organic molecules to produce water
(a) Only (B) (b) Only (D)
(c) (A) and (C) (d) All are correct

118. Metabolic water is:
(a) Used during transamination
(b) Used during photosynthesis

(c) Produced during aerobic utilisation of glucose
(d) Produced during condensation or polymerisation

119. Most of the biological energy is supplied by mitochondria through:
(a) Breaking of proteins
(b) Reduction of $NADP^+$
(c) Breaking of sugars
(d) Oxidising of TCA substrates

120. Correct sequence in electron acceptors in ATP synthesis is:
(a) Cyt b c_1 c a a_3 (b) Cyt b c c_1 a a_3
(c) Cyt c b a a_3 (d) Cyt b_c c_a a

121. What happens to the hydrogen atoms and the electrons that are removed during oxidation?
(a) They reduce NAD
(b) The oxidise NAD
(c) They are transferred to pyruvic acid
(d) They come out immediately in the form of H_2O

122. Oxygen is used by:
(a) Citric acid cycle
(b) ETC
(c) Substrate level phosphorylation
(d) ATP synthase

123. The complex concerned with oxidative phosphorylation in the inner mitochondrial membrane is:
(a) Complex IV (b) Complex V
(c) Complex III (d) Complex II

124. According to the chemiosmotic mechanism for ATP synthesis, the force/factor responsible for ATP synthesis is:
(a) Membrane potential across membrane
(b) Proton motive force
(c) Electron motive force
(d) Redox potential

125. Answer the given question with respect to the cycle given below:

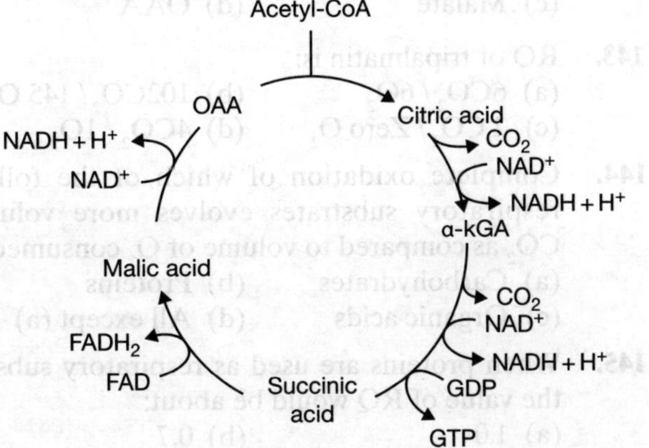

A. Total redox equivalent removed is ______
B. Condensation reaction occurs ______
C. Oxidative decarboxylation occurs ______

	A	B	C
(a)	Six	Twice	Once
(b)	Eight	Twice	Twice
(c)	Eight	Once	Once
(d)	Six	Once	Twice

126. Succinate dehydrogenase enzyme of Krebs cycle:
A. Is found dissolved in mitochondrial matrix
B. Reduces FAD to $FADH_2$
C. Catalyses the conversion of succinyl-CoA to succinate
D. Is found attached to the inner mitochondrial membrane
(a) Only (D) is incorrect
(b) Only (B) and (C) are correct
(c) Only (B) and (D) are correct
(d) All are correct except (A)

127. The continued oxidation of acetyl-CoA via the TCA cycle requires the replenishment of :
(a) α-ketoglutaric acid
(b) Succinic acid
(c) Malic acid
(d) OAA

128. Arrange the following intermediates of the TCA cycle in the correct order of their formation:
A. Citric acid
B. Malic acid
C. α-ketoglutaric acid
D. Succinic acid
(a) A, C, D, B (b) A, D, C, B
(c) A, C, B, D (d) A, B, D, C

129. If glucose is labelled with C^{14}, what molecule will become radioactive as glycolysis and the TCA cycle are completed?
(a) Water (b) NADH
(c) ATP (d) CO_2

130. In the link reaction, pyruvic acid is converted into:
(a) 5c compound (b) 2c compound
(c) 4c compound (d) 6c compound

131. If one molecule of acetyl-CoA undergoes complete oxidation, how many CO_2 molecules will be released?
(a) One (b) Three
(c) Four (d) Two

132. All of the following are steps in the aerobic phase in respiration, except:
(a) Oxidation of glucose into pyruvic acid
(b) ETS in mitochondria
(c) Conversion of pyruvic acid into acetyl-CoA
(d) TCA cycle

Respiratory Balance Sheet, Amphibolic Pathway RQ

133. Krebs cycle is considered as amphibolic pathway because:
(a) The intermediates of the cycle are used as precursors in the biosynthesis of many compounds
(b) It is the sole process to produce ATP
(c) It is the oxidation of acetyl-CoA in CO_2 and H_2O
(d) Both (a) and (c) are correct

134. RQ is the ratio of:
(a) CO_2 produced to substrate consumed
(b) CO_2 produced to O_2 consumed
(c) O_2 consumed to CO_2 produced
(d) O_2 consumed to water produced

135. The respiratory balance sheet prepared for one molecule of glucose is based on certain assumptions. Which of the following is not one of them?
(a) Orderly functioning includes glycolysis TCA and ETS
(b) Only glucose is being respired
(c) Oxidative phosphorylation of NADH
(d) The intermediates of the TCA cycle are used in different anabolic pathways

136. If fat acts as a respiratory substrate, it would need to be broken down into fatty acids and glycerol:
(i) If fatty acids would enter the respiratory pathway after being converted to ___A___
(ii) If glycerol was to be respired, it would first be degraded to ___B___

	A	B
(a)	Acetyl-CoA	Glyceraldehyde 3-phosphate
(b)	Acetyl-CoA	Pyruvic acid
(c)	Pyruvic acid	Acetyl-CoA
(d)	Pyruvic acid	Dihydroxyacetone phosphate

137. Proteins can enter the respiratory pathway by deamination of amino acids in all given forms, except:
(a) 3PGAL (b) Pyruvic acid
(c) Acetyl-CoA (d) α-ketoglutaric acid

138. First common intermediate formed by the oxidation of fat, carbohydrates and deamination of proteins is:
(a) Acetyl-CoA (b) Pyruvic acid
(c) DHAP (d) α-ketoglutarate

139. RQ can vary due to:
(a) Temperature
(b) Respiratory substrate
(c) Light and oxygen
(d) Respiratory product

140. Which of the following respiratory substrates has less than one?
(a) Glucose (b) Malic acid
(c) Sucrose (d) Tripalmatin

141. Which of the following intermediates of the Krebs cycle is used to synthesise amino acids?
(a) Succinyl-CoA
(b) Acetyl-CoA
(c) α-ketoglutaric acid
(d) OAA

142. Name the product marked A in the flow chart given below:

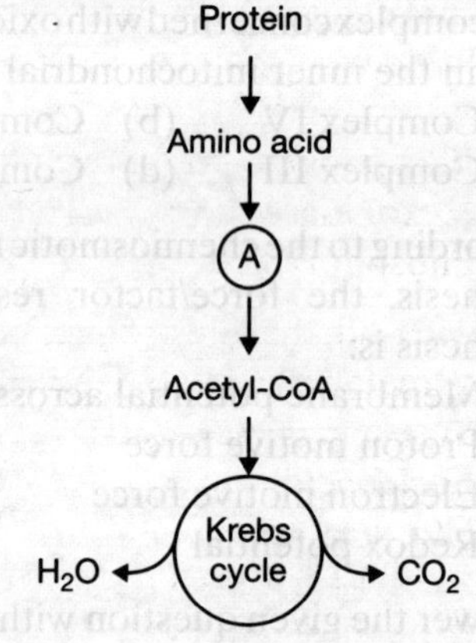

(a) Pyruvate (b) PEP
(c) Malate (d) OAA

143. RQ of tripalmatin is:
(a) $6CO_2 / 6O_2$ (b) $102CO_2 / 145 O_2$
(c) $2 CO_2$ / Zero O_2 (d) $4CO_2 / 1O_2$

144. Complete oxidation of which of the following respiratory substrates evolves more volume of CO_2 as compared to volume of O_2 consumed?
(a) Carbohydrates (b) Proteins
(c) Organic acids (d) All except (a)

145. When proteins are used as respiratory substrates, the value of RQ would be about:
(a) 1.0 (b) 0.7
(c) 0.9 (d) 1.33

146. Respiratory quotient in ethyl alcohol fermentation is:
(a) One (b) Infinity
(c) Zero (d) More than one

147. The carbon skeletons produced during respiration are used as precursors for the biosynthesis of other molecules in the cell. In light of this statement, what could be said about respiration?
(a) Respiration is a multistep process
(b) During respiration, complex sugars are synthesized from simple sugars
(c) Carbohydrates, fats, sugars or amino acids can serve as respiratory substrate
(d) Respiration is an amphibolic pathway

148. Which of the following intermediates of the Krebs cycle takes part in the synthesis of chlorophyll in the cell?
(a) Citrate (b) α-ketoglutarate
(c) OAA (d) Succinyl-CoA

149. In *Opuntia* and other succulents, night-time RQ is:
(a) One (b) More than one
(c) Zero (d) Less than one

150. Acetyl-CoA is a product of:
(a) Link reaction
(b) Deamination of amino acids
(c) β-oxidation of fatty acids
(d) All of the above

NCERT EXEMPLAR QUESTIONS

1. The ultimate electron acceptor of respiration in an aerobic organism is:
(a) Cytochrome (b) Oxygen
(c) Hydrogen (d) Glucose

2. Phosphorylation of glucose during glycolysis is catalysed by:
(a) Phosphoglucomutase
(b) Phosphoglucoisomerase
(c) Hexokinase
(d) Phosphorylase

3. Pyruvic acid, the key product of glycolysis, can have many metabolic fates. Under aerobic conditions, it forms:
(a) Lactic acid
(b) $CO_2 + H_2O$
(c) Acetyl-CoA + CO_2
(d) Ethanol + CO_2

4. Electron Transport System (ETS) is located in mitochondrial:
(a) Outer membrane
(b) Inner membrane space
(c) Inner membrane
(d) Matrix

5. Which of the following exhibits the highest rate of respiration?
(a) Growing shoot apex
(b) Germinating seed
(c) Root tip
(d) Leaf bud

6. Mitochondria are called powerhouses of the cell. Which of the following observations support this statement?
(a) Mitochondria synthesize ATP
(b) Mitochondria have a double membrane
(c) The enzymes of the Krebs cycle and cytochromes are found in mitochondria
(d) Mitochondria are found in almost all plant and animal cells

7. The end product of oxidative phosphorylation is:
(a) NADH (b) Oxygen
(c) ADP (d) ATP + H_2O

8. Match the following and choose the correct option from those give below:

Column I	Column II
A. Molecular oxygen	(i) α-ketoglutaric acid
B. Electron acceptor	(ii) Hydrogen acceptor
C. Pyruvate dehydrogenase	(iii) Cytochrome c
D. Decarboxylation	(iv) Acetyl-CoA

(a) A – (ii), B – (iii), C – (iv), D – (i)
(b) A – (iii), B – (iv), C – (ii), D – (i)
(c) A – (ii), B – (i), C – (iii), D – (iv)
(d) A – (iv), B – (iii), C – (i), D – (ii)

PRACTICE TEST PAPER

1. During respiration, three-carbon acid (3C) is formed in:
(a) Mitochondrial matrix
(b) Cytoplasm
(c) Inner mitochondrial membrane
(d) Outer mitochondrial membrane

2. Find the odd one out with respect to fermentation:
(a) Ethanol
(b) PGA
(c) OAA
(d) Dihydroxy acetone phosphate

3. TCA cycle starts with:
(a) Dehydrogenation
(b) Decarboxylation
(c) Splitting of pyruvic acid
(d) Condensation

4. Which of the following is not required for mitochondrial ETS?
(a) NADH + H^+ and $FADH_2$
(b) FMN and FAD
(c) ATP and H_2O
(d) Cytochrome and ubiquinone

5. Which of the following enzymes is required during reduction of FAD in aerobic respiration?
(a) Pyruvate dehydrogenase
(b) Succinate dehydrogenase
(c) α-ketoglutarate dehydrogenase
(d) Malate dehydrogenase

6. Respiratory substrate in which less CO_2 is released as compared to O_2 consumption during respiration is:
(a) Organic acid (b) Malic acid
(c) Sugars (d) Protein

7. Read the following statements:
A. Pyruvic acid is converted to acetyl-CoA before entering the TCA cycle
B. Succinyl-CoA is the only five-carbon intermediate in the TCA cycle
C. Fermentation is a metabolic process in which oxidation of substrate occurs without any external electron acceptor
D. In the electron transport chain, Complex IV is called cytochrome oxidase

Choose the option for correct statements:
(a) (A), (B) and (C)
(b) (B), (C) and (D)
(c) (A), (C) and (D)
(d) (A), (B), (C) and (D)

8. First decarboxylation in the TCA cycle occurs at:
(a) Pyruvic acid
(b) Malic acid
(c) Oxalo-succinic acid
(d) α-ketoglutaric acid

9. Main splitting enzyme in glycolysis is:
(a) PFK
(b) Aldolase
(c) Hexokinase
(d) Transphosphorylase

10. There are three major ways in which different cells handle pyruvic acid. Select the odd one out with respect to the fate of pyruvic acid:
(a) Aerobic respiration
(b) Lactic acid fermentation
(c) Ethanol fermentation
(d) Photo respiration

11. Match the following and choose the correct option from those given below:

Column I	Column II
A. Inner mitochondrial membrane	(i) Krebs cycle
B. Pyruvic acid is converted into ethanol and CO_2	(ii) ETC
C. Mitochondrial matrix	(iii) EMP pathway
D. Cytoplasm	(iv) Fermentation

(a) A – (ii), B – (i), C – (iii), D – (iv)
(b) A – (ii), B – (iv), C – (i), D – (iii)
(c) A – (iii), B – (iv), C – (i), D – (ii)
(d) A – (iii), B – (i), C – (iv), D – (ii)

12. A Krebs cycle intermediate which acts as raw material for chlorophyll synthesis is:
(a) Succinyl-CoA
(b) Succinic acid
(c) α-ketoglutaric acid
(d) Pyruvic acid

13. The complex concerned with oxidative phosphorylation in the inner mitochondrial membrane is:
(a) Complex III (b) Complex IV
(c) Complex V (d) Complex VI

14. Number of steps involved in decarboxylation process during aerobic respiration is:
(a) 1 (b) 3
(c) 6 (d) 14

15. In respiration, pyruvic acid is:
(a) Formed only when oxygen is available
(b) One of the products of the Krebs cycle
(c) Broken down into acetyl-CoA and CO_2
(d) Formed in the mitochondrial matrix

16. Mobile electron carriers in ETS in the mitochondrial membrane are:
(a) PQ and PC (b) Co Q, Cyt c
(c) PQ and Cyt c (d) PC, CoQ

17. Location of succinate dehydrogenase enzyme of Krebs cycle in plants is:
(a) Cytoplasm
(b) Peroxisome
(c) Mitochondrial matrix
(d) Inner mitochondrial membrane

18. What is the value of Respiratory Quotient (RQ) during germination of fatty seeds?
(a) Unity (b) Less than Unity
(c) Zero (d) More than Unity

19. After glycolysis, the fate of glucose in the mitochondrial matrix is:
(a) Oxidation
(b) Reduction
(c) Oxidative decarboxylation
(d) Hydrolysis

20. Alcoholic fermentation is carried out by:
(a) *Saccharomyces* (b) *Lactobacillus*
(c) *Clostridium* (d) *Aspergillus*

21. In glycolysis, 6-C compounds are split into 3-C compounds by the enzyme:
(a) Isomerase (b) Aldolase
(c) Hexokinase (d) Dehydrogenase

22. Nearly every living organism uses glucose as a nutrient source of energy. Why?
(a) Glucose is the only molecule capable of providing the energy to produce ATP
(b) The structure of glucose is very similar to ATP
(c) Glucose has more potential energy than any other respiratory substrate
(d) The ability to harvest energy from glucose appeared very early in biological evolution

23. Dough kept overnight in a warm place becomes soft and spongy due to:
(a) Fermentation
(b) Absorption of CO_2 from atmosphere
(c) Inhibition
(d) Capillarity

24. Respiratory Quotient of glucose is:
(a) 0.5 (b) 0.7
(c) 1.0 (d) 1.5

25. A sudden change from anaerobic to aerobic respiration is called:
(a) Emerson effect (b) Pasteur effect
(c) Blackman's law (d) Crabtree effect

26. Identify the following statements as true (T) or false (F) and select the option accordingly:
A. Glycolysis occurs in the cytoplasm of all organisms
B. Enzymes of the Krebs cycle are present in mitochondrial matrix
C. Oxidative phosphorylation takes place in the outer membrane of mitochondria
D. Through link reaction, glucose enters the aerobic respiratory pathway

	A	B	C	D
(a)	T	F	T	F
(b)	T	T	F	F
(c)	T	F	F	T
(d)	F	F	T	T

27. During oxidative phosphorylation, energy for ATP synthesis is provided by:
(a) NADH
(b) Pyruvic acid
(c) Efflux of protons to cytoplasm
(d) CoA

28. The overall goal of glycolysis, Krebs cycle and electron transport system is the formation of:
(a) Sugars
(b) ATP in one large oxidation reaction
(c) ATP in step-wise process
(d) Nucleic acid

29. ATP synthase is located in:
(a) Inner membrane of mitochondria
(b) Outer membrane of mitochondria
(c) Peri-mitochondrial space
(d) Cytoplasm

30. In which of the following processes is CO_2 not released?
(a) Aerobic respiration in eukaryotes
(b) Link reaction of aerobic respiration
(c) Lactic acid fermentation
(d) Alcoholic fermentation

31. Which of the following intermediates of the citric acid cycle is used to make chlorophyll and cytochromes?
(a) Acetyl-CoA
(b) Oxalo-acetic acid
(c) Succinyl-CoA
(d) α-ketoglutaric acid

32. Respiration of proteins is called:
(a) Floating respiration
(b) Protoplasmic respiration
(c) Salt respiration
(d) Photorespiration

33. When production of CO_2 is more than intake of oxygen, the respiratory substrate is:
(a) Glucose (b) Sucrose
(c) Organic acid (d) Protein/Fat

34. A single turn of the Krebs cycle yields three molecules of:
(a) ATP (b) CO_2
(c) $FADH_2$ (d) $NADH_2$

35. Rate of oxidation of NADH in fermentation as compared to aerobic respiration is:
(a) Slower
(b) Faster
(c) At the same rate
(d) Dependent upon the nature of substrate

36. Excess of glucose 6-phosphate inhibits:
(a) PFK (b) Hexokinase
(c) Glucokinase (d) Aldolase

37. Which of the following is a 5-carbon compound of the Krebs cycle?
(a) Citric acid
(b) Fumaric acid
(c) Oxalo-succinic acid
(d) α-ketoglutaric acid

38. How many dehydrogenation and decarboxylation reactions occur in one turn of the Krebs cycle, respectively?
(a) 3, 2 (b) 6, 4
(c) 4, 3 (d) 3, 3

39. What would be the maximum concentration of alcohol in beverages that are naturally fermented?
(a) 7% (b) 45%
(c) 13% (d) 99%

40. During ETS in mitochondria, UQ received electrons from:
(a) Complex I and II (b) Complex I only
(c) Complex III and V (d) Complex III only

41. Number of gross ATP molecules formed from complete oxidation of one molecule of glucose is:
(a) 30 (b) 36
(c) 38 (d) 40

42. The mobile electron carrier attached to the outer surface of the inner membrane of the mitochondria is:
(a) Cytochrome a (b) Ubiquinone
(c) Cytochrome a_3 (d) Cytochrome c

43. The first decarboxylation in aerobic respiration occurs when:
(a) α-ketoglutaric acid is oxidised to succinyl-CoA
(b) Phosphoenopyruvate is changed into pyruvic acid
(c) Pyruvic acid is changed into acetyl-CoA
(d) Oxalo-succinate is decarboxylated to α-ketoglutaric acid

44. Glycolysis shows partial oxidation. Identify the events not associated with this process:
(a) Ten sequence of enzymatic reactions
(b) Breakdown of C–C bonds
(c) Utilisation of reducing power
(d) Enzymes are present in the cytoplasm

45. If glucose is labelled with C^{14}, what molecule will become radioactive as the glycolysis and TCA cycles are completed?
(a) Water (b) NADH
(c) ATP (d) CO_2

ANSWER KEYS

Exercises

1. (b)	**2.** (d)	**3.** (b)	**4.** (a)	**5.** (c)	**6.** (b)	**7.** (d)	**8.** (b)	**9.** (d)	**10.** (c)
11. (a)	**12.** (c)	**13.** (c)	**14.** (d)	**15.** (c)	**16.** (d)	**17.** (d)	**18.** (d)	**19.** (a)	**20.** (c)
21. (d)	**22.** (d)	**23.** (b)	**24.** (d)	**25.** (b)	**26.** (d)	**27.** (b)	**28.** (b)	**29.** (d)	**30.** (b)
31. (a)	**32.** (c)	**33.** (d)	**34.** (d)	**35.** (c)	**36.** (b)	**37.** (b)	**38.** (d)	**39.** (c)	**40.** (a)
41. (a)	**42.** (a)	**43.** (b)	**44.** (a)	**45.** (b)	**46.** (a)	**47.** (a)	**48.** (b)	**49.** (d)	**50.** (c)
51. (d)	**52.** (c)	**53.** (a)	**54.** (d)	**55.** (d)	**56.** (b)	**57.** (c)	**58.** (b)	**59.** (c)	**60.** (a)
61. (b)	**62.** (b)	**63.** (b)	**64.** (b)	**65.** (d)	**66.** (c)	**67.** (b)	**68.** (c)	**69.** (b)	**70.** (a)
71. (a)	**72.** (c)	**73.** (c)	**74.** (c)	**75.** (c)	**76.** (d)	**77.** (b)	**78.** (a)	**79.** (a)	**80.** (b)
81. (d)	**82.** (a)	**83.** (c)	**84.** (c)	**85.** (a)	**86.** (d)	**87.** (d)	**88.** (c)	**89.** (a)	**90.** (d)
91. (d)	**92.** (b)	**93.** (d)	**94.** (c)	**95.** (c)	**96.** (d)	**97.** (a)	**98.** (a)	**99.** (d)	**100.** (b)
101. (c)	**102.** (a)	**103.** (d)	**104.** (a)	**105.** (b)	**106.** (d)	**107.** (a)	**108.** (b)	**109.** (d)	**110.** (c)
111. (d)	**112.** (a)	**113.** (a)	**114.** (b)	**115.** (c)	**116.** (a)	**117.** (b)	**118.** (c)	**119.** (d)	**120.** (a)
121. (a)	**122.** (b)	**123.** (b)	**124.** (b)	**125.** (c)	**126.** (c)	**127.** (d)	**128.** (a)	**129.** (d)	**130.** (b)
131. (d)	**132.** (a)	**133.** (d)	**134.** (b)	**135.** (d)	**136.** (a)	**137.** (a)	**138.** (a)	**139.** (b)	**140.** (d)
141. (c)	**142.** (a)	**143.** (b)	**144.** (c)	**145.** (c)	**146.** (b)	**147.** (d)	**148.** (d)	**149.** (c)	**150.** (d)

NCERT Exemplar Questions

1. (b)	**2.** (c)	**3.** (c)	**4.** (c)	**5.** (b)	**6.** (a)	**7.** (d)	**8.** (a)

Practice Test Paper

1. (b)	**2.** (c)	**3.** (d)	**4.** (c)	**5.** (b)	**6.** (d)	**7.** (c)	**8.** (c)	**9.** (b)	**10.** (d)
11. (b)	**12.** (a)	**13.** (c)	**14.** (c)	**15.** (c)	**16.** (b)	**17.** (d)	**18.** (b)	**19.** (c)	**20.** (a)
21. (b)	**22.** (d)	**23.** (a)	**24.** (c)	**25.** (b)	**26.** (b)	**27.** (a)	**28.** (c)	**29.** (a)	**30.** (c)
31. (c)	**32.** (b)	**33.** (c)	**34.** (d)	**35.** (a)	**36.** (b)	**37.** (d)	**38.** (a)	**39.** (c)	**40.** (a)
41. (c)	**42.** (d)	**43.** (c)	**44.** (c)	**45.** (d)					

SOLUTIONS

Introduction, Do Plants Breathe?

1. Cellular respiration is the mechanism of breakdown of the C–C bond of complex compounds through oxidation and release of energy.
2. Both (B) and (C) are correct. Energy is released in a series of slow, stepwise reactions controlled by enzymes.
3. Catabolic process
4. Oil and fats
5. In floating respiration, the respiratory substrates are carbohydrates and fats.
6. Glucose is the most preferred respiratory substrate.
7. (A) (B) and (C) are correct.
8. In aerobic respiration from one glucose molecule, Glycolysis = 8 ATP, Link reaction = 6 ATP, Krebs cycle = 24 ATP, Total = 38 ATP.
9. In ATP, energy is stored in pyrophosphate bonds.
10. ATP is called universal energy carrier or the energy currency of cells.
11. Only A is correct. In prokaryotes, ETS is present in mesosomes.
12. Each plant part takes care of its own gas exchange needs and there is very little transport of gases from one part to another.
13. In protoplasmic respiration, proteins act as respiratory substrate.
14. In the absence of oxygen, anaerobic respiration takes place and 1 glucose molecule yields about 2 ATP molecules. Anaerobic respiration has an efficiency of approximately 2.13%. In the presence of oxygen, one glucose molecule yields about 32 ATP molecules. So, aerobic respiration has an efficiency of approximately 40.4%. Thus, glucose consumption is about 19 times more in anaerobic respiration than in aerobic respiration.
15. Carbohydrate
16. Usable energy of respiration is stored in ATP molecules.
17. Option (d) is correct.
18. All the statements are correct.
19. 1 gm-mole of ATP contains 7.3 Kcal of energy.
20. In respiration, energy is liberated gradually, in a step-by-step process.
21. Cellular respiration converts energy stored in the chemical bonds of glucose to energy that the cell can use.

Glycolysis

22. Glycos + Lysis
23. Breakdown of glucose into two molecules of pyruvic acids
24. In glycolysis, 4 ATP are produced and 2 ATP are consumed. Hence, the net ATP production per glucose molecule is 2 ATP.
25. Invertase converts sucrose into glucose and fructose.
26. Glycolysis occurs in all living cells.
27. Hexokinase enzyme
28. Only (B) is correct. There is no decarboxylation in glycolysis.
29. Formation of BPGA
30. Zero
31. Phosphofructokinase (PEK) enzyme is called the pacemaker enzyme of glycolysis as it is the major regulatory enzyme.
32. PGAL acts as a connecting link between respiration and photosynthesis.
33. Glycolysis was discovered by Gustav Embden, Otto Meyerhof and J Parnas.
34. Acetyl-CoA is a two-carbon compound.
35. $NADH_2$ yields ATP during terminal oxidation.
36. 2 ATP

37. Glycolysis is the only source of ATP in RBCs, because they lack mitochondria.

38. Enolase enzyme catalyses the dehydration step of glycolysis.

39. End products of glycolysis are pyruvate, water, $NADH^+$, + H^+ and ATP.

40. 4 and 8 (each turn of glycolysis produces 4 ATP and consumes 2 ATP)

41. Glycolysis yields only 5% of total ATP (2 ATP out of 38 ATP) production and 2% of the total energy content of glucose.

42. Two statements (III and V) are correct. Glycolysis involves ten steps of enzymatic reactions resulting in the production of two molecules of pyruvic acid (3C). Glycolysis involves only one oxidation step.

43. Excess of glucose 6-phosphate inhibits the hexokinase enzyme.

44. Phosphofructokinase enzyme

45. Aldolase and triose phosphate isomerase are common in glycolysis and the C3 cycle.

46. Only (B) is false.

47. Two redox equivalents are removed in the form of hydrogen atoms from PGAL (triose phosphate) and transferred to a molecule of NAD^+.

48. A – Glucose 6-phosphate, B – Hexokinase

49. Each step involves a different biochemical reaction.

50. NAD^+ is a reducing agent.

51. 3-Phosphoglyceraldehyde and dehydroxyacetone phosphate are triose sugars.

Fermentation

52. Production of starch

53. Fermentation was discovered by Gay Lussac.

54. In aerobic respiration, pyruvic acid enters the mitochondria and comes under oxidative decarboxylation to produce acetyl-CoA.

55. Ethanol

56. Yeast poison themselves to death when alcoholic concentration reaches about 13%.

57. Alcohol dehydrogenase enzyme

58. Less than 7%

59. H_2O

60. A – Phosphoglyceric acid, B – Lactic acid, C – Ethanol

61. Pyruvic acid first decarboxylates and then oxidases to form ethanol.

62. (ii) is incorrect. Alcoholic fermentation involves pyruvate decarboxylase activity.

63. Option (b) is correct.

64. Decarboxylation is the first step of alcoholic fermentation.

65. All the statements are correct.

66. Zymase enzyme

67. Pyruvate decarboxylase activity

68. When O_2 is inadequate for cellular respiration in muscle, pyruvic acid is reduced to lactic acid.

69. For each molecule of glucose during fermentation, there is a net gain of 2 ATP molecules.

70. Pasteur effect is the decrease in sugar breakdown and CO_2 evolution in the shift from anaerobic to aerobic respiration in yeast.

71. $NADH_2$

72. Fermentation occurs in many prokaryotes and unicellular eukaryotes.

73. Option (c) is correct.

Aerobic Respiration (TCA Cycle, ETS, Oxidative Phosphorylation)

74. Pyruvic acid comes under oxidative decarboxylation to produce acetyl-CoA and CO_2.

75. Glycolysis, Krebs cycle, ETS

76. Oxidative decarboxylation

77. Three statements (A, B, C) are correct. The given reaction is catalysed by the enzyme pyruvate dehydrogenase. Acetyl enzyme A is involved in only aerobic respiration.

78. In the citric acid cycle, NAD^+ is reduced to form $NADH_2$.

79. Succinic dehydrogenase enzyme is present in the inner mitochondrial membrane.

80. Site of decarboxylation is the mitochondrial matrix. When one molecule of pyruvic acid in mitochondria, 3 decarboxylation occurs in the mitochondrial matrix.

81. Krebs cycle occurs in the mitochondrial matrix.

82. Bacteria (prokaryotic cells) lack mitochondria. So, the citric acid cycle occurs in the cell cytoplasm.

83. Citrate synthetase

84. In the mitochondrial matrix, Krebs cycle occurs while ETS and oxidative phosphorylation occur in the inner mitochondrial membrane.

85. Link reaction (oxidative decarboxylation)

86. Krebs cycle starts with condensation of acetyl-CoA with OAA and water to form citric acid.

87. During Krebs cycle, 1 GTP molecule is produced at substrate level phosphorylation when succinyl-CoA changes into succinic acid.

88. Oxidative decarboxylation

89. Single turn of the citric cycle yields 3 $NADH_2$, 1 $FADH_2$ and 1 GTP.

90. Alpha-ketoglutaric acid is a 5-carbon compound of the Krebs cycle.

91. 6 ATP is produced at substrate level phosphorylation (4 in Glycolysis + 2 Krebs cycle in two turns)

92. One turn of the Krebs cycle produces 12 ATP.

93. All the statements are related to the TCA cycle or Krebs cycle.

94. ETS is located in the inner mitochondrial membrane.

95. Oxidative phosphorylation

96. NADH generated in cytoplasm during glycolysis must be transported to the mitochondria and oxidised in the respiratory chain.

97. Cytochrome C is present between Complex III and Complex IV.

98. Terminal electron donor is cytochrome a_3; it transfers electrons to oxygen.

99. First decarboxylation occur swhen oxalo-succinate changes to alpha-ketoglutaric acid, which again comes under decarboxylation to produce a four-carbon compound, succinyl-CoA.

100. Succinate dehydrogenase

101. Cytochrome C and ubiquinone act as mobile electron carriers.

102. Oxygen acts as the final electron and terminal hydrogen acceptor.

103. 34

104. UQ receives electrons from Complex I and Complex II.

105. Complete oxidation of $FADH_2$ forms 2 ATP molecules; for this, four hydrogen ions are pumped into the peri-mitochondrial space for oxidation.

106. Cytochrome a_3 contains two copper centres.

107. During chemiosmosis, hydrogen ions accumulate in the peri-mitochondrial space.

108. Phenomenon of ATP synthesis is phosphorylation and endergonic.

109. For ATP synthesis, protons passes through the F_0 complex from the peri-mitochondrial space to the matrix.

110. In ETS, 12 molecules of reduced co-enzymes (10 $NADH_2$ + 2 $FADH_2$) are oxidised, which releases 12 molecules of metabolic water.

111. Headpiece of Complex V

112. 6 ATP

113. In ETC, electron carriers are arranged in order of increasing redox potential, from higher electronegative to higher electropositive redox potential.

114. Co Q and Cyt C

115. Cytochrome C is a small protein present on the outer side of the inner membrane.

116. One oxygen molecule

117. Only (D) is correct.

118. Metabolic water is one which is produced during aerobic utilisation of glucose.

119. 1 turn of Krebs cycle produces 12 ATP (1 direct + 11 through ETS)

120. Cyt b c_1 c a a_3

121. Hydrogen atom removed during oxidation is used in the reduction of NAD^+.

122. Oxygen is used by ETC.

123. Complex IV (ATP synthetase) is concerned with oxidative phosphorylation.

124. Energy released during electron transport is used in building a proton gradient or proton motive force.

125. Eight, Once, Once

126. (B) and (D) are correct.

127. OAA

128. Citric acid – α-ketoglutaric acid – Succinic acid – Malic acid

129. CO_2

130. Acetyl-CoA (2C compound)

131. One turn of Krebs cycle comes under the two decarboxylation process.

132. Glycolysis (oxidation of glucose into pyruvic acid) is a common metabolic pathway for both aerobic and anaerobic respiration.

Respiratory Balance Sheet, Amphibolic Pathway RQ

133. (A) and (C) are correct.

134. RQ is the ratio of CO_2 produced to oxygen consumed.

135. Option (d) is correct.

136. A – Acetyl-CoA, B – Glyceraldehyde 3-phosphate

137. Option (a) is correct.

138. Acetyl-CoA

139. RQ can vary due to respiratory substrate.

140. RQ of tripalmitin is less than 1 (0.7).

141. α-ketoglutaric acid reacts with NH+4 to form glutamic acid.

142. Compound X is pyruvate.

143. $102CO_2$, $145O_2$

144. Organic acids are rich in oxygen, so they require more oxygen for their oxidation.

145. RQ value of proteins is 0.9.

146. Infinity

147. Respiration is an amphibolic pathway.

148. Succinyl-CoA is the precursor of chlorophyll.

149. In succinyl xerophyte, RQ value (at night) is zero.

150. Option (d) is correct.

Plant Growth and Development

EXERCISES

Plant Growth, Measurement of Growth, Growth Rates, Conditions for Growth, Differentiation, Dedifferentiation and Redifferentiation

1. Growth:
   A. Is defined as an irreversible permanent increase in the size of an organ or its parts or even an individual cell
   B. Is accompanied by metabolic processes, both anabolic and catabolic
   (a) Only (A) is correct
   (b) Only B is correct
   (c) Both (A) and (B) are correct
   (d) Both (A) and (B) are incorrect

2. Plant growth is:
   (a) Indeterminate (b) Open ended
   (c) Localised (d) All of the above

3. The growth pattern of a plant is different from the growth pattern of an animal in having:
   (a) Localised growth centres
   (b) Indefinite growth
   (c) Diffused growth
   (d) All except (c)

4. Growth can be measured in terms of:
   (a) Fresh weight (b) Dry weight
   (c) Area (d) All of the above

5. Sum of growth and differentiation is:
   (a) Development (b) Plasticity
   (c) Redifferentiation (d) All of the above

6. Growth at cellular level is principally a consequence of the amount of:
   (a) Protoplasm (b) Cell wall
   (c) Apoplasm (d) All except (a)

7. Read the following statements with respect to arithmetic growth:
   A. It is found in root elongation at a constant rate
   B. Cells divide in such a manner that only one daughter cell remains meristematic while the others differentiate and mature
   C. It follows a mathematical progression of 2, 4, 6, 8, 16, 32…
   D. It is mathematically expressed as $L_t = L_o + rt$

   Select the correct statements:
   (a) (A) and (C)
   (b) (A), (B) and (C)
   (c) (A), (B) and (D)
   (d) (A), (B) (C) and (D)

8. Select the correct graph from the following which represent arithmetic growth:

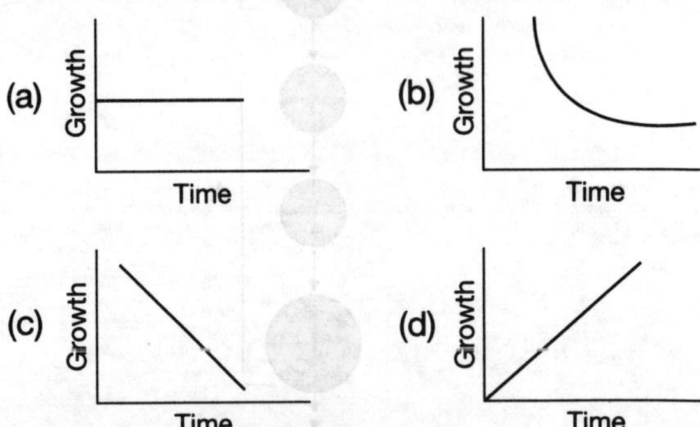

9. Mathematically, exponential growth can be expressed as:
(a) Lt = Lo + rt
(b) $W_1 = W_O e^{rt}$
(c) $W_1 = W_O - rt$
(d) $W_1 = W_O / e^{rt}$

10. Typical growth curve in plants is:
(a) Linear
(b) Parabolic
(c) Sigmoid
(d) Stair-steps shape

11. Increase in growth per unit time is called:
(a) Exponential growth
(b) Growth rate
(c) Intrinsic growth
(d) Cell elongation

12. In geometrical growth:
A. The progeny cells following mitotic division retain the ability to divide
B. The initial growth is slow, followed by the exponential phase and the stationary phase
C. Exponential growth is represented as $W_1 = W_0 e^{rt}$

Identity the correct set of statements:
(a) (A) only
(b) (B) and (C) only
(c) (A) and (C) only
(d) (A), (B) and (C)

13. Which one is the correct graph of geometric growth?

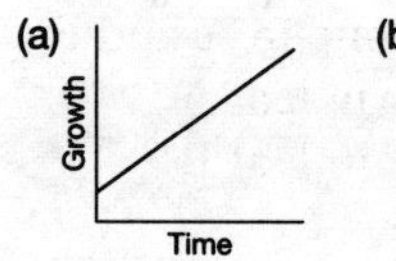

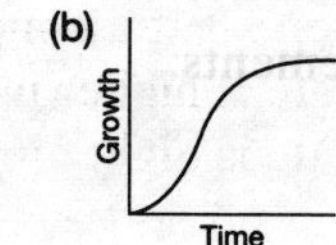

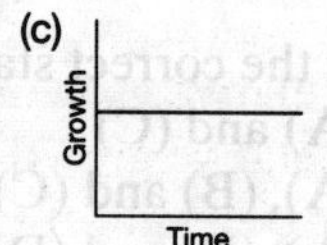

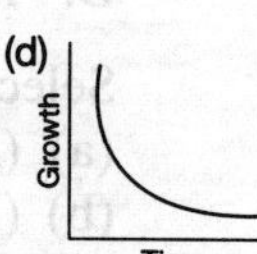

14. In the given diagram, identify the type of growth phase in (A) and (B) and choose the correct option:

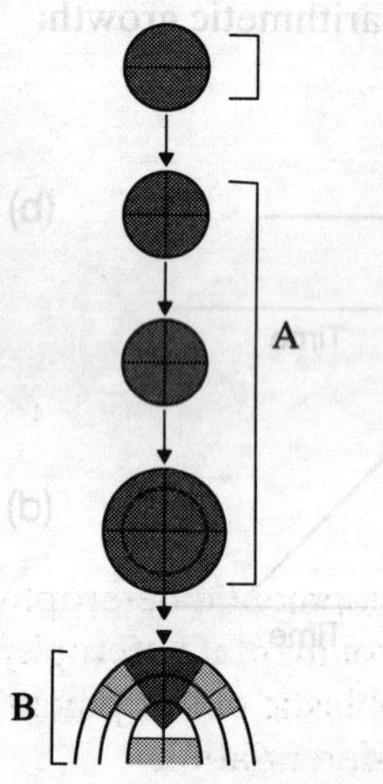

	A	B
(a)	Arithmetic phase	Geometric phase
(b)	Geometric phase	Exponential phase
(c)	Geometric phase	Arithmetic phase
(d)	Exponential phase	Geometric phase

15. One single maize root apical meristem can give rise to more than _____ new cells per hour:
(a) 3,50,000
(b) 17,500
(c) 3500
(d) 1,75,000

16. In which plant species do newly formed cells increase in size up to 3,50,000 times during its enlargement?
(a) Maize
(b) Bamboo
(c) Watermelon
(d) Cucumber

17. The growth of pollen tubes is measured in terms of:
(a) Increase in length
(b) Increase in volume
(c) Increase in surface area
(d) All of the above

18. Grand period of growth is:
(a) Lag
(b) Log
(c) Stationary
(d) Decline

19. The cells of the root apex and shoot apex:
(a) Possess dense cytoplasm and large conspicuous nucleus
(b) Possess thin cellulosic cell wall, which is primary in nature
(c) Show high respiratory rate
(d) All of the above

20. Increased vacuolation, cell enlargement and new cell wall deposition are characteristic of cells in:
(a) Formative phase
(b) Elongation phase
(c) Maturation phase
(d) Transition phase

21. Maturation phase of growth is characterised by:
(a) Presence of actively dividing cells
(b) Vacuolation and cell wall deposition
(c) Cells show structural and functional differentiation
(d) All of the above

22. For plant growth, which one(s) is(are) essential:
(a) Light
(b) Nutrients
(c) Gravity
(d) All of the above

23. During plant growth, water is needed for:
(a) Cell enlargement
(b) Providing aqueous medium for enzymatic reactions

(c) Oxydising glucose to provide energy
(d) All except (c)

24. Which of the following plays an important role in plant growth and development?
(a) Extrinsic control only
(b) Genetic control only
(c) Genetic control and PGR only
(d) Extrinsic control, genetic control and PGR

25. Select the odd one out with respect to differentiation:
(a) Secondary wall thickening in tracheids and vessels
(b) Loss of nucleus in sieve tube
(c) Loss of protoplasm in collenchyma cells
(d) Deposition of suberin in cork cells

26. Formation of interfascicular cambium and cork cambium in dicot stem is:
(a) Differentiation (b) Dedifferentiation
(c) Redifferentiation (d) Both (a) and (c)

27. Tissue culture technique, which involves formation of callus from leaf as explant, represents:
A. Growth
B. Differentiation
C. Dedifferentiation
(a) (A) and (C) (b) (A) and (B)
(c) (C) only (d) (B) only

28. Which of the following conditions is not necessary for growth at cellular level?
(a) Water (b) Nutrients
(c) Oxygen (d) Reproductive potential

29. During differentiation, the cell undergoes structural changes in its:
(a) Cell wall only (b) Protoplasm only
(c) Apoplast only (d) Cell wall and protoplast

30. The living cells lacking the capacity to divide can regain the capacity of division under some condition. This phenomenon is termed:
(a) Differentiation (b) Dedifferentiation
(c) Redifferentiation (d) Senenscence

31. Which one is an example of dedifferentiated cells?
(a) Intrafascicular cambium and cork cambium
(b) Interfascicular cambium and procambium
(c) Phellogen and phelloderm
(d) Interfascicular cambium and phellogen

32. ________ are examples of tissue formed by redifferentiation:
(a) Secondary xylem (b) Secondary phloem
(c) Phelloderm (d) All of the above

33. The ability of plants to follow different pathways and produce different structures in response to the environment and phase of life is termed:
(a) Elasticity (b) Plasticity
(c) Growth efficiency (d) Plastotron

34. Two leaves with respective growth are depicted below:

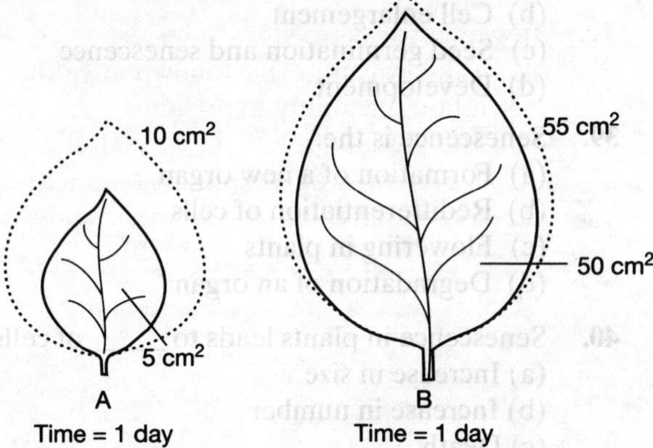

Consider A and B as the final size after growth. Which of the following statements explains the above situation appropriately:
(a) Absolute growth rate is the same in A and B
(b) Relative growth rate is higher in leaf A
(c) Relative growth rate is higher in leaf B
(d) All except (c)

35. Which of the given conditions is illustrated in the figures given below:

(a) Developmental heterophylly
(b) Environmental heterophylly
(c) Meroblastic development
(d) Dedifferentiation

36. Find the odd one out with respect to developmental heterophylly:
(a) Cotton (b) Coriander
(c) Larkspur (d) Buttercup

37. An example of plasticity in the plant *Ranunculus* is:
(a) Environmental heterophylly
(b) Developmental heterophylly
(c) Isophylly
(d) Homophylly

38. Which of the following includes all three processes?
(a) Differentiation
(b) Cell enlargement
(c) Seed germination and senescence
(d) Development

39. Senescence is the:
(a) Formation of a new organ
(b) Redifferentiation of cells
(c) Flowering in plants
(d) Degradation of an organ

40. Senescence in plants leads to _____ of cells:
(a) Increase in size
(b) Increase in number
(c) Death
(d) Differentiation

Plant Growth Regulators (The Discoveries of PGR, Physiological Effect of PGK, Auxin, Gibberellin, Cytokinin, Ethylene and Abscisic Acid)

41. Select the incorrect match:
(a) IAA – Indol compound
(b) Cytokinin – Purine derivatives
(c) Gibberellin – Precursor is tryptophan
(d) C_2H_4 – Gaseous PGR

42. Select the correct option to fill in the blanks in the following statements:
(i) The historical experiment on canary grass's coleoptile as the first step in the discovery of auxin was conducted by ___A___
(ii) ___B___ isolated auxin from coleoptile of seedlings for the first time
(iii) Auxin was first isolated from C

	A	B	C
(a)	Miller	Darwin	Coleoptile of oat
(b)	Darwin	FW Went	Human urine
(c)	Darwin	FW Went	Coleoptile of oat
(d)	FW Went	Darwin	Human urine

43. The term auxin is used for:
(a) IAA (b) IBA
(c) 2,4-D, 2,4,5 T (d) All of the above

44. The discovery of auxin has been traced back to the work of Charles Darwin and his son Francis Darwin. In their experiment they studied:
(a) Photoperiodism
(b) Phototropism
(c) Photonastic movement
(d) Photosynthesis

45. Read the following statements (A–D):
A. Promotes flowering in long-day plants under short-day conditions
B. Promotes bolting in beet and cabbage
C. Breaking seed dormancy
D. Promotes apical hook formation

Select the incorrect option with respect to the physiological role of gibberellin by selecting true (T) or false (F):

	A	B	C	D
(a)	F	F	T	T
(b)	T	F	F	T
(c)	F	T	T	F
(d)	T	T	T	F

46. Choose the incorrectly matched pair from the following:
(a) Auxin – "To grow"
(b) Gibberellins – *Gibberella fujikurai*
(c) Cytokinins – Herring sperm DNA
(d) Abscisic acid – Triple action test

47. The phenomenon of apical dominance is strengthened by:
(a) Decapitation of shoot tip
(b) Removal of leaves
(c) Auxin synthesis
(d) Cytokinin synthesis

48. The removal of shoot tips (decapitation) means loss of apical dominance and it is widely applied in:
(a) Tea plantation (b) Hedge making
(c) Bolting (d) Both (a) and (b)

49. What would be expected to happen if you cut the shoot tips of tea?
(a) Prevention of geotropism
(b) Promotion of secondary growth
(c) Growth of lateral buds
(d) Inhibition of flowering

50. Xylem differentiation is stimulated by:
(a) Auxin (b) Ethylene
(c) ABA (d) Cytokinin

51. Which includes growth promoters?
(a) Auxin, cytokinins and ABA
(b) Gibberellin, cytokinins and ethylene
(c) Auxin, gibberellins and cytokinins
(d) Ethylene, abscisic acid, salicylic acid

52. Which option includes growth inhibitors?
(a) ABA and cytocyanine
(b) GA and 1AA
(c) Inhibitor – B ethylene
(d) Ethylene, 2 4-D

53. Which of the following PGRs can fit either the growth promoter or the inhibitor?
(a) Ethylene (b) AB
(c) GAS (d) 1AA

54. Select the odd one out with respect to PGR:
(a) Inhibitor B (b) Abscission – II
(c) Dormin (d) Ethephon

55. Mark the correct option with respect to ethephon:
(a) Most widely used compound as a source of ethylene
(b) It is readily absorbed and transported within plant organs
(c) Accelerates abscission in flowers and fruits
(d) All of these are correct

56. Read the following statements:
A. Helps in stomatal closure
B. Induces seed dormancy
C. Sprouting of potato tubers
D. Decreases the tolerance of plants to various kinds of stresses

Select the incorrect statements with respect to the physiological role of ABA:
(a) (A) and (B) (b) (B), (C) and (D)
(c) (C) and (D) (d) (C) only

57. The given diagram illustrates phototropism in a plant:

Which plant hormone induces the phenomenon of photoperiodism in plants?
(a) Auxins (b) Ethylene
(c) Cytokinins (d) Gibberellin

58. Mark the incorrect statement with respect to cytokinin:
(a) Promotes adventitious shoot formation
(b) Promotes nutrient mobilisation
(c) Promotes apical dominance
(d) Produces new leaves and chloroplasts in leaves

59. Which of the following is not a function of ethylene?
A. Promotes abscission and senescence of leaves and flowers
B. Used to prepare weed-free lawns by gardeners
C. Enhances respiratory rate during fruit ripening
D. Helps to overcome the apical dominance
(a) (A) and (B) (b) (B) and (D)
(c) (C) and (D) (d) (A) and (D)

60. The PGR which is responsible for phototropism is:
(a) Auxin (b) Cytokinins
(c) Gibberellin (d) ABA

61. Which one(s) is/are acidic in nature?
(a) Auxins (b) Gibberellins
(c) ABA (d) All of the above

62 How many gibberellins have been reported from fungi and higher plants?
(a) Less than 100 (b) More than 200
(c) More than 100 (d) 100

63. Which GA was the first gibberellin to be discovered?
(a) GA1 (b) GA2
(c) GA3 (d) GA4

64. Which of the following hormones causes fruits like apple to elongate and improve its shape?
(a) GA (b) ABA
(c) NAA (d) 2, 4 D

65. The plant hormone that plays a major role in development and maturation and in inducing seed dormancy is:
(a) Cytokinin (b) ABA
(c) Auxin (d) Ethylene

66. Cytokinins are known to:
(a) Inhibit cytoplasmic movement
(b) Help in retention of chlorophyll
(c) Influence water movement
(d) Promote abscission layer formation

67. Match Column (I) and Column (II) and select the correct option:

Column I	Column II
A. C Darwin and F Darwin	(i) Gibberellin
B. Miller	(ii) Ethylene
C. Kurosawa	(iii) Auxin
D. Cousins	(iv) Cytokinin

(a) A – (iii), B – (iv), C – (i), D – (ii)
(b) A – (iii), B – (i), C – (ii), D – (iv)
(c) A – (iii), B – (ii), C – (iv), D – (i)
(d) A – (ii), B – (iii), C – (i), D – (iv)

68. Which of the following is not directly affected by auxin?
(a) Root initiation in stem cutting
(b) Parthinocarpy in tomato
(c) Used as herbicide
(d) Internodal elongation just before flowering

69. The given table lists various compounds containing plant hormones and their function in agriculture:

No	Compound	Function
(1)	2.4 D	(A)
(2)	(B)	Fruit ripening

Select the correct option to complete the table:

	A	B
(a)	Weedicide	Ethephon
(b)	Insecticide	Inhibitor – B
(c)	Weedicide	Kinetin
(d)	Insecticide	Coconut milk

70. Read the following statements and select the correct option:
A. Auxins are widely used as herbicides
B. 2.4 D is widely used to kill monocotyledonous weed, but does not affect the mature monocotyledonous plants
(a) Only (A) is correct
(b) Only (B) is correct
(c) Both (A) and (B) are correct
(d Both (A) and (B) are incorrect

71. Which of the following is not an effect of gibberellin on plants?
(a) Delay senescence
(b) Breaking dormancy
(c) Overcoming of genetic dwarfism
(d) Promote apical dominance

72. The fruits can be left on the tree longer so as to extend the market period. This is due to which function of GA?
(a) Bolting
(b) Delay senescence
(c) Internodal elongation
(d) Parthenocarpy

73. Which of the following PGRs is responsible for including horizontal growth of the seedling?
(a) Ethylene (b) ABA
(c) Auxin (d) Gibberellin

74. Ethephon in not involved in:
(a) Hastening of ripening of fruits of tomatoes and apples
(b) Thinning of cotton, cherry and walnut
(c) Increasing female flowers in cucumber
(d) Hastening the maturity period of juvenile conifers

75. Bolting is:
(a) Internodal elongation just prior to flowering in rosette plants like beet, cabbage
(b) Internodal elongation in deep water rice plants
(c) A type of grafting
(d) All except (c)

76. A: Spraying sugarcane crop with gibberellins increases the yield by as much as 20 tonnes per acre
B: Spraying juvenile conifers with cytokinins hastens the maturity period, thus leading to early seed production
(a) Only (A) is correct
(b) Only (B) is correct
(c) Both (A) and (B) are correct
(d) Both (A) and (B) are incorrect

77. Consider the statements and select the correct option:
A. Initiates germination in peanut seeds
B. Secretes hydrolyte enzyme at the time of seed germination
C. Promotes rapid internode/petiole elongation in deep water rice plant
(a) (A) and (B) are functions performed by the same PGR
(b) (A) and (C) are functions performed by ethylene
(c) (B) and (C) are effects of gibberellin
(d) (A), (B) and (C) are effects of auxin

78. Cytokinin and auxin have the opposite effect on:
(a) Leaf senescence
(b) Seed dormancy
(c) Apical dominance
(d) More than one option is correct

79. Which of the following hormones does not occur naturally in plants?
(a) Kinetin (b) Zeatin
(c) IAA (d) GA3

80. Natural cytokinins are synthesized in:
(a) Root apices (b) Developing shoot bed
(c) Young fruits (d) All of the above

81. Under what circumstances is secretion of ABA at higher concentrations?
(a) Stress (b) Differentiation
(c) Seed development (d) All except (b)

82. The term respiratory climacteric is associated with:
(a) Fruit ripening (b) Seed dormancy
(c) Seed germination (d) Malting process

83. Select the incorrect match:
(a) N6-furfurylaminopurine: Kinetin
(b) Synthetic auxin: Weed-free lawn
(c) Gaseous hydrocarbon: Synthesized by tissue undergoing senescence
(d) Terpene derivative: Root growth and root hair formation

84. Select the correct option to fill in the blanks in the following statements:
(i) (**A**) is used to speed up the process of malting in the brewing industry
(ii) (**B**) is used for synchronising fruit set in pineapples
(iii) (**C**) promotes nutrient mobilisation, which helps in the delay of leaf senescence
(iv) (**D**) is antagonistic to gibberellins

	A	B	C	D
(a)	Auxin	Cytokinin	Ethylene	Abscisic acid
(b)	Cytokinin	Gibberellins	Auxin	Ethylene
(c)	Gibberellin	Ethylene	Cytokinin	Abscisic acid
(d)	Gibberellin	Auxin	Cytokinin	Ethylene

85. A dilute solution of ______ is sprayed on litchi and pineapple, which induces flowering in them:
(a) ABA (b) GA3
(c) Ethephon (d) 2,4-D

86. Zeatin was extracted for the first time from:
(a) Maize
(b) Coconut milk
(c) Autoclaved herring sperm DNA
(d) Fungus

87. Seed dormancy is regulated by:
(a) C2H4 (b) ABA
(c) AA (d) GA3

88. Which one is a stress hormone?
(a) Abscission II (b) Inhibitor – B
(c) Dormin (d) All of the above

89. (i) "**A**" hormone helps to initiate rooting in stem cutting, an application widely used for plant propagation
(ii) '**B**' hormone promotes flowering in long-day plants under short-day conditions
(iii) "**C**" hormone helps to produce to new leaves, chloroplast in leaves and lateral shoot growth
Identify the correct name of hormones:

	A	B	C
(a)	Auxin	Giberellin	Cytokinin
(b)	Auxin	Ethylene	Cytokinin
(c)	Ethylene	Giberellin	Auxin
(d)	Giberellin	Cytokinin	Auxin

90. The most widely used PGR in agriculture is:
(a) GAS (b) 2,4-D
(c) Ethylene (d) Dormin

91. 'Richmond Lang effect' is associated with:
(a) Cytokinin (b) ABA
(c) Auxin (d) Cytokinine

92. Premature fruit fall and leaf fall is prevented by:
(a) Cytokinin (b) Gibberelin
(c) Auxin (d) Ethylene

93. All of these are growth inhibitors, except:
(a) Ethylene (b) Dormin
(c) Inhibitor-B (d) Coconut milk factor

94. Select the incorrectly matched pair:
(a) Cytokinin – Anti-ageing hormone
(b) ABA – Promote stomatal closure
(c) Auxin – Male hormone
(d) Agent orange – 2-4-D and 2,4,5-T

95. Examine the following statements:
(i) Promotes leaf senescence
(ii) Causes apical hook formation
(iii) Promotes horizontal growth of seedling
(iv) Swelling of stem axis
The above functions are carried out by:
(a) GA (b) C_2H_4
(c) Auxin (d) ABA

96. The hormone produced during adverse environmental conditions is:
(a) Abscisic acid (b) Gibberellin
(c) Cytokinin (d) 2,4-D

Photoperiodism, Vernalisation

97. The phenomenon of photoperiodism in plants was discovered by:
(a) Garner and Allard
(b) Borthwick
(c) Lysenko
(d) Blackman and Skoog

98. Pigment involved in photoperception in flowering is:
(a) Cytochrome (b) Phytochrome
(c) Carotene (d) Lycopene

99. Choose the correct option for the given figures with respect to flowering:

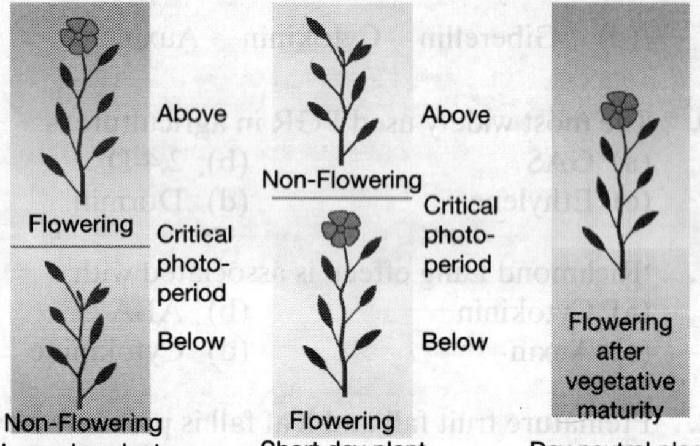

	A	B	C
(a)	LDP	SDP	DNP
(b)	LDP	DNP	SDP
(c)	SDP	LDP	DNP
(d)	SDP	DNP	LDP

100. For flowering, the critical dark period should always be exceeded in:
(a) Long-day plants
(b) Short-day plants
(c) Day-neutral plants
(d) Long-short-day plants

101. Wheat, potato, and sugar beets are:
(a) SDP (b) LDP
(c) DNP (d) LNP

102. The photoperiodic stimulus perceived by ___A___ migrates to ___B___ for inducing flowers:

	A	B
(a)	Shoot tip	Bud
(b)	Meristem	Leaves
(c)	Leaves	Shoot apex
(d)	Bud	Leaves

103. Phytochrome pigment is not involved in which of the following processes?
(a) Flowering
(b) Photomorphogenesis
(c) Seed germination
(d) Vernalisation

104. Interruption of dark period by a flash of light prevents flowering in:
(a) Short-day plants (b) Long-day plants
(c) Day-neutral plants (d) Both (b) and (c)

105. For short-day plants, the critical period is:
(a) Light (b) Dark
(c) UV ray (d) (a) and (c)

106. Choose the odd one out with respect to the photoperiod for flowering:
(a) Soya bean (b) Henbane
(c) Wheat (d) Sugar beet

107. Flowering in short-day plants is induced by:
(a) Short days and interrupted long nights
(b) Short days and uninterrupted long nights
(c) Short nights
(d) Long days with interrupted nights

108. A chemical believed to be involved in flowering is:
(a) ABA (b) Auxin
(c) Cytokinin (d) Florigen

109. Vernalisation is:
(a) Growth curve related to light
(b) Effect of duration of light on flowering
(c) Speeding up ability to flower by low temperature treatment
(d) Diurnal photoperiodicity

110. Chilling treatment prior to flowering:
(a) Induces early flowering
(b) Is essential for winter animals and biennials
(c) Promotes reproductive growth
(d) All of the above

111. Promotion of flowering by a period of low temperature is seen in biennial plants like:
(a) Sugar beet (b) Cabbage
(c) Carrots (d) All of the above

112. The stimulus of vernalisation is perceived by:
(a) Roots (b) Embryo
(c) Mature leaf (d) Flowers

113. Vernalisation can be replaced by:
(a) Gibberellin treatment
(b) Auxin treatment
(c) Cytokinin treatment
(d) Ethylene treatment

114. Most appropriate temperature for vernalisation is:
(a) 1–6°C (b) 0°C
(c) –6°C (d) 7°C

115. Sowing of winter variety of wheat in spring season after vernalisation:
(a) Helps in shortening the vegetative period of the plant and brings about early flowering
(b) Does not enable winter rice to initiate flowering and production of mature grains within the span of the flowering season
(c) Promotes delayed reproducton in the growing season
(d) Both (a) and (b)

116. Biennials are changed into annuals by:
(a) Grafting (b) Yarovisation
(c) Vernalisation (d) Both (b) and (c)

117. The diagram shows the stages of seed germination:

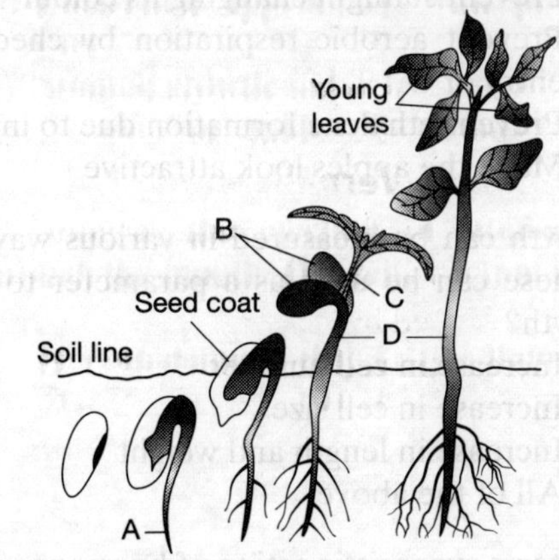

Identify A, B, C and D:

	A	B	C	D
(a)	Plumule	Cotyledons	Hypocotyl	Epicotyl
(b)	Radicle	Cotyledons	Epicotyle	Hypocotyl
(c)	Mesocotyl	Cotyledons	Hypocotyl	Epicotyl
(d)	Root hair	Cotyledons	Epicotyle	Mesocotyl

118. Steps involved in causing seed germination are:
A. Emergence of radicle
B. Formation of hypocotyl hook
C. Imbibition of water
D. Hypocotyl straightens
(a) A→B→C→D (b) A→B→C→D
(c) C→A→D→B (d) C→A→B→D

NCERT EXEMPLAR QUESTIONS

1. Ethylene is use for:
(a) Retarding ripening of tomatoes
(b) Hastening ripening of fruits
(c) Slowing down ripening of apples
(d) Both (b) and (c)

2. Coconut water contains:
(a) ABA (b) Auxin
(c) Gibberellin (d) Cytokinin

3. The effect of apical dominance can be overcome by which of the following hormones?
(a) IAA (b) Ethylene
(c) Gibberellin (d) Cytokinin

4. Match the following:

Column I	Column II
A. IAA	(i) Herring sperm DNA
B. ABA	(ii) Bolting
C. Ethylene	(iii) Stomatal closure
D. GA	(iv) Weed-free lawns
E. Cytokinins	(v) Ripening of fruits

(a) A – (iv), B – (iii), C – (v), D – (ii), E – (i)
(b) A – (v), B – (iii), C – (iv), D – (ii), E – (i)
(c) A – (iv), B – (i), C – (iv), D – (iii), E – (ii)
(d) A – (v), B – (iii), C – (ii), D – (i), E – (iv)

5. Apples are generally wrapped in waxed paper to:
(a) Prevent sunlight changing its colour
(b) Prevent aerobic respiration by checking the entry of O_2
(c) Prevent ethylene formation due to injury
(d) Make the apples look attractive

6. Growth can be measured in various ways. Which of these can be used as a parameter to measure growth?
(a) Increase in cell number
(b) Increase in cell size
(c) Increase in length and weight
(d) All of the above

7. The term synergistic action of hormones refers to:
(a) When two hormones act together but bring about opposite effects
(b) When two hormones act together and contribute to the same function
(c) When one hormone affects more than one function
(d) When many hormones bring about any one function

8. Plasticity in plant growth means that:
(a) Plant roots are extensible
(b) Plant development is dependent on the environment
(c) Stems can extend
(d) None of the above

9. To increase sugar production in sugarcanes, they are sprayed with:
(a) IAA (b) Cytokinin
(c) Gibberellin (d) Ethylene

10. ABA is antagonistic to:
(a) Ethylene (b) Cytokinin
(c) Gibberellic acid (d) IAA

11. Monocarpic plants are those which:
(a) Bear flowers with one ovary
(b) Flower once and die
(c) Bear only one flower
(d) All of the above

12. The photoperiod in plants is perceived at:
(a) Meristem (b) Flower
(c) Floral buds (d) Leaves

PRACTICE TEST PAPER

1. All of the following statements are concerned with plant growth, except:
(a) Localised (b) Open ended
(c) Determinate (d) Quantitative

2. All the changes that an organism goes through during its life cycle, from germination of the seed to senescence, is called:
(a) Development (b) Differentiation
(c) Plasticity (d) Senescence

3. In some plants, the leaves of the juvenile plant are different in shape to those in mature plants. This is heterophylly due to phases of life and seen in:
(a) Cotton, Coriander and Buttercup
(b) Buttercup and Larkspur
(c) Buttercup only
(d) Cotton, Coriander and Larkspur

4. The Avena curvature test is used for bioassay of:
(a) ABA (b) GA3
(c) Auxin (d) Cytokinin

5. Phototropic curvature is the result of uneven distribution of:
(a) Auxin (b) Gibberellin
(c) Ethylene (d) GA3

6. Most important biological effect of cytokinin is:
(a) Overcomes apical dominance
(b) Delays senescence
(c) Tissue culture
(d) Rapid cell division

7. Demand of low temperature treatment and requirement of red light can be substituted by:
(a) Cytokinin (b) GA
(c) ABA (d) Auxin

8. Which of the following is not an effect of ABA?
(a) Development of seed
(b) Promote seed dormancy
(c) Stomatal closure
(d) Delay in senescence

9. Match Column I and Column II and select the correct option:

Column I (Phytohormone)	Column II (Function)
A. Auxin	(i) Cytokinesis
B. Gibberellin	(ii) Fruit ripening
C. Cytokinin	(iii) Root initiation
D. Ethylene	(iv) Bolting hormone

(a) A – (iii), B – (iv), C – (i), D – (ii)
(b) A – (iii), B – (i), C – (iv), D – (ii)
(c) A – (ii), B – (i), C – (iii), D – (iv)
(d) A – (ii), B – (i), C – (iv), D – (iii)

10. Richmond Lang effect due to cytokinin pertains to:
(a) Root formation
(b) Morphogenesis
(c) Delay in senescence
(d) Apical dominance

11. Which accelerates abscission in flowers and fruit, like thinning of cotton, cherry and walnut?
(a) Auxin (b) Ethylene
(c) GA3 (d) ABA

12. Cell enlargement occurs due to:
(a) Florigen (b) Gibberellin
(c) Kinetin (d) Auxin

13. Plants which require short photoperiod for flowering are called:
(a) Long-day plants (b) Short-day plants
(c) Day-neutral plants (d) Both (b) and (c)

14. The height of sugarcane can be increased by the application of:
(a) IAA (b) 2 4-D
(c) Zeatin (d) Gibberellins

15. The form of growth wherein new cells are always being added to the plant body by the activity of the meristem is called:
(a) Open form of growth
(b) Closed form of growth
(c) Determinate growth
(d) Both (a) and (c)

16. Certain plants need to be exposed to low temperature to hasten flowering later in life. This treatment is known as:
(a) Stratification (b) Scarification
(c) Vernalisation (d) Photoperiodism

17. Pigment taking part in photoperception in flowering plants is:
(a) Carotene (b) Phytochrome
(c) Cytochrome (d) Lylopene

18. Natural phytohormone isolated from corn kernels and coconut milk is:
(a) Free auxin (b) Zeatin
(c) GA3 (d) Dormin

19. Which of the PGRs induces parthenocarpy in tomatoes?
(a) Auxin (b) Gibberellin
(c) Cytokinin (d) Ethylene

20. The usual shape of the growth curve is:
(a) Linear
(b) Inverted bell-shaped
(c) Sigmoid
(d) Zig-zag

21. Read the statements A, B, C and D and select the correct option with respect to gibberellins:
A. They are responsible for immature falling of leaves
B. Promote internodal elongation just prior to flowering
C. Increase sugarcane yield
D. Induce seed germination
(a) (A), (B) and (C) are correct
(b) (A), (B) and (D) are correct
(c) All are correct except (A)
(d) (A), (B) and (D) are correct

22. Cytokinesis promoting active substances obtained from autoclaved herring sperm DNA is a derivative of:
(a) Purine (b) Terpenes
(c) Indol compound (d) Pyrimidine

23. Find the incorrect match with respect to PGRs:
(a) Gibberellin – Malting process
(b) Kinetin – Overcoming apical dominance
(c) Ethylene – Apical hook formation in dicot seedling
(d) Inhibitor B – Flowering in pineapple

24. Hormonal substances responsible for inducing flowering migrates from:
(a) Shoot apex to bud
(b) Leaves to root
(c) Leaves to floral bud
(d) Root to shoot apex

25. The PGR involved in growth promoting activities are:
A. Auxin
B. Gibberellins
C. Abscisic acid
D. Cytokinin
(a) (A) and (B) only
(b) (A), (B) and (D) only
(c) (A), (B), (D) and (E)
(d) (C) and (D) only

26. Development is controlled by:
(a) Growth and differentiation
(b) Intrinsic and extrinsic factor
(c) Intracellular and intercellular factor
(d) More than one option is correct

27. One of the plant hormones which is removed along with apical bud of tea plants is:
(a) Ethylene (b) Auxin
(c) Gibberellin (d) Cytokinin

28. A synthetic auxin ____ is widely used as weedicide in cereal crop:
(a) 2,4 D (b) IAA
(c) IBA (d) NAA

29. Treatment of a plant with cytokinin enhances:
(a) Ageing of leaves
(b) Parthenocarpy in all plants
(c) Chloroplast production in leaves
(d) Apical dominance

30. The condition of viable seeds when they do not germinate even in the presence of favourable environmental conditions is called:
(a) Quiescence (b) Dormancy
(c) Vernalisation (d) Scarification

31. The plant hormone which is a derivative of carotenoids is:
(a) ABA (b) IAA
(c) $CH_2 = CH_2$ (d) Cytokinin

32. All of the following are formed from dedifferentiation of parenchyma cells, except:
(a) Wound meristem
(b) Phellogen
(c) Interfascicular cambium
(d) Intrafascicular cambium

33. Which of the following hormones plays an important role in seed development, maturation and dormancy?
(a) ABA (b) GA3
(c) Auxins (d) Cytokinins

34. Growth of pollen tube is measured in terms of:
(a) Length (b) Surface area
(c) Number of cells (d) Cell size

35. Root elongation at a constant rate shows:
(a) Geometric growth
(b) Exponential growth
(c) Arithmetic growth
(d) Determinate growth

36. Auxin and cytokinin are antagonistic in the feature:
(a) Cell division (b) Apical dominance
(c) Tropic movements (d) Both (a) and (c)

37. Sum of growth and differentiation is known as:
(a) Plasticity (b) Heterophylly
(c) Differentiation (d) Development

38. The site of perception of light/dark duration is the:
(a) Roots
(b) Stems
(c) Flowers
(d) Leaves

39. The work of Charles and Francis Darwin relates to the discovery of auxin. In their experiments, they studied:
(a) Photoperiodism (b) Phototropism
(c) Photonasty (d) Photosynthesis

40. The fruits can be left on the tree longer so as to extend the market period. It is due to which function of cytokinin?
(a) Bolting
(b) Delay in senescence
(c) Internodal elongation
(d) Parthenocarpy

41. Read the following statements and state whether true (T) or false (F):
A. GA3 is used to speed up the malting process in the brewing industry
B. Ethylene promotes root growth and root hair formation, thus increasing their absorption surface
C. Auxin is widely applied in tea plantations and hedge-making
D. During differentiation, sieve cells degenerate their nucleus and protoplasm

	A	B	C	D
(a)	T	T	T	F
(b)	T	T	F	F
(c)	F	F	T	T
(d)	T	F	F	T

42. During seed germination, the stored food is mobilised by:
(a) Cytokinin (b) ABA
(c) Gibberellin (d) Ethylene

43. Which of the following PGRs generally acts as an antagonist to gibberellins?
(a) Zeatin (b) Ethylene
(c) ABA (d) IAA

44. During ripening of fruits, rate of respiration is increased under the influence of hormones. It is called climacteric respiration. The hormone is:
(a) ABA (b) Auxin
(c) C_2H_4 (d) GA

45. Which of the following PGRs promotes stomatal closure in response to water stress?
(a) ABA (b) GA3
(c) IAA (d) C_2H_4

ANSWER KEYS

Exercises

1. (c) **2.** (d) **3.** (d) **4.** (d) **5.** (a) **6.** (a) **7.** (c) **8.** (d) **9.** (b) **10.** (c)
11. (b) **12.** (d) **13.** (b) **14.** (c) **15.** (b) **16.** (c) **17.** (a) **18.** (b) **19.** (d) **20.** (b)
21. (c) **22.** (d) **23.** (d) **24.** (d) **25.** (c) **26.** (b) **27.** (a) **28.** (d) **29.** (d) **30.** (b)
31. (d) **32.** (d) **33.** (b) **34.** (d) **35.** (a) **36.** (d) **37.** (a) **38.** (d) **39.** (d) **40.** (c)
41. (c) **42.** (b) **43.** (d) **44.** (b) **45.** (d) **46.** (d) **47.** (c) **48.** (d) **49.** (c) **50.** (a)
51. (c) **52.** (c) **53.** (a) **54.** (d) **55.** (d) **56.** (c) **57.** (a) **58.** (c) **59.** (b) **60.** (a)
61. (d) **62.** (c) **63.** (c) **64.** (a) **65.** (b) **66.** (b) **67.** (a) **68.** (d) **69.** (a) **70.** (a)
71. (d) **72.** (b) **73.** (a) **74.** (d) **75.** (a) **76.** (a) **77.** (b) **78.** (c) **79.** (a) **80.** (d)
81. (d) **82.** (a) **83.** (d) **84.** (d) **85.** (d) **86.** (a) **87.** (b) **88.** (d) **89.** (c) **90.** (c)
91. (a) **92.** (c) **93.** (d) **94.** (c) **95.** (b) **96.** (a) **97.** (a) **98.** (b) **99.** (a) **100.** (b)
101. (b) **102.** (c) **103.** (d) **104.** (a) **105.** (b) **106.** (a) **107.** (b) **108.** (d) **109.** (c) **110.** (d)
111. (d) **112.** (b) **113.** (a) **114.** (c) **115.** (a) **116.** (d) **117.** (b) **118.** (d)

NCERT Exemplar Questions

1. (b) **2.** (c) **3.** (d) **4.** (a) **5.** (b) **6.** (d) **7.** (b) **8.** (b) **9.** (c) **10.** (c)
11. (b) **12.** (d)

Practice Test Paper

1. (c) **2.** (a) **3.** (d) **4.** (c) **5.** (a) **6.** (d) **7.** (b) **8.** (d) **9.** (a) **10.** (c)
11. (b) **12.** (d) **13.** (b) **14.** (d) **15.** (a) **16.** (c) **17.** (b) **18.** (b) **19.** (a) **20.** (c)
21. (c) **22.** (a) **23.** (d) **24.** (c) **25.** (b) **26.** (d) **27.** (b) **28.** (a) **29.** (c) **30.** (a)
31. (a) **32.** (d) **33.** (a) **34.** (a) **35.** (c) **36.** (b) **37.** (d) **38.** (d) **39.** (b) **40.** (b)
41. (a) **42.** (c) **43.** (c) **44.** (c) **45.** (a)

SOLUTIONS

Plant Growth, Measurement of Growth, Growth Rates, Conditions for Growth, Differentiation, Dedifferentiation and Redifferentiation

1. Both (A) and (B) are correct.
2. Plant growth is localised, open-ended and intermediate.
3. Plant growth is unique because plants retain the capacity for unlimited growth throughout their cycle. This ability of plants is due to the presence of meristem at certain locations in their body.
4. Growth can be measured in term of increase in fresh weight, dry weight, area, etc.
5. Development includes growth and dedifferentiation.
6. Synthesis of new protoplasm is real growth.
7. (A), (B) and (D) are correct. Arithmetic growth follows mathematical progression of 1, 2, 3, 4, 5...
8. Option (d) is correct.
9. Exponential growth can be expressed as $W_1 = W_O e^{rt}$.
10. Typical growth curve in plants is sigmoid.
11. Increase in growth percentage area is called growth rate.
12. (A), (B) and (C) are correct.
13. Option (b) is correct.
14. A – Geometric phase, B – Arithmetic phase
15. One single maize root apical meristem can give rise to 1,76,500 new cells per hour.
16. Cells in a watermelon may increase in size by up to 3,50,000 times.
17. The growth of pollen tube is measured in terms of its length.
18. Grand period of growth is log phase.
19. All the statements are correct.
20. Elongation phase
21. Cells of maturation phase show structural and functional differentiation.
22. Option (d) is correct.
23. Water is needed for cell enlargement and for providing aqueous medium for enzymatic reactions.
24. Extrinsic control, genetic control and PGR
25. Collenchyma are living mechanical tissue and they are characterised by vacuolated protoplast.
26. Interfascicular cambium and cork cambium developed through dedifferentiation.
27. Cells of explant in artificial synthetic media show growth and dedifferentiation.
28. Reproductive potential
29. During differentiation, the cell undergoes structural changes in the cell wall and protoplast.
30. Dedifferentiation
31. Interfascicular cambium and phellogen (cork cambium)
32. Option (d) is correct.
33. Plants follow different pathways in response to the environment or phases of life to form different kinds of structures.
34. Relative growth rate is higher in leaf A.
35. The figure represents developmental heterophily.
36. Buttercup shows environmental heterophily.
37. Environmental heterophily
38. Development
39. Senescence is degradation of organs.
40. Senescence in plants leads to death of cells.

Plant Growth Regulators (The Discoveries of PGR, Physiological Effect of PGK, Auxin, Gibberellin, Cytokinin, Ethylene and Abscisic Acid)

41. Gibberellins are terpene derivatives.
42. A – Darwin, B – FW Went, C – Human urine
43. IAA, IBA, 24-D, 2,4,5-T all are auxins.
44. Charles Darwin and his son Francis Darwin observed that the coleoptiles of canary grass responded to unilateral illumination by growing towards the light source (photorespiration).
45. (D) is false. Apical hook formation is promoted by ethylene.
46. Triple action test is the bioassay of ethylene.
47. Auxin promotes apical dominance (in the presence of apical bud, growth of lateral buds is inhibited due to auxin secreted by apical bud).
48. Decapitation is widely applied in plantation and hedge making.
49. Removal of apical bud promotes sprouting of lateral buds.
50. Xylem differentiation is promoted by auxin.
51. Growth promoters – Auxin, Gibberellins and Cytokinin
52. Inhibitor B (Abscisic acid) is a growth inhibitor.
53. Ethylene has few growth promoting activities but largely it is a growth inhibitor.
54. Inhibitor B, dormin and abscission II are abscisic acids, while ethephon is a commercial compound which is used as a source of ethylene.
55. All the statements are correct.
56. C and D are incorrect. Ethylene promotes sprouting of potato tubers. ABA increases the tolerance of plants to various kinds of stresses.
57. Auxin is responsible for phototropism in plants.
58. Cytokinin counteracts the apical dominance of auxin.
59. (B) and (D) are not functions of ethylene. Auxin is used to prepare weed-free lawns by gardeners and cytokinin helps to overcome the apical dominance.
60. Auxin controls tropic movement.
61. Auxins, gibberellins and ABA are active in nature.
62. There are more than 100 gibberellins reported from widely different organisms such as fungi and higher plants.
63. Gibberellic acid (GA_3) was one of the first gibberellins to be discovered and remains the most intensively studied form.
64. Gibberellins cause fruits like apple to elongate and improve shape.
65. ABA plays a major role in seed development.
66. Cytokinin helps to produce new leaves and chloroplasts in leaves.
67. Charles Darwin and F Darwin – Auxin, Miller – Cytokinin, Kurosawa – Gibberellin, Cousins – Ethylene
68. Gibberellin promotes internodal elongation just prior to flowering (bolting).
69. A – Weedicide, B – Ethephon
70. Only (A) is correct. 2-4-D is widely used to kill dicotyledonous weeds, but does not affect the mature monocotyledonous plants.
71. Auxin promotes apical dominance.
72. Delay in senescence
73. Ethylene inhibits stem elongation but stimulates horizonal growth.
74. Spraying juvenile conifers with GAs hastens the maturity period, leading to early seed production.
75. Internodal elongation just prior to flowering in rosette plants like beet and cabbage
76. Only (A) is correct.
77. (A) and (C) are functions of ethylene and (B) is a function of gibberellin.
78. Cytokinin and auxins act antagonistically in the control of apical dominance.

79. Kinetin is not naturally present in plants. Miller separated it from herring sperm DNA.

80. Natural cytokinin is synthesized in regions where rapid cell division occurs.

81. Stress and seed development

82. Ethylene enhances the respiration rate during ripening of fruits; this rise is called respiratory climacteric.

83. Ethylene promotes growth of roots and root hair formation.

84. A – Gibberellin, B – Ethylene, C – Cytokinin, D – Abscisic acid

85. Auxins (2,4-D) promotes flowering in litchi and pineapple.

86. The first naturally occurring cytokinin to be chemically identified was from young maize (*Zea mays*) in 1963 and was called zeatin.

87. ABA induces bud and seed dormancy.

88. ABA is a stress hormone and helps the plant to cope with adverse environmental conditions.

89. A – Auxin, B – Gibberellin, C – Cytokinin

90. Ethylene is the most widely used PGR in agriculture.

91. Richmond Lang effect (delay in senescence) is associated with ethylene.

92. Premature leaf fall and fruit fall is presented by auxin.

93. Coconut milk factor (cytokinin) is a growth promotor.

94. Gibberellin is a male hormone that promotes the production of male flowers in female plants of cannabis.

95. All the functions are related to ethylene.

96. Abscisic acid

Photoperiodism, Vernalisation

97. Photoperiodism was discovered by Garner and Allard.

98. Phytochrome pigment

99. A – LDP, B – SDP, C – DNP

100. Short-day plants are also known as long-night plants because the requirement of darkness is above a critical length.

101. Wheat, potato and sugar beets are long-day plants.

102. A – Leaves, B – Short apex

103. Vernalisation

104. Interruption of dark period by a flash of light presents flowering in STP.

105. Darkness

106. Henbane, sugar beet and wheat are long-day plants while soya bean is a short-day plant.

107. In short-day plants, flowering is induced by short days and uninterrupted long nights.

108. When a proper photoperiod is perceived by leaves, a flowering hormone florigen is synthesized in the leaf and is transported to the bud, where flowering occurs.

109. Vernalisation is speeding up the ability to flower by low temperature treatment.

110. All the statements are correct.

111. Sugar beet, Cabbage, Carrot

112. The stimulus of vernalisation is perceived by the embryo.

113. Gibberellin treatment

114. –6°C

115. It helps in shortening the vegetative phase of the plant and brings about early flowering.

116. Vernalisation (Yarovisation)

117. A – Radicle, B – Cotyledons, C – Epicotyl, D – Hypocotyl

118. C – A – B – D

16 Digestion and Absorption

EXERCISES

Digestive System, Digestive Glands

1. Which of the following teeth are absent in milk or deciduous set of teeth?
(a) Incisors (b) Canines
(c) Premolars (d) Molars

2. The gastro-esophageal sphincter is also called:
(a) Pyloric sphincter
(b) Cardiac sphincter
(c) Sphincter of Oddi
(d) Sphincter of Boyden

3. Choose the odd one out with respect to human dentition:
(a) Thecodont, heterodont
(b) Total of 20 teeth in deciduous set
(c) Diphydont
(d) Dental formula 2123/2122

4. Which of the following processes does not occur in the buccal cavity?
(a) Breakdown of starch into maltose
(b) Mastication of food
(c) Mixing of food with saliva
(d) Absorption of maltose

5. Papillae are small projections present on the upper surface of the tongue and may also have taste buds. Which of the following papillae are the largest and present on the posterior surface of the tongue in humans?
(a) Foliate (b) Filiform
(c) Fungiform (d) Vallate

6. Identify the incorrect statement:
(a) Sphincter of Oddi guards the opening of the hepatopancreatic duct into the duodenum
(b) The primary dentition of humans differs from permanent dentition in not having only molars
(c) Vermiform appendix arises from the caecum
(d) Frenulum attaches the tongue to the floor of the buccal cavity

7. The backflow of faecal matter from the caecum into the small intestine is prevented by:
(a) Cardiac sphincter
(b) Ileo-caecal valve
(c) Pyloric sphincter
(d) Sphincter of Oddi

8. Brunner's glands are located in which layer of the wall of the small intestine?
(a) Serosa (b) Muscularis
(c) Submucosa (d) Both (b) and (c)

9. Match the following structures with their respective location in organs:

Column I	Column II
A. Crypts of Lieberkuhn	(i) Liver
B. Glisson's capsule	(ii) Small intestine
C. Islets of Langerhans	(iii) Duodenum
D. Brunner's glands	(iv) Pancreas

Choose the correct option:
(a) A – (iv), (B) – (ii), C – (i), D – (iii)
(b) A – (i), B – (iii), C – (i), D – (ii)
(c) A – (iii), B – (iv), C – (ii), D – (i)
(d) A – (ii), B – (i), C – (iv), D – (iii)

10. The structural and functional unit of the mammalian liver is:
(a) Hepatocyte (b) Glisson's capsule
(c) Hepatic lobule (d) Portal triad

11. Select the correct option with respect to the tongue:
(a) All papillae on the tongue bear taste buds
(b) Only the lower surface of the tongue has papillae
(c) It is attached to the roof of the buccal cavity by the frenulum
(d) It is a freely movable muscular organ

12. The gastrointestinal tract has the Meissner's plexus and the Auerbach's plexus in ——— and ——, respectively
(a) Submucosa and mucosa
(b) Mucosa and muscularis externa
(c) Mucosa and submucosa
(d) Submucosa and muscularis externa

13. Which of the following is an incorrect statement?
(a) The distal part of the stomach that opens into the duodenum is called the pylorus
(b) Glisson's capsule is a characteristic feature of the mammalian liver
(c) Maximum salivary secretion is contributed by the parotid glands
(d) Salivary glands are situated just outside the buccal cavity and secrete salivary juice into the buccal cavity

14. ——— layer present in the wall of the alimentary canal forms irregular folds (rugae) in the stomach and small finger-like folds called villi in the small intestine:
(a) Serosa (b) Muscularis externa
(c) Submucosa (d) Mucosa

15. ——— has an additional oblique muscle layer in the alimentary canal, besides the inner circular and outer longitudinal layer of smooth muscles in muscularis:
(a) Duodenum (b) Stomach
(c) Colon (d) Rectum

16. A compound gland which secretes hormones to regulate blood sugar level and enzymes that facilitate digestion is:
(a) Liver (b) Pancreas
(c) Adrenal gland (d) Duodenum

17. Select the incorrect match:

(a) Epiglottis	–	Guards the opening of the windpipe
(b) Pyloric sphincter		Guards the opening of the stomach into the duodenum
(c) Caecum	–	Hosts symbiotic bacteria
(d) Ascending colon	–	Opens into the rectum

18. The large intestine of humans has three parts, namely:
(a) Appendix, caecum and rectum
(b) Caecum, colon and rectum
(c) Jejunum, colon and ileum
(d) Caecum, rectum and duodenum

19. Auerbach's plexus is present:
(a) Between mucosa and submucosa
(b) In submucosa
(c) In serosa
(d) Between the longitudinal and circular muscles in muscularis

20. Choose the correct option:

(a) Cystic duct	–	Pancreas
(b) Crypts of Lieberkuhn	–	Large intestine
(c) Serosa	–	Made of thin mesothelium and some connective tissue
(d) Diphyodont	–	Teeth fitted in sockets of jaw

21. Read the following statements and choose the correct option:
I: A narrow finger-like tubular projection, the vermiform appendix, which is a vestigial organ, arises from the caecum
II: Esophagus is a thin long tube which extends posteriorly, passing through the neck, thorax and diaphragm and leading to a 'J'-shaped bag-like structure called the stomach
III: The oral cavity leads into a short larynx which serves as a common passage for food and air
(a) All the statements are correct
(b) I is correct, while II and III are incorrect
(c) I and II are correct while III is incorrect
(d) I and III are correct while II is incorrect

22. The common bile duct in humans is formed by the joining of:
(a) Hepatic duct and pancreatic duct
(b) Cystic duct and pancreatic duct
(c) Cystic duct and hepatic duct
(d) Pancreatic duct and duct of Wirsung

23. ——— duct is associated with the parotid glands:
(a) Stenson's (b) Wharton's
(c) Rivinus (d) Wirsung's

24. Read the following statements and select the correct option:
A. The intestinal mucosa has goblet cells
B. Submandibular salivary glands are also called submaxillary glands
(a) Both (A) and (B) are correct
(b) Both (A) and (B) are incorrect

(c) Only statement (A) is correct
(d) Only statement (B) is correct

25. The ——— of the colon are small pouches caused by sacculation, which give the colon its segmented appearance:
(a) Rugae (b) Haustra
(c) Crypts of Lieberkuhn (d) Fundus

26. The number of teeth in the upper jaw of a 14-year-old child is:
(a) 32 (b) 16
(c) 28 (d) 14

27. Identify the incorrect statement with respect to the liver:
(a) It is the largest gland of the body, weighing about 1.2 to 1.5 kg in adult humans
(b) It is situated in the abdominal cavity, just below the diaphragm
(c) It has two lobes
(d) The hepatic lobule has hepatic cells arranged in the form of rings

28. Which of the following is the correct location of the sphincter muscles of Boyden?
(a) Cystic duct, before combining with hepatic duct
(b) Bile duct, before combining with pancreatic duct
(c) Pancreatic duct, before it opens into the duodenum
(d) Hepatopancreatic duct, at its junction with the duodenum

29. Choose the incorrect match:

Salivary glands		Location
(a) Sub-mandibular	–	Near lower jaw
(b) Parotids	–	Near cheek
(c) Sub-lingual	–	Below tongue
(d) Sub-maxillary	–	Near cheek

30. Which of the following two structures have the same location?
(a) Gastric rugae and Auerbach's plexus
(b) Submucosal plexus and Auerbach's plexus
(c) Meissner's plexus and Brunner's glands
(d) Wharton's duct and duct of Wirsung

31. Read the following statements:
A. Esophagus opens into the fundic part of the stomach through a gastro-esophageal sphincter
B. The hepatic cells are present in the form of cords in the hepatic lobule
C. Tongue is attached to the floor of the mouth by the frenulum
D. Twelve teeth in human dentition are mono phyodont
E. Kupffer cells are phagocytic cells found in the wall of the stomach
State which are true and which are false:

	A	B	C	D	E
(a)	T	T	F	F	T
(b)	T	F	T	F	T
(c)	F	T	T	T	F
(d)	F	T	T	T	T

32. Select the mismatched pair:

Part of alimentary canal	Structures present/ function
(a) Small intestine	Lacteals
(b) Colon	Water absorption
(c) Gall bladder	Formation of bile
(d) Intestinal mucosal epithelium	Crypts of Lieberkuhn

33. Which one of the following is not a function of the liver?
(a) Detoxification of drugs and toxins
(b) Haemopoiesis in the embryonic stage
(c) Synthesis of bile juice
(d) Synthesis of secretin

34. Peyer's patches (abdominal tonsils) are lymphatic nodules present in the ——— of the alimentary canal:
(a) Submucosa (b) Mucosa
(c) Muscularis (d) Serosa

35. Achalasia cardia is the accumulation of food in the esophagus due to:
(a) Improper functioning of pyloric sphincter
(b) Improper functioning of cardiac sphincter
(c) Mechanical churning of food in stomach
(d) Improper functioning of ileo-caecal valve

36. Identify the incorrect statement:
(a) Small intestine is the longest part of the alimentary canal
(b) Parotid glands are present beneath and in front of the ears, and release secretions into the buccal cavity
(c) Swallowing is also called deglutition
(d) Mucosa is the innermost layer of the alimentary canal and forms villi in the stomach

37. The brush-border appearance of intestinal mucosa is due to:
(a) Lacteals (b) Villi
(c) Microvilli (d) Crypts of Liberkuhn

38. The long, coiled, middle part of the small intestine in humans is called:
(a) Ileum (b) Colon
(c) Duodenum (d) Jejunum

39. Select the option indicating the correct match:

Column I	Column II
A. Rectum	(i) Taste buds
B. Papillae	(ii) Anus
C. Vermiform appendix	(iii) Caecum
D. Glottis	(iv) Opening of wind pipe

(a) A – (ii), B – (i), C – (iii), D – (iv)
(b) A – (ii), B – (i), C – (iv), D – (iii)
(c) A – (iv), B – (ii), C – (iii), D – (i)
(d) A – (iii), B – (i), C – (iv), D – (ii)

40. In the alimentary canal of humans, the glands are found in:
(a) Mucosa (b) Serosa
(c) Submucosa (d) Both (a) and (c)

41. Paneth cells are a part of:
(a) Brunner's gland
(b) Crypts of Lieberkuhn
(c) Gastric pits
(d) Apical part of villi

42. Total number of premolar teeth in an adult human is:
(a) 12 (b) 8
(c) 20 (d) 6

43. The submucosal layer in the wall of the alimentary canal is formed by:
(a) Dense connective tissue
(b) Outer longitudinal and inner circular muscle layer
(c) Loose connective tissue
(d) Mesothelium

44. Which of the following are the most abundant papillae on the tongue bearing no taste buds?
(a) Filiform papillae
(b) Fungiform papillae
(c) Foliate papillae
(d) Circumvallate papillae

45. Enamel of tooth is secreted by ___X___ and is ___Y___ in origin:

	X	Y
(a)	Odontoblast	Mesodermal
(b)	Odontoblast	Ectodermal
(c)	Ameloblast	Ectodermal
(d)	Ameloblast	Mesodermal

46. Which of the following is an incorrect statement?
(a) Tusks of elephant are modified upper incisors
(b) Wisdom teeth in humans are the last molars
(c) Diastema is the toothless space in rabbits
(d) Premolars of deciduous milk set are replaced by molars of permanent set in humans

47. The largest chamber of the ruminant stomach is:
(a) Rumen (b) Reticulum
(c) Omasum (d) Abomasum

48. The oblique layer of muscles present in the muscularis of stomach are:
(a) Between the longitudinal and circular layers
(b) Between the serosa and longitudinal muscle layer
(c) Between the mucosa and submucosa
(d) Between the submucosa and circular muscle layer

49. In the given diagram of the transverse section of the gut, which layer is formed of loose connective tissues containing nerve plexus, blood and lymph vessels?

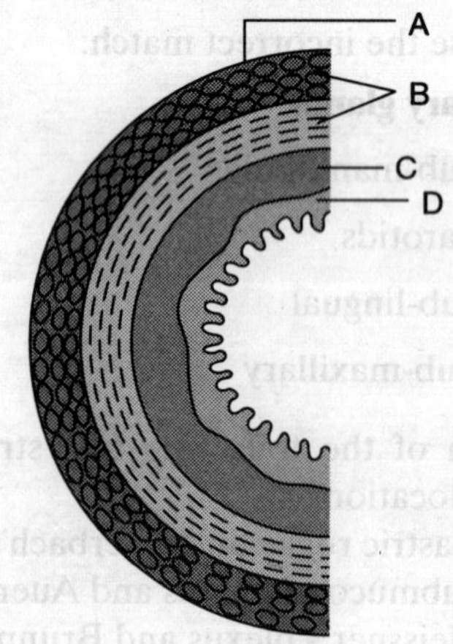

(a) A (b) B
(c) C (d) D

50. Consider the following diagram and find out the correct identification and description of the labelled structure:

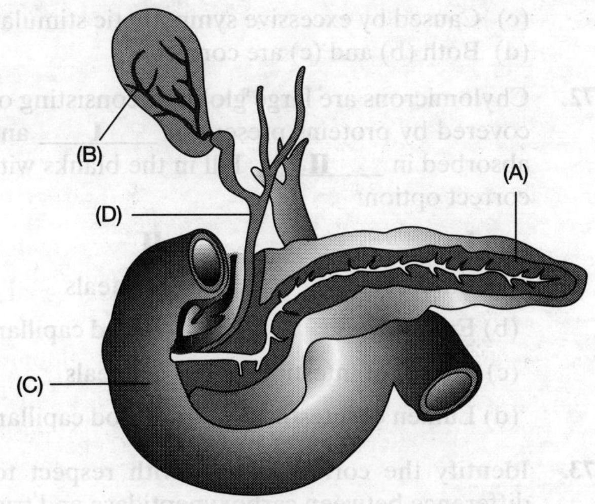

(a) A – Secretes enzymes pepsin and renin
(b) B – Contracts under influence of cholecystokinin
(c) C – Longest part of small intestine
(d) D – Duct of Wirsung formed by union of hepatic duct and cystic duct

Digestion of Food, Absorption of Digested Products, Assimilation and Defaecation

51. Protein-coated, water-soluble fat droplets called chylomicrons are formed:
(a) Inside the lumen of the intestine
(b) Inside the intestinal cells
(c) In blood capillaries
(d) In lacteals

52. Gastric juice of an infant contains:
(a) Pepsin, trypsin and renin
(b) Pepsin, renin and lipase
(c) Pepsin, chymotrypsin and lipase
(d) Pepsin, rennin and lipase

53. Which of the following will not enter the duodenum through the hepato-pancreatic ampulla?
(a) Nuclease (b) Sodium glycocholate
(c) Steapsin (d) Lactase

54. Emulsification of fats occurs in the duodenum with the help of:
(a) Bile pigments (b) Bile salts
(c) Lipase (d) Renin

55. Pepsinogen in the stomach is secreted by:
(a) Oxyntic cells (b) Parietal cells
(c) Chief cells (d) Clara cells

56. Renin is an enzyme secreted in infants. It acts on— present in milk:
(a) Fats (b) Carbohydrates
(c) Proteins (d) Lipids

57. Chymotrypsin present in pancreatic juice is activated by:
(a) Bile salts (b) Enterokinase
(c) Trypsin (d) Hydrochloric acid

58. Which of the following enzymes is not present in succus entericus?
(a) Sucrase (b) Nucleotidase
(c) Dipeptidase (d) Carboxypeptidase

59. In the process of fat emulsification, bile salts make fats more susceptible to the action of lipases by:
(a) Increasing the catalytic activity of lipase
(b) Lowering the pH of fluids in the small intestine
(c) Increasing the surface area of the fat globules by breakdown of fat globules into tiny droplets
(d) Decreasing the surface area of fat globules

60. Identify the incorrect statement:
(a) Steapsin is linked with digestion of lipids
(b) The end products of the enzyme activity of invertase are glucose and fructose
(c) Enterocytes are present in the lining of the stomach wall
(d) Intrinsic factor (IF) helps in absorption of cyanocobalamin

61. Which of the following has been correctly matched with its source cell?

(a)	Hydrochloric acid	–	Chief cells
(b)	Pepsin	–	Oxyntic cells
(c)	Chymotrypsin	–	Brush-border cells
(d)	Intrinsic factor	–	Parietal cells

62. Identify the correct option with respect to brush-border enzymes:
(a) Maltase, dipeptidase, lactase
(b) Amylase, nuclease, lipase
(c) Nucleotidase, amylase, trypsin
(d) Trypsin, sucrase, nucleotidase

63. Which of the following is an incorrect statement about the enzyme carboxypeptidase?
(a) It is an enzyme of the pancreatic juice
(b) It is an exopeptidase
(c) It uses zinc as cofactor
(d) It cleaves the peptide bond at the N-terminal end of the polypeptide chain

64. Identify the order in which the enzymes act on egg white as it moves down the alimentary canal:

(a) Pepsin → Trypsin → Dipeptidase
(b) Trypsin → Pepsin → Dipeptidase
(c) Chymotrypsin → Pepsin → Trypsin
(d) Chymotrypsin → Enterokinase → Pepsin

65. If the oxyntic cells of the gastric glands are non-functional, which of the following processes is affected?
(a) Digestion of starch
(b) Absorption of cyanocobalamin
(c) Digestion of proteins in acidic medium
(d) Both (b) and (c)

66. Identify the incorrect statement:
(a) Lysozyme present in saliva acts as anti-bacterial agent that prevents infection
(b) Mucus in saliva helps in lubricating and adhering the masticated food particles into a bolus
(c) The movement of food bolus in the esophagus is facilitated by peristaltic waves
(d) The chemical process of digestion is initiated in the oral cavity by the hydrolytic action of the proteolytic enzyme ptyalin

67. Which of the following is the correct option with respect to the enzymatic reactions occurring in the alimentary canal?
(a) Nucleic acid $\xrightarrow{\text{Nucleotidase}}$ Nucleotides
(b) Starch $\xrightarrow{\text{Amylase}}$ Disaccharides
(c) Dipeptides $\xrightarrow{\text{Pepsin}}$ Amino acids
(d) Lactose $\xrightarrow{\text{Lactase}}$ Glucose + Fructose

68. Digestive activity by pancreatic lipase yields:
(a) Amino acids
(b) Glucose
(c) Fatty acid and glycerol
(d) Nucleotides

69. The brush-bordered enzyme which activates the pancreatic proteases is:
(a) Trypsin (b) Enterokinase
(c) Enterocrinin (d) Chymosin

70. Iron deficiency anaemia is caused due to reduced secretion of:
(a) Islets of Langerhans
(b) Pancreatic acini
(c) Peptic cells of stomach
(d) Oxyntic cells of stomach

71. Achlorhydria (reduced or no HCl production) is:
(a) Caused by excessive vagus nerve stimulation
(b) Associated with iron deficiency anaemia
(c) Caused by excessive sympathetic stimulation
(d) Both (b) and (c) are correct

72. Chylomicrons are large globules consisting of fats covered by proteins present in ____I____ and are absorbed in ____II____. Fill in the blanks with the correct option:

	I	II
(a)	Enterocytes	Lacteals
(b)	Enterocytes	Blood capillaries
(c)	Lumen of intestine	Lacteals
(d)	Lumen of intestine	Blood capillaries

73. Identify the correct option with respect to the difference between carboxypeptidase and trypsin:
(a) They are released from different glands
(b) Both are released in active forms
(c) Both are released in inactive forms and are activated by hydrochloric acid
(d) Carboxypeptidase is an exopeptidase while trypsin is an endopeptidase

74. The intestinal hormone that causes gall bladder contraction and stimulates the release of enzyme-rich pancreatic juice is:
(a) Secretin
(b) Gastrin
(c) Durocrinin
(d) Cholecystokinin–Pancreozymin

75. Inhibition of proton pumps of gastric glands will result in all of the following symptoms, except:
(a) Reduced iron absorption
(b) Fall in pH of gastric juice
(c) Increased gastric infections
(d) Impaired digestion of proteins

76. Which component in the secretion of the salivary glands of the oral cavity and paneth cells of the intestine is common?
(a) Amylase (b) Lysozyme
(c) Nuclease (d) Protease

77. Which of the following is the most effective lipase enzyme that catalyses breakdown of emulsified fat?
(a) Gastric lipase (b) Pancreatic lipase
(c) Intestinal lipase (d) Lingual lipase

78. Renin is secreted in the stomach of infants for curdling of milk. It is also called:
(a) Erepsin (b) Ptyalin
(c) Chymosin (d) Steapsin

79. Non-digestive brush-bordered enzyme which does not act directly on any food source is:

(a) Enterocrinin (b) Enterogastrone
(c) Enterokinase (d) Gastrin

80. The release of water and bicarbonates from pancreatic acinar calls is stimulated by:
(a) Cholecystokinin (b) Secretin
(c) Durocrinin (d) Enterogastrone

81. The regions of the digestive system which do not secrete any digestive enzymes are
A. Liver
B. Colon
C. Pancreatic acini
D. Esophagus
(a) (A) and (B) (b) (B) and (C)
(c) (A), (B) and (D) (d) (C) and (D)

82. Which of the following is the correct statement with respect to absorption of nutrients?
(a) Carbohydrates are absorbed as disaccharides
(b) Bile pigments help in emulsification of fat droplets, facilitating their digestion and absorption
(c) Amino acids are absorbed only by passive transport
(d) Fatty acids and glycerol are first incorporated into small droplets called micelles which move into intestinal mucosa

83. The secretions of ____A____ of the mucosa along with secretions of ____B____ constitute the succus entericus.
Fill in the blanks with the correct options:

	A	B
(a)	Gastric epithelium	Goblet cells
(b)	Brush-border cells	Goblet cells
(c)	Pancreatic acini	Paneth cells
(d)	Hepatocytes	Paneth cells

84. Which of the following is incorrect about pancreas?
(a) It is a compound, leaf-like gland
(b) Islets of Langerhans in pancreas secrete insulin and glucagon
(c) The pH of pancreatic juice is 8.4
(d) Dipeptidase and lactase are enzymes of the pancreatic juice

85. Identify the correct set which shows the name of the enzyme, from where it is secreted and the substrate upon which it acts:
(a) Sucrase – Pancreas – Sucrose
(b) Steapsin – Intestine – Fat
(c) Nuclease – Pancreas – Nucleic acid
(d) Pepsin – Crypts of Liberkuhn – Peptones

86. Identify the incorrect statement with respect to enzyme-catalysed reactions occurring in the alimentary canal:
(a) Lactose $\xrightarrow{\text{Lactase}}$ Glucose + Galactose
(b) Triglyceride $\xrightarrow{\text{Lipase}}$ Glycerol + Fatty acids
(c) Nucleoside $\xrightarrow{\text{Nucleosidase}}$ Nitrogenous base + Pentose sugar
(d) Nucleotide $\xrightarrow{\text{Nucleosidase}}$ Nucleoside + Pentose sugar

87. How many of the following digestive glands secrete enzymes that optimally act at acidic pH?
Gastric glands, Salivary glands, Liver, Intestinal glands, Pancreas
(a) 5 (b) 4
(c) 3 (d) 2

88. Stomach is mainly responsible for initial digestion of ____Z____ and absorption of ____Y____.
Select the option which correctly describes Z and Y in the given blanks:
(a) Lipids, alcohol
(b) Proteins, alcohol
(c) Carbohydrates, water
(d) Lipids, amino acids

89. The percentage of starch digested into maltose by amylase in the buccal cavity of humans is:
(a) 50 (b) 30
(c) 70 (d) 90

90. Select the correct match between enzymes listed in Column I and their substrate in Column II:

Column I	Column II
(a) Dipeptidase	Amino acids
(b) Trypsin	Peptones
(c) Nuclease	Nucleosides
(d) Carboxypeptidase	Trypsinogen

91. The process of digestion is accomplished by:
(a) Mechanical process (b) Chemical process
(c) Both of these (d) None of these

92. Which of the following is/are present in gastric juice and play an important role in lubrication and protection of the mucosal epithelium from excoriation by highly concentrated hydrochloric acid?
(a) Mucus and lysozyme
(b) Lysozyme and bicarbonates
(c) Mucus and bicarbonate
(d) Renin and bicarbonates

93. The food mixed thoroughly with the acidic gastric juice of the stomach by the churning movements of its muscular wall is called:
(a) Chyle (b) Chyme
(c) Faeces (d) Lacteal

94. Identify the incorrect statement:
(a) Maximum absorption of the digested nutrients occurs in the small intestine
(b) Sphincter of Oddi is associated with opening of the hepatopancreatic duct
(c) Certain drugs when placed under the tongue are absorbed into the blood capillaries lining it
(d) Absorption of water occurs through facilitated diffusion in the alimentary canal

95. Read the following statements and select the false one among them:
(a) Large number of villi and microvilli are present for increased absorption of nutrients in the small intestine
(b) Pepsinogen is converted into its active form pepsin by the action of hydrochloric acid
(c) Bile juice makes the medium alkaline and has anti-microbial action
(d) Digestion of protein component of milk occurs by the action of lactase

96. Select the correct match of enzymes, their site of action and optimum pH:

Enzyme	Site of action	pH
(a) Trypsinogen	Duodenum	6.8
(b) Amylopsin	Pancreas	8.4
(c) Nucleotidase	Stomach	7.4
(d) Carboxypeptidase	Small intestine	7.8

97. Indigestion, diarrhea and flatulence upon consumption of milk occurs due to deficiency of:
(a) Maltase (b) Isomaltase
(c) Lactase (d) Renin

98. Which of the following statements is false with respect to digestion of food?
(a) All digestive enzymes are hydrolases
(b) Chemical process of digestion of food is initiated in the oral cavity
(c) The intestinal mucosal epithelium does not have goblet cells
(d) Pancreatic juice contains lipases and amylases

99. Identify the false statement with respect to absorption of food:
(a) Absorption is carried out by active, passive or facilitated transport mechanism
(b) Absorption is the process by which the end products of digestion pass through the intestinal mucosa into blood or lymph
(c) Maximum absorption of nutrients occurs in the small intestine
(d) Small droplets called micelles are formed for the absorption of sugars and amino acids

100. Which of the following is catalysed by the brush-border enzymes of the succus entericus?
(a) Nucleic acid $\xrightarrow{\text{Nuclease}}$ Nucleotides
(b) Starch $\xrightarrow{\text{Ptyalin}}$ Maltose
(c) Dipeptides $\xrightarrow{\text{Dipeptidase}}$ Amino acids
(d) $\left.\begin{array}{l}\text{Proteins}\\ \text{Peptones}\\ \text{Proteose}\end{array}\right\} \xrightarrow{\text{Trypsin}}$ Dipeptides

101. The amount of heat liberated from complete combustion of 1 gram of food is its gross calorific value. Gross calorific value of fat is:
(a) 5.65 kcal/g
(b) 9.45 kcal/g
(c) 4.1 kcal/g
(d) 4 kcal/g

102. The absorbed nutrients finally reach the tissues which utilise them for their activities. This process is called:
(a) Egestion (b) Assimilation
(c) Defaecation (d) Ingestion

103. The brown colour of faeces is due to:
(a) Urochrome (b) Bile salts
(c) Stercobilin (d) Roughage

104. Which of the following is correct with respect to faeces?
(a) The foul smell of faeces is because of skatole
(b) Faeces have ¾ water and ¼ solid matter
(c) Faeces have roughage, dead bacteria and undigested food
(d) All are correct

105. The dark-green mucilaginous faeces of newborns is called:
(a) Meconium (b) Lanugo
(c) Urochrome (d) Biliverdin

Disorders of the Digestive System

106. The digestive disorder in which faeces are retained within the colon as the bowel movements occur irregularly is:
(a) Diarrhea (b) Vomiting
(c) Indigestion (d) Constipation

107. Which of the following diseases result from the replacement of mother's milk by a high-calorie, low-protein diet in a child of more than one year of age?
(a) Marasmus (b) Kwashiorkar
(c) Obesity (d) Beri beri

108. Megaloblastic anaemia is caused due to deficiency of:
(a) Folic acid and ascorbic acid
(b) Cyanocobalamin and folic acid
(c) Thiamine and niacin
(d) Riboflavin and niacin

109. Which of the following is a mismatch?

Vitamin	Deficiency disorder
(a) Niacin	Pellagra
(b) Ascorbic acid	Scurvy
(c) Riboflavin	Cheilosis
(d) Tocopherol	Rickets

110. Which of the following is not a fat-soluble vitamin?
(a) Tocopherol (b) Phylloquinone
(c) Calciferol (d) Riboflavin

111. Which of the following is an incorrect statement?
(a) Calciferol is called sunshine vitamin
(b) Deficiency of iodine causes goitre
(c) Patients with scurvy are advised to include citrus fruits in their diet
(d) Intrinsic factor secreted in stomach is used for absorption of folic acid

112. Which of the following is incorrect with respect to symptoms of marasmus?
(a) Thinning of limbs, dry wrinkled skin
(b) Emaciated appearance with prominent rib cage
(c) Stunted growth and delayed development
(d) Swelling of ankles, feet and belly due to edema

113. Vomiting is the ejection of stomach contents through the mouth. This reflex action is controlled by:
(a) Hypothalamus (b) Cerebellum
(c) Medulla (d) Cerebrum

114. The abnormal frequency of bowel movement and increased liquidity of faecal discharge is called:
(a) Constipation (b) Diarrhea
(c) Indigestion (d) Jaundice

115. In patients with jaundice, the liver is affected, and skin and eyes turn yellow due to deposition of:
(a) Bile salts (b) Cholesterol
(c) Bile pigments (d) Urochrome

116. ——— is a condition in which food is not properly digested, leading to a feeling of fullness due to over-eating, spicy food, anxiety or inadequate enzyme secretion:
(a) Vomiting (b) Indigestion
(c) Constipation (d) Pellagra

NCERT EXEMPLAR QUESTIONS

1. Which of the following is not true of intestinal villi?
(a) They possess microvilli
(b) They increase the surface area
(c) They are supplied with capillaries and lacteal vessels
(d) They only participate in digestion of fats

2. Hepato-pancreatic duct opens into the duodenum and carries:
(a) Bile
(b) Pancreatic juice
(c) Both bile and pancreatic juice
(d) Saliva

3. Which of the following is not a common disorder associated with the digestive system?
(a) Tetanus (b) Diarrhea
(c) Jaundice (d) Dysentery

4. A gland not associated with the alimentary canal is:
(a) Pancreas (b) Adrenal
(c) Liver (d) Salivary glands

5. Match the two columns and select the correct options:

Column I	Column II
A. Biomacromolecules of food	(i) Alimentary canal and associated gland
B. Human digestive system	(ii) Embedded in jawbones.
C. Stomach	(iii) Outer wall of visceral organs
D. Thecodont	(iv) Converted into simple substances
E. Serosa	(v) J-shaped bag like structure

(a) A – (ii), B – (i), C – (v), D – (iii), E – (iv)
(b) A – (iv), B – (i), C – (v), D – (ii), E – (iii)
(c) A – (i), B – (ii), C – (iii), D – (iv), E – (v)
(d) A – (i), B – (iii), C – (ii), D – (iv), E – (v)

6. Match the two columns and select the right one among the options given:

Column I	Column II
A. Duodenum	(i) A cartilaginous flap
B. Epiglottis	(ii) Small blind sac
C. Glottis	(iii) 'U' shaped structure emerging from the stomach
D. Caecum	(iv) Opening of wind pipe

(a) A – (i), B – (ii), C – (iii), D – (iv)
(b) A – (iv), B – (iii), C – (ii), D – (i)
(c) A – (iii), B – (i), C – (iv), D – (ii)
(d) A – (ii), B – (iv), C – (i), D – (iii)

7. Match the enzyme with their respective substrate and choose the correct one among the options given:

Column I	Column II
A. Lipase	(i) Carbohydrates
B. Nuclease	(ii) Fats
C. Carboxypeptidase	(iii) Nucleic acids
D. Glycosidases	(iv) Proteins, peptones and proteoses

(a) A – (ii), B – (iii), C – (i), D – (iv)
(b) A – (iii), B – (iv), C – (ii), D – (i)
(c) A – (iii), B – (i), C – (iv), D – (ii)
(d) A – (ii), B – (iii), C – (iv), D – (i)

8. Dental formula in human beings is
(a) $\frac{3223}{3223}$ (b) $\frac{2123}{2123}$
(c) $\frac{1232}{1232}$ (d) $\frac{2233}{2233}$

9. Liver is the largest gland and is associated with various functions. Choose which is not correct from the following:
(a) Metabolism of carbohydrate
(b) Digestion of fat
(c) Formation of bile
(d) Secretion of hormone called gastrin

10. Mark the correct statement among the following:
(a) Trypsinogen is an inactive enzyme
(b) Trypsinogen is secreted by intestinal mucosa
(c) Enterokinase is secreted by pancreas
(d) Bile contains trypsin

PRACTICE TEST PAPER

1. Find the incorrect match:
   (a) Sucrase – Succus entericus
   (b) Nuclease – Pancreatic juice
   (c) Steapsin – Protease
   (d) Pepsin – Gastric juice

2. How many of the given structures are found in mucosa or are modifications of mucosa?
   Islets of Langerhans, Gastric glands, Villi, Crypts of Liberkuhn, Brunner's glands, Gastric rugae, Meissener's plexus
   (a) 6 (b) 5
   (c) 4 (d) 2

3. Which of the given statements is correct about enterokinase?
   (a) It is secreted by pancreatic juice
   (b) It is secreted in gastric juice
   (c) It activates trypsinogen into trypsin
   (d) It breaks down proteins into peptones and peptides

4. Curdling of milk in the stomach occurs due to the action of:
   (a) Renin (b) Steapsin
   (c) Chymotrypsin (d) Trypsin

5. Disaccharidases are found in:
   (a) Pancreatic juice (b) Gastric juice
   (c) Succus entericus (d) Bile juice

6. Which of the following is true for chylomicrons?
   (a) Chylomicrons are fat-coated protein globules
   (b) Chylomicrons are formed in intestinal cells
   (c) Chylomicrons are absorbed into blood vessels
   (d) Chylomicrons are formed in intestinal lumen

7. Find the correct match:
   (a) Rugae – The irregular folds in the submucosa of the stomach
   (b) Deglutition – The churning movements in the stomach, leading to propulsion of food
   (c) Meissner's plexus – In the muscularis layer of the gut
   (d) Ileo-caecal valve – At the junction of the small and large intestines

8. Which among the following is vestigial in humans?
   (a) Tonsils
   (b) Caecum
   (c) Vermiform appendix
   (d) Peyer's patches

9. Saliva does not have:
   (a) Mucus and water
   (b) Ptyalin
   (c) Proteolytic enzymes
   (d) Electrolytes like Na^+, K^+, Cl^-, HCO_3^-

10. The brush-bordered appearance of intestinal mucosal epithelium is because of:
    (a) Villi
    (b) Microvilli on columnar epithelial cells
    (c) Plicae circulares
    (d) Cilia on columnar epithelial cells

11. Which of the following is an incorrect statement?
    (a) The hepatic cells are present in the form of cords in the liver
    (b) No significant digestion occurs in the large intestine
    (c) Salivary glands are present just outside the buccal cavity but secrete salivary juice into the buccal cavity
    (d) The esophagus opens into the pyloric part of the stomach through the gastro-esophageal sphincter

12. Which of the following is not a component of bile juice?
    (a) Sodium bicarbonate
    (b) Cholesterol
    (c) Lipases
    (d) Degraded products of haemoglobin

13. How many teeth are common to both temporary and permanent sets of human dentition?
    (a) 12 (b) 16
    (c) 20 (d) 24

14. All of the following pass through the hepatopancreatic ampulla, except:
    (a) Trypsinogen (b) Bile salts
    (c) Dipeptidase (d) Amylopsin

15. Identify the correct statement:
(a) Lysozyme is present in gastric juice
(b) Defecation is an involuntary process
(c) Mucus in saliva helps in lubricating and adhering the masticated food particles into the bolus
(d) Maltase and lactase are present in pancreatic juice

16. Symbiotic microorganisms are normally found in:
(a) Oral cavity (b) Caecum
(c) Stomach (d) Gall bladder

17. The phosphodiester bonds in a polynucleotide chain are hydrolysed by the secretions of:
(a) Islets of Langerhans
(b) Crypts of Lieberkuhn
(c) Pancreatic acini
(d) Oxyntic cells of gastric epithelium

18. Peyer's patches are lymphatic nodules present in the mucosa of:
(a) Jejunum (b) Ileum
(c) Duodenum (d) Stomach

19. Which of the following is the correct sequence for the layers in the wall of the stomach, from inner to outer side:
(a) Mucosa — Circular muscles — Oblique muscles — Longitudinal muscles — Submucosa
(b) Mucosa — Submucosa — Oblique muscles — Circular muscles — Longitudinal muscles — Serosa
(c) Mucosa — Serosa — Submucosa — Longitudinal muscles — Circular muscles — Oblique muscles
(d) Mucosa — Submucosa — Longitudinal muscles — Oblique muscles — Circular muscles — Serosa

20. If the vagus nerve is cut, which of the following is observed?
(a) Acid production in stomach increases
(b) Reduced peristalsis
(c) Production of intestinal juices increases
(d) More than one option is correct

21. Paneth cells are found in:
(a) Gastric rugae
(b) Crypts of Liberkuhn
(c) Pancreatic acini
(d) Liver

22. Identify the incorrect statement:
(a) Liver is the largest gland of the body, weighing 1.2 to 1.5 kg in adult humans
(b) Liver is situated in the thoracic cavity, just above the diaphragm, and has two lobes
(c) Hepatic lobules are structural and functional units of the liver
(d) Each lobule is covered by thin connective tissue sheath called Glisson's capsule

23. The duct from the gall bladder (cystic duct) along with the hepatic duct from the liver forms:
(a) Duct of Wirsung (b) Duct of Santorini
(c) Common bile duct (d) Stenson's duct

24. Which of the following is an incorrect statement with respect to digestion of food?
(a) Pepsin converts proteins into proteoses and peptones
(b) Renin is a lipolytic enzyme present in the gastric juice of infants
(c) Carbohydrates in chyme are hydrolysed by pancreatic amylase into disaccharides
(d) Fats are broken down by lipases with the help of bile into di- and monoglycerides

25. Which of the following is incorrect?
(a) Jejunum is the longest portion of the small intestine
(b) Ileum is the coiled middle portion of the small intestine
(c) Jejunum opens into the large intestine
(d) All of the above are incorrect

26. Which of the following is an incorrect statement with resepct to bile juice?
(a) Bile helps in emulsification of fats
(b) Bile also activates lipases
(c) Bile is stored in gall bladder
(d) The bile released into the duodenum contains bile pigments, bile salts, cholesterol and steapsin

27. Brunner's glands are present in:
(a) Mucosa of duodenum
(b) Submucosa of duodenum
(c) Submucosa of jejunum
(d) Submucosa of ileum

28. Ptyalin and lysozyme hydrolyse:
(a) Starch and proteins, respectively
(b) Peptidoglycan and proteins, respectively
(c) Starch and peptidoglycan, respectively
(d) Proteins and peptidoglycan, respectively

29. Choose the correct statement:
(a) Chief cells secrete hydrochloric acid
(b) Parietal cells produce pepsinogen

(c) Brunner's gland secretion is acidic and rich in mucus
(d) Oxyntic cells secrete intrinsic factor

30. Which of the following is an incorrect statement?
(a) Small amount of lipases are secreted by gastric glands
(b) The pancreatic juice contains inactive enzymes – trypsinogen, chymotrypsinogen and procarboxypeptidase
(c) Gastrectomy leads to iron deficiency anaemia
(d) Absorption of water, simple sugars and alcohol takes place in the mouth

31. Haemorrhoids are enlarged rectal veins that develop in patients suffering from:
(a) Vomiting (b) Diarrhea
(c) Constipation (d) Jaundice

32. Which of the following is incorrect with respect to Kwashiorkar?
(a) It is found in children of more than one year of age
(b) Extensive oedema and swelling of body parts is seen
(c) It is caused by protein deficiency accompanied by calorie deficiency
(d) Pot belly due to fatty liver is seen

33. Which of the following is a correct set of brush-bordered enzymes?
(a) Trypsin, lipase, amylase
(b) Dipeptidase, pepsin, renin
(c) Dipeptidase, maltase, nucleotidase
(d) Nuclease, lactase, chymotrypsin

34. Identify the incorrect statement:
(a) 30% of starch is digested in the oral cavity and converted into maltose by the action of ptyalin
(b) Diarrhea is a digestive disorder in which faeces are retained within the colon as bowel movements occur irregularly
(c) Stomach receives food from the esophagus as food bolus
(d) Numerous finger-like projections called villi increase the surface area for absorption of food in the intestine

35. Which of the following is incorrect?
(a) Pepsin is the proteolytic enzyme secreted in the stomach in inactive form
(b) The mucus and bicarbonates present in gastric juice protect the mucosal epithelium from excoriation by hydrochloric acid
(c) The stomach stores the food for 4–5 hours
(d) The food mixes thoroughly with the alkaline gastric juice of the stomach by the churning movement of its muscular wall

36. Hydrolytic enzymes are absent in the secretions of:
(a) Pancreas (b) Stomach
(c) Liver (d) Small intestine

37. A muscular, glandular and voluntary organ found attached to the floor of the oral cavity by a membranous fold is:
(a) Hyoid (b) Tongue
(c) Epiglottis (d) Uvula

38. Glucose and fructose are produced by the activity of______ in the intestine:
(a) Lactase (b) Sucrase
(c) Amylase (d) Maltase

39. Select the incorrect match:

Vitamins	Deficiency disorder
(a) Thiamine	Beri beri
(b) Retinol	Night blindness
(c) Riboflavin	Scurvy
(d) Folate	Megaloblastic anaemia

40. Select the correct match of enzymes listed, their site of action, their substrate and optimum pH:

Enzyme	Site of action	Substrate	pH
(a) Nuclease	Pancreas	Nucleotide	6.8
(b) Amylopsin	Small intestine	Starch	6.8
(c) Carboxy peptidase	Small intestine	Polypeptide	7.8
(d) Trypsin	Duodenum	Peptones	1-2

41. A compound gland which secretes hormones to regulate blood sugar level and enzymes that facilitate digestion is:
(a) Liver (b) Pancreas
(c) Duodenum (d) Adrenal gland

42. Choose the incorrect match:

	Salivary glands	Location
(a)	Sub-lingual	Below tongue
(b)	Parotids	Near cheek
(c)	Sub-mandibular	Near lower jaw
(d)	Sub-maxillary	Near eyes

43. Which of the following is an incorrect statement?
(a) Trypsinogen is an inactive enzyme
(b) Aminopeptidase and carboxypeptidase are exopeptidases
(c) Small intestine is the main site for digestion and absorption of food
(d) Gastric juice contains acid which converts Fe^{2+} into Fe^{3+} and iron is absorbed as ferric ions

44. Intrinsic factor is required for absorption of:
(a) Folic acid (b) Cyanocobalamin
(c) Phylloquinone (d) Calciferol

45. Large intestine helps in absorption of:
(a) Chylomicrons
(b) Water, some drugs and minerals
(c) Glucose, fructose and amino acids
(d) Alcohol, starch and proteins

ANSWER KEYS

Exercises

1. (c)	**2.** (b)	**3.** (d)	**4.** (d)	**5.** (d)	**6.** (b)	**7.** (b)	**8.** (c)	**9.** (d)	**10.** (c)
11. (d)	**12.** (d)	**13.** (c)	**14.** (d)	**15.** (b)	**16.** (b)	**17.** (d)	**18.** (b)	**19.** (d)	**20.** (c)
21. (c)	**22.** (c)	**23.** (a)	**24.** (a)	**25.** (b)	**26.** (d)	**27.** (d)	**28.** (b)	**29.** (d)	**30.** (c)
31. (c)	**32.** (c)	**33.** (d)	**34.** (b)	**35.** (b)	**36.** (d)	**37.** (c)	**38.** (d)	**39.** (a)	**40.** (d)
41. (b)	**42.** (b)	**43.** (c)	**44.** (a)	**45.** (c)	**46.** (d)	**47.** (a)	**48.** (d)	**49.** (c)	**50.** (b)
51. (b)	**52.** (d)	**53.** (d)	**54.** (b)	**55.** (c)	**56.** (c)	**57.** (c)	**58.** (d)	**59.** (c)	**60.** (c)
61. (d)	**62.** (a)	**63.** (d)	**64.** (a)	**65.** (d)	**66.** (d)	**67.** (b)	**68.** (c)	**69.** (b)	**70.** (d)
71. (d)	**72.** (a)	**73.** (d)	**74.** (d)	**75.** (b)	**76.** (b)	**77.** (b)	**78.** (c)	**79.** (c)	**80.** (b)
81. (c)	**82.** (d)	**83.** (b)	**84.** (d)	**85.** (c)	**86.** (d)	**87.** (d)	**88.** (b)	**89.** (b)	**90.** (b)
91. (c)	**92.** (c)	**93.** (b)	**94.** (d)	**95.** (d)	**96.** (d)	**97.** (c)	**98.** (c)	**99.** (d)	**100.** (c)
101. (b)	**102.** (b)	**103.** (c)	**104.** (d)	**105.** (a)	**106.** (d)	**107.** (b)	**108.** (b)	**109.** (d)	**110.** (d)
111. (d)	**112.** (d)	**113.** (c)	**114.** (b)	**115.** (c)	**116.** (b)				

NCERT Exemplar Questions

1. (d)	**2.** (c)	**3.** (a)	**4.** (b)	**5.** (b)	**6.** (c)	**7.** (d)	**8.** (b)	**9.** (d)	**10.** (a)

Practice Test Paper

1. (c)	**2.** (c)	**3.** (c)	**4.** (a)	**5.** (c)	**6.** (b)	**7.** (d)	**8.** (c)	**9.** (c)	**10.** (b)
11. (d)	**12.** (c)	**13.** (c)	**14.** (c)	**15.** (c)	**16.** (b)	**17.** (c)	**18.** (b)	**19.** (b)	**20.** (b)
21. (b)	**22.** (b)	**23.** (c)	**24.** (b)	**25.** (d)	**26.** (d)	**27.** (b)	**28.** (c)	**29.** (d)	**30.** (d)
31. (c)	**32.** (c)	**33.** (c)	**34.** (b)	**35.** (d)	**36.** (c)	**37.** (b)	**38.** (b)	**39.** (c)	**40.** (c)
41. (b)	**42.** (d)	**43.** (d)	**44.** (b)	**45.** (b)					

SOLUTIONS

Digestive System, Digestive Glands

1. Premolars are absent in milk and deciduous teeth.
2. The gastro-esophageal sphincter is also called cardiac sphincter.
3. The dental formula of human dentition is $\frac{2123}{2123}$.
4. Maltose is not absorbed in the buccal cavity.
5. Vallate papillae are the largest and located on the posterior surface of the tongue.
6. Primary dentition differs from permanent dentition in not having premolars and last molars.
7. The ileo-caecal valve prevents backflow of faecal matter from the caecum into the small intestine.
8. Brunner's glands are located in the submucosa of the duodenum.
9. Option (d) is correct.
10. Hepatic lobule is the structural and functional unit of the liver.
11. Tongue is a freely movable muscular organ.
12. Meissner's plexus is located in the submucosa while Auerbach's plexus is located in the muscularis externa of the gastrointestinal tract.
13. Maximum salivary secretion is contributed by the sub-mandibular gland.
14. Mucosa forms irregular folds (rugae) in the stomach and small finger-like folds called villi in the small intestine.
15. Stomach has an additional oblique layer of muscles in the alimentary canal, besides a longitudinal and circular layer of smooth muscles in the muscularis.
16. Pancreas is the compound gland which secretes hormones to regulate blood sugar level and enzymes that facilitate digestion.
17. Sigmoid/Pelvic colon opens into the rectum.
18. The large intestine of humans has three parts, namely, caecum, colon and rectum.
19. Auerbach's plexus is present between the longitudinal and circular muscle layers of the muscularis externa.
20. Cystic duct is present in the gall bladder. Crypts of Lieberkuhn is present in the small intestine.
21. The oral cavity leads into the pharynx which serves as a common passage for food and air.
22. The common bile duct is formed by the joining of the cystic and hepatic ducts.
23. Stenson's duct brings the secretion of the parotid glands to the buccal cavity.
24. Option (a) is correct.
25. The haustra of colon are small pouches caused by sacculation, which gives the colon its segmented appearance.
26. Option (d) is correct.
27. The hepatic lobule has hepatic cells arranged in the form of cords.
28. The sphincter of Boyden guards the bile duct before combining with the pancreatic duct.
29. Option (d) is correct.
30. Both Meissner's plexus and Brunner's glands are present in the submucosa.
31. The esophagus opens into the cardiac part of the stomach through the gastro-esophageal sphincter. Kupffer cells are phagocytic cells present in the liver.
32. Bile is stored in the gall bladder. It is formed in the liver.
33. The hormone secretin is secreted by the small intestine.
34. Peyer's patches/Abdominal tonsils are lymphatic nodules present in the mucosa of the alimentary canal.
35. Achalasia cardia is the accumulation of food in the esophagus due to improper functioning of the cardiac sphincter. The sphincter fails

to relax to allow passage of food from the esophagus to the stomach.

36. Mucosa is the innermost layer of the alimentary canal and forms villi in the small intestine.

37. The brush-border appearance of intestinal mucosa in due to microvilli.

38. The long, coiled, middle part of the small intestine in humans is called the jejunum.

39. Option (a) is correct.

40. In the alimentary canal of humans, the glands are found in the mucosa and submucosa.

41. Paneth cells are the principal cell type in the Crypts of Lieberkuhn of the small intestine, along with goblet cells, enterocytes and enteroendocrine cells. They produce antimicrobial peptides and lysozyme.

42. Option (b) is correct.

43. The submucosal layer in the wall of the alimentary canal is formed by loose connective tissue.

44. Filiform papillae are the most abundant papillae on the tongue and lack taste buds.

45. The enamel of teeth is secreted by ameloblast and is ectodermal in origin.

46. The molars of the deciduous set are replaced by premolars of the permanent set in humans.

47. Rumen is the largest chamber of the ruminant stomach.

48. The oblique layer of muscles is present in the muscularis of the stomach, between the submucosa and circular muscle layers.

49. Option (c) is correct.

50. Option (b) is correct.

Digestion of Food, Absorption of Digested Products, Assimilation and Defaecation

51. Chylomicrons are protein-coated, water-soluble fat droplets formed inside the intestinal cells.

52. Gastric juice of infants contains pepsin, renin and lipase.

53. Lactase is secreted by the glands of the small intestine.

54. Bile salts help in emulsification of large fat droplets into small globules.

55. Pepsinogen in the stomach is secreted by peptic cells.

56. Renin is a proteolytic enzyme secreted in the stomach of infants. It acts on the milk protein casein.

57. Trypsin acts on chymotrypsinogen and forms the active chymotrypsin.

58. Carboxypeptidase is secreted by the pancreas.

59. During emulsification, the bile salts increase the surface area of fat globules by breaking them down into tiny droplets.

60. Enterocytes, or intestinal absorptive cells, are simple columnar epithelial cells which line the inner surface of the small and large intestines.

61. Intrinsic factor is a glycoprotein secreted by the parietal cells of the stomach which enables the body to absorb vitamin cyanocobalamin.

62. Maltase, dipeptidase and lactase are enzymes secreted in the succus entericus.

63. Carboxypeptidase cleaves the peptide bond at the C-terminal end of the polypeptide chain.

64. Option (a) is correct.

65. Oxyntic cells secrete hydrochloric acid and intrinsic factor (IF). HCl provides acidic medium for the action of pepsin while IF is required for absorption of cyanocobalamin.

66. The chemical process of digestion is initiated in the oral cavity by the hydrolytic action of amylase, which digests carbohydrates (starch).

67. Option (b) is correct.

68. Pancreatic lipase (steapsin) acts on fats and forms fatty acids and glycerol.

69. Enterokinase or enteropeptidase is an enzyme produced by cells of the duodenum and is involved in the digestion of proteins. Enterokinase converts trypsinogen (a zymogen) into its active form trypsin, resulting in the subsequent activation of pancreatic digestive enzymes.

70. Oxyntic cells secrete hydrochloric acid which maintains iron in ferrous form to facilitate its absorption. Thus, reduced secretion of oxyntic cells can cause iron deficiency anaemia.

71. Achlorhydria is a condition of very little or no HCl production in the stomach. It may be caused by excessive sympathetic stimulation and can cause impaired iron absorption, resulting in anaemia.

72. Chylomicrons are large globules consisting of fats covered by proteins, and formed in the intestinal cells/enterocytes. They are absorbed in lacteals.

73. Carboxypeptidase is an exopeptidase that removes amino acids of a polypeptide chain from the C-terminal end, while trypsin is an endopeptidase

74. Cholecystokinin–Pancreozymin is an intestinal hormone that causes gall bladder contraction and stimulates the release of pancreatic juice containing enzymes.

75. Inhibition of proton pumps of gastric glands will result in reduced acid production in the stomach and the increase pH of gastric juice.

76. Lysozyme is secreted by the salivary glands and paneth cells.

77. Pancreatic lipase is the most effective lipolytic enzyme.

78. Renin is also called chymosin and is secreted in the gastric juice of infants for curdling of milk.

79. Enterokinase or enteropeptidase is a brush-border enzyme that does not directly act on the food source but plays an important role in protein digestion as it converts trypsinogen (a zymogen) into the active form of trypsin. This initiates a cascade of reactions for activating other proteolytic enzymes of the pancreatic juice.

80. The release of water and bicarbonates from the pancreatic acinar cells is stimulated by secretin.

81. Option (c) is correct.

82. Carbohydrates are absorbed as monosaccharides. Bile salts help in emulsification of fat droplets. Amino acids are absorbed by active, passive and facilitated transport.

83. The secretions of brush-border cells of mucosa with goblet cells constitute the succus entericus.

84. Dipeptidase and lactase are enzymes of intestinal juice.

85. Option (c) is correct.

86. Nucleotidase catalyses breakdown of nucleotide to nucleoside and phosphate.

87. The secretions of the gastric glands and salivary glands act at acidic pH.

88. Stomach is mainly responsible for initial digestion of proteins and absorption of alcohol.

89. Thirty per cent of starch is digested by salivary amylase in the buccal cavity and converted to maltose.

90. Option (b) is correct.

91. The process of digestion is accomplished by mechanical and chemical processes.

92. Mucus and bicarbonates are present in gastric juice and play an important role in lubrication and protection of mucosal epithelium from excoriation by the highly concentrated HCl.

93. The food mixed thoroughly with acidic gastric juice of the stomach by the churning movements of its muscular wall is called chyme.

94. The absorption of water occurs by osmosis.

95. Digestion of the protein component of milk occurs by the action of renin.

96. Option (d) is correct.

97. Deficiency of lactase causes lactose intolerance, resulting in indigestion, diarrhea and flatulence.

98. Goblet cells are present in the lining of the intestinal mucosa.

99. Small droplets called micelles are formed for the absorption of fatty acids and glycerol.

100. Option (c) is correct.

101. Gross calorific value of fat is 9.45 kcal/g.

102. Assimilation is the process of utilisation of absorbed nutrients to release energy for body activities.

103. The brown colour of faeces is due to stercobilin.

104. Option (d) is correct.

105. The dark-green mucilaginous faeces of newborns is called meconium.

Disorders of the Digestive System

106. Constipation is the disorder in which faeces are retained within the colon as the bowel movements occur irregularly.

107. Kwashiorkar is a disorder of protein deficiency in children of more than one year of age, due to replacement of mother's milk with a high-calorie, low-protein diet.

108. Megaloblastic anaemia is caused by deficiency of cyanocobalamin and folic acid.

109. Rickets is caused due to deficiency of vitamin D in children.

110. Riboflavin is a water-soluble vitamin.

111. Intrinsic factor (IF) is a glycoprotein secreted by the oxyntic cells of the stomach, which enables the body to absorb vitamin cyanocobalamin.

112. Patients with Kwashiorkar have swelling of ankles, feet and belly due to oedema.

113. The vomiting centre is located in the medulla.

114. Patients with diarrhea have abnormal frequency of bowel movement and increased liquidity of faecal discharge.

115. The deposition of bile pigments causes skin and eyes to turn yellow in patients with jaundice.

116. Indigestion is a condition in which food is not properly digested, leading to a feeling of fullness due to overeating, spicy food, anxiety or inadequate enzyme secretion.

17 Breathing and Exchange of Gases

EXERCISES

Respiratory Organs, Human Respiratory System, Mechanism of Breathing

1. Identify the incorrect statement with respect to respiratory organs:
   (a) Mechanisms of breathing vary among different groups of animals depending mainly on their habitats and levels of organisation.
   (b) Insects use moist cuticles while earthworms have a network of tubes (tracheal tubes) to transport atmospheric air into the body
   (c) Special vascularised structures called gills are used by most aquatic arthropods
   (d) Reptiles, birds and mammals respire through the lungs

2. A catabolic biological process in living organisms to break organic molecules to release energy is known as:
   (a) Excretion (b) Combustion
   (c) Respiration (d) Digestion

3. The cartilaginous rings present in the trachea are:
   (a) Complete
   (b) Incomplete dorsally
   (c) Incomplete ventrally
   (d) Absent

4. Trachea is a straight tube extending up to the mid-thoracic cavity, which is divided at the level of the:
   (a) 7th thoracic vertebra
   (b) 5th cervical vertebra
   (c) 5th thoracic vertebra
   (d) 7th cervical vertebra

5. Which of the following is not a function of the conducting part of the respiratory system?
   (a) Humidification of inhaled air
   (b) Transport of atmospheric air to alveoli
   (c) Trapping dust particles and pathogens from inhaled air
   (d) Diffusion of oxygen and carbon dioxide between blood and atmospheric air

6. Incomplete cartilaginous rings support the wall of:
   (a) Trachea and primary bronchi
   (b) Secondary and tertiary bronchi
   (c) Initial bronchioles
   (d) All of the above

7. Which of the following does not constitute the conducting part of the respiratory system?
   (a) Pharynx and larynx
   (b) Primary and secondary bronchi
   (c) Tertiary bronchi and terminal bronchioles
   (d) Alveolar ducts and alveoli

8. Trachea has incomplete rings of the same cartilage that:
   (a) Forms the pinna
   (b) Forms the epiglottis
   (c) Forms the nasal septum
   (d) Forms the tip of the nose

9. In humans, the lungs are covered by a double-layered pleura with pleural fluid between them. The layer which is in close contact with the lung's surface is:
   (a) Parietal pleura (b) Visceral pleura
   (c) Peritoneal pleura (d) Fibrous pleura

10. The lungs are situated in the thoracic chamber, which is an anatomically tight chamber. The dorsal and ventral boundaries of the thoracic chamber are formed by A and B, respectively. Choose the option which correctly fills in the blanks A and B:
    (a) A – Ribs, B – Sternum
    (b) A – Vertebral column, B – Sternum
    (c) A – Sternum, B – Vertebral column
    (d) A – Sternum, B – Ribs

11. Which of the following is an incorrect statement?
(a) Alveoli are thin, irregular-walled and vascularised bag-like structures
(b) The exchange part of the respiratory tract is the site of actual diffusion of oxygen and carbon dioxide between blood and atmospheric air
(c) Larynx is a cartilaginous box which helps in sound production
(d) During swallowing, the epiglottis prevents the entry of food into the esophagus

12. Identify the incorrect statement:
(a) Cardiac notch is present in the left lung
(b) Right lung has two lobes while left lung has three lobes
(c) A pair of external nostrils is present above the upper lips
(d) The branching network of bronchi, bronchioles and alveoli comprise the lungs

13. Mark the correct option for the given statements:
A. Contraction of ______ increases the volume of the thoracic chamber in the antero-posterior axis
B. Contraction of ______ lifts up the ribs and the sternum and causes an increase in the volume of the thoracic chamber in the dorso-ventral axis
(a) A – Diaphragm,
B – External intercostal muscles
(b) A – External intercostal muscles,
B – Diaphragm
(c) A – Diaphragm,
B – Internal intercostal muscles
(d) A – External intercostal muscles,
B – Internal intercostal muscles

14. Men generally have a lower pitched voice than women because:
(a) Vocal cords in males are thinner and shorter than in females
(b) Vocal cords in males are thicker and shorter than in females
(c) Vocal cords in males are thinner and longer than in females
(d) Vocal cords in males are thicker and longer than in females

15. The expulsion of air from the lungs or expiration is caused by:
(a) Contraction of diaphragm
(b) Contraction of external intercostal muscles
(c) Relaxation of the diaphragm and intercostal muscles
(d) More than one option is correct

16. The volume of air involved in breathing movements can be estimated using a:
(a) Sphygmomanometer (b) Spirometer
(c) Kymograph (d) Barometer

17. Which of the following is an incorrect statement with respect to the mechanism of breathing?
(a) The movement of air into and out of the lungs is carried out by creating a pressure gradient between the lungs and the atmosphere
(b) Inspiration can occur if the pressure within the lungs is less than atmospheric pressure
(c) Expiration takes place when the intra-pulmonary pressure is higher than atmos-pheric pressure
(d) During inspiration, alveolar air is released out and during expiration, atmospheric air is drawn in

18. Identify the incorrect statement:
(a) Inflammation of the pleura is called pleurisy
(b) Each lung has 300–350 million alveoli
(c) The nasal chambers open into the larynx, the common passage for food and air
(d) Respiratory bronchioles, alveolar ducts and alveoli constitute the exchange part of the respiratory tract

19. Which of the following is an incorrect statement?
(a) Humans have the ability to increase the strength of inspiration and expiration with the help of additional muscles in the abdomen
(b) On an average, a healthy human breathes 12–16 times per minute
(c) Phrenic muscles are present in the diaphragm
(d) During inspiration, the diaphragm becomes dome-shaped and increases the intra-pulmonary pressure to slightly above atmospheric pressure

20. The alveoli of the lungs are lined with:
(a) Simple columnar epithelium
(b) Simple squamous epithelium
(c) Simple cuboidal epithelium
(d) Pseudostratified epithelium

21. ______ is defined as the presence of air or gas in the pleural cavity (the potential space between the visceral and parietal pleura of the lungs), causing the lungs to collapse:
(a) Pneumonia (b) Pneumothorax
(c) Emphysema (d) Hydrothorax

22. Read the following statements:
(i) The length of the trachea is 11–12 cm
(ii) Thyroid cartilage is the largest in the larynx
(iii) There are two pairs of vocal cords in the pharynx
(iv) Breathing rate is higher in adults than in children
Which of the above statements are correct?

(a) (i) and (ii) (b) (ii) and (iii)
(c) (iii) and (iv) (d) (ii) and (iv)

23. Adam's apple corresponds to:
(a) Cricoid cartilage in male
(b) Cricoid cartilage in female
(c) Thyroid cartilage in female
(d) Thyroid cartilage in male

24. Voice is produced due to the vibration of:
(a) Upper pair of vocal cords during expiration
(b) Lower pair of vocal cords during inspiration
(c) Lower pair of vocal cords during expiration
(d) Upper pair of vocal cords during inspiration

25. Lungs do not collapse between breaths and some air always remains in the lungs which can never be expelled because:
(a) Pressure in lungs is higher than atmospheric pressure
(b) There is positive intrapleural pressure
(c) There is neutral pressure in the lungs
(d) There is negative intrapleural pressure pulling at the lung walls

Respiratory Volume and Capacity, Exchange of Gases, Transport of Gases

26. The total volume of air a person can inspire after a normal expiration is called:
(a) Inspiratory reserve volume (IRV)
(b) Inspiratory capacity (IC)
(c) Expiratory capacity (EC)
(d) Vital capacity (VC)

27. The volume of air that will remain in the lungs after normal expiration is called:
(a) Vital capacity
(b) Functional residual capacity
(c) Residual volume
(d) Tidal volume

28. The volume of air involved in breathing movements can be estimated by a spirometer, except:
(a) Tidal volume
(b) Inspiratory reserve volume
(c) Residual volume
(d) Vital capacity

29. Total volume of air accommodated in the lungs at the end of forced inspiration is:
(a) Total lung capacity (TLC)
(b) VC + RV
(c) 5–6 litres
(d) All of the above

30. The correct value of pCO_2 (mm Hg) in pulmonary veins and systemic veins is:
(a) 45 and 40, respectively
(b) 40 and 45, respectively
(c) 40 in both
(d) 45 in both

31. If a person breathes in air forcibly after forceful expiration, then the volume of air breathed in is:
(a) Inspiratory capacity
(b) Expiratory capacity
(c) Vital capacity
(d) Inspiratory reserve volume

32. The amount of oxygen transported by one gram of haemoglobin is:
(a) 20 ml (b) 1.34 ml
(c) 4.7 ml (d) 15 ml

33. The haem part of haemoglobin is responsible for the transport of which of the following respiratory gas(es)?
(a) Only CO_2
(b) Only O_2
(c) Both O_2 and CO_2
(d) All atmospheric gases

34. Every 100 ml of deoxygenated blood delivers __________ of CO_2 to the alveoli:
(a) 7 ml (b) 4 ml
(c) 20 ml (d) 15 ml

35. Which of the following is incorrect with respect to carbonic anhydrase?
(a) The enzyme uses zinc as cofactor
(b) The enzyme catalyses the formation of carbonic acid
(c) The enzyme is present only in RBC and absent in plasma
(d) RBCs contain high concentration of this enzyme and minute quantities of the same are present in plasma too

36. Which of the following is incorrect?
(a) VC = IRV + TV + ERV
(b) EC = TV + ERV
(c) FRC = ERV + RV
(d) TLC = IC + RV

37. Which among the following has the lowest value?
(a) Inspiratory capacity
(b) Residual volume
(c) Expiratory reserve volume
(d) Tidal volume

38. The partial pressure of oxygen in inhaled and exhaled air is __________, respectively:
(a) 116 mm Hg and 104 mm Hg
(b) 159 mm Hg and 116 mm Hg
(c) 159 mm Hg and 104 mm Hg
(d) 116 mm Hg and 32 mm Hg

39. Hamburger phenomena involves:
(a) Diffusion of HCO_3^- into plasma and entry of Cl^- into RBC at the level of tissues
(b) Diffusion of HCO_3^- into plasma and entry of Cl^- into RBC at the level of lungs
(c) Diffusion of Cl^- into plasma and entry of HCO_3^- into RBC at the level of tissues
(d) Diffusion of Cl^- into plasma and entry of HCO_3^- into RBC at the level of lungs

40. Alveolar ventilation is:
(a) VC + RV
(b) IRV + TV + ERV
(c) TV × Respiratory rate
(d) (TV – Dead space volume) × Respiratory rate

41. Select the correct statement from the following:
(a) The pO_2 of arterial blood is more than the pO_2 of alveolar air
(b) The pO_2 of alveolar air is lower than the pO_2 of venous blood
(c) The pO_2 of alveolar air is lower than the pO_2 of exhaled air
(d) The pCO_2 of arterial blood is same as the pCO_2 of tissues

42. Which of the following is correct?
(a) Alveolar ventilation is more than pulmonary ventilation
(b) Pulmonary ventilation is equal to alveolar ventilation
(c) Pulmonary ventilation is more than alveolar ventilation
(d) Pulmonary ventilation is less than alveolar ventilation

43. In which of the following conditions does the dissociation curve of oxyhaemoglobin shift to the right of the normal curve?
(a) Fall in temperature
(b) Fall in acidity
(c) Decrease in CO_2 concentration
(d) Fall in pH

44. Chloride shift is essential for the transport of:
(a) Oxygen (b) Carbon dioxide
(c) Hydrogen (d) Carbon monoxide

45. Which of the following is incorrect with respect to functional residual capacity (FRC)?
(a) It includes ERV
(b) It is around 2 to 2.5 litres
(c) It can be exhaled totally
(d) It cannot be measured by a spirometer

46. The rate of diffusion of a gas across the respiratory membrane is affected by:
(a) Pressure gradient
(b) Solubility of gases
(c) Thickness of the membrane
(d) All of the above

47. Which of the following does not occur as blood passes through systemic capillaries in tissue?
(a) Increase in oxygen affinity of haemoglobin
(b) Shift of oxygen dissociation curve of haemoglobin to the right
(c) Unloading of oxygen by oxyhaemoglobin into tissues
(d) More than one option is correct

48. Which of the following statements is false with respect to gaseous transport?
(a) Maximum oxygen is transported in dissolved state in plasma
(b) In a resting person, the saturation of haemoglobin (with oxygen) as blood leaves the tissue capillaries is nearly 75%
(c) Carbaminohaemoglobin is formed by binding of CO_2 with the haem component of haemoglobin
(d) Both (a) and (c) are false

49. Identify the incorrect statement:
(a) Haemoglobin has the ability to bind reversibly with oxygen
(b) During transport of CO_2, blood does not become acidic due to buffering action of haemoglobin
(c) The dissociation of oxygen from oxyhaemoglobin will increase if temperature rises in the body
(d) Nearly 20–25% of CO_2 is transported in dissolved form in plasma

50. Every 100 ml of oxygenated blood delivers _____ ml of oxygen to the tissues under normal physiological conditions:
(a) 15 (b) 25
(c) 5 (d) 20

51. Maximum CO_2 is transported by deoxygenated blood to lungs as:
(a) Dissolved state in plasma
(b) Bicarbonates
(c) Carboxyhaemoglobin
(d) Carbaminohaemoglobin

52. Which of the following is incorrect with respect to oxygen transport?
(a) Each haemoglobin molecule can carry four molecules of oxygen
(b) Binding of oxygen with haemoglobin is primarily related to partial pressure of oxygen

(c) Partial pressure of CO_2, hydrogen ion concentration and temperature are the factors which interfere with binding of oxygen to haemoglobin
(d) A hyperbolic curve is obtained when percentage saturation of haemoglobin with oxygen is plotted against the pO_2

53. Which of the following is incorrect with respect to the factors that are favourable for the formation of oxyhaemoglobin in the alveoli?
(a) High pO_2
(b) Low pCO_2
(c) Lower temperature
(d) Higher H^+ concentration

54. Which of the following is an incorrect statement with respect to CO_2 transport?
(a) CO_2 is carried by haemoglobin as carbaminohaemoglobin
(b) CO_2 diffuses into blood from tissues and forms HCO_3^- and H^+, combining with water
(c) Every 100 ml of deoxygenated blood delivers approximately 4 ml of CO_2 to the alveoli
(d) Chloride shift is essential for CO_2 transport and refers to the movement of chloride ions from RBCs into plasma and the simultaneous transfer of HCO_3^- from plasma into RBCs

55. The extra amount of air that remains in the lungs after normal expiration and can be expired forcefully is:
(a) Tidal volume
(b) Residual volume
(c) Expiratory reserve volume
(d) Vital capacity

56. During gaseous exchange, oxygen moves from alveolar air to blood of the pulmonary capillaries by moving across I and II.
Choose the option which fills in the blanks I and II correctly in the above statement:
(a) 1 columnar cell layer, 2 squamous cell layers
(b) 2 squamous cell layers, 1 non-cellular layer
(c) 1 cuboidal cell layer, 2 squamous cell layers
(d) 1 squamous cell layer, 1 non-cellular layer

57. Partial pressure of CO_2 in deoxygenated blood is equivalent to the partial pressure of:
(a) Oxygen in oxygenated blood
(b) Carbon dioxide in alveoli
(c) Carbon dioxide in tissues
(d) Oxygen in alveolar air

58. State true (T) or false (F) for the following statements and select the correct option:
A. Lungs do not collapse between breaths and some air always remains in the lungs called residual volume
B. Under normal physiological conditions, percentage of CO_2 transported through blood plasma is 77%
C. Vital capacity (VC) = Inspiratory capacity (IC) + Expiratory reserve volume (ERV)
D. Oxygen diffuses more easily than carbon dioxide across the diffusion boundary in alveoli because solubility of oxygen is higher than carbon dioxide
(a) A(T), B(F), C(T), D(F)
(b) A(T), B(T), C(T), D(F)
(c) A(F), B(T), C(F), D(T)
(d) A(F), B(F), C(T), D(T)

59. In methaemoglobin, iron is present in the form of:
(a) Ferrous ions
(b) Ferric ions
(c) Carboxyhaemoglobin
(d) None of the above

60. Which of the following is observed if human blood becomes acidic?
(a) RBC count falls
(b) WBC count rises
(c) Oxygen carrying capacity of haemoglobin decreases
(d) Oxygen carrying capacity of haemoglobin increases

61. Which of the following is observed during acclimatisation of body at high altitude?
(a) Increase in respiratory rate
(b) Increase in levels of 2.3-diphosphoglycerate
(c) Increase in RBC count in blood
(d) All of the above

62. The diffusion membrane in the lungs is made of:
(a) Thin squamous epithelium of alveoli
(b) Endothelium of alveolar capillaries
(c) The basement substance between alveolar epithelium and capillaries
(d) All of the above

63. Which of the following is an incorrect statement?
(a) Lecithin acts as a lung surfactant and prevents collapse of the alveoli
(b) Dead space refers to the volume of air present in the alveoli and is used for gaseous exchange
(c) Pressure contributed by an individual gas in a mixture of gases is called partial pressure
(d) Oxygen in expired air is 16%

64. Carbon monoxide combines with haemoglobin to form:
(a) Carbaminohaemoglobin
(b) Methaemoglobin
(c) Carboxyhaemoglobin
(d) Oxyhaemoglobin

65. ________ refers to the deficiency of oxygen in tissues:
(a) Hypoxia (b) Apnea
(c) Eupnea (d) Hypercapnia

66. Identify the incorrect statement:
(a) pO_2 and pCO_2 of alveolar air is 104 mm Hg and 40 mm Hg, respectively
(b) In lungs, gaseous exchange takes place by diffusion
(c) The dissociation of oxygen from oxyhaemoglobin will increase if CO_2 levels in blood fall
(d) In comparison with adult haemoglobin, foetal haemoglobin has higher oxygen affinity

67. Apnea refers to:
(a) Normal breathing
(b) Absence of breathing
(c) Difficulty in breathing
(d) Rapid shallow breathing

68. Yawning occurs due to:
(a) Excess of oxygen
(b) Deficiency of oxygen
(c) Excess of nitrogen
(d) Deficiency of carbon dioxide

69. Which of the following is an incorrect statement?
(a) pO_2 of alveolar air is more than pO_2 of arterial blood
(b) pCO_2 of systemic veins is more than pCO_2 of alveolar air
(c) Sum of partial pressure of all gases at sea level is 760 mm Hg
(d) pCO_2 is more than pO_2 in exhaled air

70. Involuntary spasms of the diaphragm causing sudden jerky inspiration is called:
(a) Asphyxia (b) Hiccup
(c) Yawning (d) Heartburn

71. Partial pressure of O_2 and CO_2 in atmospheric air compared to those in alveolar air is:

	pO_2	pCO_2
(a)	Lower	Higher
(b)	Higher	Higher
(c)	Higher	Lower
(d)	Lower	Lower

72. Match the following columns and choose the correct option:

Column I (% of CO_2 transported)	Column II (Form of CO_2)
A. 70%	(i) Dissolved in plasma
B. 20–25%	(ii) Carbaminohaemoglobin
C. 5–10%	(iii) Bicarbonates

(a) A – (i), B – (ii), C – (iii)
(b) A – (ii), B – (i), C – (iii)
(c) A – (iii), B – (ii), C – (i)
(d) A – (ii), B – (iii), C – (i)

73. The oxygen dissociation curve is:
(a) Hyperbolic (b) 'J' shaped
(c) Sigmoid (d) Bell shaped

74. In which of the following conditions does dissociation of oxygen from oxyhaemoglobin occur?
(a) High H^+ (b) Low pO_2
(c) High pCO_2 (d) All of the above

75. The solubility of CO_2 in blood is:
(a) Slightly lower than that of O_2
(b) 20–25 times higher than that of O_2
(c) 50–60 times higher than that of O_2
(d) Almost equal to that of O_2

76. pO_2 in humans is similar in:
(a) Alveoli and atmospheric air
(b) Oxygenated and deoxygenated blood
(c) Alveoli and deoxygenated blood
(d) Deoxygenated blood and tissue

77. Match the following columns and choose the correct option:

Column I	Column II
A. Vital capacity	(i) ERV + TV
B. Function residual capacity	(ii) IRV + TV
C. Inspiratory capacity	(iii) TV + IRV + ERV
D. Expiratory capacity	(iv) ERV + RV

(a) A – (i), B – (iii), C – (ii), D – (iv)
(b) A – (iv), B – (iii), C – (i), D – (ii)
(c) A – (i), B – (iii), C – (iv), D – (ii)
(d) A – (ii), B – (i), C – (iii), D – (iv)

78. The given graph is a demonstration of ________ on the 'A' axis and ________ on the 'B' axis:

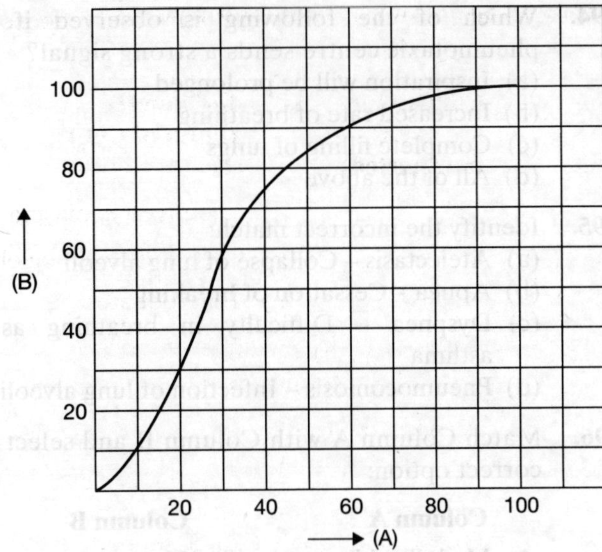

(a) (A) % Saturation of haemoglobin with O_2
  (B) Partial pressure of O_2
(b) (A) Partial pressure of O_2
  (B) % Saturation of Hb with O_2
(c) (A) Solubility of O_2
  (B) Saturation percentage of O_2 with haemoglobin
(d) (A) Saturation percentage of O_2 with haemoglobin
  (B) Solubility of oxygen

79. **In the given diagram, certain parts have been labelled A, B and C. Mark the correct option with respect to identification of the labelled structures:**

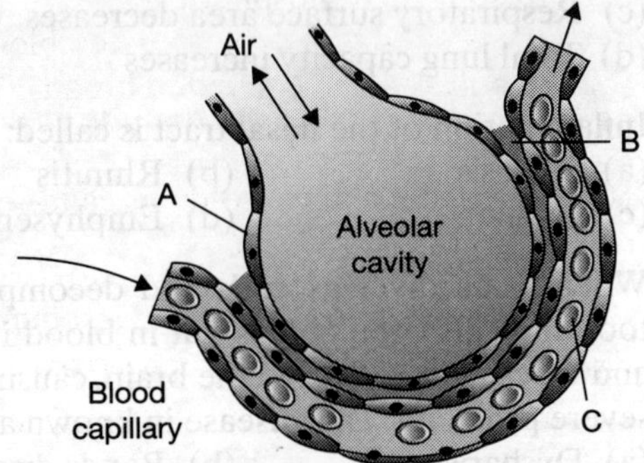

(a) A – Alveolar wall
  B – Basement substance
  C – Red blood cell
(b) A – Blood capillary
  B – Alveolar wall
  C – White blood cell
(c) A – Blood capillary
  B – Basement membrane
  C – Red blood cell
(d) A – Alveolar wall
  B – Blood capillary
  C – Red blood cell

80. **Consider the following table of respiratory gases oxygen and carbon dioxide along with their partial pressure (in mm Hg) in atmospheric air, alveoli, oxygenated blood and tissue. Choose the correct option that fills in the blanks A, B, C and D in the given table:**

Respiratory gas	Atmospheric air	Alveoli	Oxygenated blood	Tissues
O_2	A	104	95	D
CO_2	0.3	B	C	45

(a) A – 95
  B – 45
  C – 40
  D – 40
(b) A – 159
  B – 40
  C – 40
  D – 40
(c) A – 104
  B – 40
  C – 45
  D – 40
(d) A – 159
  B – 45
  C – 40
  D – 95

81. **Hering–Breuer reflexes help to:**
(a) Increase respiratory rate
(b) Trap dust particles from inhaled air
(c) Prevent over-inflation of lungs
(d) Transport gases in blood

82. **__________ is a chronic disorder in which alveolar walls are damaged, due to which respiratory surface is decreased:**
(a) Asthma (b) Emphysema
(c) Bronchitis (d) Pneumonia

83. **A strong pneumotaxic signal will:**
(a) Increase duration of inspiration
(b) Decrease duration of inspiration
(c) Cause partial filling of lungs
(d) Both (b) and (c) are correct

84. **Peripheral chemoreceptors that recognise changes in CO_2 and H^+ concentration and send signals to the rhythm centre for remedial action are located in:**
(a) Brachial artery and aortic arch
(b) Pulmonary artery and carotid artery
(c) Carotid artery and aortic arch
(d) Renal artery and abdominal aorta

85. Respiratory centre of the brain is stimulated by:
(a) O_2 content in deoxygenated blood
(b) CO_2 content in deoxygenated blood
(c) CO_2 content in oxygenated blood
(d) O_2 content in oxygenated blood

86. Asthma may be attributed to:
(a) Accumulation of fluid in the lungs
(b) Inflammation of trachea
(c) Allergic reaction of the mast cells in the lungs
(d) Bacterial infection of the lungs

87. Lack of pulmonary surfactant can cause:
(a) Asthma
(b) Pneumonia
(c) Emphysema
(d) Respiratory distress syndrome

88. Pneumotaxic centre for respiration is located in:
(a) Medulla (b) Pons
(c) Cerebellum (d) Hypothalamus

89. Which of the following is/are example(s) of occupational lung disease?
(a) Silicosis (b) Pneumoconiosis
(c) Asbestosis (d) All of the above

90. The neural signals from the pneumotaxic centre can ________ the duration of inspiration and thereby alter the respiratory rate:
(a) Enhance (b) Reduce
(c) Not change (d) Increase

91. A chemosensitive area is situated adjacent to the rhythm centre, which is highly sensitive to:
(a) CO_2 and O_2 (b) O_2 and H^+
(c) CO_2 and H^+ (d) Only O_2

92. When blood contains high percentage of CO_2 and very low percentage of O_2, the condition is called:
(a) Emphysema (b) Asphyxia
(c) Eupnea (d) Pleurisy

93. Workers in the grinding and stone-breaking industries may suffer from:
(a) Emphysema
(b) Bacterial infection of alveoli
(c) Pneumothorax
(d) Lung fibrosis

94. Which of the following is observed if the pheumotaxic centre sends a strong signal?
(a) Inspiration will be prolonged
(b) Increased rate of breathing
(c) Complete filling of lungs
(d) All of the above

95. Identify the incorrect match:
(a) Atelectasis – Collapse of lung alveoli
(b) Apnea – Cessation of breaking
(c) Dyspnea – Difficulty in breathing as in asthma
(d) Pneumoconiosis – Infection of lung alveoli

96. Match Column A with Column B and select the correct option:

Column A	Column B
A. Medulla oblongata	(i) Chemoreceptors
B. Pons Varolii	(ii) Switch off centre for breathing
C. Aortic arch and carotid artery	(iii) Normal respiratory rhythm centre

(a) A – (i), B – (ii), C – (iii)
(b) A – (iii), B – (ii), C – (i)
(c) A – (ii), C – (i), C – (iii)
(d) A – (ii), B – (i), C – (iii)

97. Emphysema is a chronic disorder in which the walls of the alveoli are damaged; as a result:
(a) Vital capacity increases
(b) Respiratory surface area increases
(c) Respiratory surface area decreases
(d) Total lung capacity increases

98. Inflammation of the nasal tract is called:
(a) Silicosis (b) Rhinitis
(c) Asthma (d) Emphysema

99. When a sea diver ascends and decompression is too rapid, nitrogen comes out in blood in bubbles and blocks blood flow to the brain, causing pain or severe problems. This disease in known as:
(a) Dysbarism (b) Bends disease
(c) Caisson's disease (d) All of the above

100. Which of the following respiratory diseases is called whooping cough?
(a) Tuberculosis (b) Pertussis
(c) Diphtheria (d) Pneumonia

NCERT EXEMPLAR QUESTIONS

1. Respiration in insects is called direct because:
   (a) The cells exchange O_2/CO_2 directly with the air in the tubes
   (b) The tissues exchange O_2/CO_2 directly with coelomic fluid
   (c) The tissues exchange O_2/CO_2 directly with the air outside through body surface
   (d) Tracheal tubes exchange O_2/CO_2 directly with the haemocoel which then exchanges with tissues

2. Regarding the functions of our nasal chamber, mark the incorrect entry:
   (a) Humidifies the air (b) Warms up the air
   (c) Exchange of gases (d) Cleans up the air

3. A person suffers punctures in his chest cavity in an accident without any damage to the lungs. Its effect could be:
   (a) Reduced breathing rate
   (b) Rapid increase in breathing rate
   (c) No change in respiration
   (d) Cessation of breathing

4. It is known that exposure to carbon monoxide is harmful to animals because:
   (a) It reduces CO_2 transport
   (b) It reduces O_2 transport
   (c) It increases CO_2 transport
   (d) It increases O_2 transport

5. Mark the true statement among the following with reference to normal breathing:
   (a) Inspiration is a passive process whereas expiration is active
   (b) Inspiration is a active process whereas expiration is passive
   (c) Inspiration and expiration are active processes
   (d) Inspiration and expiration are passive processes

6. A person breathes in some volume of air by forced inspiration after having a forced expiration. This quantity of air taken in is:
   (a) Total lung capacity (b) Tidal volume
   (c) Vital capacity (d) Inspiratory capacity

7. Mark the incorrect statement in the context of O_2 binding to Hb:
   (a) Higher pH (b) Lower temperature
   (c) Lower pCO_2 (d) Higher PO_2

8. Which of the following statements is incorrect regarding the respiratory system?
   (a) Each terminal bronchiole gives rise to a network of bronchi
   (b) The alveoli are highly vascularised
   (c) The lungs are covered by a double-layered membrane
   (d) The pleural fluid reduces friction on the lung surface

9. Incidence of emphysema, a respiratory disorder, is high in cigarette smokers. In such cases:
   (a) The bronchioles are found damaged
   (b) The alveolar walls are found damaged
   (c) The plasma membrane is found damaged
   (d) The respiratory muscles are found damaged

10. Respiratory process is regulated by certain specialised centres in the brain. One of the following centres can reduce the inspiratory duration upon stimulation:
    (a) Medullary inspiratory centre
    (b) Pneumotaxic centre
    (c) Apneustic centre
    (d) Chemosensitive centre

11. CO_2 dissociates from carbaminohaemoglobin when:
    (a) pCO_2 is high and pO_2 is low
    (b) pO_2 is high and pCO_2 is low
    (c) pCO_2 and pO_2 are equal
    (d) None of the above

12. In breathing movements, air volume can be estimated by:
    (a) Stethoscope
    (b) Hygrometer
    (c) Sphigmomanometer
    (d) Spirometer

13. From the following relationships between respiratory volume and capacities and mark the correct answer:
    (i) Inspiratory capacity (IC) = Tidal Volume + Residual Volume
    (ii) Vital Capacity (VC) = Tidal Volume (TV) + Inspiratory Reserve Volume (IRV) + Expiratory Reserve Volume (ERV)
    (iii) Residual Volume (RV) = Vital Capacity (VC) – Inspiratory Reserve Volume (IRV)
    (iv) Tidal Volume (TV) = Inspiratory Capacity (IC) – Inspiratory Reserve Volume (IRV)

(a) (i) Incorrect, (ii) Incorrect, (iii) Incorrect, (iv) Correct
(b) (i) Incorrect, (ii) Correct, (iii) Incorrect, (iv) Correct
(c) (i) Correct, (ii) Correct, (iii) Incorrect, (iv) Correct
(d) (i) Correct, (ii) Incorrect, (iii) Correct, (iv) Incorrect

14. The oxygen–haemoglobin dissociation curve will show a right shift in the case of:
(a) High pCO_2
(b) High pO_2
(c) Low pCO_2
(d) Less H^+ concentration

15. Match the following and mark the correct options:

Animal	Respiratory organ
A. Earthworm	(i) Moist cuticle
B. Insects	(ii) Gills
C. Fishes	(iii) Lungs
D. Birds/Reptiles	(iv) Trachea

(a) A – (ii), B – (i), C – (iv), D – (iii)
(b) A – (i), B – (iv), C – (ii), D – (iii)
(c) A – (i), B – (iii), C – (ii), D – (iv)
(d) A – (i), B – (ii), C – (iv), D – (iii)

PRACTICE TEST PAPER

1. Peripheral chemoreceptors are located in:
(a) Lung alveoli
(b) Bronchioles
(c) Carotid artery and aortic arch
(d) Peripheral nervous system

2. Pneumotaxic centre is related to ___I___ and located in ___II___. Fill in the blanks with the correct option:

	I	II
(a)	Increasing the respiratory rate	Cerebral cortex
(b)	Stimulating of exhalation	Medulla oblongata
(c)	Switching off inspiration	Pons Varolii
(d)	Inhibition of inhalation	Lungs themselves

3. A chronic disorder in which the alveolar walls are damaged due to which respiratory surface is decreased is:
(a) Bronchitis (b) Emphysema
(c) Pneumonia (d) Asthma

4. Which of the following is a function of the respiratory/exchange part of the respiratory system?
(a) Humidification of inhaled air
(b) Clearing the inhaled air from dust particles
(c) Conduction of air to alveoli
(d) Diffusion of O_2 and CO_2 between atmospheric air and blood

5. Identify the incorrect match:
(a) Residual volume → 1100–1200 ml
(b) Tidal volume → 500 ml
(c) Inspiratory reserve volume → 1000–1100 ml
(d) Vital capacity → 3.4–4.8 litres

6. Which of the following shifts the oxygen dissociation curve to the left?
(a) High pCO_2
(b) High temperature
(c) High pH
(d) High H^+ concentration

7. Spirometry cannot measure the following volumes and capacities, except:
(a) Total lung capacity
(b) Residual volume
(c) Expiratory reserve volume
(d) Functional residual capacity

8. Mark the correct match:
(a) Cockroach – Gills
(b) Scorpion – Book lungs
(c) Earthworm – Tracheal respiration
(d) Lizard – Skin

9. The partial pressure of oxygen in oxygenated blood is:
(a) 104 mm Hg (b) 95 mm Hg
(c) 40 mm Hg (d) 45 mm Hg

10. Which of the following is correct?
(a) Trachea measures 15–16 cm in length
(b) Cartilaginous rings of trachea are incomplete ventrally
(c) Larynx has a pair of thyroid cartilages
(d) Vocal cords are longer and thicker in males due to the effect of androgens

11. Read the following statements and choose the correct option:
I: Hypoxia refers to deficiency of oxygen in tissue
II: Chemoreceptors in medulla respond only to low levels of oxygen in blood
(a) Both statements are correct
(b) Both statements are incorrect
(c) Only statement I is correct
(d) Only statement II is correct

12. Which of the following is an incorrect statement with respect to the human respiratory system?
(a) Right lung is larger than the left
(b) Cardiac notch is present in right lung
(c) Muscles of the diaphragm contract during inspiration
(d) Respiratory bronchioles, alveolar ducts and alveoli form the exchange part of the respiratory tract

13. Carbon monoxide forms a stable complex called ________ with haemoglobin:
(a) Carbaminohaemoglobin
(b) Carboxyhaemoglobin
(c) Methaemoglobin
(d) Haemocyanin

14. Identify the incorrect statement with respect to the regulation of respiration:
(a) The rhythm centre of respiration is located in the medulla
(b) Neural signal from pneumotaxic centre reduces the duration of inspiration
(c) Receptors in the carotid artery and aortic arch are sensitive to CO_2 and H^+
(d) The role of oxygen in regulation of respiratory rhythm is very significant

15. During swallowing, entry of food into the larynx is prevented by a thin cartilaginous flap called:
(a) Uvula (b) Gullet
(c) Glottis (d) Epiglottis

16. Total volume of air that a person can expire after normal inspiration is:
(a) Residual volume
(b) Expiratory reserve volume
(c) Expiratory capacity
(d) Functional residual capacity

17. Select the correct statement:
(a) CO_2 is transported as carboxyhaemoglobin in blood
(b) Dissolution of O_2 from haemoglobin is favoured by high pH at the tissue site
(c) Carbonic anhydrase is present mainly in plasma and in small amounts in RBCs
(d) CO_2 combines with globin component of haemoglobin to form carbaminohaemoglobin

18. Cartilaginous rings are present in the wall of all of the following, except:
(a) Primary bronchi
(b) Trachea
(c) Respiratory bronchioles
(d) Secondary bronchi

19. Every 1000 ml of oxygenated blood delivers approximately ________ ml of O_2 to the tissues under normal physiological conditions. Choose the option which fills in the blank correctly:
(a) 5 (b) 4
(c) 50 (d) 40

20. The sigmoid curve representing the percentage saturation of haemoglobin with O_2 will shift to the left due to:
(a) High temperature at the level of tissues
(b) High H^+ concentration in the blood
(c) High pCO_2 and low pO_2 of tissues
(d) Low metabolic rate of the tissues

21. The total amount of CO_2 transported through plasma in dissolved state is:
(a) 7% (b) 20–25%
(c) 70% (d) 25–35%

22. Which of the following does not occur as blood passes through systemic capillaries in tissues:
(a) Reduced affinity of haemoglobin for oxygen
(b) Unloading of oxygen
(c) Shift of the oxyhaemoglobin dissociation curve to the right
(d) Shift of the oxyhaemoglobin dissociation curve to the left

23. Which of the following respiratory systems is not closely associated with blood supply?
(a) Skin of earthworm
(b) Gills of fish
(c) Alveoli of human lungs
(d) Trachea of insects

24. Select the incorrect statement:
(a) One haemoglobin transports four oxygen molecules
(b) The pO_2 of inhaled air is 159 mm Hg

(c) CO and O_2 bind on the same site on haemoglobin
(d) CO_2 and O_2 bind on the same site on haemoglobin

25. The sum total of inspiratory reserve volume, tidal volume and expiratory reserve volume is:
(a) Functional residual capacity
(b) Residual volume
(c) Vital capacity
(d) Expiratory capacity

26. ______ is defined as the volume of gas that diffuses through the respiratory membrane per minute for a partial pressure difference of 1 mm Hg:
(a) Inspiratory capacity
(b) Diffusing capacity
(c) Vital capacity
(d) Functional residual capacity

27. Which of the following is an incorrect match with respect to the animal and its corresponding respiratory organs?
(a) Prawn – Gills
(b) Frog – Lungs
(c) Earthworm – Moist skin
(d) *Planaria* – Trachea

28. Chloride shift is essential for the transport of:
(a) Oxygen (b) Nitrogen
(c) Hydrogen (d) Carbon dioxide

29. Which of the following is the maximum?
(a) Inspiratory capacity
(b) Vital capacity
(c) Expiratory capacity
(d) Functional residual capacity

30. The thoracic volume in the antero-posterior axis mainly increases by:
(a) Contraction of diaphragm
(b) Contraction of external intercostal muscles
(c) Relaxation of diaphragm
(d) Relaxation of external intercostal muscles

31. The additional volume of air a person can breathe in after normal inspiration is:
(a) Tidal volume
(b) Inspiratory reserve volume
(c) Expiratory reserve volume
(d) Residual volume

32. Hiccup occurs due to muscular spasm of:
(a) Trachea (b) Larynx
(c) Diaphragm (d) Esophagus

33. Select the incorrect statement:
(a) Nasal chambers are responsible for cleaning and humidification of air
(b) Trachea is lined by ciliated epithelium
(c) Vocal cords are present in the pharynx
(d) Trachea divides into a pair of primary bronchi at the level of the 5th thoracic vertebra

34. The volume of air accommodated in the lungs at the end of forced inspiration is:
(a) Vital capacity
(b) Residual volume
(c) Total lung capacity
(d) Functional residual capacity

35. Read the following statements and choose the correct option:
I: Spirometer cannot measure the volume of air remaining in the lungs after forced expiration
II: Functional residual capacity is around 2–2.5 litres and cannot be totally exhaled
(a) Both statements are correct
(b) Both statements are incorrect
(c) Only statement I is correct
(d) Only statement II is correct

36. In haemoglobin, iron is present in ______ form, while in methaemoglobin, iron is present as ______:
(a) Ferrous, ferric (b) Ferric, ferrous
(c) Ferric, ferric (d) Ferrous, ferrous

37. Select the incorrect statement:
(a) Oxygen is totally absent in exhaled air
(b) Carbonic anhydrase uses zinc as cofactor
(c) Haemoglobin has the ability to bind reversibly with oxygen
(d) The pO_2 of alveolar air is 95 mm Hg

38. The chemoreceptors of the respiratory centre are activated by:
(a) Level of O_2 in arterial blood
(b) Level of O_2 in venous blood
(c) Level of CO_2 in arterial blood
(d) Level of CO_2 in venous blood

39. Ordinary or normal breathing is called:
(a) Dyspnea (b) Tachypnea
(c) Eupnea (d) Apnea

40. The partial pressure of oxygen in the alveoli of the lungs is:
(a) Equal to that in the blood
(b) More than that in the blood
(c) Less than that in the blood
(d) None of the above

41. Bowman's glands are found in:
(a) Olfactory epithelium
(b) External auditory canal
(c) Cortical nephrons
(d) Renal pelvis

42. Amphibians like frogs respire through:
(a) Lungs and moist skin
(b) Gills and moist cuticle
(c) Tracheal tubes and gills
(d) Gills and lungs

43. Surfactant is:
(a) Phospholipid
(b) Deficient in premature infant
(c) Lacking in individuals suffering from acute respiratory distress syndrome
(d) All of the above

44. The factor that affects the rate of alveolar diffusion is:
(a) Solubility of gases
(b) Thickness of the membrane
(c) Pressure gradient
(d) All of the above

45. _______ is difficulty in breathing causing wheezing due to inflammation of bronchi and bronchioles:
(a) Emphysema (b) Diphtheria
(c) Asthma (d) Silicosis

ANSWER KEYS

Exercises

1. (b) **2.** (c) **3.** (b) **4.** (c) **5.** (d) **6.** (d) **7.** (d) **8.** (c) **9.** (b) **10.** (b)
11. (d) **12.** (b) **13.** (a) **14.** (d) **15.** (c) **16.** (b) **17.** (d) **18.** (c) **19.** (d) **20.** (b)
21. (b) **22.** (a) **23.** (d) **24.** (c) **25.** (d) **26.** (b) **27.** (b) **28.** (c) **29.** (d) **30.** (b)
31. (c) **32.** (b) **33.** (b) **34.** (b) **35.** (c) **36.** (d) **37.** (d) **38.** (b) **39.** (a) **40.** (d)
41. (c) **42.** (c) **43.** (d) **44.** (b) **45.** (c) **46.** (d) **47.** (a) **48.** (d) **49.** (d) **50.** (c)
51. (b) **52.** (d) **53.** (d) **54.** (d) **55.** (c) **56.** (b) **57.** (c) **58.** (b) **59.** (b) **60.** (c)
61. (d) **62.** (d) **63.** (b) **64.** (c) **65.** (a) **66.** (c) **67.** (b) **68.** (b) **69.** (d) **70.** (b)
71. (c) **72.** (c) **73.** (c) **74.** (d) **75.** (b) **76.** (d) **77.** (b) **78.** (b) **79.** (a) **80.** (b)
81. (c) **82.** (b) **83.** (d) **84.** (c) **85.** (c) **86.** (c) **87.** (d) **88.** (b) **89.** (d) **90.** (b)
91. (c) **92.** (b) **93.** (d) **94.** (b) **95.** (d) **96.** (b) **97.** (c) **98.** (b) **99.** (d) **100.** (b)

NCERT Exemplar Questions

1. (a) **2.** (c) **3.** (d) **4.** (b) **5.** (b) **6.** (c) **7.** (a) **8.** (c) **9.** (b) **10.** (b)
11. (b) **12.** (d) **13.** (b) **14.** (a) **15.** (b)

Practice Test Paper

1. (c) **2.** (c) **3.** (b) **4.** (d) **5.** (c) **6.** (c) **7.** (c) **8.** (b) **9.** (b) **10.** (d)
11. (c) **12.** (b) **13.** (b) **14.** (d) **15.** (d) **16.** (c) **17.** (d) **18.** (c) **19.** (c) **20.** (d)
21. (a) **22.** (d) **23.** (d) **24.** (d) **25.** (c) **26.** (b) **27.** (d) **28.** (d) **29.** (b) **30.** (a)
31. (b) **32.** (c) **33.** (c) **34.** (c) **35.** (a) **36.** (a) **37.** (a) **38.** (c) **39.** (c) **40.** (b)
41. (a) **42.** (a) **43.** (d) **44.** (d) **45.** (c)

SOLUTIONS

Respiratory Organs, Human Respiratory System, Mechanism of Breathing

1. Insects have a network of tubes (tracheal tubes) to transport atmospheric air into the body, while earthworms have moist cuticles for gaseous exchange.

2. Respiration is a catabolic biological process in living organisms to break organic molecules to release energy.

3. The rings of cartilage in the trachea are incomplete dorsally.

4. Trachea divides into a pair of primary bronchi at the level of the 5th thoracic vertebra.

5. The conducting part of the respiratory system does not participate in diffusion/exchange of gases.

6. Option (d) is correct.

7. Option (d) is correct.

8. Hyaline cartilage is present in the nasal septum and trachea. Epiglottis, pinna and tip of the nose have elastic cartilage.

9. Visceral pleura is in contact with the lung's surface.

10. The thoracic chamber is bounded dorsally by the vertebral column and ventrally by the sternum.

11. During swallowing, the epiglottis prevents the entry of food into the trachea.

12. The right lung has three lobes while the left lung has two lobes.

13. Contraction of the diaphragm increases the size of the thoracic chamber in the antero-posterior axis, while contraction of the external intercostal muscles lifts up the ribs and increases the volume of the thoracic chamber in the dorso-ventral axis.

14. Men have a lower pitched voice than women because vocal cords in males are thicker and longer than in females.

15. The expulsion of air from the lungs is caused by relaxation of the diaphragm and external intercostal muscles.

16. Pulmonary function and pulmonary volumes/ capacities are measured by spirometer.

17. During inspiration, atmospheric air is drawn in while during expiration, alveolar air is released out.

18. The nasal chamber opens into the pharynx, the common passage for food and air.

19. During inspiration, the diaphragm becomes flat and lowers the intra-pulmonary pressure slightly below atmospheric pressure.

20. The alveoli of the lungs are lined with simple squamous epithelium.

21. Presence of air or gas in the pleural cavity is called pneumothorax.

22. There are two pairs of vocal cords in the larynx. Breathing rate is higher in children than in adults.

23. Adam's apple corresponds to the thyroid cartilage in males; it enlarges at puberty under the influence of androgens.

24. Voice is produced by vibration of the lower pair of vocal cords during expiration.

25. Lungs do not collapse between breaths because there is negative intrapleural pressure pulling at the lung walls.

Respiratory Volume and Capacity, Exchange of Gases, Transport of Gases

26. Inspiratory capacity (IC) refers to the total volume of air a person can inspire after a normal expiration.

27. The volume of air that will remain in the lungs after the normal expiration is called function residual capacity (FRC).

28. Residual volume (RV) cannot be estimated by a spirometer.

29. Option (d) is correct.

30. The correct value of pCO_2 (mm Hg) in pulmonary veins and systemic veins is 40 and 45, respectively.

31. If a person breathes in air forcibly after forceful expiration, it is called vital capacity.

32. The volume of oxygen transported by one gram of haemoglobin is 1.34 ml.

33. Only oxygen is transported by the haem part of haemoglobin.

34. Every 100 ml of deoxygenated blood delivers 4 ml of CO_2 to the alveoli.

35. RBCs contain very high concentration of the enzyme carbonic anhydrase and minute quantities of the same in plasma too.

36. TLC = IC + FRC

37. Option (d) is correct.

38. The partial pressure of oxygen in inhaled and exhaled air is 159 mm Hg and 116 mm Hg, respectively.

39. Hamburger phenomena involves diffusion of HCO_3^- into plasma and entry of Cl^- into the RBCs at the level of tissues.

40. Option (d) is correct.

41. The pO_2 of alveolar air (104 mm Hg) is lower than exhaled air (116 mm Hg).

42. Pulmonary ventilation is more than alveolar ventilation.

43. Fall in pH shifts the dissociation curve of oxyhaemoglobin to the right.

44. Chloride shift is essential for CO_2 transport.

45. FRC cannot be totally exhaled.

46. Option (d) is correct.

47. As the blood passes through systemic capillaries, the affinity for oxygen in haemoglobin decreases.

48. Maximum oxygen is transported combined with haemoglobin. Carbaminohaemoglobin is formed by binding of CO_2 with the globin part of haemoglobin.

49. Nearly 5–10% of CO_2 is transported in dissolved state in plasma.

50. Every 100 ml of oxygenated blood delivers 5 ml of oxygen to tissues under normal physiological conditions

51. Maximum (70%) CO_2 is transported by deoxygenated blood to the lungs as bicarbonates.

52. A sigmoid curve is obtained when percentage saturation of haemoglobin with oxygen is plotted against the pO_2.

53. High H^+ concentration lowers the affinity of haemoglobin to oxygen.

54. Chloride shift is essential for CO_2 transport and refers to the movement of HCO_3^- from RBCs to plasma and that of chloride ions from plasma to RBCs.

55. Expiratory reserve volume (ERV) refers to the extra amount of air that remains in the lungs after normal expiration and can be forcefully exhaled.

56. During gaseous exchange, oxygen moves from alveolar air to the blood of the pulmonary capillaries by moving across two squamous cell layers (alveoli and capillary endothelium) and one non-cellular layer (basement membrane).

57. The partial pressure of CO_2 in deoxygenated blood (40 mm Hg) is equivalent to the pCO_2 in tissues.

58. CO_2 diffuses more easily than O_2 across the diffusion boundary because the solubility of CO_2 is higher than O_2.

59. In methaemoglobin, iron is present in ferric form.

60. If blood becomes acidic (pH falls), the oxygen affinity of haemoglobin decreases and more oxygen is delivered to tissues by haemoglobin. Thus, the oxygen carrying capacity of haemoglobin decreases.

61. At high altitude, the RBC count and respiratory rate increase. The rise in levels of 2, 3-diphosphoglycerate (DPG) lowers the oxygen affinity of haemoglobin.

62. The diffusion membrane in the lungs is made up of the squamous epithelium of alveoli, endothelium of alveolar capillaries and the basement substance between them.

63. Dead space refers to the air present in large airways like the trachea and bronchi, and is not used/involved in gaseous exchange in the lungs.

64. Carbon monoxide (CO) combines with haemoglobin to form carboxyhaemoglobin.

65. Hypoxia refers to deficiency of oxygen in tissues.

66. The dissociation of oxygen from haemoglobin will decrease if CO_2 level in blood falls.

67. Apnea refers to absence of breathing.

68. Yawning occurs due to deficiency of oxygen.

69. pO_2 of exhaled air is 116 mm Hg while pCO_2 of exhaled air is 32 mm Hg.

70. Hiccup refers to involuntary spasms of the diaphragm causing sudden jerky inspiration.

71. Option (c) is correct.

72. Option (c) is correct.

73. The oxygen dissociation curve of haemoglobin is sigmoid shaped.

74. The dissociation of oxygen from haemoglobin increases in conditions like high H^+, low pO_2 and high pCO_2.

75. The solubility of CO_2 in blood is 20–25 times higher than that of O_2.

76. The pO_2 in deoxygenated blood and tissue is 40 mm Hg.

77. Option (b) is correct.

78. Option (b) is correct.

79. Option (a) is correct.

80. Option (b) is correct.

81. Hering–Breur reflex prevents over-inflation of the lungs. Pulmonary stretch receptors present on the wall of bronchi and bronchioles of the airway respond to excessive stretching of the lungs during large inspirations.

82. Emphysema is a chronic disorder in which alveolar walls are damaged, due to which the respiratory surface is decreased.

83. Pneumotaxic centre reduces the duration of inspiration. Thus, a strong pneumotaxic signal will decrease the duration of inspiration and cause partial filling of the lungs.

84. Peripheral chemoreceptors located in the carotid artery and aorta recognise the changes in CO_2 and H^+ concentration and send signals to the rhythm centre for remedial action.

85. The respiratory (chemosensitive) area of the brain is stimulated by CO_2 content in oxygenated blood.

86. Asthma may be attributed to allergic reaction of the mast cells in the lungs.

87. Deficiency of pulmonary surfactant can cause respiratory distress syndrome.

88. Pneumotaxic centre is located in pons and reduces the duration of inspiration.

89. Occupational lung diseases are caused in workers of industries involving grinding or stone-breaking, giving rise to inflammation leading to fibrosis (proliferation of fibrous tissue) and thus causing serious lung damage. For example, silicosis, asbestosis and pneumoconiosis.

90. The neural signals from the pneumotaxic centre can reduce the duration of inspiration.

91. The chemosensitive area, situated adjacent to the rhythm centre, is highly sensitive to CO_2 and H^+.

92. Asphyxia is a condition where blood contains a high percentage of CO_2 and low percentage of O_2.

93. The workers of industries related to grinding and stone-breaking suffer from lung fibrosis.

94. A strong pneumotaxic signal increases the rate of breathing.

95. Pneumoconiosis is an occupational lung disease seen in workers of coal mines.

96. Option (b) is correct.

97. Emphysema is a disorder of smokers, where the walls of the alveoli are damaged and the respiratory surface area for gaseous exchange is reduced.

98. Inflammation of the nasal tract is called rhinitis.

99. Option (d) is correct.

100. Pertussis is called whooping cough and is caused by the bacteria *Bordetella pertussis*.

Body Fluids and Their Circulation

EXERCISES

Body Fluids

1. The most numerous leucocytes in blood are:
(a) Eosinophils (b) Neutrophils
(c) Lymphocytes (d) Basophils

2. Total leucocyte count (TLC) in the blood of a healthy human is:
(a) 12,000–14,000 per cu mm
(b) 6000–8000 per cu mm
(c) 1000–2000 per cu mm
(d) 20,000–22,000 per cu mm

3. Select the incorrect statement:
(a) The haemoglobin level in blood is 12–16 g/dL
(b) The lifespan of RBCs in humans is 60 days
(c) Nucleus, mitochondria and Golgi bodies are absent in erythrocytes
(d) The size of the human erythrocyte is 7–8 μm

4. The phagocytic WBCs of blood are:
(a) Neutrophil and lymphocyte
(b) Neutrophil and monocyte
(c) Monocyte and lymphocyte
(d) Acidophil and lymphocyte

5. The leucocyte that mediates inflammatory reactions is:
(a) Acidophil (b) Basophil
(c) Lymphocyte (d) Neutrophil

6. Select the plasma protein that chiefly helps in maintaining osmotic balance:
(a) Albumin (b) Globulin
(c) Fibrinogen (d) Transferrin

7. Select the human organ often referred to as the 'graveyard' of RBCs:
(a) Thymus (b) Liver
(c) Bone marrow (d) Spleen

8. Serum can be described as:
(a) Plasma with formed elements
(b) Blood without clotting factors
(c) Plasma without clotting factors
(d) Blood without formed elements

9. Select the correct match:

	Leucocytes	Per cent of total leucocytes
(a)	Monocytes	0.5–1%
(b)	Basophils	6–8%
(c)	Eosinophils	2–3%
(d)	Neutrophils	20–25%

10. Special cells in the bone marrow produce non-nucleated colourless cell fragments called:
(a) Megakaryocytes (b) Lymphocytes
(c) Acidophils (d) Thrombocytes

11. Body fluid often referred to as "middle man of the body" is:
(a) Blood (b) Serum
(c) Plasma (d) Lymph

12. Conversion of inactive prothrombin into thromb in occurs in the presence of:
(a) Fibrin and Ca^{2+}
(b) Ca^{2+} only
(c) Thromboplastin only
(d) Thrombokinase and Ca^{2+}

13. Erythroblastosis foetalis can be prevented by administering anti-Rh antibodies to:
(a) Rh–ve mother after delivery of Rh+ve baby
(b) Rh+ve mother after delivery of Rh+ve baby
(c) Rh+ve mother after delivery of Rh–ve baby
(d) Rh+ve father

14. Person with blood group 'O⁻' can donate blood to any person with any blood group under the ABO system because they:
(a) Lack antigens of RBCs and have no antibody in plasma
(b) Lack A, B and Rh antigens of RBCs
(c) Have A, B and Rh antigens on RBCs
(d) Lack anti-A, anti-B and anti-Rh antibodies in plasma

15. Which of the following converts inactive fibrinogens of blood into fibrins during cascade of reactions of blood clotting?
(a) Thrombin (b) Thromboplastin
(c) Prothrombin (d) Thrombokinase

16. Enucleated cells present in blood are:
(a) Monocytes (b) Neutrophils
(c) B-lymphocytes (d) Erythrocytes

17. Monocytes and neutrophils differ from each other in all of the following, except:
(a) The shape of their nuclei in mature form
(b) Presence of granules in cytoplasm
(c) Their mode of action against foreign organisms
(d) Their contribution to total WBC content of blood

18. Select the incorrect statement:
(a) Blood is a fluid connective tissue consisting of about 45% formed elements and 55% plasma
(b) Formed elements represent the cellular portion
(c) The ratio of RBC:WBC is 600:1 in blood
(d) Blood has a straw-coloured viscous fluid called plasma, 90–92% of which is made up of proteins

19. Choose the incorrect statement:
(a) Fibrinogen is produced by the liver
(b) Globulins are primarily involved in defense mechanisms of the body
(c) Blood is prevented from clotting inside blood vessels due to the presence of fibrinogen
(d) Plasma contains 6–8% proteins

20. The proteolytic enzyme which causes lysis of fibrin during fibrinolysis is:
(a) Plasmin (b) Thrombin
(c) Fibrin (d) Thrombokinase

21. Bursa of Fabricius is an important organ of birds, linked with:
(a) Formation of nitrogen waste
(b) Generation of B lymphocytes
(c) Production of lipoproteins
(d) Storage of glycogen

22. Erythroblastosis foetalis can occur if the blood group of the father and mother is, respectively:
(a) Rh–ve and Rh+ve
(b) Rh–ve and Rh–ve
(c) Rh+ve and Rh–ve
(d) Rh+ve and Rh+ve

23. Select the incorrect statement:
(a) Rhesus antigen is present in RBCs
(b) Patients with 'AB' blood group do not have any antibodies in plasma
(c) The granules of eosinophils stain with methylene blue
(d) Hirudin is an anticoagulant found in the saliva of leech

24. Blood groups A, B and O were discovered by:
(a) William Harvey
(b) Landsteiner
(c) Weiner
(d) de Castello and Sturli

25. Ethylenediaminetetra-acetic acid (EDTA) strongly and irreversibly chelates ___________ and prevents blood from clotting:
(a) Fibrin (b) Calcium ions
(c) Thrombin (d) Thromboplastin

26. Vitamin essential for blood clotting is:
(a) Retinol (b) Phylloquinone
(c) Tocopherol (d) Riboflavin

27. Christmas disease is caused by deficiency of factor:
(a) IX (b) VIII
(c) III (d) II

28. Which of the following is/are the function(s) of blood?
(a) Transportation of gases
(b) Transportation of food
(c) Defense against infection
(d) All of the above

29. The role of platelets is:
(a) Transport of oxygen
(b) Providing immunity
(c) Blood clotting
(d) Maintain osmotic pressure

30. Select the incorrect statement:
(a) Lifespan of platelets is one week
(b) Size of platelets is 2–3 μm
(c) Abnormal fall in platelet count is thrombocytosis
(d) Breakdown of megakaryocytes in bone marrow forms platelets

31. Which of the following is correct with respect to the lymph?
(a) It is rich in RBCs and thrombocytes
(b) It has WBCs and mostly lymphocytes
(c) It is red in colour due to haemoglobin
(d) It is richer in proteins compared to blood

32. Find the mismatch:
(a) Factor I – Fibrinogen
(b) Factor III – Thromboplastin
(c) Factor VIII – Anti-haemophilic globulin (AHG)
(d) Factor IV – Prothrombin

33. Deficiency of ________ and ________ causes megaloblastic/macrocytic anaemia:
(a) Folic acid, cyanocobalamin
(b) Phylloquinone, ascorbic acid
(c) Riboflavin, thiamine
(d) Folic acid, thiamine

34. Which of the following is correct for lymph?
(a) It transports oxygen from the heart to different organs
(b) It brings tissue fluid back to systemic circulation
(c) It brings metabolic waste from cells into the venous system
(d) Both (b) and (c)

35. Eosinophils:
(a) Constitute 60–65% of total WBCs
(b) Are called polymorphonuclear leucocytes
(c) Secrete antibodies and provide humoral immunity
(d) Resist infections and are associated with allergic reactions

36. The movement of leucocytes out of the circulatory system and towards the site of infection or tissue damage is called:
(a) Diapedesis (b) Synapsis
(c) Haemopoiesis (d) Leucopoiesis

37. Compared to blood, lymph has:
(a) More waste and less proteins
(b) More waste and more proteins
(c) Less waste and less proteins
(d) Less waste and more proteins

38. Select the true statement with respect to leucocytes:
(a) They are all non-phagocytic
(b) Their normal range is 60,000–80,000 per μmm of blood
(c) They are nucleated
(d) They all have granules that stain either acidic or basic dyes

39. Which of the following does not show coagulation/clotting?
(a) Plasma (b) Lymph
(c) Blood (d) Serum

40. Identify the correct statement:
(a) Neutrophils are the most numerous WBCs
(b) A reduction in the number of RBCs causes clotting disorders
(c) Neutrophils are called polymorphonuclear leucocytes
(d) More than one option is correct

Human Circulatory System

41. Incomplete double circulation is seen in:
(a) Aves (b) Mammals
(c) Pisces (d) Amphibians

42. Heart pumps out only deoxygenated blood in:
(a) Pisces (b) Amphibians
(c) Reptiles (d) Aves

43. Heart is protected by a double-walled membranous bag called:
(a) Myocardium (b) Endocardium
(c) Pericardium (d) Epicardium

44. Stroke volume of the human heart under normal physiological conditions is:
(a) 5000 ml (b) 70 ml
(c) 130 ml (d) 700 ml

45. Chordae tendineae in the heart are:
(a) Thin fibres of ventricular nodal tissue
(b) Connection between pulmonary trunk and aorta during embryonic stage
(c) Connection between the cuspid valves and the papillary muscles of ventricles
(d) Valve between left ventricle and aorta

46. The opening present in the atrioventricular septum between the left atrium and left ventricle is guarded by:
(a) Tricuspid valve (b) Thebesian valve
(c) Eustachian valve (d) Bicuspid valve

47. Maximum filling of blood in ventricles occurs during:
(a) Atrial systole (b) Ventricular systole
(c) Joint diastole (d) None of the above

48. Select the incorrect statement with respect to the nodal tissue of the heart:
(a) Sino-atrial node acts as the pacemaker of the heart
(b) Both sino-atrial node and atrio-ventricular node are located in the right atrium
(c) All parts of nodal tissue can generate the same number of action potentials per minute
(d) The entire nodal tissue from the sino-artial node to Purkinje fibres has the ability to generate action potentials

49. The total duration of phase of the cardiac cycle in which atria receive blood from their respective veins is:
(a) 0.1 seconds (b) 0.7 seconds
(c) 0.8 seconds (d) 0.3 seconds

50. The opening present on the inter-atrial septum of the heart during foetal life is called:
(a) Ductus arteriosus
(b) Fossa ovalis
(c) Foramen ovale
(d) Ligamentum arteriosum

51. Select the incorrect statement:
(a) Eustachian valve guards the opening of the coronary sinus into the right atrium
(b) The wall of the left ventricle is the thickest among the four chambers of the human heart
(c) Coronary sulcus is the groove separating the atria from the ventricles
(d) Mitral valve is also called bicuspid valve

52. The thick muscular layer present in the wall of heart is:
(a) Endocardium (b) Myocardium
(c) Pericardium (d) Epicardium

53. Bundle of His is a network of:
(a) Nerve fibres found only in atria
(b) Muscle fibres found only in the ventricles
(c) Muscle fibres found only in atria
(d) Nerve fibres found only in the ventricles

54. Which of the following has the shortest duration in the cardiac cycle?
(a) Atrial systole (b) Atrial diastole
(c) Ventricular systole (d) Ventricular diastole

55. Semilunar valve is present at:
(a) Junction of right ventricle and pulmonary trunk
(b) Junction of left ventricle and aorta
(c) Junction of left atrium and left ventricle
(d) Both (a) and (b)

56. Systemic heart refers to:
(a) Right atrium and right ventricle in human heart
(b) Left atrium and left ventricle in human heart
(c) Both right and left ventricles in human heart
(d) Entire heart in lower vertebrates

57. Select the incorrect statement:
(a) Left atrium receives oxygenated blood from the lungs through two pairs of pulmonary arteries
(b) All four chambers of the heart are relaxed during joint diastole
(c) Volume of blood pumped out by each ventricle per minute is called stroke volume
(d) Both (a) and (c) are incorrect

58. Open vascular system is found in:
(a) Apple snail and earthworm
(b) Earthworm and cockroach
(c) Housefly and prawn
(d) Frog and apple snail

59. The opening of which blood vessel is not guarded by any valves in human beings?
(a) Inferior vena cava (b) Pulmonary artery
(c) Pulmonary veins (d) Both (b) and (c)

60. Find the incorrect statement:
(a) Reptilian heart is arteriovenous heart that shows incomplete double circulation
(b) Heart of fish receives only venous blood which is pumped into lungs for oxygenation
(c) Heart of vertebrates is enclosed in pericardium
(d) Human heart is nearly of the size equivalent to a clenched fist

61. Sino-atrial node has high rhythmicity and is called the pacemaker. It is moderated by:
(a) Cardiac muscles present in the atrio-ventricular node
(b) Cardiac centre present in the medulla oblongata
(c) It is self-regulating and self-stimulating, not moderated
(d) Pons Varolii

62. The time interval between closing of the semilunar valve and the atrio-ventricular valve is:
(a) 0.3 seconds (b) 0.5 seconds
(c) 0.8 seconds (d) 0.1 seconds

63. All of the following valves are in contact with deoxygenated blood, except:
(a) Eustachian valve
(b) Thebesian valve
(c) Mitral valve
(d) Pulmonary semilunar valve

64. First heart sound ___A___ is produced by ___B___ valves.
Choose the option which fills in the blanks correctly:

	A	B
(a)	Lub	Opening of atrioventricular valve
(b)	Lub	Closing of atrioventricular valve
(c)	Dub	Closing of semilunar valve
(d)	Dub	Opening of semilunar valve

65. Identify the incorrect statement:
(a) Atrioventricular bundle is a tract of conducting fibres present in the interventricular bundle
(b) Impulse generated by sino-atrial node stimulates the atria and ventricles to contract simultaneously
(c) SA node has inherent ability to generate nerve impulse at higher rate compared to the rest of the cardiac muscle fibres
(d) Purkinje fibres are conduction fibres distributed in the ventricular muscles

66. Which of the following is an incorrect statement?
(a) P-wave represents depolarisation of the atria
(b) Ventricular systole begins shortly after Q-wave
(c) T-wave represents the return of the ventricles from excited to normal state
(d) The end of T-wave marks the end of atrial systole

67. Human heart is myogenic because:
(a) Impulse for heart beat is generated by cardiac centre in medulla
(b) Myocardium is present in heart wall
(c) Impulse for heart beat is generated in sino-atrial node
(d) Parasympathetic nerves reduce the heart rate

68. In the foetal heart, ductus arteriosus connects:
(a) Right and left atria
(b) Pulmonary artery and pulmonary vein
(c) Pulmonary artery and systemic aorta
(d) Pulmonary vein and systemic aorta

69. Compared to the second heart sound, the first heart sound is:
(a) Less loud (b) Longer
(c) Lower pitched (d) All of the above

70. Upon stimulation of the vagus nerve, the heart rate:
(a) Increases
(b) Decreases
(c) Remains unchanged
(d) First increases and then decreases

71. If a person goes into cardiac arrest, his ECG shows:
(a) Flattened P-wave
(b) Flattened QRS complex
(c) Flattened T-wave
(d) All of the above

72. Select the incorrect statement:
(a) ECG is a graphical representation of the electrical activity of the heart during the cardiac cycle
(b) To obtain a standard ECG, the patient is connected with two leads, one to the left wrist and one to the ankle
(c) The "T-wave" represent the return of the ventricle from excited to normal state
(d) The end of "T-wave" marks the end of systole

73. Which of the following is the correct statement?
(a) Blood pressure is lower in a closed circulatory system compared to an open circulatory system
(b) Molluscs and arthropods have closed circulatory system
(c) The flow of blood in a closed circulatory is slower than in an open circulatory system
(d) Blood flow to organs is precisely regulated in a closed circulatory system

74. In joint diastole:
(a) Atria contract, while ventricles relax
(b) Atria relax, while ventricles contract
(c) Both the atria and ventricles contract simultaneously
(d) All four chambers of the heart are in relaxed state

75. Which among the following is the last to be depolarised in a cardiac cycle?
(a) AV node (b) SA node
(c) Purkinje fibres (d) Bundle of His

76. Read the given statements and choose the correct option:
I: The sino-atrial node generates an action potential which stimulates both the ventricles to undergo simultaneous relaxation
II: Ventricular diastole increases ventricular pressure, causing closure of the tricuspid and bicuspid valves due to attempted backflow of blood into the atria
(a) Both statements are correct
(b) Both statements are incorrect
(c) Only statement I is correct
(d) Only statement II is correct

77. Parasympathetic stimulation results in all, except:
(a) Decrease in cardiac output
(b) Decrease in rate of heart beat
(c) Decrease in the conduction of action potential in heart muscle fibres
(d) Increase in strength of ventricular contraction

78 Which of the following is correct with respect to second heart sound during a cardiac cycle?
(a) It is associated with closure of the semilunar valves at the beginning of ventricular diastole
(b) It is caused due to opening of atrioventricular valves

(c) It is caused due to closure of atrioventricular valves
(d) It is associated with closure of the semilunar valves at the beginning of ventricular systole

79. The increase in rate of heart beat, strength of ventricular contraction and hence the cardiac output can be achieved through:
A. Action of adrenal medullary hormones
B. Neural signals from sympathetic nerves
C. Neural signals from parasympathetic nerves
D. Action of neurotransmitters like catecholamines
Which of the above statements are correct?
(a) (A) and (C) (b) (B), (C) and (D)
(c) (A), (B) and (D) (d) (C) and (D)

80. The semilunar valves open during:
(a) Ventricular systole
(b) Ventricular diastole
(c) Atrial systole
(d) Joint diastole

81. The body has the ability to:
(a) Alter the stroke volume but not the heart rate
(b) Alter the heart rate but not the stroke volume
(c) Alter the heart rate as well as the stroke volume
(d) Alter neither the heart rate nor the stroke volume

82. Identify the incorrect statement:
(a) 72 cardiac cycles are performed by the human heart per minute
(b) Heart sounds are heard using a stethoscope
(c) The heart sounds 'lub' and 'dub' are of no clinical significance
(d) The cardiac output of an athlete is higher than that of an ordinary person

83. Relaxation of the atria coincides with:
(a) Atrial systole
(b) Ventricular systole
(c) Blood flow into atria from large veins
(d) More than one option is correct

84. Heart murmurs:
(a) Are heard due to defective heart valves
(b) Are generated by turbulent flow of blood
(c) Are heard by stethoscope
(d) All of the above are correct

85. Select the incorrect statement:
(a) Purkinje fibres are modified cardiac muscle fibres
(b) Papillary muscles are found in ventricles
(c) Both SA node and AV node are found in the right atrium
(d) Impulse from SA node is spread to the atria via bundle of His while AV bundle spreads the impulse in ventricles

86. The thick muscular projections on the wall of the ventricles are called:
(a) Chordae tendineae (b) Conus arteriosus
(c) Sinus venosus (d) Columnae carneae

87. Choose the option which correctly fills in the given blank:
By the end of joint diastole, the ventricles are almost ________ filled.
(a) One-third (b) Two-third
(c) Half (d) One-fourth

88. The part of the conducting system present in the ventricular wall of the human heart is:
(a) Sino-atrial node
(b) Atrioventricular node
(c) Purkinje fibres
(d) Internodal pathways

89. Read the following statements and select the correct option:
A. First heart sound is 'lub'
B. 'Dub' sound is produced during joint diastole
(a) Only (A) is correct
(b) Only (B) is correct
(c) Both (A) and (B) are correct
(d) Both (A) and (B) are incorrect

90. Given are the events in a cardiac cycle. Mark the incorrect option:
(a) The atrioventricular valves open due to contraction of atria during systole
(b) Bicuspid and tricuspid valves close during ventricular systole
(c) Bundle of His transmits stimuli received from AV node through ventricular musculature
(d) During ventricular relaxation, closure of the semilunar valve precedes opening of the bicuspid and tricuspid valves

Blood Vessels, Circulation through Special Regions, Disorders of the Circulatory System

91. Arteries differ from veins in having:
(a) Wider lumen (b) Thicker wall
(c) Valves (d) Low pressure

92. Which one of the following is absent in humans?
(a) Hepatic vein
(b) Renal portal vein
(c) Hepatic portal vein
(d) Hypophyseal portal vein

93. Myocardial ischaemia leads to a symptom of acute chest pain which is called:
(a) Heart block (b) Heart failure
(c) Angina pectoris (d) Arteriosclerosis

94. Tunica media of blood vessels is rich in:
(a) Fibrous tissue
(b) Adipose tissue
(c) Smooth muscles and elastic fibres
(d) Areolar tissue

95. Maximum urea levels are present in:
(a) Hepatic vein (b) Hepatic portal vein
(c) Renal vein (d) Hepatic artery

96. Gnarled, enlarged veins, mostly appearing in the legs and feet due to abnormal valves, are called:
(a) Varicose veins (b) Aneurysm
(c) Stroke (d) Portal veins

97. Hypophyseal portal veins carry blood from the hypothalamus to:
(a) Anterior pituitary (b) Neurohypophysis
(c) Posterior pituitary (d) Kidneys

98. The endothelial lining of blood vessels is made up of:
(a) Simple cuboidal epithelium
(b) Simple columnar epithelium
(c) Simple squamous epithelium
(d) Compound squamous epithelium

99. The wall of artery is ________ and blood moves at ________ pressure:
(a) Thick, low (b) Thin, low
(c) Thick, high (d) Thin, high

100. Which of the following heart disorders is characterised by sudden damage to heart muscles due to inadequate blood supply?
(a) Heart failure (b) Heart attack
(c) Heart block (d) Cardiac arrest

101. Read the following statements and choose the correct option:
I: A portal vein does not carry blood directly to the heart but forms a network of capillaries in another or intermediate organ before reaching the heart
II: Humans have hepatic and hypophyseal portal systems
(a) Both statements are correct
(b) Both statements are incorrect
(c) Only statement I is correct
(d) Only statement II is correct

102. In which of the following condition(s) is the artificial pacemaker not implanted for treatment?
(a) Heart block
(b) Irregularity of heart rhythm
(c) Both (a) and (b)
(d) Atherosclerosis

103. Small blood vessels supplying blood to the walls of large blood vessels are called:
(a) Varicose veins (b) Vasa vasorum
(c) Retia mirabilia (d) Portal vessels

104. Heart attack is also called:
(a) Myocardial infarction
(b) Heart block
(c) Arrhythmia
(d) Stroke

105. Coronary artery disease (CAD) is often referred to as:
(a) Arteriosclerosis (b) Atherosclerosis
(c) Heart failure (d) Hypertension

106. Identify the incorrect statement:
(a) Hypertension is the term for blood pressure that is higher than normal: 120/80 mm Hg
(b) Regular exercise and low cholesterol levels in blood increase the risk of hypertension
(c) If repeated checks of blood pressure of an individual shows that it is 140/90 mm Hg or higher, it shows hypertension
(d) High blood pressure leads to heart disease and also affects vital organs like the brain and kidneys

107. Which of the following is an incorrect statement?
(a) Blood pressure is measured using a sphygmomanometer
(b) Atherosclerosis is caused by deposits of calcium, fat, cholesterol and fibrous tissues, which makes the lumen of arteries narrower
(c) Angina is more common in younger people than in the elderly
(d) Heart failure causes congestion in the lungs

108. When the heart suddenly stops beating, it is called:
(a) Tachycardia (b) Bradycardia
(c) Cardiac arrest (d) Heart failure

109. Hardening and loss of elasticity of the arteries is called:
(a) Atherosclerosis (b) Arteriosclerosis
(c) Angina pectoris (d) Oedema

110. The state of the heart when it is not pumping blood effectively to meet the needs of the body is called:
(a) Angina (b) Coronary artery disease
(c) Heart failure (d) Atherosclerosis

111. A special ________ system of blood vessels is present in our body exclusively for circulation of blood to and from cardiac musculature:
(a) Cerebral (b) Coronary
(c) Portal (d) Renal

112. Irregular heart beat is called:
(a) Arrhythmia (b) Tachycardia
(c) Bradycardia (d) Heart failure

113. Abnormal ballooning in the weakened area in an artery is called:
(a) Aneurysm (b) Varicosity
(c) Fibrillation (d) Angitis

114. The characteristic common to both blood capillary and lymph capillary is:
(a) Presence of valves
(b) Presence of muscle layer in their wall
(c) Presence of RBCs and WBCs
(d) Lined by single layer of endothelium

115. Identify the incorrect statement:
(a) Hepatic portal system is found in all vertebrates
(b) Veins have valves to maintain blood flow at low pressure
(c) Carotid artery carries oxygenated blood to the brain
(d) Both pulmonary and renal arteries contain oxygenated blood

NCERT EXEMPLAR QUESTIONS

1. Which of the following cells does not exhibit phagocytic activity?
(a) Monocyte (b) Neutrophil
(c) Basophil (d) Macrophage

2. One of the common symptoms observed in people infected with dengue fever is:
(a) Significant decrease in RBC count
(b) Significant decrease in WBC count
(c) Significant decrease in platelet count
(d) Significant increase in platelet count

3. Which among the following is correct during each cardiac cycle?
(a) The volume of blood pumped out by the right and left ventricles is the same.
(b) The volume of blood pumped out by the right and left ventricles is different
(c) The volume of blood received by each atrium is different
(d) The volume of blood received by the aorta and pulmonary artery is different

4. Cardiac activity could be moderated by the autonomous neural system.
Tick the correct answer:
(a) The parasympathetic system stimulates heart rate and stroke volume
(b) The sympathetic system stimulates heart rate and stroke volume
(c) The parasympathetic system decreases the heart rate but increases stroke volume
(d) The sympathetic system decreases the heart rate but increases stroke volume

5. Mark the pair of substances among the following which is essential for coagulation of blood:
(a) Heparin and calcium ions
(b) Calcium ions and platelet factors
(c) Oxalates and citrates
(d) Platelet factors and heparin

6. ECG depicts the depolarisation and repolarisation processes during the cardiac cycle. In the ECG of a normal healthy individual, one of the following waves is not represented:
(a) Depolarisation of atria
(b) Repolarisation of atria
(c) Depolarisation of ventricles
(d) Repolarisation of ventricles

7. Which one of the following types of cells lack nucleus in humans?
(a) Erythrocytes (b) Neutrophils
(c) Eosinophils (d) Monocytes

8. Which one of the following blood cells is involved in antibody production?
(a) B-lymphocytes (b) T-lymphocytes
(c) RBC (d) Neutrophils

9. The cardiac impulse is initiated and conducted further up to the ventricle.
The correct sequence of conduction of impulse is:
(a) S A Node → A V Node → Purkinje fibre → A V Bundle

(b) S A Node → Purkinje fibre → A V Node → A V Bundle
(c) S A Node → A V Node → A V Bundle → Purkinje fibre
(d) S A Node → Purkinje fibre → A V Bundle → A V Node

10. The cells involved in inflammatory reactions are:
(a) Basophils (b) Neutrophils
(c) Eosinophils (d) Lymphocytes

11. The second heart sound (dubb) is associated with the closure of:
(a) Tricuspid valve
(b) Semilunar valves
(c) Bicuspid valve
(d) Tricuspid and bicuspid valves

12. Which of the following correctly explains a phase/event in the cardiac cycle in a standard electrocardiogram?
(a) QRS complex indicates atrial contraction
(b) QRS complex indicates ventricular contraction
(c) Time between S and T represents atrial systole
(d) P-wave indicates beginning of ventricular contraction

13. Which of the following statements is incorrect?
(a) A person of 'O' blood group has anti 'A' and anti 'B' antibodies in his blood plasma
(b) A person of 'B' blood group cannot donate blood to a person of 'A' blood group
(c) Blood group is designated on the basis of the presence of antibodies in the blood plasma
(d) A person of AB blood group is universal recipient

14. What would be the cardiac output of a person having 72 heart beats per minute and a stroke volume of 50 ml?
(a) 360 mL (b) 3600 mL
(c) 7200 mL (d) 5000 mL

15. Match the terms given under Column 'A' with their functions given under Column 'B' and select the answer from the options given below:

Column I	Column II
A. Lymphatic system	(i) Carries oxygenated blood
B. Pulmonary vein	(ii) Immune response
C. Thrombocytes	(iii) To drain back the tissue fluid to the circulatory system
D. Lymphocytes	(iv) Coagulation of blood

(a) A – (ii), B – (i), C – (iii), D – (iv)
(b) A – (iii), B – (i), C – (iv), D – (ii)
(c) A – (iii), B – (i), C – (iii), D – (iv)
(d) A – (ii), B – (i), C – (iii), D – (iv)

16. Read the following statements and choose the correct option:
Statement 1: Atria receive blood from all parts of the body which subsequently flows to ventricles
Statement 2: Action potential generated at sino-atrial node passes from atria to ventricles
(a) Action mentioned in Statement 1 is dependent on action mentioned in Statement 2
(b) Action mentioned in Statement 2 is dependent on action mentioned in Statement 1
(c) Action mentioned in Statements 1 and 2 are independent of each other
(d) Action mentioned in Statements 1 and 2 are synchronous

PRACTICE TEST PAPER

1. The haemoglobin content per 100 ml of blood of a normal healthy human adult is:
(a) 12–16 mg (b) 12–16 g
(c) 25–30 g (d) 17–40 mg

2. Person with blood group AB is considered as universal recipient because he has:
(a) Both A and B antigens on RBC but no antibodies in plasma
(b) Both A and B antibodies in plasma
(c) No antigen on RBC and no antibody in plasma
(d) Both A and B antigens in plasma but no antibodies

3. Bundle of His in heart is/are:
(a) Nerve tissue supplied to ventricles
(b) Muscular tissue supplied to atria
(c) Muscular tissue supplied to ventricles
(d) Nerves regulating opening and closing of semilunar valves

4. Which of the following blood corpuscles helps in blood coagulation?
(a) RBCs (b) Lymphocytes
(c) Thrombocytes (d) Basophils

5. Identify the incorrect statement:

(a) Formation of RBCs is called erythropoiesis
(b) Megakaryocytes of bone marrow produce cell fragments called thrombocytes
(c) Albumin is the most abundant protein in plasma
(d) Deficiency of folate and cyanocobalamin causes microcytic anaemia

6. Choose the incorrect match with respect to leucocytes and their percentage in blood?
(a) Monocytes → 6–8%
(b) Neutrophils → 60–65%
(c) Eosinophils → 2–3%
(d) Lymphocytes → 0.5–1%

7. Which of the following is an incorrect statement?
(a) Blood group AB is the universal recipient
(b) In Pisces, there is single circuit of blood flow through the heart
(c) Fibrinogen and prothrombin are present in blood serum
(d) The oxidation state of iron in haemglobin is Fe^{2+}

8. Which of the flowing plasma proteins are primarily involved in defense mechanisms of the body?
(a) Albumin (b) Prothrombin
(c) Globulin (d) Fibrinogen

9. Which of the following are paired structures of the human circulatory system?
(a) Superior vena cava
(b) Inferior vena cava
(c) Systemic aorta
(d) Pulmonary vein

10. Which of the following is the correct sequence of the formed elements of blood in increasing order of their count in blood?
(a) Platelets, RBCs, WBCs
(b) WBCs, Platelets, RBCs
(c) RBCs, WBCs, Platelets
(d) RBCs, Platelets, WBCs

11. Which layer of an artery is lined by endothelium?
(a) Tunica media (b) Tunica interna
(c) Tunica externa (d) Tunica adventitia

12 Choose the incorrect statement with respect to blood clotting:
(a) Calcium is important for blood clotting
(b) Tissue injury causes release of clotting factor thromboplastin
(c) Thrombin converts fibrin to fibrinogen
(d) Coagulum is a reddish-brown scum formed at the site of injury

13. Bursa of Fabricius is an important organ of birds. This organ is associated with:
(a) Metabolism of fatty acids
(b) Formation of urea from ammonia
(c) Generation of metabolic waste
(d) Generation of B cells

14. Which of the following is an incorrect statement?
(a) Heart sounds are heard using stethoscope
(b) Heart of humans is neurogenic
(c) Heart is found in mediastinum
(d) Arthropods and molluscs have open circulatory system

15. All of the following statements are correct, except:
(a) Smaller the organism, higher the metabolic rate
(b) Angina pectoris is caused by hypoxia of cardiac muscles
(c) Hypertension refers to persistent increased blood pressure
(d) The average weight of the human heart is 80–100 grams

16. Chordae tendineae in the heart are:
(a) Present in the interventricular septum
(b) Channels connecting pulmonary trunk and the aorta during embryonic stages
(c) Foramen between right and left atria in embryonic stages
(d) Connections between the atrio-ventricular valves and the papillary muscles of ventricles

17. The blood group that has antigens 'A' and 'B' on RBCs is __I__ and humans with this blood group can receive blood from donors whose blood group is/are __II__. The option that fills in both blanks correctly is:

	I	II
(a)	O	O
(b)	AB	A, B, AB and O
(c)	A	A and O
(d)	B	B and O

18. Which of the following converts inactive fibrinogens of blood into fibrins during cascade of reactions of blood clotting?
(a) Thrombokinase (b) Thrombin
(c) Prothrombinase (d) Thromboplastin

19. _____ is a symptom of acute chest pain when enough oxygen is not reaching the heart muscles:
(a) Heart block (b) Heart failure
(c) Angina pectoris (d) Arteriosclerosis

20. The opening present in the atrioventricular septum between the right atrium and right ventricle is guarded by:
(a) Bicuspid valve (b) Thebesian valve
(c) Eustachian valve (d) Tricuspid valve

21. Maximum filling of blood in ventricles occurs during:
(a) Atrial systole (b) Ventricular systole
(c) Joint diastole (d) Both (a) and (b)

22. Which of the following are enucleated cells present in blood?
(a) Monocytes and neutrophils
(b) Erythrocytes and thrombocytes
(c) Lymphocytes and basophils
(d) Thrombocytes and acidophils

23. Blood from the intestine is carried to the liver via ________ before it is returned to systemic circulation. The correct option that fills in the blank is:
(a) Hepatic vein
(b) Hepatic artery
(c) Hepatic portal vein
(d) Renal vein

24. The increase in the rate of heart beat, strength of ventricular contraction and hence the cardiac output can be achieved through all of the following, except:
(a) Neural signals from sympathetic
(b) Neural signals from parasympathetic
(c) Action of adrenal medullary hormones
(d) Action of neurotransmitters like catecholamines

25. Neutrophils and monocytes are similar in:
(a) Shape of their nuclei
(b) Their cytoplasmic contents
(c) Their contribution to total WBC content of blood
(d) Their phagocytic action against foreign organisms

26. Which of the following is correct with respect to second heart sound during a cardiac cycle?
(a) It is caused by closing of atrioventricular valves at the beginning of ventricular systole
(b) It is cause by closing of semilunar valve at the beginning of ventricular diastole
(c) It is caused by closing of semilunar valve at the beginning of ventricular systole
(d) It is caused by the closing of atrioventricular valves at the beginning of ventricular diastole

27. Which of the following is the correct statement?
(a) Pulmonary circulation starts with the pumping of deoxygenated blood from the right ventricle into the lungs and its return in oxygenated form to the right atrium
(b) Angina pectoris is more common in younger persons
(c) During heart attack, heart muscle is suddenly damaged by inadequate blood supply
(d) Systemic circulation starts with pumping of oxygenated blood from the left ventricle through the aorta to all organs and return of deoxygenated blood into the left atrium through the veins

28. Select the incorrect statement with respect to nodal tissue of the heart:
(a) Sino-atrial node is located at the upper right corner of the right atrium
(b) Atrio-ventricular node is located at the lower left corner of the right atrium
(c) The entire nodal tissue has the ability to generate action potentials
(d) Atrio-ventricular node normally acts as the pacemaker

29. A special coronary system of blood vessels is present in our body exclusively for the circulation of ____I____ to and from ____II____. Fill in the blanks with the appropriate option:

I	II
(a) Blood	Skeletal musculature
(b) Lymph	Liver
(c) Blood	Cardiac musculature
(d) Serum	Cardiac musculature

30. Which part of the brain moderates cardiac functions through the autonomic nervous system?
(a) Cerebrum
(b) Cerebellum
(c) Pons Varolii
(d) Medulla oblongata

31. Identify the incorrect match:
(a) Left ventricle – Pumps blood to aorta
(b) Right atrium – Receives blood from pulmonary veins
(c) Right ventricle – Pumps blood to pulmonary artery
(d) Mitral valve – Left artio-ventricular valve

32. If the vagus nerve supply to the human heart is cut, the heart will:
(a) Beat faster (b) Beat slowly
(c) Stop beating (d) Show no change

33. Which of the following layers is thickest in the heart wall?

(a) Endocardium (b) Myocardium
(c) Pericardium (d) Epicardium

34. A vein differs from an artery in that:
(a) The wall of the vein is non-collapsible
(b) Internal valves are absent in veins
(c) The tunica media of the vein is very thick
(d) The lumen of the vein is wider

35. Identify the incorrect statement:
(a) Neutrophils are the most abundant WBCs in blood
(b) Eosinophils are also called acidophils
(c) Lymphocytes and monocytes are granulocytes
(d) Basophils mediate inflammatory reactions

36. Choose the option that fills in the given blanks appropriately. When systolic blood pressure is ___**I**___ mm Hg and diastolic blood pressure is ___**II**___ mm Hg, it indicates hypertension.
(a) 90 and 140 (b) 140 and 90
(c) 120 and 80 (d) 80 and 110

37. The condition where the heart is not pumping blood effectively to meet the needs of the body is called:
(a) Heart failure (b) Heart block
(c) Heart attack (d) Angina pectoris

38. The heart is protected by a double-walled membranous bag called:
(a) Epicardium (b) Endocardium
(c) Pericardium (d) Myocardium

39. Deposition of fats, cholesterol and fibres in the lining of arteries is called:
(a) Arthritis (b) Atherosclerosis
(c) Angiogenesis (d) Arteriosclerosis

40. Which blood vessel in mammals would normally carry the maximum amount of urea?
(a) Renal vein
(b) Hepatic vein
(c) Hepatic artery
(d) Hepatic portal vein

41. The T-wave in an ECG represents:
(a) Electrical excitation of the atria
(b) Depolarisation of ventricles
(c) Return of ventricles from the excited state
(d) Beginning of systole

42. Pulse is a direct measure of:
(a) Cardiac output (b) Stroke volume
(c) Blood pressure (d) Heart rate

43. Which of the following are required for blood clotting?
(a) Na^+, K^+
(b) Ca^{2+}, thromboplastin
(c) K^+, prothrombin
(d) Na^+, Ca^{2+}

44. Compared to blood, our lymph has:
(a) More RBCs and less WBCs
(b) More WBCs and no RBCs
(c) Plasma without proteins
(d) More RBCs and platelets

45. The duration of atrial systole in each cardiac cycle is:
(a) 0.7 sec (b) 0.1 sec
(c) 0.3 sec (d) 0.5 sec

ANSWER KEYS

Exercises

1. (b) **2.** (b) **3.** (b) **4.** (b) **5.** (b) **6.** (a) **7.** (d) **8.** (c) **9.** (c) **10.** (d)
11. (d) **12.** (d) **13.** (a) **14.** (b) **15.** (a) **16.** (d) **17.** (c) **18.** (d) **19.** (c) **20.** (a)
21. (b) **22.** (c) **23.** (c) **24.** (d) **25.** (b) **26.** (b) **27.** (a) **28.** (d) **29.** (c) **30.** (c)
31. (b) **32.** (d) **33.** (a) **34.** (d) **35.** (d) **36.** (a) **37.** (a) **38.** (c) **39.** (d) **40.** (d)
41. (d) **42.** (a) **43.** (c) **44.** (b) **45.** (c) **46.** (d) **47.** (c) **48.** (c) **49.** (b) **50.** (c)
51. (a) **52.** (b) **53.** (b) **54.** (a) **55.** (d) **56.** (b) **57.** (d) **58.** (c) **59.** (c) **60.** (b)
61. (b) **62.** (b) **63.** (c) **64.** (b) **65.** (b) **66.** (d) **67.** (c) **68.** (c) **69.** (d) **70.** (b)
71. (d) **72.** (b) **73.** (d) **74.** (d) **75.** (c) **76.** (b) **77.** (d) **78.** (a) **79.** (c) **80.** (a)
81. (c) **82.** (c) **83.** (d) **84.** (d) **85.** (d) **86.** (d) **87.** (b) **88.** (c) **89.** (c) **90.** (a)
91. (b) **92.** (b) **93.** (c) **94.** (c) **95.** (a) **96.** (a) **97.** (a) **98.** (c) **99.** (c) **100.** (b)
101. (a) **102.** (d) **103.** (b) **104.** (a) **105.** (b) **106.** (b) **107.** (c) **108.** (c) **109.** (b) **110.** (c)
111. (b) **112.** (a) **113.** (a) **114.** (d) **115.** (d)

NCERT Exemplar Questions

1. (c)	2. (c)	3. (a)	4. (b)	5. (b)	6. (b)	7. (a)	8. (a)	9. (c)	10. (a)
11. (b)	12. (b)	13. (c)	14. (b)	15. (b)	16. (d)				

Practice Test Paper

1. (b)	2. (a)	3. (c)	4. (c)	5. (d)	6. (d)	7. (c)	8. (c)	9. (d)	10. (b)
11. (b)	12. (c)	13. (d)	14. (b)	15. (d)	16. (d)	17. (b)	18. (b)	19. (c)	20. (d)
21. (c)	22. (b)	23. (c)	24. (b)	25. (d)	26. (b)	27. (c)	28. (d)	29. (c)	30. (d)
31. (b)	32. (a)	33. (b)	34. (d)	35. (c)	36. (b)	37. (a)	38. (c)	39. (b)	40. (b)
41. (c)	42. (d)	43. (b)	44. (b)	45. (b)					

SOLUTIONS

Body Fluids

1. Neutrophils are the most numerous (60–65%) leucocytes in blood.
2. Total leucocyte count (TLC) in blood is 6000–8000 per cu mm.
3. The lifespan of a human RBC is 120 days.
4. Neutrophil and monocyte are phagocytic WBCs of blood.
5. Basophil is the WBC that mediates inflammatory reaction.
6. Albumins are the plasma proteins that help in maintaining osmotic balance in blood.
7. Spleen is called the "graveyard of RBCs".
8. Serum is defined as plasma without clotting factors.
9. Option (c) is correct.
10. Thrombocytes are non-nucleated colourless cell fragments formed by the breakdown of special cells called megakaryocytes.
11. Lymph is often referred to as "middle man of the body".
12. Conversion of inactive prothrombin into thrombin occurs in the presence of thrombokinase and calcium.
13. Erythroblastosis foetalis can be prevented by administering anti–Rh antibodies to Rh–ve mother after delivery of Rh+ve baby.
14. Blood group O–ve is the universal donor because it lacks A, B and Rh antigens on RBCs.
15. Thrombin converts inactive fibrinogen of blood into fibrin during blood clotting.
16. Erythrocytes are enucleated cells of blood.
17. Both monocytes and neutrophils are phagocytic WBCs.
18. Plasma is a straw-coloured, viscous fluid constituting nearly 55% of the blood. 90–92% of plasma is water and proteins contribute 6–8% of it.
19. Blood is prevented from clotting inside blood vessels due to the presence of heparin.
20. Plasmin/Fibrinolysin is the proteolytic enzyme which causes lysis of fibrin during fibrinolysis.
21. Bursa of Fabricius is present in the cloaca of birds and is necessary for generation of B lymphocytes.
22. Erythroblastosis foetalis can occur if the blood group of the father and mother is Rh+ve and Rh–ve, respectively.
23. The granules of eosinophil stain with eosin.
24. Blood groups A, B and O were discovered by the Austrian biologist Karl Landsteiner.
25. EDTA strongly and irreversibly chelates calcium and prevents blood from clotting.
26. Vitamin essential for blood clotting is phylloquinone (vitamin K) since it is required for activation of clotting factors like prothrombin.
27. Christmas disease is caused by deficiency of clotting factor IX/anti-haemophilic factor B. It is also called haemophilia B and the disease is

named for Stephen Christmas, who was the first person diagnosed with the condition in 1952.

28. Option (d) is correct.

29. Platelets/Thrombocytes are required for normal blood clotting.

30. Abnormal fall in platelet count is called thrombocytopenia.

31. Lymph has WBCs and mostly lymphocytes.

32. Prothrombin or clotting factor II is a vitamin K–dependent coagulation factor. On activation, prothrombin is cleared to form thrombin, which in turn acts as a protease that converts fibrinogen to fibrin.

33. Deficiency of folic acid and cyanocobalamin causes megaloblastic/macrocytic anaemia.

34. Lymph brings tissue fluid back to systemic circulation. It also brings metabolic waste from cells into the venous system.

35. Eosinophils constitute 2–3% of leucocytes, resist infections and are associated with allergic reactions.

36. The movement of leucocytes out of the blood vessels towards the site of infection or tissue damage is called diapedesis.

37. Compared to blood, lymph has more wastes and less proteins.

38. Leucocytes or WBCs are nucleated cells and their number averages 6000–8000 mm^{-3} of blood. They are short-lived and are colourless due to lack of haemoglobin. The two main categories of WBCs are granulocytes and agranulocytes. Among them, neutrophils and monocytes are phagocytic.

39. Serum is the fluid and solute component of blood which does not play any role in clotting. It may be defined as blood plasma without fibrinogens. Serum includes all proteins not used in blood clotting, all electrolytes, antibodies and hormones.

40. Neutrophils are the most abundant cells of the WBCs (60–65%). They have a multilobed nucleus and called polymorphonuclear leucocytes.

Human Circulatory System

41. Amphibians and reptiles (except crocodiles) have a three-chambered heart with two atria and a single ventricle. The left atrium receive oxygenated blood from the gills/lungs/skin and the right atrium gets deoxygenated blood from other parts of the body. However, they get mixed in a single ventricle and it pumps mixed blood. So, it is called incomplete double circulation.

42. In fish, the heart pumps out deoxygenated blood which is oxygenated by the gills and supplied to the body parts, from where deoxygenated blood returns to the heart, and is therefore called single circulation.

43. The heart is protected by a double-walled membranous bag called the pericardium.

44. Stroke volume refers to the volume of blood pumped by each ventricle in a cardiac cycle and averages 70 ml.

45. Chordae tendineae (tendinous cords), colloquially known as heart strings, are tendons resembling fibrous cords of connective tissue that connect the papillary muscles to the tricuspid valve and the mitral valve in the heart.

46. Bicuspid/Mitral valve guards the opening present in the atrio-ventricular septum between the left atrium and left ventricle.

47. Maximum ventricular filling (70%) occurs during joint diastole.

48. The nodal musculature has the ability to generate action potentials without an external stimuli, that is, it is autoexcitable. However, the number of action potentials that could be generated in a minute vary at different parts of the nodal system. The sino-atrial node (SAN) can generate the maximum number of action potentials, 70–75 per minute, and is responsible for initiating and maintaining the rhythmic contractile activity of the heart.

49. Atria receive blood from the respective veins for 0.7 seconds during each cardiac cycle.

50. Foramen ovale is present on the inter-atrial septum of the heart during foetal life. It is an anatomic adaptation in the foetus to allow oxygenated blood from the umbilical vein via the inferior vena cava to bypass pulmonary circulation.

51. Eustachian valve guards the opening of the inferior vena cava into the right atrium. Thebesian valve guards the opening of the coronary sinus into the right atrium.

52. Myocardium is the thick muscle layer present in the wall of the heart.

53. Bundle of His is the network of heart muscle cells specialised for electrical conduction that transmits electrical impulses from the AV node to the point of the apex of the fascicular branches via the bundle branches.
54. Atrial systole has the shortest duration of 0.1 seconds in a cardiac cycle.
55. The openings of the right and left ventricles into the pulmonary artery and the aorta, respectively, are provided with semilunar valves.
56. Systemic heart refers to the left atrium and left ventricle in the human heart.
57. Left atrium receives oxygenated blood from the lungs through two pairs of pulmonary veins. Volume of blood pumped out by each ventricle per minute is called cardiac output.
58. Housefly and prawn are both arthropods and have open circulatory system.
59. Due to oblique opening of pulmonary veins into the right atrium, there are no valves at this junction.
60. In fish, the heart pumps out deoxygenated blood which is oxygenated by the gills and supplied to the body parts, from where deoxygenated blood is returned to the heart.
61. A special neural centre in the medulla oblongata can moderate cardiac function through the autonomous nervous system (ANS).
62. The time interval between the opening of the semilunar valve and the atrio-ventricular valve is 0.5 seconds.
63. A bicuspid or mitral valve guards the opening between the left atrium and the left ventricle and is therefore in contact with oxygenated blood only.
64. The first heart sound is called 'lub' and is produced by the closing of the atrioventricular valve.
65. The impulse generated by the sino-atrial node (SAN) produces action potential which stimulates both the atria to undergo simultaneous contraction – the atrial systole. This increases the flow of blood into the ventricles by about 30%.
66. The end of the T-wave marks the end of ventricular systole.
67. Human heart is myogenic since the normal activities of the heart are regulated intrinsically that is, autoregulated by specialised muscles (nodal tissues).
68. In the foetal heart, the ductus arteriosus is present between the pulmonary artery and systemic aorta. It is also called ductus Botalli and allows most of the blood from the right ventricle to bypass the foetus's fluid-filled non-functioning lungs.
69. Option (d) is correct.
70. The vagus nerve brings parasympathetic neural signals and decreases the heart rate, speed of action potential and thereby the cardiac output.
71. Option (d) is correct.
72. To obtain a standard ECG, the patient is connected to the machine with three electrical leads (one to each wrist and to the left ankle) that continuously monitor the heart activity.
73. Blood pressure is lower in an open circulatory system compared to a closed circulatory system. Molluscs and arthropods have an open circulatory system. The flow of blood in an open circulatory system is comparatively slower.
74. In joint diastole, all four chambers of the heart are in relaxed state.
75. Purkinje fibres are the last to be depolarised in a cardiac cycle.
76. The sino-atrial node (SAN) generates an action potential which stimulates both the atria to undergo simultaneous contraction. Ventricular systole increases ventricular pressure, causing the closure of the bicuspid and tricuspid valves due to attempted backflow of blood into atria.
77. Parasympathetic stimulation decreases the strength of ventricular contraction and thereby the cardiac output.
78. The second heart sound is associated with the closure of the semilunar valves at the beginning of ventricular diastole.
79. The neural signals from the parasympathetic nerves decrease the heart rate, the strength of ventricular contraction and thereby the cardiac output.
80. The semilunar valves open during ventricular systole. The ventricular systole increases the ventricular pressure to force open the semilunar valves guarding the pulmonary artery and aorta, allowing the blood in these ventricles to flow through these vessels into the circulatory pathways.
81. Out body has the ability to alter heart rate as well stroke volume.

82. Option (c) is correct.

83. Relaxation of atria coincides with ventricular systole and the flow of blood into the atria from large veins.

84. Heart murmurs are hissing/abnormal heart sounds heard due to defective heart valves. They are generated due to turbulent flow of blood and heard by a stethoscope.

85. An impulse from the SA node is spread to the atria via internodal fibres/tract. A bundle of nodal fibres, atrio-ventricular bundle (AV bundle) continues from the AV node which passes through the atrio-ventricular septa to emerge at the top of the interventricular septum and immediately divides into right and left bundles. These branches give rise to minute fibres throughout the ventricular musculature of the respective sides and are called Purkinje fibres.

86. The thick muscular projections on the wall of the ventricles are called columnae carneae.

87. By the end of joint diastole, the ventricles are almost two-third filled.

88. Purkinje fibres are distributed in the ventricular walls.

89. Option (c) is correct.

90. The atrio-ventricular valves are pushed open during ventricular diastole due to decline in ventricular pressure, and due to rising pressure in atria exerted by the blood which was emptied into them by the veins.

Blood Vessels, Circulation through Special Regions, Disorders of the Circulatory System

91. The wall of arteries are thicker than veins due to prominent tunica media.

92. Renal portal system is absent in humans.

93. Myocardial ischaemia leads to a symptom of acute chest pain called angina pectoris. It appears when not enough oxygen is reaching the heart muscle. It is more common among middle-aged people and the elderly.

94. The tunica media of blood vessels is rich in smooth muscles and elastic fibres.

95. The maximum urea levels are present in the hepatic vein.

96. Varicose veins are gnarled, enlarged veins appearing mostly in the legs due to abnormal valves.

97. Hypophyseal portal veins carry blood from the hypothalamus to the anterior pituitary.

98. Simple squamous epithelium forms the endothelial lining of blood vessels.

99. The wall of the artery is thick and blood moves at high pressure.

100. Heart attack occurs due to sudden damage to heart muscles due to inadequate blood supply.

101. Option (a) is correct.

102. Artificial pacemaker is used for treatment of heart block and irregularity of heart rhythm.

103. Vasa vasorum are small blood vessels supplying blood to the walls of large blood vessels.

104. Heart attack is also called myocardial infarction.

105. Coronary artery disease (CAD) is also called atherosclerosis, and affects the vessels that supply blood to the heart muscle. It is caused by deposition of calcium, fat, cholesterol and fibrous tissue, which makes the lumen of arteries narrower.

106. Regular exercise and low cholesterol levels in blood decrease the risk of hypertension.

107. Angina is more common among the middle-aged and the elderly.

108. When the heart suddenly stops beating, it is called cardiac arrest.

109. Hardening and loss of elasticity of the arteries is called arteriosclerosis.

110. Heart failure means the state of heart when it is not pumping blood effectively enough to meet the needs of the body. It is sometimes called congestive heart failure because congestion of the lungs is one of the main symptoms.

111. A special coronary system of blood vessels is present in our body exclusively for circulation of blood to and from cardiac musculature.

112. Arrhythmia refers to irregularity of heart beat. It occurs when electrical impulses in the heart do not work properly.

113. Aneurysm refers to abnormal ballooning in the weakened area in an artery. A ruptured aneurysm can result in internal bleeding and stroke.

114. Both blood and lymph capillaries are lined by a single layer of endothelium.

115. Pulmonary artery carries deoxygenated blood while renal artery carries oxygenated blood.

Excretory Products and Their Elimination

EXERCISES

Modes of Excretion

1. Which of the following nitrogenous waste products is most toxic and requires a large amount of water for its elimination?
(a) Urea (b) Ammonia
(c) Uric acid (d) Guanine

2. Mark the odd one out with respect to examples of ureotelic animals:
(a) Marine fishes (b) Mammals
(c) Aquatic insects (d) Terrestrial amphibians

3. Identify the incorrect statement:
(a) Ammonia, urea and uric acid are the main forms of nitrogenous waste excreted by animals
(b) Ammonia is generally excreted by diffusion across body surfaces or through gill surfaces as ammonium ions
(c) Aquatic amphibians excrete nitrogenous waste as uric acid in the form of pellets or paste with minimum loss of water
(d) Uric acid is the least toxic of the three nitrogenous waste products

4. Malpighian tubules are the excretory structures of:
(a) Rotifers (b) *Amphioxus*
(c) Insects (d) Flatworms

5. Crustaceans like prawn perform excretion using:
(a) Antennal glands (b) Nephridia
(c) Green glands (d) Both (a) and (c)

6. Small amount of ______________ may be retained in the kidney matrix of some terrestrial animals to maintain the desired osmolarity.
Fill in the blank with the correct option:
(a) Ammonia (b) Urea
(c) Uric acid (d) Ornithuric acid

7. Find the mismatched pair:
(a) Aquatic insects – Ammonotelic
(b) Land snails – Ureotelic
(c) Birds – Uricotelic
(d) Reptiles Uricotelic

8. Which of the following is an incorrect statement?
(a) Kidneys play a significant role in the excretion of ammonia in ammonotelic animals
(b) Minimum loss of water occurs during excretion of uric acid
(c) Protonephridia are excretory structures in Platyhelminthes
(d) Terrestrial adaption necessitated the production of lesser toxic nitrogenous waste like urea and uric acid for conservation of water

9. Select the option which shows the correct matching of the animal with its excretory organ and excretory product:

Animal	**Excretory organ**	**Excretory product**
(a) Shark	Kidneys	Uric acid
(b) Vulture	Kidneys	Urea
(c) Housefly	Malpighian tubules	Uric acid
(d) Salamander	Kidney	Urea

10. Identify the incorrect statement:
(a) Ammonia is readily soluble and diffuses easily
(b) In ureotelic animals, ammonia is not a product of metabolism
(c) Earthworm excretes ammonia in the rainy season and urea in a dry environment
(d) Vertebrates have complex tubular organs called kidneys for excretion

11. The principal nitrogenous excretory waste in mammals are synthesized:
(a) In the kidneys and eliminated by the liver
(b) In the liver and eliminated by the kidneys
(c) In the liver and eliminated by the intestine
(d) In the kidneys and eliminated by the kidneys

12. The excretory structures in rotifers are:
(a) Green glands
(b) Gills
(c) Malpighian tubules
(d) Protonephridia

13. The characteristic(s) shared by urea, uric acid and ammonia is/are:
A. They are nitrogenous waste
B. They all need a large amount of water for excretion
C They are equally toxic
D. They are produced in the kidneys
(a) (A) and (C) (b) (A) and (D)
(c) (A) only (d) (A), (B) and (D)

14. Freshwater bony fish maintain water balance by:
(a) Excreting waste in the form of uric acid
(b) Drinking a large amount of water
(c) Excreting ammonia and hypotonic urine
(d) Retaining osmolytes like trimethylamine oxide

15. Which one of the following options gives the correct categorisation of animals according to the type of nitrogenous waste they give out?

	Ammonotelic	Ureotelic	Uricotelic
(a)	Humans, cockroach	Birds, reptiles	Frog, aquatic insects
(b)	Marine fishes, pigeon	Land snails, land insects	Humans, earthworm
(c)	Many bony fishes, aquatic insects	Mammals, marine fishes	Birds, land snails
(d)	Reptiles, birds	Cockroach, aquatic amphibians	Many bony fishes, aquatic insects

16. Which of the following amino acids play a role in the ornithine cycle?
(a) Ornithine, methionine
(b) Glycine, ornithine
(c) Ornithine, citrulline
(d) Phenylalanine, methionine

17. Which of the following wastes are removed from blood during the Krebs–Henseleit cycle?
(a) CO_2 and urea
(b) Ammonia and CO_2
(c) Urea and uric acid
(d) CO and uric acid

18. In humans, uric acid is produced due to metabolism of:
(a) Protein (b) Purine
(c) Pyrimidine (d) Fats

19. In the human body, the blood vessel carrying maximum urea is:
(a) Hepatic vein (b) Hepatic portal vein
(c) Hepatic artery (d) Renal artery

20. In the ornithine cycle, urea is synthesised in the liver from:
(a) Breakdown of citrulline
(b) Breakdown of arginine
(c) Breakdown of ornithine
(d) Breakdown of purines

Human Excretory System

21. Which of the following is an incorrect statement?
(a) Each kidney weighs 120 –170 g
(b) Kidneys are present from T_{12} to L_3 vertebrae
(c) Kidneys are located in the pelvis
(d) Adipose tissue forms a protective cushion over the kidneys

22. The section of kidneys shows extension of the cortex into the medullary regions between the pyramids and is called:
(a) Major calyx (b) Minor calyx
(c) Columns of Bertini (d) Ducts of Bellini

23. Bowman's capsule and glomerulus are together called:
(a) Malpighian body
(b) Malpighian tubule
(c) Renal corpuscle
(d) Both (a) and (c)

24. Each kidney of an adult human measures **A** in length, **B** in width and **C** in thickness.
Fill in the blanks with the correct options:

	A	B	C
(a)	10–12 cm	2–3 cm	5–7 cm
(b)	10–12 cm	5–7 cm	2–3 cm
(c)	5–7 cm	10–12 cm	2–3 cm
(d)	5–7 cm	2–3 cm	10-12 cm

25. Identify the incorrect statement:
(a) Each kidney has nearly one million complex tubular structures called nephrons
(b) Each nephron has two parts – the glomerulus and the renal tubule
(c) Glomerulus is a tuft of capillaries formed by the efferent arteriole – a fine branch of the renal artery
(d) The renal tubule begins with a double-walled cup-like structure called Bowman's capsule, which encloses the glomerulus

26. In cortical nephrons, the loop of Henle is:
(a) Very long and runs deep into the medulla
(b) Too short and extends very little into the medulla
(c) Absent
(d) Very long but found only in the cortex of the kidneys

27. The distal convoluted tubules (DCT) of many nephrons open into a straight tube called:
(a) Henle's loop
(b) Collecting duct
(c) Proximal convoluted tubule
(d) Ureter

28. Which of the following is the correct statement?
(a) PCT and DCT are present in the cortical region of the kidneys whereas the malpighian corpuscle and loop of Henle dip into the medulla
(b) Malpighian corpuscle, PCT and DCT of the nephron are situated in the cortical region of the kidney whereas the loop of Henle dips into the medulla
(c) Malpighian corpuscle, PCT and DCT are present in the medulla while the loop of Henle is present in the cortical region of the kidneys
(d) Malpighian corpuscle and loop of Henle are present in the cortical region of the kidney, while PCT and DCT are present in the medulla

29. ______ emerging from the glomerulus forms a fine capillary network around the renal tubule, called the peritubular capillaries:
(a) Afferent arteriole (b) Efferent arteriole
(c) Renal artery (d) Vasa recta

30. Which of the following is the correct statement?
(a) Afferent arteriole is narrower than efferent arteriole
(b) Efferent arteriole is narrower than afferent arteriole
(c) PCT is lined by brush-bordered columnar epithelium
(d) Vasa recta is absent or highly reduced in juxtamedullary nephrons

31. Which of the following is the correct statement?
(a) The glomerulus encloses the Bowman's capsule
(b) Blood enters the glomerulus through the efferent arteriole
(c) Hilum is a notch towards the centre of the inner concave surface of the kidney through which the ureter, blood vessels and nerves enter
(d) Blood from the glomerulus is carried away by the renal artery

32. Match the options given in Column I with Column II:

Column I	Column II
A. Bowman's capsule	(i) Brush-bordered cuboidal epithelium
B. PCT	(ii) Simple squamous epithelium
C. Urinary bladder	(iii) Transitional epithelium

(a) A – (i), B – (ii), C – (iii)
(b) A – (ii), B – (i), C – (iii)
(c) A – (iii), B – (ii), C – (i)
(d) A – (ii), B – (iii), C – (i)

33. Inside the hilum of the kidney is a broad funnel-shaped space called the ___I___ with projections called ___II___.
Fill in the blanks:

	I	II
(a)	Cortex	Calyces
(b)	Calyces	Cortex
(c)	Renal pelvis	Calyces
(d)	Renal medulla	Cortex

34. Vasa recta:
(a) Runs parallel to Henle's loop
(b) Is 'U' shaped
(c) Is highly developed in juxtamedullary nephron
(d) All of the above are correct

35. Identify the correct statement:
(a) The left kidney is slightly below the right
(b) Podocytes are present in the inner wall of Bowman's capsule
(c) DCT receives filtrate directly from Bowman's capsule
(d) The number of nephrons is equal to twice the number of Bowman's capsules

36. Which of the following is not a part of the medullary pyramid?
(a) Proximal convoluted tubule
(b) Distal convoluted tubule
(c) Renal corpuscle
(d) All of the above

37. Bowman's capsule is:
(a) A single-walled cup-like structure
(b) Also called renal corpuscle
(c) Lined by columnar epithelium
(d) Present in the cortical region of the kidney

38. The retroperitoneal kidney is:
(a) Kidney present in thorax
(b) Kidney present in pelvic cavity
(c) Kidney covered by peritoneum on ventral side
(d) Kidney covered by peritoneum on dorsal side

39. The dotted appearance of the cortex of the kidney is due to:
(a) Loop of Henle (b) Collecting duct
(c) Calyces (d) Malpighian body

40. The total length of each nephron in humans is about:
(a) 5 mm (b) 30 mm
(c) 90 mm (d) 150 mm

Urine Formation, Function of Tubules, Regulation of Kidney Function

41. Which of the following is absorbed passively in uriniferous tubules?
(a) Glucose (b) Sodium ions
(c) Nitrogenous waste (d) Amino acids

42. Glomerular filtration rate (GFR) in a healthy individual is approximately:
(a) 180 litres per day
(b) 180 litres per hour
(c) 125 ml per hour
(d) Both (a) and (c)

43. Which of the following is incorrect with respect to the proximal convoluted tubule (PCT)?
(a) It is lined by simple cuboidal brush-bordered epithelium
(b) 70–80 per cent of electrolytes and water are reabsorbed by this segment
(c) PCT helps to maintain pH and ionic balance
(d) It allows passage of small amounts of urea into the medullary interstitium to keep up the osmolarity

44. Which of the following is incorrect with respect to the function of renal tubules?
(a) Tubular reabsorption occurs by active or passive mechanisms
(b) 99 per cent of the filtrate is reabsorbed by the renal tubules
(c) Tubular secretion helps in maintenance of ionic and acid base balance of the body fluids
(d) Reabsorption of water occurs only by active mechanism

45. Conditional reabsorption of Na^+ and water takes place in:
(a) Proximal convoluted tubule
(b) Distal convoluted tubule
(c) Henle's loop
(d) Collecting duct

46. The descending limb of the loop of Henle is:
(a) Permeable to water but almost impermeable to electrolytes
(b) Impermeable to water but allows transport of electrolytes actively or passively
(c) Permeable to water and electrolytes
(d) Impermeable to water and electrolytes

47. The maintenance of pH and ionic balance of blood by selective secretion of H^+ and K^+ ions occurs in:
(a) PCT (b) DCT
(c) Collecting duct (d) All of the above

48. Glomerular filtrate consists of:
(a) All constituents of blood
(b) All constituents of blood, except proteins
(c) All constituents of blood, except blood corpuscles
(d) All constituents of blood, except blood corpuscles and proteins

49. In which of the following parts of the nephron does reabsorption of sodium take place under the influence of aldosterone?
(a) PCT (b) Loop of Henle
(c) DCT (d) Collecting duct

50. On an average, ____A____ of blood is filtered by the kidneys per minute, which consistutes nearly ____B____ of the blood pumped out by each ventricle of the heart in a minute.
Fill in the blanks for options A and B:
(a) 500–600 ml, 1/4th
(b) 1100–1200 ml, 1/5th
(c) 2000–2500 ml, 1/5th
(d) 1100–1200 ml, 1/3rd

51. Identify the incorrect statement:
(a) The epithelial cells of Bowman's capsule are called podocytes
(b) The filtrate has almost all the constituents of plasma, except proteins

(c) Glomerular capillary blood pressure causes filtration of blood through the endothelium of glomerular blood vessels and epithelium of PCT
(d) The first step in urine formation is the filtration of blood, which is carried out by the glomerulus and is called glomerular filtration

52. If Henle's loop were absent from the mammalian nephron, which of the following is to be expected?
(a) There will be no urine formation
(b) Urine will be more dilute
(c) Urine will be more concentrated
(d) No change in quantity or quality of urine

53. A fall in glomerular filtration rate causes:
(a) Release of cortisol from the adrenal cortex
(b) Juxtaglomerular cells to release renin
(c) Heart to release atrial natriuretic factor
(d) Adrenal medulla to release adrenaline

54. __________ is released by the walls of atria in response to increase in blood volume and blood pressure:
(a) Renin
(b) Angiotensinogen
(c) Atrial natriuretic factor
(d) Melatonin

55. Excessive loss of fluid from the body can activate the osmoreceptors to release __________ from the neurohypophysis:
(a) Atrial natriuretic factor
(b) Vasopressin
(c) Renin
(d) Angiotensin

56. The function of renin is:
(a) To convert angiotensin I to angiotensin II
(b) To convert angiotensinogen to angiotensin I
(c) To cause vasodilation
(d) To release aldosterone from the adrenal medulla

57. Which of the following correctly describes the function of the atrial natriuretic factor (ANF)?
(a) Vasodilation
(b) Decrease blood pressure
(c) Keeping a check on the renin-angiotensin mechanism
(d) All of the above

58. Which of the following is incorrect with respect to the functions of ADH?
(a) Facilitates water reabsorption from the latter parts of the renal tubule
(b) Increases blood pressure
(c) Constrictor effect on blood vessels
(d) Causes reabsorption of Na^+ and water from PCT

59. Human kidneys can produce urine nearly __________ times more concentrated than the initial filtrate formed:
(a) Six (b) Four
(c) Ten (d) Two

60. The renin-angiotensin mechanism activates the adrenal cortex to release:
(a) Renin (b) Aldosterone
(c) Rennin (d) Vasopressin

61. Deficiency of vasopressin causes:
(a) Glycosuria
(b) Diabetes insipidus
(c) Diabetes mellitus
(d) Proteinuria

62. Angiotensinogen is secreted by:
(a) Heart (b) Liver
(c) Kidneys (d) Adrenal cortex

63. ADH controls the water permeability of:
(a) Loop of Henle
(b) Proximal convoluted tubule
(c) Distal convoluted tubule
(d) Bowman's capsule

64. An increase in body fluid volume can:
(a) Suppress the release of vasopressin
(b) Promote the release of vasopressin
(c) Switch off the osmoreceptors
(d) Both (a) and (c) are correct

65. The osmolarity of interstitial fluid in kidneys is ___I___ in the cortex and ___II___ in the inner medulla:

	I	II
(a)	300 mOsm/L	600 mOsm/L
(b)	600 mOsm/L	1000 mOsm/L
(c)	300 mOsm/L	1200 mOsm/L
(d)	1200 mOsm/L	300 mOsm/L

66. The chemicals which increase urine volume are called:
(a) Monouretics (b) Diuretics
(c) Triuretics (d) Tetrauretics

67. The reabsorption of Na^+ and water from the distal parts of the renal tubule is caused by:
(a) ANF (b) Angiotensin II
(c) Aldosterone (d) Renin

68. The osmolarity gradient, that is, an increasing osmolarity towards the inner medullary interstitium, is caused by:
(a) NaCl and urea (b) Ammonia and urea
(c) Urea and K^+ (d) HCO_3^- and K^+

69. Mark the incorrect match:
(a) Henle's loop – Concentration of urine
(b) PCT – Absorption of Na^+ and K^+
(c) Bowman's capsule – Glomerular filtration
(d) DCT – Absorption of glucose

70. Which one of the following is not secreted into the filtrate by the tubules to maintain the pH and ionic balance of body fluids?
(a) K^+ (b) H^+
(c) HCO_3^- (d) NH_3

71. Which of the following is an incorrect statement?
(a) The flow of filtrate in the two limbs of Henle's loop is in opposite directions and thus forms a counter current
(b) The flow of blood through the two limbs of vasa recta is also in a counter current pattern
(c) The proximity between Henle's loop and vasa recta, as well as the counter current in them, helps in maintaining decreasing osmolarity towards the inner medullary interstitium
(d) An increasing osmolarity gradient is maintained from the cortex to the medulla by NaCl and urea

72. The filtrate becomes __________ in the descending limb of Henle's loop:
(a) Hypotonic (b) Hypertonic
(c) Isotonic (d) All of the above

73. The part of the nephron impermeable to water is:
(a) PCT
(b) DCT
(c) Descending limb of Henle's loop
(d) Ascending limb of Henle's loop

74. The ability to produce concentrated urine in vertebrates depends on:
(a) Presence of fenestrae in glomerular capillaries
(b) Length of Henle's loop
(c) Presence of brush-bordered epithelium in Henle's loop
(d) Presence of podocytes in Bowman's capsule

75. Identify the incorrect statement:
(a) The flow of filtrate in the two limbs of Henle's loop is in opposite directions and forms a counter-current
(b) Small amount of urea enters the descending limb of Henle's loop and returns to the interstitium through the collecting tubule
(c) Filtrate in ascending limb of Henle's loop becomes hypotonic to blood plasma
(d) Vasopressin is a vasoconstrictor

76. The presence of brush-bordered simple cuboidal epithelium in PCT helps in:
(a) Counter-current mechanism
(b) Passing out dilute urine
(c) Increasing the surface area for reabsorption
(d) Selective secretion

77. Which of the following is an incorrect statement?
(a) Renin is secreted in response to low blood volume
(b) In the dehydrated state, maximum water is reabsorbed from DCT
(c) Angiotensin II increases glomerular blood pressure
(d) When a person drinks a lot of water, the release of vasopressin is suppressed

78. __________ g of urea is excreted per day by human kidneys:
(a) 25–30 (b) 45–50
(c) 60–70 (d) 90–100

79. The reabsorption of ions by the tubules of the kidneys is enhanced by:
(a) Cortisol (b) Erythropoietin
(c) Aldosterone (d) Calciferol

80. Increased ADH secretion is likely after:
(a) Eating a sugary snack
(b) Ingestion of alcoholic drinks
(c) Drinking lots of pure water
(d) Sweating-induced dehydration increasing plasma osmolarity

Disorders of the Excretory System, Accessory Excretory Organs, Artificial Kidney

81. Increase in frequency of urination is called:
(a) Oliguria (b) Polyuria
(c) Nephritis (d) Pyuria

82. In patients with renal failure, the blood urea levels are raised. This condition is called:
(a) Cystitis (b) Uremia
(c) Ketosis (d) Oliguria

83. Identify the mismatched pair:
(a) Hematuria – Blood in urine
(b) Oliguria – Urine output below normal
(c) Pyuria – Pus cells in urine
(d) Cystitis – Inflammation of glomeruli

84. Presence of glucose and ketone bodies in urine indicates:
(a) Diabetes insipidus (b) Diabetes mellitus
(c) Phenylketonuria (d) Renal calculi

85. Identify the incorrect statement:
(a) Lungs remove 200 ml of CO_2 per hour
(b) Liver secretes bile containing substances like bilirubin, biliverdin, cholesterol, degraded steroid hormones, vitamins and drugs
(c) Sweat produced by sweat glands contains substances like sterols, hydrocarbons and waxes
(d) Both (a) and (c)

86. Stone or insoluble mass of crystallised salts (oxalates) formed within the kidneys is called:
(a) Renal calculi
(b) Cholelithiasis
(c) Pyelonephritis
(d) Glomerulonephritis

87. Which of the following is/are found in the secretions of sweat glands?
(a) NaCl
(b) Lactic acid
(c) Small amount of urea
(d) All of the above

88. Ketone bodies are observed in the urine of patients with:
(a) Starvation (b) Diabetes mellitus
(c) Diabetes insipidus (d) Both (a) and (b)

89. Discomfort, pain or burning sensation while urinating is called:
(a) Enuresis (b) Dysuria
(c) Anuria (d) Proteinuria

90. Lack of voluntary control over micturition is called:
(a) Cystitis
(b) Urinary incontinence
(c) Nephrolithiasis
(d) Pyuria

91. When the kidney of a person is damaged, the patient suffers from anaemia because:
(a) Haemoglobin is not synthesized sufficiently
(b) RBCs pass through the glomerulus
(c) Sufficient erythropoietin is not produced
(d) Iron and cyanocobalamin are not able to bind with haemoglobin

92. Which of the following is/are removed during haemodialysis?
(a) Glucose (b) Urea
(c) Amino acids (d) All of the above

93. Which of the following is added to blood during haemodialysis?
(a) Thrombin (b) Fibrin
(c) Heparin (d) Thromboplastin

94. Which of the following is an incorrect statement?
(a) pH of urine is around 6
(b) The yellow colour of urine is due to urochrome
(c) Dialysis is carried out when a patient suffers from uremia
(d) Glycosuria occurs if a person eats more sugar

95. Glomerulonephritis refers to:
(a) Inflammation of urethra
(b) Inflammation of glomeruli
(c) Precipitation of kidney stones
(d) Presence of uric acid in urine

96. Which of the following is incorrect with respect to haemodialysis?
(a) Blood drained from a convenient artery is pumped into a dialysing unit after adding an anticoagulant like heparin
(b) The dialysing fluid has the same composition as that of plasma, except the nitrogenous waste
(c) Nitrogenous wastes are removed by active transport
(d) The cleared blood is pumped back to the body through a vein after adding anti-heparin to it

97. In humans, gout is caused due to:
(a) High levels of ammonia
(b) High levels of urea
(c) High levels of uric acid
(d) High levels of amino acids

98. Which of the following helps in elimination of excretory wastes?
(a) Liver only (b) Skin only
(c) Kidneys only (d) Lungs, liver and skin

99. Choose the correct statement:
(a) Small amount of nitrogenous waste can be eliminated through saliva
(b) On an average, 1–1.5 litres of urine are excreted out by adult human kidneys
(c) Diuretics are drugs used to pass out excess volume of urine
(d) All are correct

100. In animals, nitrogenous wastes are produced mostly by catabolism of:
(a) Nucleic acid (b) Protein
(c) Both (a) and (b) (d) Fatty acids

NCERT EXEMPLAR QUESTIONS

1. The following substances are the excretory products in animals. Choose the least toxic form among them?
(a) Urea (b) Uric acid
(c) Ammonia (d) Carbon dioxide

2. Filtration of blood takes place at
(a) PCT (b) DCT
(c) Collecting ducts (d) Malpighian body

3. Which of the following statements is correct?
(a) ADH – prevents conversion of angiotensinogen in blood to angiotensin
(b) Aldosterone – facilitates water reabsorption
(c) ANF – enhances sodium reabsorption
(d) Renin – causes vasodilation

4. Which of the following is removed from our body by lungs?
(a) CO_2 only (b) H_2O only
(c) CO_2 and H_2O (d) Ammonia

5. The pH of human urine is approximately:
(a) 8.5 (b) 7
(c) 6 (d) 7.5

6. Different types of excretory structures and animals are given below. Match them appropriately and mark the correct answer from among those given below:

Excretory structure/organ	Animals
A. Protonephridia	(i) Prawn
B. Nephridia	(ii) Cockroach
C. Malpighian tubules	(iii) Earthworm
D. Green gland or Antennal gland	(iv) Flatworms

(a) A – (iv), B – (iii), C – (ii), D – (i)
(b) A – (iii), B – (i), C – (iv), D – (ii)
(c) A – (iii), B – (iv), C – (ii), D – (i)
(d) A – (iii), B – (i), C – (ii), D – *(iv)*

7. Which one of the following statements is incorrect?
(a) Birds and land snails are uricotelic animals
(b) Mammals and frogs are ureotelic animals
(c) Aquatic amphibians and aquatic insects are ammonotelic animals
(d) Birds and reptiles are ureotelic

8. Which of the following pairs is wrong?
(a) Uricotelic ---------- Birds
(b) Ureotelic ---------- Insects
(c) Ammonotelic ---------- Tadpole
(d) Ureotelic ---------- Elephant

9. Which one of the following statements is incorrect?
(a) The medullary zone of kidney is divided into a few conical masses called medullary pyramids projecting into the calyces
(b) Inside the kidney, the cortical region extends between the medullary pyramids as the renal pelvis
(c) Glomerulus along with Bowman's capsule is called the renal corpuscle
(d) Renal corpuscle, proximal convoluted tubule (PCT) and distal convoluted tubule (DCT) of the nephron are situated in the cortical region of the kidney

10. The condition of accumulation of urea in the blood is termed as:
(a) Renal calculi (b) Glomerulonephritis
(c) Uremia (d) Ketonuria

11. Which one of the following is also known as antidiuretic hormone?
(a) Oxytocin (b) Vasopressin
(c) Adrenaline (d) Calcitonin

12. Match the terms given in Column I with their physiological processes given in Column II and choose the correct answer:

Column I	Column II
A. Proximal convoluted tubule	(i) Formation of concenterated urine
B. Distal convoluted tubule	(ii) Filtration of blood
C. Henle's loop	(iii) Reabsorption of 70–80% of electrolytes
D. Counter-current mechanism	(iv) Ionic balance
E. Renal corpuscle	(v) Maintenance of concentration gradient in medulla

(a) A – (iii), B – (v), C – (iii), D – (ii), E – (i)
(b) A – (iii), B – (iv), C – (i), D – (v), E – (ii)
(c) A – (i), B – (iii), C – (ii), D – (v), E – (iv)
(d) A – (iii), B – (i), C – (iv), D – (v), E – (ii)

13. Match the abnormal conditions given in Column I with their explanations given in Column II and choose the correct option:

Column I		Column II
A. Glycosuria	(i)	Accumulation of uric acid in joints
B. Renal calculi	(ii)	Inflammation in glomeruli
C. Glomerular nephritis	(iii)	Mass of crystallised salts within the kidney
D. Gout	(iv)	Presence of glucose in urine

(a) A – (i), B – (iii), C – (ii), D-iv
(b) A – (iii), B – (ii), C – (iv), D – (i)
(c) A – (iv), B – (iii), C – (ii), D-i
(d) A – (iv), B – (ii), C – (iii), D – (i)

14. We can produce concentrated/dilute urine. This is facilitated by a special mechanism. Identify the mechanism:
(a) Reabsorption from PCT
(b) Reabsorption from collecting duct
(c) Reabsorption/Secretion in DCT
(d) Counter-current mechanism in Henle's loop/ vasa recta

15. Dialysing unit (artificial kidney) contains a fluid which is almost the same as plasma, except that it has:
(a) High glucose (b) High urea
(c) No urea (d) High uric acid

PRACTICE TEST PAPER

1. The component of blood that does not enter the nephron during ultrafiltration is:
(a) Glucose (b) Urea
(c) Sodium ions (d) Albumin

2. Which of the following is the correct statement with respect to vasopressin?
(a) It promotes sodium reabsorption from PCT and DCT
(b) It is synthesized in the neurohypophysis
(c) It promotes vasodilation of blood vessels
(d) It increases the permeability of the collecting duct and DCT for water

3. Select the incorrect match between the organism and its corresponding excretory structure:
(a) Protonephridia – *Planaria*
(b) Antenal gland – Crab
(c) Nephridia – Earthworm
(d) Malpighian tubules – Tapeworm

4. Increase in blood volume and blood pressure stimulates the atrial wall to release a peptide hormone called:
(a) Renin
(b) Aldosterone
(c) Atrial natriuretic factor
(d) Anti diuretic hormone

5. Identify the correct match:
(a) DCT – Impermeable to water
(b) Afferent arteriole – Carries blood away from the glomerulus
(c) PCT – Conditional reabsorption of water and electrolytes
(d) Descending limb of Henle's loop – Permeable to water and impermeable to electrolytes

6. Tubular reabsorption of water is minimum in:
(a) PCT (b) Collecting duct
(c) DCT (d) Henle's loop

7. Vasa rectae refers to:
(a) Capillaries surrounding kidneys in the lumbar region
(b) Capillaries present in Bowman's capsule
(c) Capillaries running parallel to Henle's loop
(d) Capillaries formed by breakdown of portal vein

8. Which of the following is incorrect with respect to haemodialysis?
(a) For dialysis, blood is taken out from the body through an artery
(b) Anti-heparin is added to blood prior to haemodialysis
(c) Nitrogenous wastes are removed from blood by simple diffusion
(d) Haemodialysis is used for patients suffering from uremia

9. An increase in body fluid due to excess water intake:
(a) Causes release of vasopressin
(b) Inhibits release of vasopressin
(c) Causes release of aldosterone
(d) Decreases glomerular blood flow

10. ____________ allows passage of a small amount of urea into the medullary interstitium to keep up the osmolarity:

(a) Henle's loop (b) PCT
(c) DCT (d) Collecting duct

11. Proximal convoluted tubule of nephron differs from distal convoluted tubule, as the former:
(a) Has modified cells called macula densa
(b) Is the active site for glucose reabsorption
(c) Allows reabsorption of electrolytes only by passive transport
(d) Both (a) and (c)

12. Identify the correct statement:
(a) Renin converts angiotensinogen to angiotensin I
(b) Urea is reabsorbed in the entire loop of Henle
(c) Water reabsorption is minimum in proximal convoluted tubule
(d) Excessive loss of water from the body stimulates the osmoreceptors which stimulate the hypothalamus to release vasopressin from the neurohypophysis

13. The osmolarity of fluid in the inner medullary interstitium is nearly ______ times the osmolarity in the cortex.
Choose the option which correctly fills in the given blank:
(a) Two (b) Three
(c) Four (d) Six

14. Which of the following is a ureotelic organism?
(a) Pigeon (b) Land snail
(c) Shark (d) Salamander

15. A special sensitive region called juxtaglomerular apparatus (JGA) is formed by certain cells of:
A. Proximal convoluted tubule
B. Afferent arteriole
C. Distal convoluted tubule
(a) (A) and (B) only (b) (B) and (C) only
(c) (A), (B) and (C) (d) (A) and (C) only

16. Which of the following is an incorrect match with respect to secretion in human body and the excretory wastes it contains?
(a) Sebum – Sterols, waxes and hydrocarbons
(b) Sweat – NaCl, small amount of urea, lactic acid
(c) Bile – Bilirubin, biliverdin
(d) Saliva – Lysozyme, ptyalin, thiocyanate

17. Extensions of the cortex between the medullary pyramids in the kidneys are known as:
(a) Renal pelvis
(b) Renal calyces
(c) Columns of Bertini
(d) Column of Bellini

18. Which of the following ions is not eliminated from the human kidney by tubular secretion?
(a) H^+ (b) K^+
(c) Na^+ (d) NH_4^+

19. Select the option to the fill in the blank correctly:
(a) Reabsorption of water from distal parts of the tubules is facilitated by the hormone ______
(b) Dialysing fluid contains all the constituents as in plasma, except ______
(c) A healthy adult human excretes (on an average) ______ g of urea per day
(a) (A) Aldosterone
(B) Nitrogenous wastes
(C) 150–180
(b) (A) Renin
(B) Urea
(C) 5–10
(c) (A) Vasopressin
(B) Urea
(C) 100–120
(d) (A) Vasopressin
(B) Nitrogenous wastes
(C) 25–30

20. Select the odd one out with respect to active reabsorption in the kidneys:
(a) Na^+ (b) Glucose
(c) Urea (d) Amino acids

21. Substance used to evaluate glomerular filtration rate (GFR) is:
(a) Insulin (b) Inulin
(c) Glucose (d) Sodium ions

22. The part of the nephron that is primarily helpful in maintaining osmolarity gradient within the renal interstitium is:
(a) DCT (b) Henle's loop
(c) PCT (d) Bowman's capsule

23. A juxtamedullary nephron is different from a cortical nephron in all features, except:
(a) Length of Henle's loop
(b) Presence of vasa recta
(c) Number
(d) Functions like filtration and reabsorption

24. Identify the incorrect statement:
(a) Each nephron has two parts – the glomerulus and the renal tubule
(b) Peritubular capillaries arise from the efferent arteriole
(c) Vasopressin increases blood pressure by a constrictor effect on blood vessels
(d) Appearance of ketone bodies in urine is indicative of diabetes insipidus

25. Glomerular filtrate is:
(a) Hypertonic to blood plasma
(b) Deproteinised plasma
(c) Blood without excretory wastes
(d) Plasma with blood corpuscles

26. Which of the following is an incorrect statement?
(a) Urinary bladder is lined with transitional epithelium
(b) Kidneys are situated between the levels of the last thoracic and the third lumbar vertebra, close to the dorsal inner wall of the abdominal cavity
(c) Right kidney lies at a level slightly higher than the left
(d) Each kidney weighs 120–170 g

27. Excretory structures in rotifers are:
(a) Malpighian tubules
(b) Flame cells
(c) Green glands
(d) Kidneys

28. Uricotelism is found in:
(a) Frogs and toads
(b) Sharks and aquatic insects
(c) Birds and reptiles
(d) Mammals and marine fishes

29. Human urine is acidic because:
(a) Acidic proteins are excreted with urine
(b) Hydrogen ions are actively secreted into the filtrate
(c) Potassium and sodium exchange generates acidity
(d) Reabsorptions of H^+ occurs actively in DCT

30. Effective filtration pressure in the glomerulus is caused due to:
(a) Secretion of adrenaline
(b) Vacuum develops in PCT and sucks the blood
(c) Pumping action of heart
(d) Afferent arteriole is slightly larger than efferent arteriole

31. Glucose and amino acids are reabsorbed in:
(a) Henle's loop
(b) Collecting duct
(c) Proximal convoluted tubule
(d) Distal convoluted tubule

32. Juxtaglomerular cells secrete the enzyme renin. It performs directly/indirectly all of the following functions, except:
(a) Regulate blood pressure
(b) Regulate GFR
(c) Moderate renal blood flow
(d) Increase RBC synthesis

33. The greater the ability of an animal to excrete hypertonic urine:
(a) Longer is Henle's loop
(b) Smaller is Henle's loop
(c) More the number of juxtamedullary nephrons
(d) Both (a) and (c)

34. Which of the following is incorrect with respect to the body's response to increased blood osmolarity in humans?
(a) Increased thirst
(b) Release of ADH by neurohypophysis
(c) Production of dilute urine
(d) Increased permeability of the collecting duct to water

35. Proximal convoluted tubule is not involved in:
(a) Active transport of glucose and amino acids
(b) Secretion of H^+ and NH_3
(c) Selective reabsorption of HCO_3^-
(d) Active reabsorption of urea

36. Which is the first nitrogenous catabolite of amino acids?
(a) Urea (b) Uric acid
(c) Ammonia (d) Hippuric acid

37. Sudoriferous glands produce sweat, whose main function is to facilitate cooling of the body surface. Which of the following is not part of sweat secretion?
(a) Urea (b) Glucose
(c) Lactic acid (d) Uric acid

38. Angiotensinogen is produced by:
(a) Lungs (b) Liver
(c) Kidneys (d) Intestine

39. A decrease in blood pressure/volume will not cause the release of:
(a) Vasopressin
(b) Renin
(c) Aldosterone
(d) Atrial natriuretic factor

40. Identify the incorrect statement:
(a) Urea is synthesised in the liver
(b) Uric acid is the chief nitrogenous waste of cockroach

(c) Earthworm excretes urea in rainy season and ammonia in dry conditions
(d) Aquatic amphibians are ammonotelic

41. Select the correct statement:
(a) Vasa recta is well developed in cortical nephrons
(b) PCT and DCT are situated in the medullary region of the kidney
(c) The ascending limb of Henle's loop extends as the DCT
(d) The juxtamedullary nephrons have short Henle's loop

42. Read the following statements:
(i) Podocytes are arranged in an intricate manner in Bowman's capsule to leave filtration slits or slit pores
(ii) The collecting duct of the nephron is located in the cortical region of the kidney
(iii) Tubular secretion helps in the maintenance of ionic and acid-base balance of body fluids
(iv) Blood vessels, nerves and urethra enter the kidney through the hilum
Which of the above statements are correct?
(a) (i) and (ii)
(b) (ii) and (iii)
(c) (ii) and (iv)
(d) (i), (iii) and (iv)

43. The ascending limb of Henle's loop:
(a) Is permeable to water
(b) Is impermeable to electrolytes
(c) Allows transport of electrolytes actively or passively
(d) Does not reabsorb water or electrolytes

44. The diagram given below represents the longitudinal section of a kidney with certain parts labelled A, B, C and D.

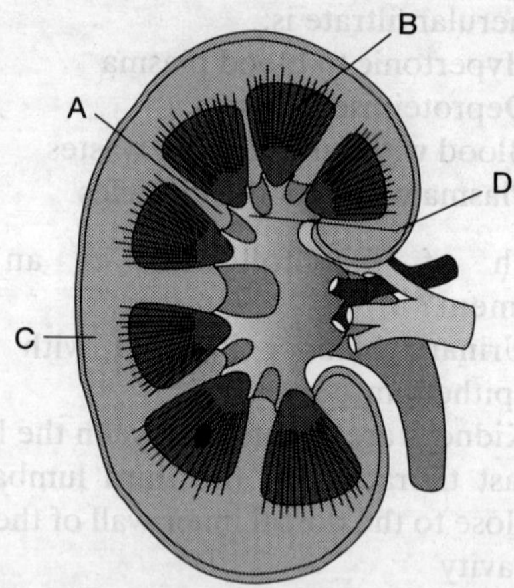

Which of the following is the correct explanation of the labelled parts?
(a) A – Formed by Bowman's capsule
(b) B – Contains loop of Henle and collecting ducts
(c) C – Called renal pelvis
(d) D – Called medullary pyramids

45. Following is a diagram of the renal corpuscle:

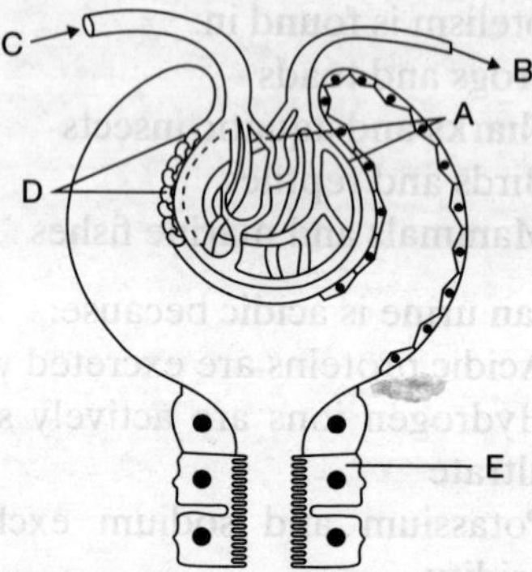

Identify the region from which podocytes arise:
(a) E only
(b) B and C
(c) D only
(d) A and B

ANSWER KEYS

Exercises

1. (b)	**2.** (c)	**3.** (c)	**4.** (c)	**5.** (d)	**6.** (b)	**7.** (b)	**8.** (a)	**9.** (c)	**10.** (b)
11. (b)	**12.** (d)	**13.** (c)	**14.** (c)	**15.** (c)	**16.** (c)	**17.** (b)	**18.** (b)	**19.** (a)	**20.** (b)
21. (c)	**22.** (c)	**23.** (d)	**24.** (b)	**25.** (c)	**26.** (b)	**27.** (b)	**28.** (b)	**29.** (b)	**30.** (b)
31. (c)	**32.** (b)	**33.** (c)	**34.** (d)	**35.** (b)	**36.** (d)	**37.** (d)	**38.** (c)	**39.** (d)	**40.** (b)
41. (c)	**42.** (d)	**43.** (d)	**44.** (d)	**45.** (b)	**46.** (a)	**47.** (d)	**48.** (d)	**49.** (c)	**50.** (b)
51. (c)	**52.** (b)	**53.** (b)	**54.** (c)	**55.** (b)	**56.** (b)	**57.** (d)	**58.** (d)	**59.** (b)	**60.** (b)
61. (b)	**62.** (b)	**63.** (c)	**64.** (d)	**65.** (c)	**66.** (b)	**67.** (c)	**68.** (a)	**69.** (d)	**70.** (c)
71. (c)	**72.** (b)	**73.** (d)	**74.** (b)	**75.** (b)	**76.** (c)	**77.** (b)	**78.** (a)	**79.** (c)	**80.** (d)
81. (b)	**82.** (b)	**83.** (d)	**84.** (b)	**85.** (d)	**86.** (a)	**87.** (d)	**88.** (d)	**89.** (b)	**90.** (b)
91. (c)	**92.** (b)	**93.** (c)	**94.** (d)	**95.** (b)	**96.** (c)	**97.** (c)	**98.** (d)	**99.** (d)	**100.** (c)

NCERT Exemplar Questions

1. (b)	**2.** (d)	**3.** (b)	**4.** (c)	**5.** (c)	**6.** (a)	**7.** (d)	**8.** (b)	**9.** (b)	**10.** (c)
11. (b)	**12.** (d)	**13.** (c)	**14.** (b)	**15.** (c)					

Practice Test Paper

1. (d)	**2.** (d)	**3.** (d)	**4.** (c)	**5.** (d)	**6.** (d)	**7.** (c)	**8.** (b)	**9.** (b)	**10.** (d)
11. (b)	**12.** (d)	**13.** (c)	**14.** (c)	**15.** (b)	**16.** (d)	**17.** (c)	**18.** (c)	**19.** (d)	**20.** (c)
21. (b)	**22.** (b)	**23.** (d)	**24.** (d)	**25.** (b)	**26.** (c)	**27.** (b)	**28.** (c)	**29.** (b)	**30.** (d)
31. (c)	**32.** (d)	**33.** (d)	**34.** (c)	**35.** (d)	**36.** (c)	**37.** (d)	**38.** (b)	**39.** (d)	**40.** (c)
41. (c)	**42.** (d)	**43.** (c)	**44.** (b)	**45.** (c)					

SOLUTIONS

Modes of Excretion

1. Ammonia is the most toxic nitrogenous waste and requires a large amount of water for its elimination.
2. Aquatic insects are ammonotelic in nature. Mammals, many terrestrial amphibians and marine fishes mainly excrete urea and are called ureotelic.
3. Aquatic amphibians excrete ammonia and require a large amount of water for its elimination.
4. Malpighian tubules are the excretory structures of most insects including cockroaches.
5. Antennal glands or green glands perform the excretory function in crustaceans like prawns.
6. Small amount of urea may be retained in the kidney matrix of some terrestrial animals to maintain the desired osmolarity.
7. Land snails excrete nitrogenous wastes as uric acid in the form of pellets or paste.
8. Kidneys do not play a significant role in the excretion of ammonia in ammonotelic animals. Ammonia, as it is readily soluble, is generally excreted by diffusion across body surfaces or through gill surfaces (in fish) as ammonium ions.
9. Option (c) is correct.

10. In ureotelic animals, ammonia produced by metabolism is converted into urea in the liver of animals like mammals, marine fishes and many terrestrial amphibians, and released into blood which is filtered and excreted out by the kidneys.
11. The principal nitrogenous excretory waste in mammals is synthesized in the liver and eliminated by the kidneys.
12. Protonephridia or flame cells are the excretory structures in Platyhelminthes (for example, *Planaria*), rotifers, some annelids and the Cephalochordates – *Amphioxus*.
13. Ammonia, urea and uric acid are all nitrogenous wastes excreted by animals. Ammonia is the most toxic form and requires a large amount of water for its elimination, whereas uric acid, being the least toxic, can be removed with minimum loss of water.
14. Freshwater bony fishes maintain water balance by excreting ammonia and hypotonic urine.
15. Option (c) is correct.
16. Ornithine and citrulline are amino acids that participate in ornithine/urea cycle.
17. Wastes like ammonia and CO_2 are removed from blood during the Krebs–Henseliet cycle.
18. In humans, uric acid is a byproduct of purine metabolism.
19. Hepatic vein is the blood vessel carrying maximum urea.
20. In the ornithine cycle, urea is synthesised in the liver from breakdown of arginine.

Human Excretory System

21. Kidneys are reddish brown, bean-shaped structures situated between the levels of the last thoracic and third lumbar vertebra, close to the dorsal inner wall of the abdominal cavity.
22. In kidneys, the cortex extends between the medullary pyramids as renal columns called columns of Bertini.
23. Glomerulus along with Bowman's capsule is called the Malpighian body or renal corpuscle.
24. Each kidney of an adult human measures 10–12 cm in length, 5–7 cm in width, 2–3 cm in thickness.
25. Glomerulus is a tuft of capillaries formed by the afferent arteriole – a fine branch of the renal artery.
26. In cortical nephrons, the loop of Henle is too short and extends very little into the medulla.
27. The DCT of many nephrons open into a straight tube called the collecting duct.
28. Malpighian corpuscle, PCT and DCT of the nephron are situated in the cortical region of the kidney whereas the loop of Henle dips into the medulla.
29. Efferent arteriole emerging from the glomerulus forms a fine capillary network around the renal tubule called the peritubular capillaries.
30. Efferent arteriole is narrower than afferent arteriole.
31. Towards the centre of the inner concave surface of the kidneys is a notch called hilum through which ureter, blood vessels and nerves enter.
32. Option (b) is correct.
33. Inner to the hilum of the kidney is a broad funnel-shaped space called the renal pelvis with projections called calyces.
34. Vasa recta is a 'U'-shaped minute vessel that runs parallel to Henle's loop and is highly developed in the juxtamedullary nephron.
35. The inner wall of Bowman's capsule has epithelial cells called podocytes that are arranged in an intricate manner so as to leave minute spaces called filtration slits or slit pores.
36. Option (d) is correct.
37. Bowman's capsule is present in the cortical region of the kidneys. It is a double-walled cup-like structure. It is lined by simple squamous epithelium.
38. The retroperitoneal kidney is covered by peritoneum on the ventral side.
39. The dotted appearance of the cortex of the kidneys is due to Malpighian body.
40. The total length of each nephron in humans is about 30 mm.

Urine Formation, Function of Tubules, Regulation of Kidney Function

41. Nitrogenous wastes are absorbed passively in uriniferous tubules.
42. Glomerular filtration rate (GFR) in a healthy individual is approximately 180 litres per day or 125 ml per minute.

43. Collecting duct allows passage of small amounts of urea into the medullary interstititum to keep up the osmolarity.
44. Reabsorption of water occurs passively in the initial segments of the nephron.
45. Conditional reabsorption of Na^+ and water takes place in DCT.
46. The descending limb of Henle's loop is permeable to water but almost impermeable to electrolytes.
47. The maintenance of pH and ionic balance of blood occurs by selective secretion of H^+ and K^+.
48. Glomerular filtrate consists of all constituents of blood except blood corpuscles and proteins.
49. In DCT, reabsorption of sodium takes place under the influence of aldosterone.
50. On an average 1100–1200 ml of blood is filtered by the kidneys per minute which constitute nearly one-fifth of blood pumped out by each ventricle of the heart in a minute.
51. The glomerular capillary blood pressure causes filtration of blood through three layers: the endothelium of glomerular blood vessels, the epithelium of Bowman's capsule and a basement membrane between the two layers.
52. If Henle's loop were absent from the mammalian nephron, urine would be more dilute.
53. A fall in glomerular blood flow/blood pressure causes juxtaglomerular cells (JG cells) to release renin. Renin catalyses conversion of angiotensinogen to angiotensin I, which forms angitensin II and increases blood pressure.
54. Atrial natriuretic factor (ANF) is released by the walls of atria in response to increase in blood volume and blood pressure.
55. An excessive loss of fluid from the body can activate osmoreceptors in the body to release vasopressin/ADH/pitressin from the neurohypophysis.
56. Renin is released from juxtaglomerular (JG) cells and converts angiotensinogen to angiotensin I.
57. ANF is a vasodilator and lowers blood pressure. It is released in response to a rise in blood pressure/blood volume by the atrial walls. It inhibits the renin-angiotensin mechanism.
58. ADH/Vasopressin causes water reabsorption from the distal tubules.
59. Human kidney can produce urine nearly four times more concentrated than the initial filtrate formed.
60. In the renin-angiotensin mechanism, angiotensin-II activates the adrenal cortex to release aldosterone.
61. Deficiency of vasopressin causes diabetes insipidus.
62. Angiotensinogen is secreted by the liver.
63. ADH controls the water permeability of DCT and promotes it reabsorption when water is scarce and osmolarity of body fluids is raised.
64. An increase in body fluid volume can suppress the release of vasopressin. It also switches off the osmoreceptors present in the hypothalamus.
65. The osmolarity of the interstitial fluid in kidney is 300 mOsm/L in the cortex and 1200 mOsm/L in the inner medulla.
66. The chemicals which increase the urine volume are called diuretics.
67. The reabsorption of Na^+ and water from the distal parts of the renal tubule is caused by aldosterone.
68. The osmolarity gradient from the cortex to the inner medulla is maintained by NaCl and urea.
69. The reabsorption of glucose occurs primarily in PCT.
70. pH and ionic balance of the body fluids is maintained by selective secretion of hydrogen ions and ammonia potassium ions into the filtrate and by absorption of HCO_3^- from it.
71. Minimum water reabsorption occurs in Henle's loop.
72. The filtrate becomes hypertonic in the descending limb of Henle's loop.
73. Ascending limb of Henle's loop is impermeable to water.
74. The ability to produce concentrated urine depends on the length of Henle's loop.
75. Small amount of urea enters the thin segment of the ascending limb of Henle's loop which is transported back to the interstitium by the collecting tubule.

76. The presence of brush-bordered simple cuboidal epithelium in PCT helps in increasing the surface area for reabsorption.

77. Nearly 70–80% of water is reabsorbed from PCT.

78. 25–30 grams of urea are excreted out per day by human kidneys.

79. Aldosterone causes reabsorption of Na^+ and water from the distal parts of the tubule.

80. Increased ADH secretion is likely after sweating-induced dehydration, increasing plasma osmolarity. The osmoreceptors in the body are activated and stimulate the hypothalamus to release ADH from the neurohypophysis.

Disorders of the Excretory System, Accessory Excretory Organs, Artificial Kidney

81. Polyuria refers to increase in frequency of urination.

82. Uremia refers to raised blood urea levels in patients with renal failure.

83. Cystitis refers to inflammation of the urinary bladder.

84. Presence of glucose and ketone bodies in urine indicates diabetes mellitus.

85. Lungs remove nearly 200 ml of CO_2 per minute. Sweat produced by sweat glands contain NaCl, urea, glucose, amino acid, lactic acid and water.

86. Renal calculi refers to stone or crystallised salts (oxalates) within the kidneys.

87. Option (d) is correct.

88. Ketone bodies are byproducts of fat breakdown and are observed in the urine of patients with starvation and diabetes mellitus.

89. Dysuria refers to pain or burning sensation while passing urine.

90. Lack of voluntary control over micturition is called urinary incontinence.

91. Anaemia is seen in patients with renal failure because sufficient erythropoietin is not produced.

92. Nitrogenous wastes like urea, ammonia and uric acid are removed during haemodialysis.

93. An anticoagulant heparin is added to blood during haemodialysis.

94. Glycosuria refers to the presence of glucose in urine. It is observed in patients with diabetes mellitus. It occurs because the blood glucose levels exceed the tubular maxima/renal threshold (180 mg/dl).

95. Glomerulonephritis refers to inflammation of glomeruli.

96. During haemodialysis, nitrogenous wastes are removed by passive transport/diffusion.

97. In humans, gout is caused due to high levels of uric acid and its precipitation in joints.

98. Option (d) is correct.

99. Option (d) is correct.

100. In animals, nitrogenous wastes are produced by catabolism of proteins and nucleic acid.

Locomotion and Movement

EXERCISES

Types of Muscles, Mechanism of Muscle Contraction

1. Identify the incorrect match:
   (a) Amoeboid movement – Macrophage
   (b) Flagellar movement – Trachea
   (c) Ciliary movement – Fallopian tube
   (d) Muscular movement – Tongue

2. Select the incorrect statement:
   (a) All locomotion is movement, but all movement is not locomotion
   (b) Ciliary movement is shown by *Euglena*
   (c) Spermatozoa swim using their flagella
   (d) Protoplasmic streaming in *Amoeba* is a simple form of movement

3. A common connective tissue layer that holds the muscle bundles or fascicles together is called:
   (a) Myofibril (b) Fascia
   (c) Perimysium (d) Endomysium

4. In meromyosin, the globular head:
   A. Has the binding site for ATP
   B. Has the binding site for actin
   C. Is a component of light meromyosin
   D. Has an active ATPase enzyme
   E. Is a component of heavy meromyosin
   (a) (A), (B) and (C) only
   (b) (B), (C) and (E) only
   (c) (A), (B), (D) and (E) only
   (d) (C), (D) and (E) only

5. Release of which of the following helps in relaxation of skeletal muscle fibre?
   (a) Acetylcholine (b) Calcium
   (c) Dopamine (d) Sodium

6. Skeletal muscle fibres are different from cardiac muscle fibres as:
   (a) Former lack actin and myosin
   (b) Latter have thick and thin filaments
   (c) Latter have intercalated discs
   (d) Former have longer refractory period than the latter

7. Which of the following properties is not common between muscle fibres and neurons?
   (a) Conductivity (b) Excitability
   (c) Depolarisation (d) Elasticity

8. Absence of ATP after the contracted sarcomere begins to relax would result in:
   (a) Quick release of myosin from actin
   (b) Inability of myosin heads to detach from actin
   (c) Binding of troponin to actin
   (d) No effect on muscle relaxation

9. Which of the following statements is incorrect?
   (a) Origin of muscle is mesodermal
   (b) The muscles of the limbs are striated and voluntary
   (c) Heart muscles are unstriated and unbranched
   (d) Repeated contraction of muscle results in accumulation of lactic acid

10. A single isolated contraction of a muscle fibre is called muscle:
    (a) Tonus (b) Twitch
    (c) Tetany (d) Spasm

11. In patients with myasthenia gravis:
    (a) Lactic acid is accumulated in muscle
    (b) Deficiency of acetylcholine is observed
    (c) Antibodies are formed against neuromuscular junction
    (d) Calcium does not enter synaptic knob

12. Sarcomere is the length of muscle between two:
    (a) M-lines (b) Z-lines
    (c) Lines of Krause (d) Both (b) and (c)

13. The characteristic feature of smooth muscle is/are:
(a) Uninucleated
(b) Fusiform
(c) Unbranched and involuntary
(d) All of the above

14. Each sarcomere is formed by:
(a) 1 A-band and 2 I-bands
(b) 2 A-bands and 1 I-band
(c) 1 A-band and 2 half I-bands
(d) 2 A-bands and 2 I-bands

15. The junction between a motor neuron and the sarcolemma of the muscle fibre is called:
(a) Neuromuscular junction
(b) Motor end plate
(c) Cross-bridge
(d) Both (a) and (b)

16. Anisotropic band is made of:
(a) Actin filaments
(b) Myosin filaments
(c) Calmodulin
(d) Both (a) and (b)

17. Ions used for muscle contraction are:
(a) Na^+ and Ca^{++} (b) Ca^{++} and Mg^{++}
(c) Mg^{++} and Na^+ (d) Na^+ and K^+

18. Which of the following is not a function of muscles?
(a) Voiding out urine during micturition
(b) Propelling of food through gut
(c) Exchange of gases in alveoli
(d) Bending of forearm over arm

19. Which of the following is correct for cardiac muscles?
(a) Gap junctions are present
(b) They show auto-rhythmicity
(c) They are branched and voluntary
(d) More than one option is correct

20. The difference between a red and a white muscle fibre in skeletal muscles can be based on all of the following, except:
(a) Thickness of muscle fibre
(b) Myoglobin content
(c) Amount of sarcoplasmic reticulum
(d) Presence of sarcomere

21. Which of the following is not observed during skeletal muscle contraction?
(a) A-band length is unchanged
(b) Sarcomere lengthens
(c) H-zone disappears
(d) Z-lines come closer

22. Which of the following is correct with respect to the events of muscle contraction?
(a) Change in size of A-band
(b) Change in length of myosin filament
(c) Change in length of actin filament
(d) Sliding of actin filament over myosin filament

23. During contraction of skeletal muscles:
(a) Size of I-band decreases, while A-band remains the same
(b) Size of both I-band and A-band increases
(c) Size of I-band remains the same while that of A-band decreases
(d) Size of both I-band and A-band decreases

24. Rapid spasms (wild contractions) in muscle due to low calcium in body fluid is called:
(a) Muscular dystrophy (b) Tetany
(c) Tonus (d) Gout

25. Rigor mortis is the stiffening of the body muscles a few hours after death due to:
(a) Deposition of lactic acid
(b) Breakdown of acto-myosin complex
(c) Lack of ATP and phosphocreatine
(d) Relaxation of muscle fibres

26. Identify the incorrect statement:
(a) Each 'F actin' is a polymer of 'G actin'
(b) Muscles possess excitability, elasticity, contractility and extensibility
(c) Skeletal muscles are classified as red and white primarily based on the amount of red myoglobin pigment in them
(d) The region H-zone has only actin filaments

27. Fill in the blanks with correct options:
A. Each actin (thin) filament is made of ___**I**___ 'F' (filamentous) actins helically wound to each other
B. Two filaments of another protein ___**II**___ also run close to 'F' actins throughout its length.
C. In the resting state, a subunit of ___**III**___ masks the active binding sites for myosin on actin filaments
D. Many monomeric proteins called ___**IV**___ constitute one thick filament

	I	II	III	IV
(a)	Four	Troponin	Tropomyosin	Actin
(b)	Two	Tropomyosin	Troponin	Meromyosin
(c)	Three	Tropomyosin	Troponin	Actin
(d)	Two	Troponin	Tropomyosin	Meromyosin

28. The mitochondria present in muscle fibres are called:

(a) Sarcomere
(b) Sarcosomes
(c) Sarcolemma
(d) Sarcoplasmic reticulum

29. The process of muscle contraction continues till ________ are pumped back to ________ resulting in the masking of actin filaments which cause the return of 'Z-lines' back to their original position:
(a) Calcium, sarcosomes
(b) Calcium, sarcoplasmic reticulum
(c) Magnesium, sarcosomes
(d) Sodium, sarcoplasmic reticulum

30. Which of the following describes the role of calcium during contraction of skeletal muscles?
(a) Binding to the acetylcholine receptors at the neuromuscular junction
(b) Binding with troponin so that myosin head binding sites on actin are exposed
(c) Binding with actin to unmask the active sites for myosin
(d) Binding with tropomyosin to activate troponin

31. Cross-arms of the myosin monomer consist of:
(a) Trimers of troponin
(b) Outward projection of the head region of meromyosin
(c) Outward projection of the tail region of meromyosin
(d) Polymer of G actin to form F actin

32. When an action potential reaches the axon terminal of a motor neuron:
(a) Physical contact between the motor neuron and the muscle fibre occurs
(b) Acetylcholine is released into the neuromuscular junction
(c) Calcium is released inside the muscle fibre
(d) Acetylcholinesterase cleaves the acetycholine at the neuromuscular junction

33. Find out the correct sequence of events that take place during muscle contraction:
A. Hydrolysis of ATP
B. Formation of cross-bridge
C. Release of calcium ions in sarcoplasm
D. Release of neurotransmitter at neuromuscular junction
E. Generation of action potential in the sarcolemma
F. Binding of calcium with subunit of troponin
(a) (A), (C), (D), (F), (E), (B)
(b) (C), (D), (E), (A), (B), (F)
(c) (D), (E), (C), (F), (A), (B)
(d) (D), (E), (C), (A), (F), (B)

34. During muscle contraction:
(a) Hydrolysis of ATP does not occur
(b) Temperature of muscle decreases
(c) Thick filaments slide over thin filaments
(d) Chemical energy is changed into mechanical energy

35. Study the given figure and identify its parts:

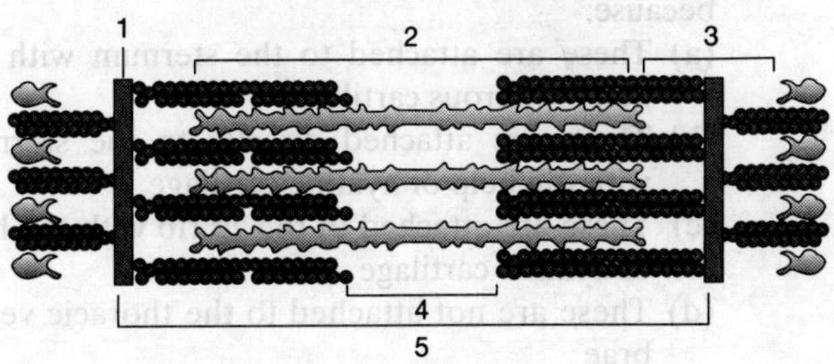

	1	2	3	4	5
(a)	I-band	A-band	Z-line	Sarcomere	H-zone
(b)	Z-line	A-band	I-band	H-zone	Sarcomere
(c)	A-band	I-band	Z-line	H-zone	Sarcomere
(d)	Z-line	I-band	A-band	Sarcomere	H-zone

Axial and Appendicular Skeleton

36. Select the incorrect match:
(a) Coxal bone – Innominate bone
(b) Clavicle – Collar bone
(c) Femur – Thigh bone
(d) Scapula – Pelvic girdle

37. Read the following statements and select the option that correctly classifies them as (T) or false (F):
A. There are 12 pairs of ribs in humans
B. Sternum is present on the dorsal side in the rib cage
C. Glenoid cavity is present in the pelvic girdle
D. There are 8 cranial bones in the skull

	(A)	(B)	(C)	(D)
(a)	T	F	F	T
(b)	T	T	F	F
(c)	F	T	F	T
(d)	F	F	F	T

38. The head of femur articulates with ________ of coxal bone to form the hip joint:
(a) Glenoid cavity
(b) Acetabulum
(c) Obturator foramen
(d) Foramen magnum

39. Select the correct match with respect to the human bone and its feature:
(a) Sternum – Neural canal
(b) Coxal bone – Acromion process
(c) Ribs – Bicephalic
(d) Hyoid – L-shaped bone

40. The 8th, 9th and 10th pairs of ribs are called false ribs because:
(a) These are attached to the sternum with the help of fibrous cartilage
(b) These are attached directly to the sternum with the help of hyaline cartilage
(c) These are attached to the 7th rib with the help of hyaline cartilage
(d) These are not attached to the thoracic vertebrae

41. Consider the following statements with certain blanks. Find out the option which correctly fills in these blanks:
A. The human face is made of ____I____ bones
B. All mammals (except a few) have ____II____ cervical vertebrae
C. The number of phalanges in each limb of humans is ____III____
D. A cup-shaped bone called ____IV____ covers the knee ventrally

	I	II	III	IV
(a)	12	8	12	Scapula
(b)	8	5	8	Patella
(c)	8	7	15	Tibia
(d)	14	7	14	Patella

42. Scapula is a large ____I____ bone situated in the ____II____ part of the thorax between ____III____ and ____IV____ ribs.
Fill in the blanka with correct options:

	I	II	III	IV
(a)	Circular	Dorsal	First	Seventh
(b)	Rectangular	Dorsal	Second	Sixth
(c)	Triangular	Dorsal	Second	Seventh
(d)	Triangular	Ventral	Second	Eighth

43. The number of bones is the same in which set listed in the options below?
(a) Face and tarsals of one limb
(b) Carpals and tarsals
(c) Cranium and carpals of one limb
(d) Pelvic girdle and pectoral gridle

44. Each rib is called bicephalic because:
(a) It is connected dorsally to the vertebral column and ventrally to the sternum
(b) It is connected dorsally to the sternum and ventrally to the vertebral column
(c) It has two articulation surfaces on its dorsal end
(d) It has two articulation surfaces on its ventral end

45. Mark the odd one out with respect to the bones of the axial skeleton:
(a) Ribs (b) Atlas
(c) Sternum (d) Fibula

46. Femur, a bone of the appendicular skeleton, articulates with all, except:
(a) Patella (b) Tibia
(c) Fibula (d) Coxal

47. Which of the following is a correct match?
(a) Cranial bone – Mandible
(b) Ossicle – Hyoid
(c) Flat bones – Ribs
(d) Pelvic girdle – Scapula

48. Tibia is:
(a) Thinner than the fibula
(b) Thicker than the fibula
(c) Present in the forearm
(d) Of the same thickness as the fibula

49. 11th and 12th pairs of ribs are called:
(a) True ribs (b) False ribs
(c) Floating ribs (d) None of the above

50. Acromion process is a part of:
(a) Humerus (b) Femur
(c) Pectoral girdle (d) Pelvic girdle

51. Which of the following is incorrect with respect to the functions of the vertebral column?
(a) Protects the spinal cord
(b) Supports the head
(c) Serves as the point of attachment for the ribs and musculature of the back
(d) Protects the brain

52. Each vertebra has a central hollow portion called ________ through which the spinal cord passes:
(a) Foramen magnum
(b) Neural canal
(c) Obturator foramen
(d) Foramen ovale

53. The two halves of the pelvic girdle meet ____I____ to form the ____II____ containing ____III____ cartilage.
Fill in the blanks with correct options:

	I	II	III
(a)	Dorsally	Pubic symphysis	Fibrous
(b)	Dorsally	Pectoral girdle	Calcified
(c)	Ventrally	Pubic symphysis	Fibrous
(d)	Ventrally	Suture	Hyaline

54. Read the following statements and choose the correct option:

I: The dorsal, flat, triangular body of the scapula has a slightly elevated ridge called the spine which projects as a flat, expanded process called the acromion.

II: Below the acromion is a depression called the acetabulum, which articulates with the head of the humerus to form the shoulder joint

(a) Both statements are correct
(b) Both statements are incorrect
(c) Only statement I is correct
(d) Only statement II is correct

55. Each half of the pectoral girdle consists of:
(a) Clavicle and sternum
(b) Two coxal bones
(c) Clavicle and scapula
(d) Ilium, ischium and pubis

56. The rib cage is formed by:
(a) Cervical vertebrae, ribs and sternum
(b) Thoracic vertebrae, ribs and sternum
(c) Lumbar vertebrae, ribs and sternum
(d) Sacrum, ribs and sternum

57. Select the incorrect statement:
(a) Each limb is made of 30 bones
(b) Vertebral column is formed by 28 serially arranged bones and is ventrally placed
(c) Pelvic girdle is formed by two coxal bones
(d) Wrist of each forelimb has 8 carpal bones

58. The number of lumbar vertebrae in the human body is:
(a) 12 (b) 7
(c) 5 (d) 4

59. The number of metacarpals in each hand is:
(a) 4 (b) 6
(c) 5 (d) 8

60. Olecranon process is present in:
(a) Radius (b) Ulna
(c) Humerus (d) Tibia

61. Appendicular skeleton is formed by all, except:
(a) Pectoral girdle (b) Skull
(c) Limb bones (d) Pelvic girdle

62. _____ is an example of sesamoid bone:
(a) Patella (b) Scapula
(c) Femur (d) Sacrum

63. Knee joint is formed between:
(a) Humerus and radius–ulna
(b) Femur and coxal bone
(c) Femur and fibula
(d) Femur and tibia

64. Each middle ear has _____ ossicles:
(a) 2 (b) 3
(c) 4 (d) 5

65. Ribs attached to sternum are:
(a) First seven pairs
(b) Last three pairs
(c) Vertebrochondral
(d) More than one option is correct

66. Which of the following is the correct sequential order of vertebrae in the vertebral column of humans?
(a) Cervical – Lumbar – Sacral – Coccygeal – Thoracic
(b) Cervical – Coccygeal – Lumbar – Sacral – Thoracic
(c) Cervical – Thoracic – Lumbar – Sacral – Coccygeal
(d) Cervical – Sacral – Thoracic – Lumbar – Coccygeal

67. Skull of human beings is:
(a) Monocondylic (b) Dicondylic
(c) Tricondylic (d) Tetracondylic

68. The occipital bone has a large opening called _____ through which the brain communicates with the spinal cord:
(a) Foramen magnum
(b) Foramen caecum
(c) Foramen rotundus
(d) Foramen spinosum

69. The strongest and largest vertebra is:
(a) Atlas (b) Thoracic
(c) Lumbar (d) Coccyx

70. Coccyx is formed by fusion of _____ vertebrae:
(a) Five (b) Four
(c) Three (d) Two

71. The first cervical vertebra is:
(a) Atlas
(b) Axis
(c) Vertebra prominens
(d) Coccyx

72. Odontoid process is present on:
(a) Atlas
(b) First lumbar vertebra
(c) Axis
(d) First thoracic vertebra

73. The unpaired bones of the face are:
(a) Vomer and maxilla
(b) Vomer and mandible
(c) Nasal and lacrimal
(d) Lacrimal and zygomatic

74. The first vertebra is ____I____ and articulates with ____II____.
Fill in the blanks for options I and II:
(a) Atlas, vomer
(b) Atlas, occipital condyles
(c) Axis, temporal
(d) Atlas, glenoid cavity

75. Which of the following is an incorrect statement?
(a) Cranium is formed of eight bones of which two are paired
(b) Clavicle is a link between the axial and appendicular skeletons
(c) The coxal bone of the pelvic girdle is formed by fusion of ilium, ischium and pubis
(d) Axial skeleton has 126 bones while appendicular skeleton has 80 bones

76. Identify the correct option for A, B, C and D:

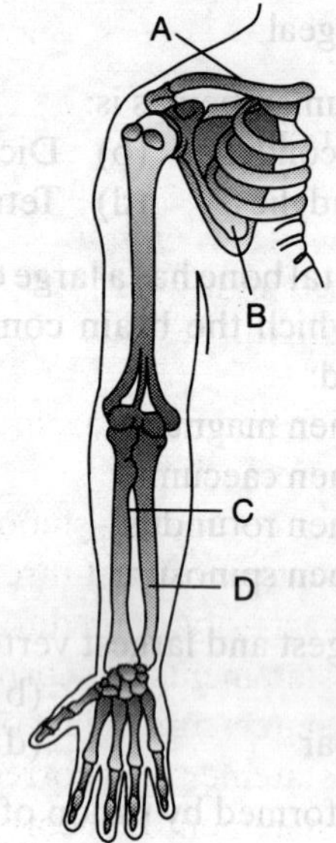

	A	B	C	D
(a)	Scapula	Clavicle	Radius	Ulna
(b)	Clavicle	Scapula	Radius	Ulna
(c)	Coxal bone	Scapula	Ulna	Radius
(d)	Clavicle	Scapula	Ulna	Radius

77. The figure given below shows the pelvic girdle and lower limb bones. Identify the parts labelled A, B, C, D and E and select the correct option:

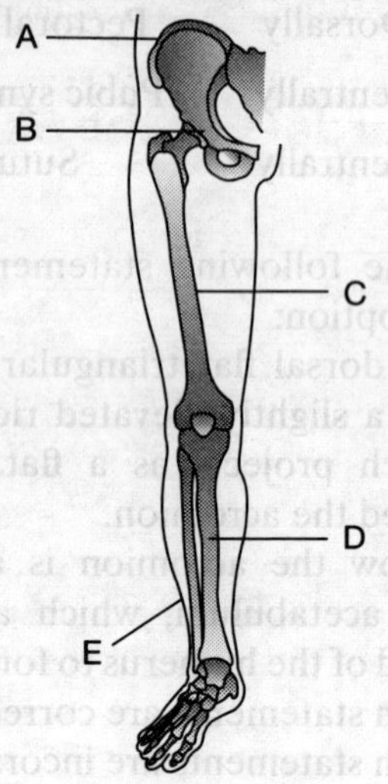

	A	B	C	D	E
(a)	Sacrum	Pubis	Patella	Tibia	Fibula
(b)	Ilium	Ischium	Femur	Tibia	Fibula
(c)	Ilium	Ischium	Femur	Fibula	Tibia
(d)	Ischium	Ilium	Patella	Tibia	Tarsal

78. Select the incorrect statement:
(a) Pectoral and pelvic girdle bones help in the articulation of the upper and lower limbs, respectively, with the axial skeleton
(b) Scapula has a slightly elevated ridge called spine which projects an expanded process called the acromion
(c) Greater and lesser trochanters are present on the femur
(d) Appendicular skeleton has bones distributed along the main axis of the body

79. Read the following statements:
(i) Haversian canal is present in bone
(ii) Bones are attached to each other by tendons
(iii) Upper jaw is formed of two zygomatic bones
(iv) Thoracic vertebrae are the strongest and heaviest
Which of the above statements are incorrect?
(a) (i) and (ii)
(b) (ii) and (iv) only
(c) (i) and (iv)
(d) (ii), (iii) and (iv)

80. Identify the incorrect statement:
(a) 11th and 12th pairs of ribs are vertebral ribs
(b) The lumbar region of the vertebral column in humans is made up of 5 vertebrae

(c) The skeletal system has two principle divisions: the exial skeleton and the appendicular skeleton
(d) Each foot has eight tarsal bones

Joints, Disorders of the Skeletal System

81. ______ joints do not allow any movement:
(a) Fibrous (b) Cartilaginous
(c) Synovial (d) Hinge

82. The joint between atlas and axis is an example of:
(a) Hinge joint
(b) Pivot joint
(c) Saddle joint
(d) Ball and socket joint

83. Inflammation of joints due to accumulation of uric acid crystals is called:
(a) Rheumatoid arthritis
(b) Osteoarthritis
(c) Gout
(d) Osteoporosis

84. Nodding movement of head occurs due to the joint between:
(a) Second and third cervical vertebrae
(b) First and second cervical vertebrae
(c) Atlas and occipital condyles
(d) Cervical and thoracic vertebrae

85. An auto-immune disorder in which there is inflammation of synovial membrane, secreting abnormal granules and causing erosion of articular cartilage is:
(a) Myasthenia gravis
(b) Gouty arthritis
(c) Rheumatoid arthritis
(d) Osteoarthritis

86. Which of the following is an example of cartilaginous joint?
(a) Between the carpals
(b) Between carpal and metacarpal of thumb
(c) Pubic symphysis
(d) Sutures of skull

87. The type of joint in which flat skull bones fuse end-to-end with the help of dense fibrous connective tissues in the form of sutures is:
(a) Cartilaginous (b) Fibrous
(c) Synovial (d) Saddle

88. Which type of joint occurs between carpals in humans?
(a) Pivot joint (b) Hinge joint
(c) Saddle joint (d) Gliding joint

89. Select the correct matching of the type of the joint with the example in the human skeletal system:

Type of joint	Example
(a) Pivot joint	Between the third and fourth cervical vertebrae
(b) Hinge joint	Between humerus and pectoral girdle
(c) Cartilaginous joint	Between frontal and parietal
(d) Saddle joint	Between carpal and metacarpal of thumb

90. Glenoid cavity articulates:
(a) Clavicle with scapula
(b) Humerus with scapula
(c) Scapula with acromion
(d) Clavicle with acromion

91. Which of the following is correctly matched?
(a) Hinge joint – Between vertebrae
(b) Gliding joint – Between zygapophyses of successive vertebrae
(c) Fibrous joint – Between phalanges
(d) Cartilaginous joint – Skull bones

92. ______ joint is present between the ribs and sternum:
(a) Cartilaginous (b) Fibrous
(c) Gliding (d) Angular

93. Joints are points of:
(a) Contact between bones and cartilages
(b) Contact between bones
(c) Both (a) and (b)
(d) Crack in the bones

94. Read the following statements and choose the correct option:
I: Force generated by smooth muscles is used to carry out movement through joints
II: Joints act as fulcrum during movement
(a) Both statements are correct
(b) Both statements are incorrect
(c) Only statement I is correct
(d) Only statement II is correct

95. In cartilaginous joints, the bones involved are joined together with the help of:
(a) Dense fibrous connective tissues
(b) Cartilages
(c) Fluid-filled synovial cavity
(d) Adipose tissue

96. Which of the following is incorrect with respect to osteoporosis?
(a) It is an age-related disorder
(b) Bone mass is decreased
(c) Patients have reduced chance of fractures
(d) Decreased levels of estrogen is a common cause

97. Fibrous joints are also called:
(a) Synarthroses (b) Synovial joints
(c) Diarthroses (d) Amphiarthroses

98. Mark the odd one out with respect to examples of hinge joints:
(a) Knee joint
(b) Interphalangeal joint
(c) Elbow joint
(d) Intercarpal joint

99. Identify the incorrect statement:
(a) Deficiency of vitamin D causes rickets in children and osteomalacia in adults
(b) Dislocation refers to displacement of the articular surface out of the joint cavity
(c) The primary movement at the pivot joint is rotation
(d) Ball-and-socket joint is a mono-axial joint that allows movement in one plane only

100. Which of the following is a degenerative joint disease characterised by the progressive erosion of articular cartilage in old age?
(a) Gout (b) Rheumatoid arthritis
(c) Osteoarthritis (d) Osteoporosis

NCERT EXEMPLAR QUESTIONS

1. Match the following and mark the correct option:

Column I	Column II
A. Fast muscle fibres	(i) Myoglobin
B. Slow muscle fibres	(ii) Lactic acid
C. Actin filament	(iii) Contractile unit
D. Sarcomere	(iv) I-band

(a) A – (i), B – (ii), C – (iv), D – (iii)
(b) A – (ii), B – (i), C – (iii), D – (iv)
(c) A – (ii), B – (i), C – (iv), D – (iii)
(d) A – (iii), B – (ii), C – (iv), D – (i)

2. Ribs are attached to:
(a) Scapula (b) Sternum
(c) Clavicle (d) Ilium

3. What is the type of movable joint present between the atlas and axis?
(a) Pivot (b) Saddle
(c) Hinge (d) Gliding

4. ATPase of the muscle is located in:
(a) Actinin (b) Troponin
(c) Myosin (d) Actin

5. Intervertebral disc is found in the vertebral column of:
(a) Birds (b) Reptiles
(c) Mammals (d) Amphibians

6. Which one of the following is showing the correct sequential order of vertebrae in the vertebral column of human beings?
(a) Cervical — lumbar — thoracic — sacral — coccygeal
(b) Cervical — thoracic — sacral — lumbar — coccygeal
(c) Cervical — sacral — thoracic — lumbar — coccygeal
(d) Cervical — thoracic — lumbar — sacral — coccygeal

7. Which one of the following pairs is incorrect?
(a) Hinge joint: between humerus and pectoral girdle
(b) Pivot joint: between atlas and axis
(c) Gliding joint: between the carpals
(d) Saddle joint: between carpals and metacarpals of thumb

8. Knee joint and elbow joint are examples of:
(a) Saddle joint
(b) Ball and socket joint
(c) Pivot joint
(d) Hinge joint

9. Macrophages and leucocytes exhibit:
(a) Ciliary movement
(b) Flagellar movement
(c) Amoeboid movement
(d) Gliding movement

10. Which one of the following is not a disorder of bones?
(a) Arthritis (b) Osteoporosis
(c) Rickets (d) Atherosclerosis

11. Which one of the following statements is incorrect?
(a) Heart muscles are striated and involuntary
(b) The muscles of hands and legs are striated and voluntary
(c) The muscles located in the inner walls of the alimentary canal are striated and involuntary
(d) Muscles located in the reproductive tracts are unstriated and involuntary

12. Which one of the following statements is true?
(a) Head of humerus bone articulates with acetabulum of pectoral girdle
(b) Head of humerus bone articulates with glenoid cavity of pectoral girdle
(c) Head of humerus bone articulates with acetabulum of pelvic girdle
(d) Head of humerus bone articulates with glenoid cavity of pelvic girdle

13. Muscles with characteristic striations which are involuntary are:
(a) Muscles in the wall of the alimentary canal
(b) Muscles of the heart
(c) Muscles assisting locomotion
(d) Muscles of the eyelids

14. Match the followings and mark the correct option:

Column I	Column II
A. Sternum	(i) Synovial fluid
B. Glenoid cavity	(ii) Vertebrae
C. Freely movable joint	(iii) Pectoral girdle
D. Cartilaginous joint	(iv) Flat bones

(a) A – (ii), B – (i), C – (iii), D – (iv)
(b) A – (iv), B – (iii), C – (i), D – (ii)
(c) A – (ii), B – (i), C – (iv), D – (iii)
(d) A – (iv), B – (i), C – (ii), D – (iv)

PRACTICE TEST PAPER

1. Which of the following remains unchanged during contraction of skeletal muscle fibres?
(a) Length of H-zone
(b) Length of isotropic band
(c) Length of actin filament
(d) Length between two consecutive Z-lines

2. Which of the following bones is not a part of the axial skeleton?
(a) Cervical vertebra (b) Malleus
(c) Femur (d) Frontal

3. Which type of movement is shown by some specialised cells like macrophages and leucocytes found in blood in our body?
(a) Ciliary movement
(b) Amoeboid movement
(c) Flagellar movement
(d) Muscular movement

4. Each organised skeletal muscle in our body is made up of a number of muscle bundles or fascicles. These muscle bundles are held together by a common collagenous connective tissue layer called:
(a) Epimysium (b) Epineurium
(c) Endomyosium (d) Fascia

5. ATP binding site and actin binding site are located in which part of the myosin filament?
(a) Light meromyosin
(b) Short arm of meromyosin
(c) Globular head of meromyosin
(d) Tail of meromyosin

6. Mark the odd one out with respect to paired bones in the human body:
(a) Femur (b) Ribs
(c) Mandible (d) Patella

7. Progressive degeneration of skeletal muscle mostly due to genetic disorder is:
(a) Tetany
(b) Myasthenia gravis
(c) Muscular dystrophy
(d) Gout

8. Complete the following analogy:
I-band : Actin :: H-zone: _____
Choose the correct option:
(a) Myosin (b) Actin
(c) Actin and myosin (d) Actinin

9. Locomotion requires perfect, coordinated activity of:
(a) Muscular system (b) Skeletal system
(c) Neural system (d) All of the above

10. Identify the incorrect statement with respect to red muscle fibres and white muscle fibres:
(a) Red muscle fibres are rich in myoglobin

(b) Amount of sarcoplasmic reticulum is high in white muscle fibres
(c) Red muscle fibres contract very fast and for very short duration
(d) Red muscle fibres have plenty of mitochondria

11. Which of the following is a single U-shaped bone present at the base of the buccal cavity, and does not articulate with any other bone?
(a) Mandible (b) Hyoid
(c) Atlas (d) Zygomatic

12. Identify the incorrect statement:
(a) 80 bones are present in the axial skeleton
(b) 12 bones are present in the face
(c) Cranium is formed of 8 bones
(d) Each middle ear has 3 ossicles

13. Humerus and radius–ulna articulate by ______ joint:
(a) Pivot (b) Hinge
(c) Ball and socket (d) Fibrous

14. Sesamoid bones are formed by ossification of:
(a) Smooth muscles (b) Skeletal muscles
(c) Tendons (d) Cartilage

15. Dorsally, the 8th, 9th and 10th pairs of ribs articulate with:
(a) Sternum (b) 7th rib
(c) Cartilage of 7th rib (d) Vertebral column

16. Muscles have the property of:
(a) Contractility (b) Excitability
(c) Elasticity (d) All of the above

17. Gliding joint is present between:
(a) Humerus and scapula
(b) Femur and tibia
(c) Radius and ulna
(d) Zygapophyses of vertebrae

18. Choose the correct option:
(a) True ribs – Vertebrochondral ribs
(b) Floating ribs – Vertebral ribs
(c) False ribs – Vertebrosternal ribs
(d) True ribs – 8th, 9th and 10th pairs

19. The continued state of contraction caused by rapid succession of many stimuli is called:
(a) Muscle twitch (b) Tetanus
(c) Fatigue (d) Cramp

20. The 'thick' filament in muscles is a polymerised protein of:
(a) G actin (b) F actin
(c) Meromyosin (d) Troponin

21. Acromion process is a part of:
(a) Pelvic girdle (b) Pectoral girdle
(c) Vertebral column (d) Skull

22. Scapula is a large triangular flat bone situated in the dorsal part of the thorax between:
(a) Second and tenth ribs
(b) Third and ninth ribs
(c) Second and seventh ribs
(d) First and fifth ribs

23. Pectoral girdle is formed by:
(a) Two coxal bones
(b) Scapula and clavicle
(c) Maxilla and mandible
(d) Radius and ulna

24. In resting muscle fibre, a subunit of troponin covers:
(a) Actin binding sites on myosin
(b) Myosin binding sites on actin
(c) Calcium binding sites on tropomyosin
(d) Calcium binding sites on actin

25. Pubic symphysis is an example of:
(a) Fibrous joint (b) Cartilaginous joint
(c) Hinge joint (d) Synovial joint

26. Read the following statements and mark the incorrect one among them:
(a) Tetany is caused by rapid spasm in muscles due to low calcium in body fluids
(b) Bone matrix has calcium salts while cartilage has chondroitin sulphate
(c) Fibrous joints play a major role in locomotion
(d) Osteoporosis is an age-related disease of the skeletal system that may occur due to oestrogen deficiency

27. Fill in the blanks with correct options:
The skull is composed of two sets of bones: cranial and facial, that total ___**I**___ bones. Cranial bones are ___**II**___ whereas facial bones are ___**III**___ in number.

	I	II	III
(a)	22	8	12
(b)	22	14	8
(c)	22	8	14
(d)	24	10	14

28. Which of the following is an incorrect statement?
(a) Floating ribs are attached only to the vertebral column
(b) A motor neuron along with the muscle fibres connected to it constitutes a motor unit
(c) F actin is a polymer of monomeric G actin
(d) Many monomeric proteins called tropomyosin constitute one thick filament

29. Identify the incorrect statement:
(a) Each half of the pectoral gridle consists of a clavicle and a scapula
(b) Axial skeleton comprises 126 bones distributed along the main axis of the body
(c) Bone and cartilage are specialised connective tissue
(d) Skeletal muscles are classified as red and white fibres based primarily on the amount of myoglobin pigment in them

30. Read the following statements and choose the correct option:
I: Force generated by the muscles is used to carry out movement through joints, where joints act as fulcrum.
II: In cartilaginous joints, the bones involved are joined with the help of dense fibrous connective tissues.
(a) Both statements are correct
(b) Both statements are incorrect
(c) Only statement I is correct
(d) Only statement II is correct

31. Passage of ova through the female reproductive tract is facilitated by:
(a) Flagellar movements
(b) Ciliary movements
(c) Amoeboid movements
(d) Protoplasmic streaming

32. Scapula is a flat triangular bone with a slightly elevated ridge called the spine which projects as a flat, expanded process called ___I___. The ___II___ articulates with this.

	I	II
(a)	Acromion	Sternum
(b)	Olecranon	Clavicle
(c)	Acromion	Clavicle
(d)	Coracoid	Sternum

33. Sternum is connected to ribs by _____ cartilage:
(a) Hyaline (b) Elastic
(c) Fibrous (d) Calcified

34. The central part of the thick filament, not overlapped by thin filament, is called:
(a) Z-line (b) H-zone
(c) Line of Krause (d) Sarcomere

35. White muscles have:
(a) Abundant blood supply
(b) Aerobic respiration only
(c) High myoglobin pigment
(d) Fewer mitochondria

36. Which of the following is an incorrect statement?
(a) Each limb is made up of 30 bones
(b) Skeleton system consists of a framework of bones and a few cartilages
(c) According to the sliding filament theory, contraction of muscle fibre takes place by sliding of thin filaments over thick filaments
(d) Each myofibril has dark I-bands and light A-bands

37 During muscular contraction:
(a) Chemical energy is converted to mechanical energy
(b) Mechanical energy is converted to chemical energy
(c) Physical energy is converted to mechanical energy
(d) Potential energy is converted to chemical energy

38. Which of the statements about the mechanism of muscle contraction are correct?
(i) During muscle contraction, isotropic band becomes elongated
(ii) Repeated activation of the muscles can lead to lactic acid accumulation
(iii) Acetycholine is released when the neural signal reaches the motor end plate
(iv) Muscle contraction is initiated by a signal sent by the CNS via a sensory neuron
(a) (i) and (ii) (b) (iii) and (iv)
(c) (i) and (iii) (d) (ii) and (iii)

39. On simulation, a muscle cell releases calcium from:
(a) Mitochondria
(b) Nucleus
(c) Sarcoplasmic reticulum
(d) Synaptic vesicle

40. The pivot joint between atlas and axis is a type of:
(a) Saddle joint
(b) Synovial joint
(c) Fibrous joint
(d) Cartilaginous joint

41. Muscles of the heart are:
(a) Striated and voluntary
(b) Non-striated and involuntary
(c) Striated and involuntary
(d) Striated, unbranched and syncytial

42. Which of the following is not a function of the skeletal system?
(a) Locomotion
(b) Storage of minerals
(c) Production of erythrocytes
(d) Production of body heat

43. The matrix of bone and cartilage can be distinguished by the presence of:
(a) Haversian canal (b) Lacuna
(c) Adipose cells (d) Chromatophores

44. Knee joint is an example of:
(a) Pivot joint
(b) Hinge joint
(c) Gliding joint
(d) Ball and socket joint

45. Pubic symphysis:
(a) Contains fibrous cartilage
(b) Is a fibrous joint
(c) Is a cartilaginous joint
(d) Both (a) and (c) are correct

ANSWER KEYS

Exercises

1. (b)	2. (b)	3. (b)	4. (c)	5. (b)	6. (c)	7. (d)	8. (b)	9. (c)	10. (b)
11. (c)	12. (d)	13. (d)	14. (c)	15. (d)	16. (d)	17. (b)	18. (c)	19. (d)	20. (d)
21. (b)	22. (d)	23. (a)	24. (b)	25. (c)	26. (d)	27. (b)	28. (b)	29. (b)	30. (b)
31. (b)	32. (b)	33. (c)	34. (d)	35. (b)	36. (d)	37. (a)	38. (b)	39. (c)	40. (c)
41. (d)	42. (c)	43. (c)	44. (c)	45. (d)	46. (c)	47. (c)	48. (b)	49. (c)	50. (c)
51. (d)	52. (b)	53. (c)	54. (c)	55. (c)	56. (b)	57. (b)	58. ()	59. (c)	60. (b)
61. (b)	62. (a)	63. (d)	64. (b)	65. (a)	66. (c)	67. (b)	68. (a)	69. (c)	70. (b)
71. (a)	72. (c)	73. (b)	74. (b)	75. (d)	76. (b)	77. (b)	78. (d)	79. (d)	80. (d)
81. (a)	82. (b)	83. (c)	84. (c)	85. (c)	86. (c)	87. (b)	88. (d)	89. (d)	90. (b)
91. (b)	92. (a)	93. (c)	94. (d)	95. (b)	96. (c)	97. (a)	98. (d)	99. (d)	100. (c)

NCERT Exemplar Questions

1. (c)	2. (b)	3. (a)	4. (c)	5. (c)	6. (d)	7. (a)	8. (d)	9. (c)	10. (d)
11. (c)	12. (b)	13. (b)	14. (b)						

Practice Test Paper

1. (c)	2. (c)	3. (b)	4. (d)	5. (c)	6. (c)	7. (c)	8. (a)	9. (d)	10. (c)
11. (b)	12. (b)	13. (b)	14. (c)	15. (d)	16. (d)	17. (d)	18. (b)	19. (b)	20. (c)
21. (b)	22. (c)	23. (b)	24. (b)	25. (b)	26. (c)	27. (c)	28. (d)	29. (b)	30. (c)
31. (b)	32. (c)	33. (a)	34. (b)	35. (d)	36. (d)	37. (a)	38. (d)	39. (c)	40. (b)
41. (c)	42. (d)	43. (a)	44. (b)	45. (d)					

SOLUTIONS

Types of Muscles, Mechanism of Muscle Contraction

1. Trachea is lined with ciliated epithelium.
2. Flagellar movement is seen in *Euglena*.
3. A common connective tissue layer that holds the muscle bundles or fascicles together is called fascia.
4. Each meromyosin has two important parts, a globular head with a short arm and a tail, the former called heavy meromyosin (HMM) and the latter, light meromyosin (LMM). The globular head is an active ATPase enzyme and has binding sites for ATP and active sites for actin.
5. For relaxation in skeletal muscle, calcium is pumped back into sarcoplasmic cisternae, resulting in masking of actin filament.
6. Cardiac muscle fibres can be distinguished from skeletal muscles due to the presence of intercalated discs.
7. Elasticity is a property of muscle fibres and not seen in neurons.
8. Myosin head detaches from the actin molecule when a new ATP molecule binds to it. Thus, absence of ATP molecule would cause inability of myosin head to detach from actin.
9. Cardiac muscle fibres are branched and striated.
10. A single isolated contraction of a muscle fibre is called muscle twitch.
11. In patients with myasthenia gravis, antibodies are formed against the neuromuscular junction.
12. Sarcomere is the length of muscle between two Z-lines/Lines of Krause.
13. Smooth muscle fibres are uninucleated, unbranched, fusiform and involuntary.
14. Each sarcomere is formed by 1 A-band and two half I-bands.
15. The junction between a motor neuron and sarcolemma of muscle fibre is called motor end plate or neuromuscular junction.
16. Anisotropic band (A-band) is made of both myosin and actin filaments.
17. Ca^{2+} and Mg^{2+} are used for muscle contraction.
18. Option (c) is correct.
19. Cardiac muscles are characterised by gap junctions and also show auto-rhythmicity.
20. Both red and white muscle fibres have sarcomere.
21. The length of the sarcomere reduces during skeletal muscle contraction.
22. During muscle contraction, the actin filaments slide over myosin filaments.
23. During contraction of skeletal muscles, the size of the I-band decreases while that of the A-band remains the same.
24. Tetany refers to the wild spasm of muscles due to low calcium levels.
25. Rigor mortis is stiffening of the body muscles a few hours after death due to lack of ATP and phosphocreatine.
26. The region of the H-zone has only myosin filaments.
27. Option (b) is correct.
28. The mitochondria present in muscle fibres are called sarcosomes.
29. The process of muscle contraction continues till calcium ions are pumped back to the sarcoplasmic reticulum, resulting in the masking of actin filaments, which cause the return of Z-lines to their original position.
30. During muscle contraction, calcium binds with troponin so that myosin head binding sites on actin are exposed.
31. The heavy meromyosin (HMM) component, that is, the head and short arm, projects outwards at a regular distance and angle from each other from the surface of a polymerised myosin filament and is known as cross-arm.
32. When an action potential reaches the axon terminal of the motor neuron, acetylcholine is released into the neuromuscular junction.

33. Option (c) is correct.

34. During muscle contraction, chemical energy is changed into mechanical energy.

35. Option (b) is correct.

Axial and Appendicular Skeleton

36. Scapula is a part of the pectoral girdle.

37. Option (a) is correct.

38. The head of the femur articulates with the acetabulum to form the hip joint.

39. Each rib has two articulation surfaces on its dorsal end and hence is called bicephalic.

40. The 8th, 9th and 10th pairs of ribs do not articulate directly with the sternum but join the seventh rib with the help of hyaline cartilage, and are called false ribs or vertebrochondral ribs.

41. Option (d) is correct.

42. Option (c) is correct.

43. There are eight bones in the cranium and also in the carpals of one limb.

44. Each rib is called bicephalic because it has two articulation surfaces on its dorsal end.

45. Fibula is a bone of the appendicular skeleton.

46. Femur does not articulate with fibula. At the knee joint, the lower end of the femur articulates with the tibia.

47. Option (c) is correct.

48. Tibia is thicker and longer than fibula.

49. The last two pairs (11th and 12th) of ribs are not connected ventrally and are therefore called floating ribs.

50. Acromion process is a part of the pectoral girdle.

51. Option (d) is correct.

52. Each vertebra has a central hollow portion called neural canal through which the spinal cord passes.

53. Option (c) is correct.

54. Below the acromion is a depression called the glenoid cavity which articulates with the head of the humerus to form the shoulder joint.

55. Each half of the pectoral girdle consists of a clavicle and scapula.

56. Thoracic vertebrae, ribs and sternum together form the rib cage.

57. Our vertebral column is formed of 26 serially arranged units called vertebrae and is dorsally placed.

58. The number of lumbar vertebrae in the human body is 5.

59. There are five metacarpals in each hand.

60. Olecranon process is present in the ulna.

61. Appendicular skeleton is formed by limb bones, and pectoral and pelvic girdles.

62. Patella is an example of sesamoid bone.

63. Knee joint is formed between the femur and tibia.

64. There are three ossicles/ear bones in each middle ear.

65. The first seven pairs of ribs are called true ribs. Dorsally, they are attached to thoracic vertebrae and ventrally connected to the sternum with the help of hyaline cartilage.

66. Option (c) is correct.

67. Skull of humans is dicondylic, since it has two occipital condyles with the help of which it is connected to the vertebral column.

68. The occipital bone has a large opening called foramen magnum through which the brain communicates with the spinal cord.

69. Lumbar vertebra are the strongest and largest.

70. Coccyx is formed by fusion of four coccygeal vertebrae.

71. Atlas refers to the first cervical vertebrae.

72. Odontoid process is present on the axis.

73. The unpaired bones of the face are vomer and mandible.

74. The first vertebra is atlas and articulates with the occipital condyles.

75. The axial skeleton has 80 bones while the appendicular skeleton has 126 bones.

76. Option (b) is correct.

77. Option (b) is correct.

78. Axial skeleton has bones distributed along the main axis of the body. The bones of the limbs along with their girdles constitute the appendicular skeleton.

79. Bones are attached to each other by ligaments. Upper jaw is formed by two maxillary bones. Lumbar vertebrae are the strongest and heaviest.

80. Each foot has seven tarsal bones.

Joints, Disorders of the Skeletal System

81. Fibrous joints do not allow any movement. This type of joint is shown by the flat skull bones which fuse end-to-end with the help of dense fibrous connective tissues in the form of sutures, to form the cranium.

82. The joint between the atlas and axis is an example of pivot joint.

83. Gout refers to inflammation of joints due to accumulation of uric acid crystals.

84. Nodding movement of head occurs due to joint between atlas and occipital condyles.

85. Rheumatoid arthritis is an autoimmune disease in which there is inflammation of the synovial membrane, secreting abnormal granules and causing erosion of articular cartilage.

86. In cartilaginous joints, the bones involved are joined together with the help of cartilages, for example, pubic symphysis between the two coxal bones.

87. Fibrous joints are present between the bones of the cranium to form sutures.

88. Gliding joints are present between the carpal bones.

89. Option (d) is correct.

90. Humerus articulates with the glenoid cavity of the scapula to form the shoulder joint.

91. Option (b) is correct.

92. Cartilaginous joint is present between the ribs and sternum.

93. Joints are points of contact between bones, or between bones and cartilage.

94. Force generated by skeletal muscles is used to carry out movement through joints, where the joint acts as a fulcrum.

95. In cartilaginous joints, the bones involved are joined together with the help of cartilage.

96. Osteoporosis is an age-related disorder characterised by decreased bone mass and increased chance of fractures.

97. Fibrous joints are also called synarthroses.

98. Intercarpal joints are example of gliding joints.

99. Ball-and-socket joint is the one in which a partially spherical end lies in a socket, allowing multidirectional movement and rotation.

100. Osteoarthritis is a degenerative joint disease characterised by the progressive erosion of articular cartilage in old age.

Neural Control and Coordination

EXERCISES

Neurons, Nerve Impulse

1. The resting neural membrane has poor permeability for ____________, although it has a higher permeability for ____________:
(a) Na^+ and K^+ (b) K^+ and Na^+
(c) Na^+ and Mg^{2+} (d) Ca^{2+} and Na^+

2. Which of the following is an advantage of electrical synapses?
(a) They may allow two-way transmission of impulses in contrast to chemical synapses
(b) They are faster than chemical synapses
(c) They can synchronise the activity of a group of neurons or muscle fibres
(d) All of the above

3. The functional unit of the nervous system is:
(a) Nephron (b) Neuron
(c) Nekton (d) *Neophron*

4. 'Neurons are excitable cells'. This statement is best illustrated as:
(a) Membrane is permeable to negatively charged ions and impermeable to positively charged ions
(b) Ion channels on the neural membrane are selectively permeable
(c) Different ion channels are present on the neural membrane
(d) The neuronal membrane exists in a polarised state and responds to stimulus

5. Find the incorrect combination of type of neuron and its location:

Neuron	Location
(a) Multipolar	Cerebral cortex
(b) Bipolar	Retina
(c) Unipolar	Cerebellum
(d) Pseudounipolar	Dorsal root ganglia

6. Consider the following statements and choose the correct option:
A. In *Hydra*, the neural organisation is composed of a network of neurons
B. In insects, the neural system is organised as a brain along with a number of ganglion and neural tissue
(a) Statement (A) is incorrect but Statement (B) is correct
(b) Statement (A) is correct but Statement (B) is incorrect
(c) Both statements are correct
(d) Both statements are incorrect

7. A neuron with one axon and one dendrite is found in:
(a) Embryonic stage
(b) Retina
(c) Olfactory epithelium
(d) Both (b) and (c)

8. Which of the following statements are correct for chemical synapses?
A. Pre- and post-synaptic neurons are separated by fluid-filled synaptic cleft
B. The membranes of the pre-synaptic and post-synaptic neurons are in close proximity
C. Bidirectional movement of impulse
D. Presence of neurotransmitter vesicles in the axon terminals
(a) (A) and (B)
(b) (B) and (C)
(c) (A) and (D)
(d) (B) and (D)

9. In resting neurons, the axonal membrane:
(a) Has high concentration of K^+ in the axoplasm along with low concentration of Na^+
(b) Is more permeable to Na^+ and impermeable to K^+

(c) Is more permeable to K^+ and impermeable to Na^+
(d) Both (a) and (c)

10. Nerve impulse transmission within a nerve fibre is independent of:
(a) Diameter of neuron
(b) Presence or absence of myelin
(c) Concentration of Na^+ and K^+ ions inside and outside the neuron
(d) Chemical nature of neurotransmitter

11. Saltatory conduction nerve impulse takes place through:
(a) Dendrite
(b) Cyton
(c) Myelinated neuron
(d) Non-myelinated neuron

12. During a nerve impulse, repolarisation occurs due to:
(a) Influx of Na^+
(b) Influx of K^+
(c) Efflux of K
(d) Efflux of Na^+

13. Which of the following statements is correct for the nodes of Ranvier?
(a) Neurilemma is discontinuous
(b) Myelin sheath is discontinuous
(c) Covered by myelin
(d) Both neurilemma and myelin sheath are discontinuous

14. Myelin sheath is secreted in the central nervous system (CNS) by:
(a) Microglia
(b) Ependymal cells
(c) Schwann cells
(d) Oligodendrocytes

15. Nissl granules are absent in:
(a) Cyton
(b) Axon
(c) Dendrite
(d) Cell body

16. What used to be described as Nissl bodies is now defined as:
(a) Mitochondria
(b) Golgi bodies
(c) Centrioles
(d) Ribosomes

17. Influx of ______ ions is responsible for the release of neurotransmitters from the synaptic vesicles into the synaptic cleft:
(a) Sodium
(b) Calcium
(c) Potassium
(d) Magnesium

18. In a neuron, the impulse is conducted from:
(a) Cell body to dendrite
(b) Cell body to axon
(c) Dendrite to cell body
(d) Both (b) and (c)

19. Which of the following is the role of Na^+–K^+ pumps in the axonal membrane?
(a) To maintain the polarised state in the resting state of the neuron
(b) To pump $2Na^+$ out for every $3K^+$ pumped into the axon
(c) To pump 3 K^+ out for every 2 Na^+ pumped into the axon
(d) More than one option is correct

20. The bulbous structure present at the terminal end of an axon is:
(a) Schwann cell
(b) Synaptic knob
(c) Synaptic vesicle
(d) Dendrite

21. Which of the following neurotransmitters produces inhibitory post-synaptic potential?
(a) Acetylcholine
(b) Epinephrine
(c) Gamma-aminobutyric acid (GABA)
(d) Norepinephrine

22. Myelinated neurons are abundant in:
(a) Grey matter of brain
(b) Grey matter of spinal cord
(c) Both (a) and (c)
(d) White matter of brain and spinal cord

23. During the conduction of nerve impulses, the action potential is the result of the movement of:
(a) Na^+ from extracellular fluid into axon
(b) K^+ from extracellular fluid into axon
(c) Na^+ from axon to extracellular fluid
(d) Ca^{2+} from extracellular fluid into axon

24. The binding of neurotransmitters with receptors opens ion channels that generate a new potential in:
(a) Synaptic vesicles
(b) Pre-synaptic membrane
(c) Post-synaptic membrane
(d) Synaptic cleft

25. Short fibres which branch repeatedly and project out of the cell body and contain Nissl granules are called:
(a) Cyton
(b) Axon

(c) Perikaryon
(d) Dendrite

26. Which of the following is correct?
(a) A sub-threshold stimulus can generate an action potential
(b) Electrical synapse works by releasing a chemical neurotransmitter
(c) Action potential generation in neuron follows the all-or-none principle
(d) The resting membrane of the neuron is permeable to negatively charged proteins present in the axoplasm

27. There is free flow of electrical impulses in myelinated axons at the nodes of Ranvier due to the:
(a) Lack of axon
(b) Lack of cell body
(c) Lack of myelin sheath
(d) Lack of neurilemma

28. Identify the incorrect statement:
(a) The neural system provides an organised network of point-to-point connections for quick coordination
(b) The endocrine system provides chemical integration through hormones
(c) In our body, the neural system and endocrine system jointly coordinate and integrate all the activities of the organs so that they function in a synchronised fashion
(d) Coordination is the process through which organs function independently without complementing each other's functions

29. Find the incorrect match with respect to the parts labelled A, B, C and D in the given diagram:

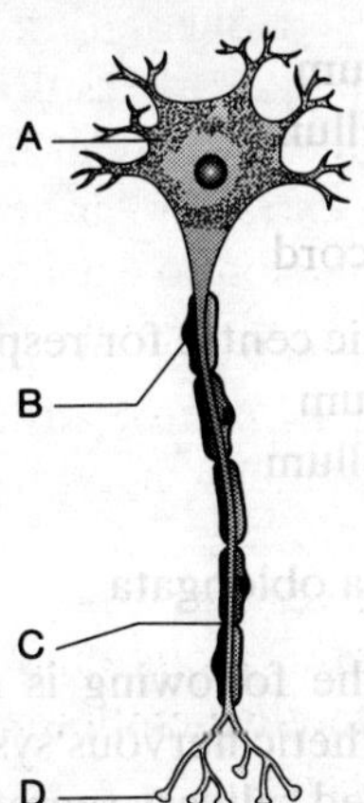

(a) A - Nissl granules – Present in cyton and dendrites
(b) B - Schwann cells - Forms myelin sheath in CNS
(c) C - Nodes of Ranvier – Gap between two adjacent myelin sheaths
(d) D - Synaptic knob – Contains neurotransmitter vesicles

30. Which of the following statements is correct with respect to electrical synapses?
(a) The impulse can flow directly from one neuron to the other
(b) Transmission of impulse is very similar to impulse conduction along a single axon
(c) The pre- and post-synaptic membranes lie in close proximity
(d) All of the above are correct

31. All of the following are correct about non-medullated nerve fibres, except:
(a) It is enclosed by Schwann cells that do not form a myelin sheath around axons
(b) It is found in autonomous and somatic neural systems
(c) It is found in spinal and cranial nerves
(d) Both (a) and (b)

32. Which of the following is an incorrect match?
(a) Autonomic neural system – Controls smooth muscles
(b) Somatic neural system – Controls skeletal muscles
(c) Sympathetic neural system – Controls involuntary organs
(d) Parasympathetic neural system – Controls voluntary organs

33. The Na^+–K^+ pump in the axonal membrane:
(a) Is operative during the resting stage and transports 3 Na^+ into the cell and 2 K^+ outwards
(b) Is operative during depolarisation and transports 3 Na^+ outwards and 2 K^+ into the cell
(c) Is operative during repolarisation and transports 3Na^+ and 2 K^+ into the cell
(d) Is operative during the resting stage and transports 3 Na^+ outwards and 2 K^+ into the cell

34. Identify the incorrect statement:
(a) The CNS includes the brain and the spinal cord and is the site of information processing and control
(b) The PNS comprises all the nerves of the body associated with the CNS
(c) The afferent nerve fibres transmit impulses from the CNS to the concerned peripheral tissues while efferent fibres transmit impulses from the tissues to the CNS
(d) The PNS is divided into two divisions, called somatic neural system and autonomic neural system

35. Identify the site where the following type of neuron is found:

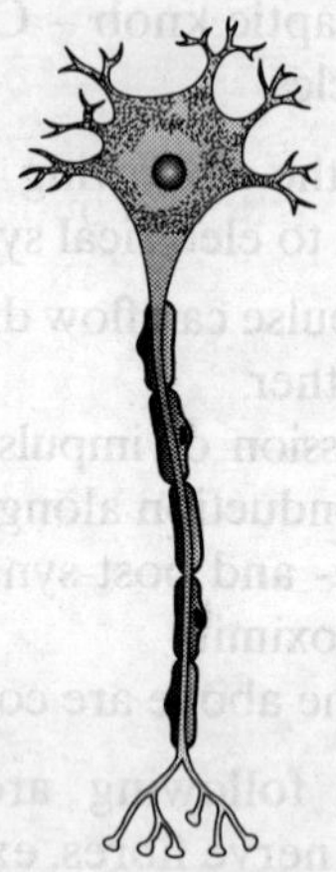

(a) *Hydra* (b) Cerebral cortex
(c) Embryonic stage (d) Retina

Neural System (CNS, PNS and ANS)

36. Which of the following is not a part of the forebrain?
(a) Cerebrum
(b) Thalamus
(c) Hypothalamus
(d) Cerebellum

37. Which of the following part of the human brain is paired?
(a) Pons
(b) Medulla
(c) Cerebral hemisphere
(d) Thalamus

38. The cerebrum wraps around a structure called ________, which is a major coordinating centre for sensory and motor signalling:
(a) Epithalamus
(b) Thalamus
(c) Hypothalamus
(d) Midbrain

39. The inner parts of the cerebral hemisphere and a group of associated deep structures like amygdala and hippocampus form a complex structure called:
(a) Basal ganglia
(b) Corpus striatum
(c) Limbic system
(d) Hypothalamus

40. The midbrain is located between ___I___ of the forebrain and ___II___ of the hindbrain. Fill in the blanks:
(a) Thalamus, pons
(b) Pons, thalamus
(c) Thalamus, cerebellum
(d) Hypothalamus, thalamus

41. The brainstem is formed by:
(a) Midbrain
(b) Pons
(c) Medulla oblongata
(d) All of the above

42. Find the correct match:
(a) Thalamus – Controls body temperature, urge to eat and drink
(b) Association areas – Neither clearly sensory nor motor in function
(c) Corpus callosum – Tract of nerve fibres which connects the two cerebellar hemispheres
(d) Pons – Controls gastric secretions and cardiovascular reflexes

43. All of the following are functions of cerebrospinal fluid (CSF), except:
(a) It provides a shock absorbing cushion to the CNS
(b) It provides buoyancy to the brain
(c) It is present in the subdural space between the duramater and the arachnoid mater
(d) It is present in the ventricles of the brain and the central canal of the spinal cord

44. Which of the following are correctly paired?
(a) Forebrain and pons
(b) Midbrain and thalamus
(c) Hindbrain and medulla oblongata
(d) Brainstem and cerebrum

45. ________ is responsible for coordinating voluntary movements such as posture, balance and speech, resulting in smooth and balanced muscular activity:
(a) Cerebrum
(b) Cerebellum
(c) Pons
(d) Spinal cord

46. Pneumotaxic centre for respiration is present in:
(a) Cerebrum
(b) Cerebellum
(c) Pons
(d) Medulla oblongata

47. Which of the following is not a function of the parasympathetic nervous system?
(a) Increased salivary secretion
(b) Increased heart rate
(c) Increased peristalsis
(d) Pupillary constriction

48. Mammalian brain differs from amphibian brain in having:

(a) Cerebrum
(b) Cerebellum
(c) Medulla
(d) Corpus callosum

49. Damage to the cerebellum causes:
(a) Slurred speech
(b) Stumbling
(c) Ataxia
(d) All of the above

50. Identify the incorrect match:
(a) Prosencephalon – Forebrain
(b) Mesencephalon – Midbrain
(c) Rhombencephalon – Hindbrain
(d) Diencephalon – Pons

51. Which part of the brain commands "how and when to move the body parts"?
(a) Only the cerebellum's conscious directions
(b) Only the cerebrum's unconscious instructions
(c) The cerebellum's unconscious directions and the cerebrum's conscious instructions
(d) The cerebellum's conscious directions and the cerebrum's unconscious instructions

52. Mark the key communication link between the senses and the cortex that receives the majority of incoming signals and evaluates their importance before passing them to the cerebrum:
(a) Epithalamus
(b) Hypothalamus
(c) Thalamus
(d) Pons

53. Autonomic neural system relays impulses from:
(a) CNS to glands
(b) CNS to involuntary muscles
(c) Both (a) and (b)
(d) CNS to skeletal muscles

54. Decoding of visual information occurs in:
(a) Temporal lobe
(b) Parietal lobe
(c) Frontal lobe
(d) Occipital lobe

55. In the human spinal cord, the ciliated cells present inside it are called:
(a) Podocyte
(b) Porocyte
(c) Schwannn cells
(d) Ependymal cells

56. Corpus callosum is found in which part of the human brain?
(a) Forebrain
(b) Midbrain
(c) Hindbrain
(d) Rhombencephalon

57. Corpora quadrigemina consists of four round swellings (lobes) which are present on:
(a) Dorsal portion of midbrain
(b) Ventral portion of midbrain
(c) Dorsal portion of hindbrain
(d) Dorsal portion of forebrain

58. Which lobe of the cerebral hemisphere registers feelings of hot, cold, touch and pain?
(a) Frontal
(b) Occipital
(c) Temporal
(d) Parietal

59. Bulges present between the adjacent grooves of the cerebrum are called:
(a) Sulci
(b) Fissures
(c) Gyri
(d) Cortex

60. The outermost meninx made of fibrous tissue surrounding the human brain is:
(a) Duramater
(b) Piamater
(c) Arachnoid membrane
(d) Tunica vaginalis

61. Which of the following is an incorrect statement?
(a) All spinal nerves arise from the spinal cord
(b) Fifty pairs of spinal nerves arise from the spinal cord
(c) All spinal nerves are mixed
(d) Spinal cord runs through the neural canal of the vertebral column

62. A canal called the cerebral aqueduct passes through:
(a) Midbrain
(b) Pons
(c) Medulla
(d) Thalamus

63. Which of the following is not a function of the hypothalamus?
(a) Control body temperature
(b) Urge for eating and drinking
(c) Secrete neurohormones
(d) Vomiting

64. Purkinje cells are characteristic of:
(a) Cerebrum
(b) Pons

(c) Medulla
(d) Cerebellum

65. Which of the following functions is not regulated by the medulla oblongata?
(a) Respiration
(b) Gastric secretion
(c) Cardiovascular reflexes
(d) Expression of emotional reactions

66. Which part of the limbic system converts information from short-term to long-term memory, essential in learning:
(a) Amygdala
(b) Hippocampus
(c) Hypothalamus
(d) Septum

67. Which of the following is an incorrect statement?
(a) Neural cell bodies are concentrated in the cerebral cortex
(b) Centre for coughing and vomiting are present in the medulla
(c) Cerebrum forms the major part of the human brain
(d) The cerebral cortex is referred to as white matter due to its pale appearance

68. Which of the following is the longest cranial nerve?
(a) Olfactory (b) Vagus
(c) Facial (d) Optic

69. Which of the following are purely sensory cranial nerves?
(a) Olfactory, optic and occulomotor
(b) Facial, vagus and hypoglossal
(c) Occulomotor, trochlear and abducens
(d) Olfactory, optic and auditory

70. How many sympathetic ganglia are found in humans on either side of the spinal cord?
(a) 10 (b) 21
(c) 7 (d) 14

71. Which of the following cranial meninges is the innermost and has the Latin meaning "tender mother"?
(a) Duramater
(b) Arachnoid mater
(c) Piamater
(d) Choroid plexus

72. Find the correct match:
(a) Cerebral cortex – Inner part of cerebrum
(b) Cerebral aqueduct – Present in spinal cord
(c) Hypothalamus – Lies at the base of the thalamus
(d) Pons – Thermoregulation

73. Corpora quadrigemina comprises:
(a) Tracts which connect the cerebellum with pons
(b) Tracts which connect the forebrain with the hindbrain
(c) Two superior and two inferior colliculi
(d) Ascending tracts of the spinal cord

74. The lateral ventricles are located in:
(a) Cerebrum (b) Midbrain
(c) Pons (d) Medulla

75. The movement of tongue muscles is controlled by:
(a) Facial nerve
(b) Vagus nerve
(c) Hypoglossal nerve
(d) Trigeminal nerve

76. Sympathetic nerves arise from:
(a) Craniosacral region
(b) Thoracolumbar region
(c) Craniolumbar region
(d) Lumbosacral region

77. Third and fourth ventricles of the brain are connected by:
(a) Foramen of Monro
(b) Aqueduct of Sylvius
(c) Foramen of Magendie
(d) Foramen of Luschka

78. Which of the following statements is correct with respect to the spinal cord?
(a) It is composed of grey matter only
(b) It is composed of outer grey matter and inner white matter
(c) It is composed of outer white matter and inner grey matter
(d) It is composed of white matter only

79. Comprehension of spoken and written words takes place in the region of:
(a) Broca's area
(b) Wernicke's area
(c) Association area
(d) Motor area

80. Which of the following cranial nerves innervates the heart, lungs and stomach?
(a) Trigeminal nerve
(b) Vagus nerve
(c) Auditory nerve
(d) Trochlear nerve

81. In the brain, lateral ventricles are connected to the third ventricle by:
(a) Foramen of Monro
(b) Foramen of Luschka
(c) Foramen of Magendie
(d) Foramen of Rotundus

82. The part of the hindbrain that is responsible for hand–eye coordination is:
(a) Cerebellum
(b) Thalamus
(c) Pons
(d) Medulla oblongata

83. Anterior choroid plexus is present on:
(a) Floor of diencephalon
(b) Floor of medulla oblongata
(c) Roof of diencephalon
(d) Roof of medulla oblongata

84. Which of the following is an incorrect statement?
(a) Vagus is a mixed cranial nerve
(b) Parasympathetic nerves arise from the craniosacral region
(c) The thermal regulation centre of the brain is located in the hypothalamus
(d) All motor neurons are afferent neurons

85. Identify the label 'X' in the given diagrammatic representation of a reflex action:

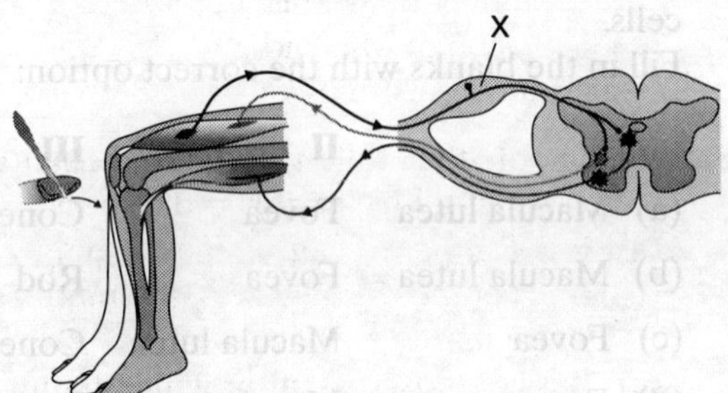

(a) Muscle spindle (b) Dorsal root ganglia
(c) Motor end plate (d) Interneuron

86. Arbor vitae is found in:
(a) Cerebrum
(b) Cerebellum
(c) Midbrain
(d) Hypothalamus

87. The length of the spinal cord is:
(a) 60–65 cm (b) 42–45 cm
(c) 90–95 cm (d) 20–22 cm

88. ________ serves as a neuronal link between the cerebral cortex and the cerebellum:
(a) Spinal cord (b) Medulla
(c) Pons (d) Epithalamus

89. Identify the incorrect statement with respect to the autonomic nervous system (ANS):
(a) The fibres of the parasympathetic efferent neurons are called craniosacral outflow
(b) The ganglia of the parasympathetic nervous system are situated close to, or within, the effector organ
(c) The neurotransmitter within the ganglia is acetylcholine for both sympathetic and parasympathetic nerves
(d) In the sympathetic system, the neurotransmitter at the terminal synapse is acetylcholine

90. Which lobe of the cerebrum is responsible for inner monitoring of complex thoughts and actions, creative ideas, intellectual insight, reality testing by judgement, and translation of perceptions and memories into plans of muscle movement?
(a) Frontal lobe (b) Parietal lobe
(c) Temporal lobe (d) Occipital lobe

Sensory Reception and Processing

91. The outermost layer of the eyeball called ________ is a tough but elastic sheath of fibrous connective tissue:
(a) Choroid (b) Sclera
(c) Retina (d) Conjunctiva

92. ________ is a reflex mechanism by which light rays from objects at various locations in the near visual field are brought to focus on the retina:
(a) Adaptation
(b) Accommodation
(c) Reflection
(d) Refraction

93. The fluid-filled spiral chamber of the inner ear is shaped like a narrow snail shell and is called:
(a) Malleus (b) Cochlea
(c) Maculae (d) Saccule

94. Human ear is responsible for:
(a) Static equilibrium
(b) Dynamic equilibrium
(c) Hearing
(d) All of the above

95. Which of the following eye defects can be cured by surgical removal of the defective lens?
(a) Cataract (b) Glaucoma
(c) Presbyopia (d) Myopia

96. Hyposmia is:
(a) Reduced ability to hear
(b) Reduced ability to smell

(c) Reduced ability to respire
(d) Reduced ability to see

97. Which of the following ossicles is not present in the human middle ear?
(a) Incus
(b) Malleus
(c) Weberian ossicle
(d) Stapes

98. Waxy material which entraps dust and also lubricates the tympanum is secreted by:
(a) Meibomian gland
(b) Ceruminous gland
(c) Lacrimal gland
(d) Harderian gland

99. The sensory receptors responsible for detecting heat or high temperature are:
(a) Merkel's discs
(b) Golgi–Mazzoni organs
(c) Ruffini's organs
(d) End bulbs of Krause

100. For accommodation, the shape of the lens is regulated by:
(a) Intraocular pressure of aqueous humor
(b) Intraocular pressure of vitreous humor
(c) Smooth muscles of iris
(d) Smooth muscles of ciliary body

101. The colour of the eye is due to the colour of:
(a) Sclera (b) Choroid
(c) Iris (d) Ciliary body

102. Vestibular apparatus consists of:
(a) Middle ear + Cochlea
(b) Semicircular canals + Otolith organ
(c) Semicircular canals + Cochlea
(d) Otolithic membrane + Cochlea

103. Which of the following is not a refractive abnormality?
(a) Myopia
(b) Glaucoma
(c) Hypermetropia
(d) Astigmatism

104. Which of the following is incorrect with respect to ear ossicles?
(a) Handle of malleus is attached to tympanum
(b) Stapes pushes against the oval window
(c) Ear ossicles act in concert as a lever system
(d) They reduce the frequency of vibration

105. If we accidently focus on intense light sources such as the sun, it will damage _______ part of eye:
(a) Blind spot (b) Yellow spot
(c) Macula lutea (d) Both (b) and (c)

106. Tension of which of the following regions of the eye determines the shape of the lens for focusing distant objects?
(a) Ciliary muscle
(b) Circular iris muscle
(c) Suspensory ligament
(d) Longitudinal iris muscle

107. The receptor cells for both hearing and balancing are:
(a) Chemoreceptor (b) Thermoreceptor
(c) Mechanoreceptor (d) Photoreceptor

108. Eustachian tube connects:
(a) Middle ear with external ear
(b) Middle ear with internal ear
(c) External ear with pharynx
(d) Middle ear with pharynx

109. Which of the following disorders is caused due to blockage in the canals of Schlemm and raised intraocular pressure?
(a) Cataract (b) Presbyopia
(c) Squint (d) Glaucoma

110. At the posterior pole of the eye, lateral to th e blind spot, there is a yellowish pigmented spot called ___I___ with a central pit called the ___II___, having densely packed ___III___ cells.
Fill in the blanks with the correct option:

	I	II	III
(a)	Macula lutea	Fovea	Cone
(b)	Macula lutea	Fovea	Rod
(c)	Fovea	Macula lutea	Cone
(d)	Fovea	Macula lutea	Rod

111. The receptors for touch in our skin are:
(a) Ruffini corpuscle
(b) Krause's bulbs
(c) Meissner's corpuscle
(d) Johnston's organs

112. If the ciliary muscles fail to contract:
(a) We cannot see distant objects
(b) We cannot focus nearer objects
(c) We cannot focus both distant and nearer objects
(d) We cannot distinguish between red and green colours

113. Photopigments in the human eyes are composed of opsin (a protein) and retinal (aldehyde of Vitamin A) and light induces:

(a) Dissociation of retinal from opsin, resulting in changes in the structure of opsin
(b) Dissociation of retinal from opsin, resulting in changes in the structure of retinal
(c) Dissociation of retinal from opsin, resulting in changes in the structure of both retinal and opsin
(d) Association of retinal with opsin, resulting in changes in the structure of retinal

114. All of the following statements are correct with respect to the cornea in the human eye, except:
(a) It is highly vascular
(b) It can be easily transplanted
(c) It obtains nutrition from aqueous humor
(d) Major refraction takes place at the cornea

115. Consider the following statements A, B, C and D with certain blanks. Find the option which correctly fills in these blanks:
A. The membranous labyrinth is filled with a fluid called ___I___
B. The space within the cochlea called ___II___ is filled with endolymph
C. At the base of the cochlea, the scala vestibuli ends at the ___III___
D. The organ of Corti is a structure located on ___IV___

	I	II	III	IV
(a)	Endolymph	Scala tympani	Oval window	Basilar membrane
(b)	Perilymph	Scala vestibule	Round window	Tectorial membrane
(c)	Endolymph	Scala media	Oval window	Basilar membrane
(d)	Perilymph	Scala media	Round window	Tectorial membrane

116. Consider the following events that take place during vision and find the correct sequence with respect to the mechanism of vision:
A. Action potential transmitted by optic nerve
B. Light induces dissociation of retinal from opsin
C. Potential differences generated in photoreceptor cell
D. Light rays of visible wavelength focused on retina
E. Action potential generated in ganglion cells through bipolar cells
(a) (A), (B), (D), (E), (C)
(b) (B), (C), (D), (E), (A)
(c) (D), (A), (E), (C), (B)
(d) (D), (B), (C), (E), (A)

117. Which of the following parts of the human eye are incorrectly matched with its identification and feature/function?

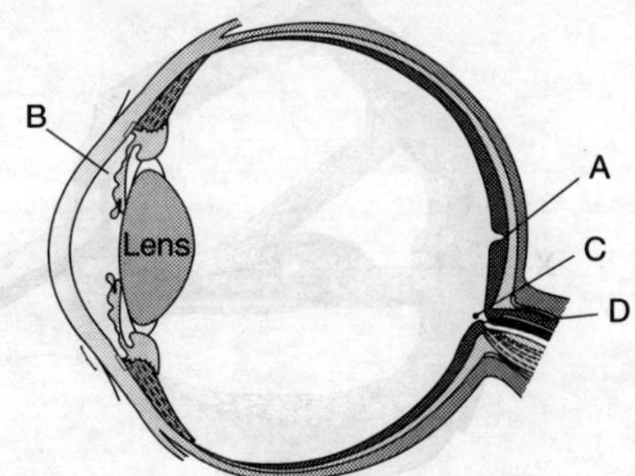

(a) D – Optic nerve – Transmits impulses to visual cortex of the brain
(b) C – Blind spot – Rods are absent, while cones are present
(c) A – Fovea – Point of greatest visual activity
(d) B – Aqueous humor – Present in the space between the cornea and lens

118. Which of the following is the correct transmission of visual impulse in the retina of the human eye?
(a) Bipolar cells → ganglion cells → cone
(b) Cone → bipolar cells → ganglion cells
(c) Ganglion cells → cone → bipolar cells
(d) Cone → ganglion cells → bipolar cells

119. Which of the following is an incorrect statement?
(a) Optic nerve arises from the blind spot
(b) The space between the lens and retina is called aqueous chamber, which is filled with thin watery fluid
(c) Membranous labyrinth is surrounded by a fluid called perilymph
(d) Rods contain a purplish-red protein called rhodopsin or visual purple

120. Which of the following labelled parts in the given diagram of the human ear help in amplification of sound waves?

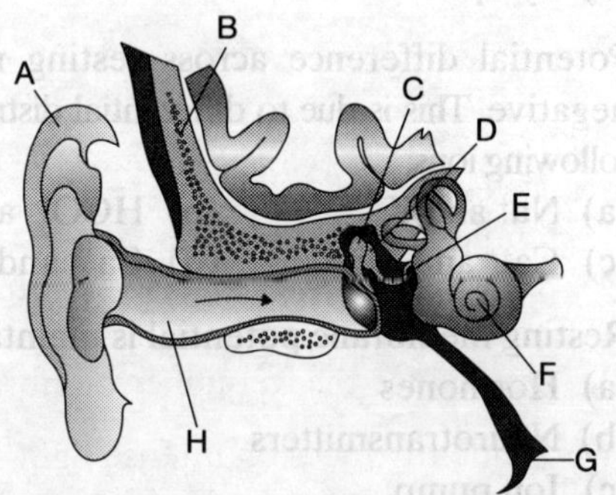

(a) B, C, D (b) C, D, E
(c) D, E, F (d) F, G, H

121. Identify part 'X' in the sectional view of the cochlea given below:

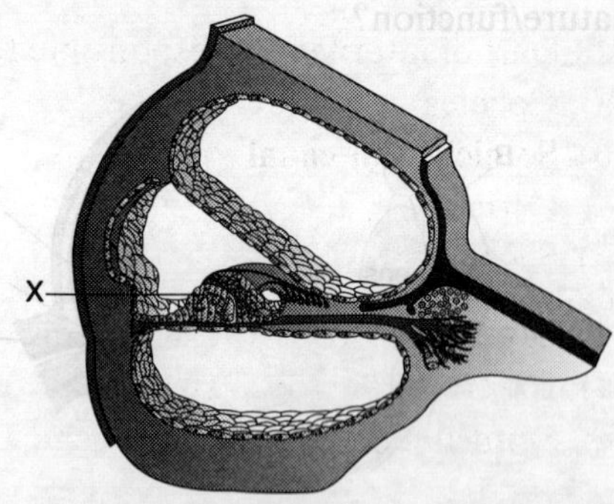

(a) Scala tympani (b) Scala vestibuli
(c) Organ of Corti (d) Reissner's membrane

122. Which of the following is correctly labelled in the given diagram of the human ear?

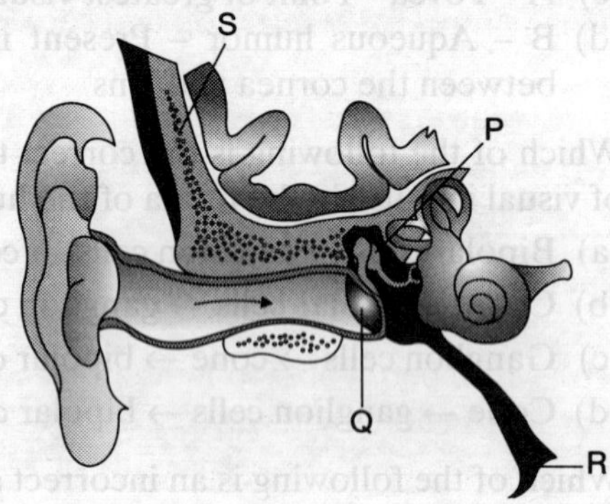

(a) Q – Stapes (b) P Cochlea
(c) R – Malleus (d) S – Temporal bone

123. Identify the incorrect statement:
(a) Rods are responsible for scotopic vision
(b) There are no photoreceptors in the blind spot
(c) Cones are responsible for twilight vision
(d) Rods and cones contain light-sensitive proteins called photopigments

124. Which of the following is/are correct with respect to blind spot?
(a) The optic nerves leave the eye
(b) Retinal blood vessels enter the eye
(c) It is located at a point medial to and slightly above the posterior pole of the eyeball
(d) All are correct

125. Which of the following is incorrect with respect to the ear?
(a) The outer ear consists of pinna and external auditory meatus
(b) Tympanum is composed of connective tissues covered with mucous membrane outside and skin inside
(c) The pinna collects vibrations in the air which produce sound
(d) There are very fine hairs and wax secreting sebaceous glands in the skin of the pinna and the meatus

NCERT EXEMPLAR QUESTIONS

1. Chemicals which are released at the synaptic junction are called:
(a) Hormones
(b) Neurotransmitters
(c) Cerebrospinal fluid
(d) Lymph

2. Potential difference across resting membrane is negative. This is due to differential distribution of the following ions:
(a) Na^+ and K^+ (b) HCO_3^- and Cl^-
(c) Ca^{++} and Mg^{++} (d) Ca^{++} and Cl^-

3. Resting membrane potential is maintained by:
(a) Hormones
(b) Neurotransmitters
(c) Ion pump
(d) None of the above

4. The function of our visceral organs is controlled by:
(a) Sympathetic and somatic neural system
(b) Sympathetic and parasympathetic neural system
(c) Central and somatic nervous system
(d) None of the above

5. Which of the following is not involved in knee-jerk reflex?
(a) Muscle spindle
(b) Motor neuron
(c) Brain
(d) Spinal cord

6. An area in the brain which is associated with strong emotions is:
(a) Cerebral cortex
(b) Cerebellum
(c) Limbic system
(d) Medulla

7. Mark the vitamin present in rhodopsin:
(a) Vitamin A (b) Vitamin B
(c) Vitamin C (d) Vitamin D

8. Human eyeball consists of three layers and it encloses:
(a) Lens, iris, optic nerve
(b) Lens, aqueous humor and vitreous humor
(c) Cornea, lens, iris
(d) Cornea, lens, optic nerve

9. Wax secreting gland present in the ear canal is called:
(a) Sweat gland (b) Prostate gland
(c) Cowper's gland (d) Ceruminous gland

10. The part of internal ear responsible for hearing is:
(a) Cochlea
(b) Semicircular canal
(c) Utriculus
(d) Sacculus

11. The organ of Corti is a structure present in:
(a) External ear (b) Middle ear
(c) Semicircular canal (d) Cochlea

PRACTICE TEST PAPER

1. Which of the following structures form the brainstem?
(a) Cerebrum, thalamus and hypothalamus
(b) Midbrain, pons and medulla
(c) Hypothalamus and cerebellum
(d) Cerebrum and cerebellum

2. Corpus callosum, a tract of nerve fibres, connects:
(a) Two cerebellar hemispheres
(b) Medulla and spinal cord
(c) Two cerebral hemispheres
(d) Midbrain and hindbrain

3. Speed of propagation of an action potential can be determined by:
(a) Presence or absence of myelin sheath
(b) Diameter of axon
(c) Both (a) and (b)
(d) Length of the axon

4. Which of the following is not regulated by the hypothalamus?
(a) Body temperature
(b) Hunger and satiety
(c) Release of hormones by anterior pituitary
(d) Coughing, sneezing and swallowing

5. If the ventral root of the spinal cord is damaged, then
(a) Afferent impulses are affected
(b) Efferent impulses are affected
(c) Both motor and sensory impulses are damaged
(d) There will be no overall effect

6. Mark the incorrect statement:
(a) Autonomous nervous system is self-governed without conscious control
(b) Centres for respiration and cardiac activity are present in the medulla
(c) Neurotransmitters are released by dendrites
(d) Hippocampus converts short-term memory to long-term memory

7. Nerve fibres in resting stage:
(a) More permeable to Na^+ than K^+
(b) More permeable to K^+ than Na^+
(c) Permeable to negatively charged proteins
(d) Permeable to proteins and Na^+

8. Sexual behaviour and emotional reactions of the human body are chiefly controlled by:
(a) Cerebellum (b) Midbrain
(c) Limbic system (d) Pons

9. A polarised neuronal membrane is one:
(a) That does not respond to threshold stimulus
(b) That is incapable of generating action potential
(c) That is in resting state
(d) That cannot conduct stimulus

10. Identify the incorrect statement:
(a) Electrical synapses are rare in the human body
(b) Neurotransmitters are used in chemical synapses
(c) Impulse transmission across an electrical synapse is always slower than that across a chemical synapse
(d) Synapse comprises a pre-synaptic and post-synaptic membrane that may or may not be separated by a synaptic cleft

11. Which of the following is an incorrect statement?
(a) Cranial and spinal nerves are made of myelinated nerve fibres

(b) Myelinated nerve fibres are enveloped by Schwann cells, which form a myelin sheath around the axon
(c) Non-myelinated nerve fibres are enclosed by Schwann cells that do not form a myelin sheath
(d) Myelinated nerve fibres are found in autonomous and somatic neural systems

12. Nissl granules are not found in:
(a) Cell body (b) Cyton
(c) Axon (d) Dendrites

13. Complex functions like intersensory associations, memory and communication are due to ________________ areas of the cerebral cortex:
(a) Motor (b) Sensory
(c) Association (d) Limbic

14. Cerebral aqueduct is a canal that passes through:
(a) Pons (b) Cerebellum
(c) Medulla (d) Midbrain

15. Identify the incorrect statement:
(a) Cerebrum forms the major part of the human brain
(b) The grey matter of the cerebral cortex has tracts of fibres covered with myelin sheath
(c) Hypothalamus has several groups of neurosecretory cells
(d) The midbrain is located between the thalamus/hypothalamus of the forebrain and pons of the hindbrain

16. The dorsal portion of the ________ consists mainly of four round swellings or lobes called corpora quadrigemina:
(a) Cerebrum
(b) Midbrain
(c) Thalamus
(d) Hypothalamus

17. Which of the following is an incorrect statement with respect to the ear?
(a) The ear is divided into three major sections called the outer ear, middle ear and inner ear
(b) The outer ear consists of the pinna and external auditory canal
(c) The middle ear contains three ossicles called malleus, incus and stapes, which are attached to one another in a chain-like fashion
(d) Eustachian tube connects the inner ear cavity to the pharynx

18. At the base of the cochlea, the scala vestibuli ends at the ____I____, while the scala tympani terminates at the ____II____ which opens to the middle ear.
Fill in I and II with correct options:

	I	II
(a)	Round window	Oval window
(b)	Oval window	Round window
(c)	Round window	Tympanum
(d)	Oval window	Auditory meatus

19. In the internal ear, the organ of Corti is a structure located on the ____________ which contains hair cells that act as auditory receptors:
(a) Tectorial membrane
(b) Basilar membrane
(c) Reissner's membrane
(d) Basement membrane

20. The visible coloured portion of the eye is called:
(a) Pupil (b) Iris
(c) Choroid (d) Ciliary body

21. Rods are photoreceptors of the retina used for:
(a) Photopic vision
(b) Scotopic vision
(c) Colour vision
(d) Both (a) and (c)

22. The visual cortex of the brain, where neural impulses are analysed and images formed on the retina are recognised based on earlier memory and experience, is located in ____________ lobe of the cerebrum:
(a) Parietal (b) Occipital
(c) Temporal (d) Frontal

23. Choose the correct match:

Column I	Column II
A. Pons	(i) Midbrain
B. Cerebrum	(ii) Hindbrain
C. Superior colliculi	(iii) Forebrain
D. Vestibular apparatus	(iv) Equilibrium

(a) A – (ii), B – (i), C – (iii), D – (iv)
(b) A – (ii), B – (iii), C – (i), D – (iv)
(c) A – (iv), B – (i), C – (ii), D – (iii)
(d) A – (iii), B – (i), C – (iv), D – (ii)

24. In the retina of the human eye, visual acuity is greatest at:
(a) Fovea (b) Yellow spot
(c) Macula lutea (d) Blind spot

25. Which of the following is not a part of the forebrain?
(a) Cerebrum (b) Thalamus
(c) Hypothalamus (d) Cerebellum

26. Cerebrospinal fluid (CSF) is present:
(a) In extradural space
(b) Between arachnoid mater and piamater
(c) Between duramater and archanoid mater
(d) Between piamater and the brain

27. The fluid between the eye lens and retina is called:
(a) Aqueous humor
(b) Vitreous humor
(c) Serum
(d) Lymph

28. Identify the incorrect statement:
(a) Membranous labyrinth is filled with endolymph
(b) Middle ear is filled with perilymph
(c) Malleus is the largest ear ossicle
(d) Membranous labyrinth floats within a bony labyrinth in a fluid called perilymph

29. Unipolar neuron has a cell body with one axon only and is found in:
(a) Cerebral cortex
(b) Embryonic stage
(c) Retina of eye
(d) Olfactory membrane

30. Mark the incorrect statement for the choroid layer of the eyeball:
(a) It contains many blood vessels and looks bluish in colour
(b) It is thin over the posterior two-thirds of the eyeball, but becomes thick in the anterior part to form the ciliary body
(c) It is pigmented and converts the interior of the eyeball into a dark chamber
(d) Its posterior part has a yellowish spot called macula lutea with a central pit called the fovea

31. Endolymph is present in:
(a) Scala tympani
(b) Scala vestibuli
(c) Scala media
(d) Middle ear

32. The vestibular apparatus of the internal ear is composed of:
(a) Pinna and auditory meatus
(b) Cochlea and saccule
(c) Cochlea and utricle
(d) Three semicircular canals and otolith organ

33. The outermost protective covering of the human brain is:
(a) Piamater
(b) Duramater
(c) Arachnoid mater
(d) Tela choroidea

34. The rods contain:
(a) Visual purple
(b) Rhodopsin
(c) Purplish-red protein
(d) All of the above

35. Which of the following is the correct statement?
(a) The malleus is attached to the oval window of the cochlea
(b) The stapes is attached to the tympanic membrane
(c) The ear ossicles increase the efficiency of transmission of sound waves to the inner ear
(d) All the statements are correct

36. All spinal nerves are:
(a) Motor
(b) Sensory
(c) Mixed
(d) None of the above

37. Cornea transplant in human beings is almost never rejected, because:
(a) It is a non-living layer
(b) It has no blood supply
(c) It has enucleated cells
(d) Its cells are resistant to bacteria

38. Injury localised to the hypothalamus would most likely disrupt:
(a) Short-term memory
(b) Executive functions, such as decision making
(c) Regulation of body temperature
(d) Coordination during locomotion

39. Which of the following regions of the brain is incorrectly paired with its function?
(a) Medulla oblongata – Cardiovascular and respiratory centre
(b) Cerebrum – Calculation and contemplation
(c) Cerebellum – Language comprehension
(d) Corpus callosum – Communication between the two cerebral hemispheres

40. Which one of the following pairs of structures distinguishes a nerve cell from other types of cells?
(a) Vacuoles and fibres
(b) Perikaryon and dendrites

(c) Nucleus and mitochondria
(d) Flagellum and medullary sheath

41. Which of the following pairs of ions are required for conduction of nerve impulses?
(a) Na^+ and Ca^{2+}
(b) Mg^{2+} and Ca^{2+}
(c) Na^+ and K^+
(d) K^+ and Ca^{2+}

42. Layers in the wall of eyeball from inside outwards are:
(a) Sclera, retina, choroid
(b) Retina, choroid, sclera
(c) Choroid, retina, sclera
(d) Retina, sclera, choroid

43. Identify the incorrect statement:
(a) Humans have 12 pairs of cranial nerves
(b) All cranial nerves are mixed
(c) Humans have 31 pairs of spinal nerves
(d) Vagus is the longest cranial nerve

44. A gymnast is able to balance his body upside down even in darkness because of:
(a) Cochlea
(b) Basilar membrane
(c) Vestibular apparatus
(d) Reissner's membrane

45. ____________ lens is used for correction in myopia:
(a) Convex
(b) Cylindrical
(c) Concave
(d) None of the above

ANSWER KEYS

Exercises

1. (a) **2.** (d) **3.** (b) **4.** (d) **5.** (c) **6.** (c) **7.** (d) **8.** (c) **9.** (d) **10.** (d)
11. (c) **12.** (c) **13.** (b) **14.** (d) **15.** (b) **16.** (d) **17.** (b) **18.** (d) **19.** (a) **20.** (b)
21. (c) **22.** (d) **23.** (a) **24.** (c) **25.** (d) **26.** (c) **27.** (c) **28.** (d) **29.** (b) **30.** (d)
31. (c) **32.** (d) **33.** (d) **34.** (c) **35.** (b) **36.** (d) **37.** (c) **38.** (b) **39.** (c) **40.** (a)
41. (d) **42.** (b) **43.** (c) **44.** (c) **45.** (b) **46.** (c) **47.** (b) **48.** (d) **49.** (d) **50.** (d)
51. (c) **52.** (c) **53.** (c) **54.** (d) **55.** (d) **56.** (a) **57.** (a) **58.** (d) **59.** (c) **60.** (a)
61. (b) **62.** (a) **63.** (d) **64.** (d) **65.** (d) **66.** (b) **67.** (d) **68.** (b) **69.** (d) **70.** (b)
71. (c) **72.** (c) **73.** (c) **74.** (a) **75.** (c) **76.** (b) **77.** (b) **78.** (c) **79.** (b) **80.** (b)
81. (a) **82.** (a) **83.** (c) **84.** (d) **85.** (b) **86.** (b) **87.** (b) **88.** (c) **89.** (d) **90.** (a)
91. (b) **92.** (b) **93.** (b) **94.** (d) **95.** (a) **96.** (b) **97.** (c) **98.** (b) **99.** (c) **100.** (d)
101. (c) **102.** (b) **103.** (b) **104.** (d) **105.** (d) **106.** (c) **107.** (c) **108.** (d) **109.** (d) **110.** (a)
111. (c) **112.** (b) **113.** (a) **114.** (a) **115.** (c) **116.** (d) **117.** (b) **118.** (b) **119.** (b) **120.** (b)
121. (c) **122.** (d) **123.** (c) **124.** (d) **125.** (b)

NCERT Exemplar Questions

1. (b) **2.** (a) **3.** (c) **4.** (b) **5.** (c) **6.** (c) **7.** (a) **8.** (b) **9.** (d) **10.** (a)
11. (d)

Practice Test Paper

1. (b) **2.** (c) **3.** (c) **4.** (d) **5.** (b) **6.** (c) **7.** (b) **8.** (c) **9.** (c) **10.** (c)
11. (d) **12.** (c) **13.** (c) **14.** (d) **15.** (b) **16.** (b) **17.** (d) **18.** (b) **19.** (b) **20.** (b)
21. (b) **22.** (b) **23.** (b) **24.** (a) **25.** (d) **26.** (b) **27.** (b) **28.** (b) **29.** (b) **30.** (d)
31. (c) **32.** (d) **33.** (b) **34.** (d) **35.** (c) **36.** (c) **37.** (b) **38.** (c) **39.** (c) **40.** (b)
41. (c) **42.** (b) **43.** (b) **44.** (c) **45.** (c)

SOLUTIONS

Neurons, Nerve Impulse

1. The resting neural membrane has poor permeability for Na^+, although it has higher permeability for K^+.

2. Option (d) is correct.

3. The functional unit of the nervous system is the neuron.

4. Neurons are excitable cells and, in the resting state, they exist in polarised state and respond to stimulus.

5. Unipolar neurons are found in the early embryonic stages.

6. Option (c) is correct.

7. Bipolar neurons are found in the retina and olfactory epithelium. They have an axon and a dendrite attached to the cell body.

8. In chemical synapse, the pre- and post-synaptic membranes are separated by a fluid-filled cleft. Impulses are conducted by chemical neurotransmitters. These chemicals are present in vesicles in the axon terminals. Only unidirectional movement of impulses takes place across these synapses.

9. In the resting neuron, the axonal membrane has high concentration of K^+ in the axoplasm along with low concentration of Na^+. Also, the axonal membrane is more permeable to K^+ and impermeable to Na^+ in the resting neuron.

10. Nerve impulse transmission through nerve fibres is independent of the chemical nature of the neurotransmitter.

11. Saltatory conduction of nerve impulse takes place through myelinated neurons.

12. During a nerve impulse, repolarisation occurs due to efflux of K^+.

13. Myelin sheath is discontinuous at the nodes of Ranvier.

14. Myelin sheath is secreted in the CNS by oligodendrocytes.

15. Nissl granules are absent in axon.

16. Nissl bodies or granules are granules of rough endoplasmic reticulum (RER) with rosettes of free ribosomes and are the site of protein synthesis.

17. Influx of calcium ions is responsible for the release of neurotransmitters from synaptic vesicles into the synaptic cleft.

18. In a neuron, the impulse is conducted from the dendrite to the cell body and from the cell body to the axon.

19. Na^+–K^+ pumps in the axonal membrane help in maintaining the polarised state of the neuron in the resting state.

20. Synaptic knob is the bulbous structure present at the terminal end of the axon.

21. Gamma aminobutyric acid (GABA) is a neurotransmitter that produces inhibitory post-synaptic potential.

22. Myelinated neurons are present in the white matter of the brain and spinal cord.

23. During conduction of nerve impulse, the action potential is the result of the movement of Na^+ from extracellular fluid into the axon.

24. The binding of neurotransmitters generates new potential in the post-synaptic membrane.

25. Dendrites are short fibres that branch repeatedly and project out of the cell body and contain Nissl granules.

26. An impulse equal to or greater than the threshold impulse can only generate action potential. Chemical synapses work by releasing chemical neurotransmitters. Resting membrane in the neuron is impermeable to negatively charged proteins present in the axoplasm.

27. The free flow of impulses at the nodes of Ranvier in myelinated neurons is due to lack of myelin sheath.

28. Coordination is the process through which two or more organs interact and complement the

functions of one another.

29. Option (b) is correct.

30. Option (d) is correct.

31. Myelinated nerve fibres are found in the spinal and cranial nerves. Unmyelinated nerve fibres are commonly found in the autonomous and somatic neural systems.

32. The autonomic neural system (sympathetic and parasympathetic neural system) transmits impulses from the CNS to the involuntary organs and smooth muscles of the body.

33. Na^+–K^+ pumps of the axonal membrane are operative during resting state and transport $3Na^+$ outwards and $2K^+$ into the cell.

34. The efferent nerve fibres transmit impulses from the CNS to the concerned peripheral tissues while afferent fibres transmit impulses from the tissues to the CNS.

35. Option (b) is correct.

Neural System (CNS, PNS and ANS)

36. The forebrain consists of the cerebrum, thalamus and hypothalamus.

37. Cerebrum is the major part of the human brain and a deep cleft divides the cerebrum longitudinally into two halves, which are termed as left and right cerebral hemispheres.

38. The cerebrum wraps around a structure called the thalamus, which is a major coordinating centre for sensory and motor signalling.

39. The inner parts of the cerebral hemispheres and a group of associated deep structures like amygdala and hippocampus form a complex structure called the limbic system.

40. The midbrain is located between the thalamus of the forebrain and pons of the hindbrain.

41. The brainstem is formed by midbrain, pons and medulla oblongata.

42. Hypothalamus controls body temperature, and the urge to eat and drink. Corpus callosum is a tract of nerve fibres which connects the two cerebral hemispheres. Medulla oblongata controls gastric secretion and cardiovascular reflexes.

43. Cerebrospinal fluid (CSF) is present between arachnoid mater and piamater in the subarachnoid space.

44. Hindbrain comprises pons, medulla and cerebellum.

45. Cerebellum is responsible for coordinating voluntary movements such as posture, balance and speech, and maintains smooth and balanced muscular activity.

46. Pneumotaxic centre is present in pons and regulates respiration.

47. Parasympathetic nervous system decreases the heart rate.

48. Mammalian brains are characterised by the presence of corpus callosum.

49. Damage to the cerebellum causes slurred speech, unsteady gait, frequent stumbling, impaired coordination, uncontrolled eye movements and ataxia.

50. Diencephalon is part of the forebrain and comprises the epithalamus, thalamus and hypothalamus. Pons is a part of hindbrain.

51. Cerebellum's unconscious directions and cerebrum's conscious instructions from the brain command how and when to move the body parts.

52. Thalamus is the key communication link between the senses and the cortex that receives the majority of incoming signals and evaluates their importance before passing them to the cerebrum.

53. Autonomic neural system relays impulses from the CNS to glands and involuntary muscles.

54. The visual cortex is the primary cortical region of the brain that receives, integrates and processes visual information relayed from the retinas. It is the occipital lobe of the primary cerebral cortex, which is the most posterior region of the brain.

55. Ependymal cells form the epithelial lining of the ventricles in the brain and the central canal of the spinal cord. They have cilia to push cerebrospinal fluid (CSF).

56. Corpus callosum is present in the forebrain and links the two cerebral hemispheres.

57. Corpora quadrigemina consists of four round swellings which are present on the dorsal portion of the midbrain.

58. Parietal lobe of cerebral hemisphere registers feelings of hot, cold, touch and pain.

59. Gyri are the bulges present between adjacent grooves in the cerebrum.

60. Duramater is the outermost meninx made of fibrous tissue surrounding the human brain.

61. A total of 31 pairs of spinal nerves arise from the spinal cord.

62. A canal called cerebral aqueduct/Iter/aqueduct of Sylvius passes through the midbrain.

63. Centre for vomiting is located in the medulla.

64. Purkinje cells, also called Purkinje neurons, are found in the cerebellar cortex of the brain. Purkinje cell bodies are shaped like a flask and have many thread-like extensions called dendrites, which receive impulses from other neurons called granule cells.

65. Medulla oblongata contains centres which control respiration, cardiovascular reflexes and gastric secretions.

66. Hippocampus converts information from short-term to long-term memory, essential in learning.

67. The cerebral cortex is referred to as grey matter due to the presence of neuronal cell bodies and unmyelinated axons, that give a dull grey appearance to it.

68. Vagus/Pneumogastric nerve is the longest cranial nerve.

69. Olfactory, optic and auditory are purely cranial nerves.

70. The sympathetic ganglia are nerve cell bodies that run along either side of the spinal cord. There are 21 pairs of these ganglia and present mainly in the thoracic and lumbar regions and called thoraco-lumbar outflow.

71. Piamater is the innermost meninx and has the Latin meaning 'tender mother'.

72. Cerebral cortex appears dull grey in colour; the cerebral aqueduct passes through the midbrain. Hypothalamus has centres for thermoregulation.

73. Corpora quadrigemina comprises two superior and two inferior colliculi.

74. The lateral ventricles are present in the cerebrum.

75. The movement of tongue muscles is controlled by the hypoglossal nerve.

76. Sympathetic nerves arise from the thoracolumbar region.

77. Third and fourth ventricles of the brain are connected by aqueduct of Sylvius.

78. Spinal cord comprises outer white matter and inner grey matter.

79. Wernicke's area is located in the temporal lobe and is associated with comprehension of the spoken and written word.

80. Vagus nerve innervates the heart, lungs, stomach and liver. It is the 10th cranial nerve.

81. In the brain, the lateral ventricles are connected to the third ventricle by the foramen of Monro.

82. The part of the hindbrain responsible for hand–eye coordination is the cerebellum.

83. Anterior choroid plexus is present on the roof of the diencephalon and secretes cerebrospinal fluid (CSF).

84. All motor neurons are efferent and carry information from the CNS to muscles.

85. Option (b) is correct.

86. Arbor vitae is cerebellar white matter, so called for its branched, tree-like appearance.

87. The length of the spinal cord is 42–45 cm.

88. Pons serves as a neuronal link between the cerebral cortex and cerebellum.

89. In the sympathetic nervous system, the neurotransmitter at the terminal synapse is adrenaline.

90. Frontal lobe is responsible for inner monitoring of complex thoughts and actions, creative

ideas, intellectual insight and translates perception and memories into plans of muscle movement.

Sensory Reception and Processing

91. The outermost layer of the eyeball called sclera is a tough but elastic sheath of fibrous connective tissue.

92. Accommodation is a reflex mechanism by which light rays from objects at various locations in the near visual field are brought to focus on the retina.

93. Cochlea is the fluid-filled spiral chamber of the inner ear and is shaped like a narrow snail shell.

94. The human ear is responsible for hearing as well as maintaining equilibrium/balance (both static and dynamic). So it is called a stato-acoustic organ.

95. Cataract is controlled by surgical removal of defective lens.

96. Hyposmia is the reduced ability to smell.

97. Weberian ossicles are present in the air bladder of bony fish.

98. Ceruminous glands are specialised sudoriferous glands located subcutaneously in the external auditory canal. They produce waxy material which entrap dust particles and lubricate the tympanum.

99. Ruffini's organs are sensory receptors responsible for detecting heat or high temperature.

100. For accommodation, the shape of the lens is regulated by smooth muscles of the ciliary body.

101. The colour of the eye is due to the colour of the iris.

102. Vestibular apparatus is composed of three semicircular canals and the otolithic organ consisting of the utricle and saccule.

103. Glaucoma is a disease of the eye in which fluid pressure within the eye rises. If left untreated, the patient may lose vision and even become blind.

104. The auditory ossicles are located in the middle ear and serve to transmit sounds from the air to the fluid-filled labyrinth (cochlea). As the sound waves vibrate the tympanum, it in turn moves the nearest ossicle, the malleus, to which it is attached. The malleus then transmits the vibration via lever action to the incus and ultimately to the stapes. Stapes pushes against the membrane of fenestra ovalis/oval window. Due to their presence, the conduction of the impulse is amplified many times compared to if the sound waves were transmitted directly from the external ear to the oval window.

105. If we accidentally focus on intense light sources such as the sun, it will damage the yellow spot/macula lutea.

106. The suspensory ligament becomes taut/stretched and makes the lens more flat (less curved) to focus on distant objects.

107. The receptors for hearing (phonoreceptor) and those used for maintaining balance (statoreceptor) are mechanoreceptors.

108. Eustachian tube connects the middle ear to the pharynx.

109. Glaucoma is caused due to blockage in canals of Schlemm and raised intraocular pressure.

110. Option (a) is correct.

111. Meissner's corpuscles are receptors for touch (tangoreceptors) in our skin.

112. If ciliary muscles fail to contract, the lens cannot be pulled to increase convexity. Therefore, the patient will experience difficulty in focusing on nearby objects.

113. Light induces dissociation of retinal from opsin, resulting in change in the structure of opsin.

114. Cornea is avascular and its cells obtain nourishment from aqueous humor.

115. Option (c) is correct.

116. Option (d) is correct.

117. Option (b) is correct.

118. Option (b) is correct.

119. The space between the lens and retina is called vitreous chamber and is filled with a a transparent gel called vitreous humor.

120. Option (b) is correct.

121. Option (c) is correct.

122. Option (d) is correct.

123. Daylight (photopic) vision and colour vision are functions of cones and twilight (scotopic) vision is a function of rods.

124. Option (d) is correct.

125. Tympanum is composed of connective tissues covered with mucous membrane inside and skin outside.

Chemical Control and Integration

EXERCISES

Endocrine Glands and Hormones, Hormone Action

1. Hormones are ____I____ chemicals which act as ____II____ messengers and are produced in ____III____ amounts.
Fill in the blanks with correct options:

	I	II	III
(a)	Nutrient	Intercellular	Large
(b)	Non-nutrient	Intracellular	Trace
(c)	Nutrient	Intracellular	Large
(d)	Non-nutrient	Intercellular	Trace

2. Endocrine glands:
(a) Have ducts
(b) Lack ducts
(c) Secrete hormones
(d) Both (b) and (c) are correct

3. Which of the following is an incorrect statement?
(a) Neural coordination is fast but short-lived
(b) The nerve cells innervate all the cells of the body
(c) Cellular functions need to be continuously regulated
(d) The neural and endocrine systems jointly coordinate and regulate the physiological functions in the body

4. Identify the incorrect statement:
(a) Hormones are transported to distantly located target organs
(b) Hormones act as chemical messengers
(c) A large number of chemicals act as hormones and provide coordination in invertebrates
(d) The neural system provides point-to-point rapid coordination among organs

5. Hormones are chemically:
(a) Steroids (b) Proteins
(c) Amines (d) All of the above

6. Hormone receptors:
(a) Are always present in the cytoplasm of target cells
(b) Are present on the plasma membrane of target cells for steroidal hormones
(c) Are specific proteins to which hormones bind to produce their effects
(d) Are non-specific and can bind to several hormones of diverse chemical nature

7. Steroid hormones and iodothyronines:
(a) Are water soluble
(b) Generate secondary messengers
(c) Bind with the membrane receptors
(d) Pass through the plasma membrane of target tissues and are lipophilic

8. Mark the mismatch with respect to the chemical nature of hormones:
(a) Peptide hormones – Pituitary hormones
(b) Amino acid derivatives – Adrenal cortex hormones
(c) Steroidal hormones – Testosterone, estradiol
(d) Protein hormones – Hypothalamic hormones

9. Hormones which interact with membrane-bound receptors:
(a) Are steroidal in nature
(b) Enter the target cells
(c) Generate secondary messengers
(d) Regulate gene expression by directly interacting with genome

10. Hormones which interact with intracellular receptors:
(a) Regulate gene expression
(b) Regulate chromosomal function by interaction of the hormone–receptor complex with the genome
(c) Generate secondary messengers
(d) Both (a) and (b) are correct

11. Mark the odd one out with respect to the chemical nature of hormones:
(a) Cortisol (b) Testosterone
(c) Adrenaline (d) Estradiol

12. Which of the following hormones is not derived from the amino acid tyrosine?
(a) Thyroxine (b) Epinephrine
(c) Erythropoietin (d) Dopamine

13. Mark the odd one out with respect to the peptide hormones:
(a) Insulin (b) Glucagon
(c) Vasopressin (d) Aldosterone

14. All hormones:
(a) Are produced by exocrine grands
(b) Are lipid-soluble molecules
(c) Are proteins and elicit some response
(d) Are carried to target cells via blood

15. Which of the following do not act as secondary messengers that regulate cellular metabolism in response to interaction of hormones with membrane-bound receptors?
(a) Ca^{++}
(b) Cyclic AMP
(c) Inositol triphosphate
(d) ATP

16. ______ is an example of a hormone that acts through cAMP (secondary messenger):
(a) Progesterone (b) Corticosterone
(c) Testosterone (d) Epinephrine

17. Testosterone and estradiol are steroidal hormones and hence they bind to:
(a) Lipid receptors on cell membrane
(b) Enzyme–linked membrane receptors
(c) Intracellular receptors
(d) G–protein linked membrane receptors

18. On the basis of similarity in nature of receptors (membrane-bound receptors or intracellular receptors), which set is correct?
(a) Thyroxine and insulin
(b) Insulin and growth hormone
(c) Glucagon and cortisol
(d) Oxytocin and testosterone

19. Following is a diagrammatic representation of the mechanism of hormone action. Which of the following hormones represent/perform the given mechanism of action?

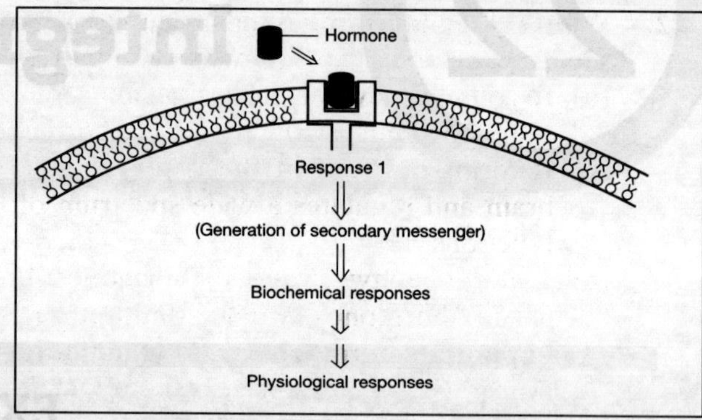

(a) Insulin, Progesterone, Estradiol
(b) Adrenaline, Glucagon, LH
(c) Cortisol, Thyroxine, Noradrenaline
(d) Aldosterone, Insulin, FSH

20. Identify the correct statement with respect to the mechanism of hormone action given below:

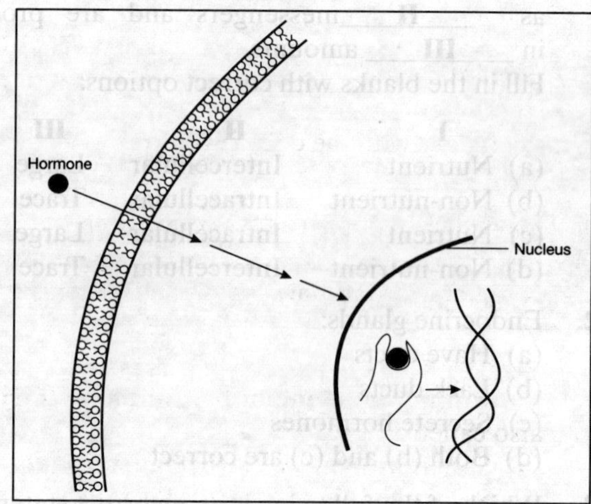

(a) Hormones mediating such mechanisms generate secondary messengers
(b) Such hormones bind with membrane-bound receptors
(c) Such hormones bind with intracellular receptors and mostly regulate gene expression
(d) Peptide hormones and catecholamines exhibit such mechanisms

Hypothalamus and Pituitary Gland

21. Which of the following is a neuroendocrine structure?
(a) Thymus (b) Pancreas
(c) Testis (d) Hypothalamus

22. Which of the following is incorrect with respect to the hypothalamus?
(a) It contains several groups of neurosecretory cells called nuclei which produce hormones
(b) It forms the roof of the diencephalon, fore-brain and regulates a wide spectrum of body functions
(c) It produces two types of hormones: the releasing hormones (which stimulate secretion of pituitary hormones) and the inhibiting hormones (which inhibit secretions of pituitary hormones).
(d) None of the above

23. The hypothalamic hormones:
(a) Reach the anterior pituitary through axons
(b) Reach the posterior pituitary through the portal vein
(c) Reach the adenohypophysis through the portal circulatory system
(d) Are steroidal in nature

24. Gonadotropin releasing hormone (GnRH) stimulates ____I____ to release ____II____. Fill in the blanks with the correct option:

	I	II
(a)	Neurohypophysis	Androgens
(b)	Testis	Luteinizing hormone
(c)	Anterior pituitary	Gonadotropins
(d)	Ovary	Progesterone

25. Growth hormone inhibitory hormone (GHIH) is also called
(a) Somatotropin (b) Somatostatin
(c) Somatomedin (d) Secretin

26. Which of the following is absent in the portal vein linking the hypothalamus with the anterior pituitary?
(a) Prolactin releasing hormone
(b) Thyrotropin releasing hormone
(c) Somatotropin releasing hormone
(d) Somatotropin

27. Identify the correct match:
(a) Prolactin – Causes milk ejection by contracting mammary gland ducts
(b) Luteinizing hormone – Induces ovulation from Graafian follicle
(c) Vasopressin – Assists parturition by contraction of uterine muscles
(d) Thyroid stimulating hormones – Promotes secretion of thymosin from the thymus

28. Which of the following hormones is not produced by the pars distalis region of the pituitary?
(a) Follicle stimulating hormone (FSH)
(b) Oxytocin
(c) Prolactin
(d) Growth hormone

29. Which of the following statements is correct with respect to the growth hormone (GH)?
(a) Excessive secretion of GH in middle age causes gigantism
(b) Low secretion of GH leads to stunted growth in children resulting in pituitary dwarfism
(c) Oversecretion of GH stimulates growth of hands and feet in children causing acromegaly, along with protrusion of lower jaw
(d) All of the above

30. Identify the incorrect statement with respect to the pituitary hormones:
(a) Luteinizing hormone stimulates release of androgens from testis in males
(b) Prolactin regulates the growth of mammary glands and formation of milk in them
(c) In females, FSH maintains the corpus luteum formed from remnants of the Graafian follicle after ovulation
(d) MSH acts on melanocytes and regulates pigmentation of skin

31. ________ stimulates the synthesis and secretion of steroid hormones called glucocorticoids (cortisol) from the adrenal cortex:
(a) Corticotropin releasing hormone
(b) Adrenocorticotropic hormone (ACTH)
(c) Corticosterone
(d) Cortisone

32. Which of the following is incorrect with respect to the neurohypophysis?
(a) It is called pars nervosa
(b) Secretes oxytocin and vasopressin
(c) Releases hormones that are actually synthesized in the hypothalamus
(d) None of the above

33. Pars intermedia of the pituitary gland secretes only one hormone called:
(a) Growth hormone
(b) Melanocyte stimulating hormone
(c) Prolactin
(d) Thyroid stimulating hormone

34. Which of the following is incorrect with respect to the pituitary gland?
(a) It is located in a bony cavity called sella turcica
(b) It is attached to the hypothalamus by a stalk
(c) It is anatomically divided into an adenohypophysis and neurohypophysis
(d) Adenohypophysis consists of two portions, pars nervosa and pars intermedia

35. Which of the following is the correct option with respect to the function of the following gonadotropins?

	FSH	LH
(a)	Release of ovum from Graafian follicle	Development of follicle in menstrual cycle
(b)	Release of progestrone from corpus luteum	Release of oestrogen from developing follicle
(c)	Release of androgens from Leydig cells	Promotes spermatogenesis
(d)	Stimulates growth of ovarian follicles	Induces ovulation of mature follicles

36. Which of the following statements is incorrect?
(a) In males, FSH regulates spermatogenesis
(b) Oxytocin acts on the smooth muscles of our body and stimulates their contraction
(c) Hypersection of GH does not further increase height because the epiphyseal plates close after adolescence
(d) Oxytocin and vasopressin are synthesized in the neurohypophysis

37. Fill in the blanks with the correct options:

In females, ___I___ stimulates contraction of the uterus at the time of child birth, and ___II___ from the mammary gland. ___III___ mainly acts on the kidney and stimulates reabsorption of water by the ___IV___ and thereby reduces water loss through urine.

	I	II	III	IV
(a)	Vasopressin	Milk ejection	Oxytocin	Distal tubules
(b)	Oxytocin	Milk ejection	Vasopressin	Distal tubules
(c)	Oxytocin	Milk synthesis	Vasopressin	Proximal tubules
(d)	Vasopressin	Milk synthesis	Oxytocin	Proximal tubules

38. Deficiency of vasopressin results in reduced ability of the kidneys to conserve water, leading to water loss and dehydration. This condition is called:
(a) Diabetes mellitus (b) Diabetes insipidus
(c) Oliguria (d) Proteinuria

39. The posterior lobe of the pituitary gland is connected to the hypothalamus through:
(a) Neurosecretory cells
(b) Hypophyseal veins
(c) Hypophseal arteries
(d) Both (b) and (c)

40. Patients with diabetes insipidus have:
(a) Glycosuria and polyuria
(b) Polyuria and ketonuria
(c) Glycosuria and ketonuria
(d) Only polyuria

Thyroid Gland, Parathyroid Gland

41. Identify the incorrect statement with respect to the functions of the thyroid gland:
(a) Regulation of basal metabolic rate
(b) Supports the process of RBC formation
(c) Controlling the metabolism of carbohydrates, proteins and fats
(d) Suppressing the immune system and anti-inflammatory action

42. Thyroid gland is composed of two lobes that are interconnected by a thin flap of connective tissue called:
(a) Vermis (b) Isthmus
(c) Stalk (d) Infundibulum

43. Which of the following is incorrect with respect to the thyroid gland?
(a) It is the largest endocrinal gland
(b) The gland is composed of follicles and stromal tissue
(c) The gland has the exceptional feature of storing its secretions
(d) The gland is located on the dorsal side of the forebrain

44. Which of the following is not seen in the growing baby if the mother develops hypothyroidism during pregnancy?
(a) Mental retardation
(b) Low intelligence quotient and deaf-mutism
(c) Stunted growth
(d) Weight loss and emaciated appearance

45. Thyroid gland secretes a protein hormone called ___I___ which regulates ___II___:

	I	II
(a)	Thyroxine	Blood calcium levels
(b)	Thyrocalcitonin	Blood calcium levels
(c)	Tri-iodothyronine	Basal metabolic rate
(d)	Thyrocalcitonin	Basal metabolic rate

46. Which of the following is a function of the parathyroid hormone?
(a) Maintenance of water and electrolyte balance
(b) Increases blood calcium levels
(c) Hypocalcemic effect, by promoting calcium deposition in bones
(d) Maintains normal rhythm of sleep–wake cycle

47. Parathyroid hormone:
(a) Acts on bones and stimulates the process of bone resorption
(b) Stimulates resorption of calcium by the renal tubules
(c) Increases calcium absorption from digested food
(d) Performs all of the above functions

48. Which of the following is an autoimmune disease of the thyroid glands in which antibodies mimic TSH and cause hyperthyroidism?
(a) Gull's disease (b) Myxedema
(c) Grave's disease (d) Addison's disease

49. Deficiency of iodine in diet causes:
(a) Parry's disease (b) Goitre
(c) Tetany (d) Diabetes mellitus

50. Which of the following are antagonistic hormones?
(a) Thyrocalcitonin and thyroxine
(b) Thyroxine and parathyroid hormone
(c) Thyrocalcitonin and thymosin
(d) Parathyroid hormone and thyrocalcitonin

51. Which of the following is an incorrect statement?
(a) Protruding tongue, mental retardation and deaf-mutism are symptoms of cretinism
(b) Tri-iodothyronine is also called thyroxine
(c) In humans, four parathyroid glands are present on the back of the thyroid gland
(d) Thyroid gland is composed of two lobes which are located on either side of the trachea

52. Hypocalcemic tetany is caused by:
(a) Hypothyroidism
(b) Hyperthyroidism
(c) Hypoparathyroidism
(d) Hyperparathyroidism

53. Identify the incorrect statement:
(a) Blood calcium levels are regulated by parathyroid hormone and thyrocalcitonin
(b) Toxic agents present in food which interfere with thyroxine synthesis lead to development of simple goitre
(c) Parathyroid hormone is released in response to raised blood calcium levels
(d) Cretinism results from hypothyroidism in children

54. Patients with hypothyroidism in adulthood suffer from myxedema. These patients do not have:
(a) Weight gain
(b) Slow heart rate
(c) Swollen puffy appearance of face and limbs
(d) Exophthalmic goitre

55. Parathyroid hormone is not a/an:
(a) Peptide hormone
(b) Steroidal hormone
(c) Hypercalcemic hormone
(d) Antagonist to thyrocalcitonin

Adrenal Gland, Pineal Gland, Thymus

56. ________ is also called supra-renal gland:
(a) Thyroid (b) Adrenal
(c) Thymus (d) Pineal

57. The hormones of ________ are called catecholamines:
(a) Adrenal cortex (b) Adrenal medulla
(c) Pineal (d) Thymus

58. ________ is the main glucocorticoid:
(a) Cortisol (b) Aldosterone
(c) Adrenaline (d) Noradrenaline

59. Corticoids which regulate the balance of water and electrolytes in our body are called:
(a) Glucocorticoids
(b) Sex corticoids
(c) Mineralocorticoids
(d) Catecholamines

60. Which of the following is not a function of glucocorticoids?
(a) They have anti-inflammatory action and suppress the immune response
(b) Maintenance of cardio-vascular system as well as kidney function
(c) They stimulate glycogen and protein synthesis
(d) They stimulate production of erythrocytes

61. Aldosterone acts mainly at the renal tubules and stimulates:
(a) Reabsorption of K^+ and phosphate
(b) Excretion of Na^+ and water
(c) Reabsorption of Na^+, water and excretion of K^+, phosphate
(d) None of the above

62. The mineralocorticoid in our body is:
(a) Epinephrine (b) Cortisol
(c) Noradrenaline (d) Aldosterone

63. The emergency hormones or hormones of fight or flight are:
(a) Thyrocalcitonin and parathyroid
(b) Insulin and glucagon
(c) Cortisol and aldosterone
(d) Adrenaline and noradrenaline

64. Glucocorticoids are secreted by ________:
(a) Zona glomerulosa
(b) Zona fasciculata
(c) Zona reticularis
(d) Zona polysoma

65. ________ gland degenerates in old age, causing weakening of immune response:
(a) Thymus (b) Pancreas
(c) Parathyroid (d) Pineal

66. Which of the following is not a function of catecholamines?
(a) Breakdown of glycogen
(b) Pupillary dilation
(c) Slowdown of heart rate
(d) Increased alertness and sweating

67. The innermost layer of the adrenal cortex which is responsible for secreting sex corticoids is:
(a) Zona glomerulosa
(b) Zona reticularis
(c) Zona fasciculata
(d) Medulla

68. Identify the incorrect statement:
(a) Cortisol increases WBC count in blood and stimulates high immune response
(b) Aldosterone maintains electrolyte balance in the body
(c) Adrenal cortex hormones play a role in growth of axial hair, pubic hair and facial hair during puberty
(d) Cortisol stimulates gluconeogenensis and lipolysis

69. Our body has __**I**__ pair of adrenal glands located at __**II**__ part of kidney:
(a) I – One, II – Posterior
(b) I – One, II – Anterior
(c) I – Two, II – Posterior
(d) I – Two, II – Anterior

70. Which of the following diseases is caused due to destruction of adrenal cortex, for example, tuberculosis?
(a) Conn's syndrome
(b) Addison's disease
(c) Adrenal virilism
(d) Cushing's syndrome

71. Identify the incorrect match:
(a) Hirsutism – Abnormal growth of hair in female's face and body
(b) Gynaecomastia – Development of breasts in female at puberty
(c) Thymus – Throne of immunity
(d) Zona reticularis – Sex corticoids

72. Adrenal cortex is derived from __**A**__ while adrenal medulla is derived from __**B**__:
(a) A – Ectoderm, B – Endoderm
(b) A – Mesoderm, B – Ectoderm
(c) A – Ectoderm, B – Mesoderm
(d) A – Mesoderm, B – Endoderm

73. All of the following are functions of cortisol, except:
(a) Gluconeogenesis
(b) Proteolysis
(c) Increases cellular uptake and utilisation of amino acid
(d) Lipolysis

74. Excessive secretion of androgen in females causes excessive facial hair, deepening of voice, male pattern baldness and irregular menstruation. This condition is called:
(a) Addison's disease
(b) Adrenal virilism
(c) Conn's syndrome
(d) Pheochromocytoma

75. Match the options given columns I and II:

	Column I		Column II
A.	Melatonin	(i)	Milk ejection
B.	Cortisol	(ii)	Sleep–wake cycle
C.	Thyrocalcitonin	(iii)	Immunosuppression
D.	Oxytocin	(iv)	Hypocalcemic hormone

(a) A – (ii), B – (iii), C – (iv), D – (i)
(b) A – (iii), B – (i), C – (iv), D – (ii)

(c) A – (iv), B – (i), C – (ii), D – (iii)
(d) A – (i), B – (iv), C – (iii), D – (ii)

76. Hassall's corpuscles are found in:
(a) Thyroid (b) Thymus
(c) Pineal gland (d) Adrenal cortex

Pancreas

77. Identify the incorrect statement:
(a) Pancreas is a composite gland
(b) The endocrine pancreas consists of 'Islets of Langerhans'
(c) There are 10 to 20 million Islets of Langerhans in a normal human pancreas
(d) The exocrinal pancreas consists of pancreatic acini

78. Identify the incorrect statement with respect to glucagon:
(a) It plays an important role in maintaining normal blood glucose levels
(b) It stimulates glycogenolysis in the liver
(c) It stimulates process of gluconeogenesis
(d) It increases cellular glucose uptake and utilisation

79. Fill in the blanks with the correct option: Insulin acts mainly on liver cells and stimulates ____I____ resulting in ____II____ blood sugar:

	I	II
(a)	Glycogenesis	Decreases
(b)	Glycogenesis	Increases
(c)	Glycogenolysis	Decreases
(d)	Glycogenolysis	Increases

80. Which of the following is an incorrect statement?
(a) Insulin and glucagon are antagonistic hormones
(b) Glucose homeostasis in blood is maintained by insulin and glucagon
(c) Glucagon acts mainly on liver cells and stimulates breakdown of glycogen
(d) Insulin is a hyperglycemic hormone

81. Patients with diabetes mellitus have:
(a) Hyperglycemia
(b) Polyuria
(c) Loss of glucose in urine
(d) All of the above

82. The α-cells of the pancreas secrete a hormone which:
(a) Increases cellular glucose uptake and utilisation
(b) Stimulates glycogenesis
(c) Stimulates gluconeogenesis
(d) Causes hypoglycemia

83. Somatostatin is secreted by the delta cells of the Islets of Langerhans. It acts to:
(a) Inhibit release of glucagon and promote release of insulin
(b) Inhibit release of insulin and promote release of glucagon
(c) Inhibit release of both insulin and glucagon
(d) Stimulate release of insulin and glucagon

84. Which of the following hormones causes hyperglycemia?
(a) Adrenaline (b) Glucagon
(c) Growth hormone (d) All of the above

85. Which of the following symptoms is/are common in patients with diabetes mellitus and diabetes insipidus?
(a) Polyuria and ketonuria
(b) Glycosuria and polyuria
(c) Glycosuria and ketonuria
(d) Only polyuria

86. Select the incorrect statement:
(a) Insulin has 2 polypeptides A and B with 21 and 30 amino acids, respectively
(b) Patients with diabetes mellitus exhibit excessive thirst and hunger
(c) Beta cells of pancreas secrete insulin
(d) Glucagon is a steroidal hormone

Testis, Ovary and Hormones of the Heart, Kidneys and Gastrointestinal Tract

87. The atrial walls of our heart secrete ____I____ which decreases ____II____. Fill in the blanks with the correct options:

	I	II
(a)	Atrial natriuretic factor	Blood glucose
(b)	Melatonin	Serum calcium
(c)	Somatomedin	Electrolyte imbalance
(d)	Atrial natriuretic factor	Blood pressure

88. Juxtaglomerular cells (JG cells) of the kidneys secrete:
(a) Estradiol
(b) Erythropoietin
(c) Gastric inhibitory peptide
(d) Calcitonin

89. Which of the following is not a function of progesterone?
(a) Maintenance of endometrium
(b) Prepares uterus for pregnancy
(c) Growth of alveoli in mammary glands and milk secretion
(d) Responsible for ovulation

90. Select the incorrect statement:
(a) Progesterone is secreted by the corpus luteum
(b) Oestrogen is secreted by the theca layer of the developing follicle
(c) Oestrogen is called pregnancy hormone
(d) FSH is responsible for the development of Graafian follicle

91. Interstitial cells of Leydig in the testis secrete androgen under the influence of:
(a) Follicle stimulating hormone
(b) Leuteinizing hormone
(c) GnRH
(d) Aldosterone

92. Select the correct option which matches the endocrine gland with its hormone and its function:

Endocrine gland	Hormone	Function
(a) Corpus luteum	Progesterone	Responsible for ovulation
(b) Ovary	FSH	Stimulates follicle development
(c) Adrenal cortex	Aldosterone	Gluconeogenesis
(d) Leydig cells	Androgen	Spermatogenesis

93. Decreased levels of oestrogen in females, particularly after menopause, cause:
(a) Rheumatoid arthritis
(b) Osteoporosis
(c) Tetany
(d) Gout

94. Identify the incorrect match:
(a) Melanocyte stimulating hormone – Skin pigmentation
(b) Atrial natriuretic factor – Increases blood pressure
(c) Gastric inhibitory peptide – Inhibits gastric secretion
(d) Oxytocin – Parturition

95. ________ increases secretion of hydrochloric acid from the oxyntic cells of the stomach:
(a) Secretin (b) Enterogastrone
(c) Enterokinase (d) Gastrin

96. Identify the correct set of antagonistic hormones:
(a) Gastrin and secretin
(b) Gastrin and enterogastrone
(c) Cholecystokinin and secretin
(d) Enterokinase and enterogastrone

97. Which of the following is a function of cholecystokinin?
(a) Stimulates release of gastric juice
(b) Stimulates contraction of gall bladder
(c) Converts angiotensin–I to angiotensin-II
(d) Promotes sodium loss with urine

98. Secretin:
(a) Causes gall bladder contraction
(b) Increases production of enzymes in pancreatic juice
(c) Stimulates secretion of bicarbonate from pancreas
(d) Increases gastric peristalsis

99. Identify the incorrect statement:
(a) Enterogastrone reduces production of hydrochloric acid in stomach
(b) ANF is a vasodilator
(c) Relaxin is secreted by the corpus luteum towards the end of pregnancy
(d) Angiotensinogen is secreted by the kidneys

100. Which of the following is a correct match with respect to hormones, its source and target organ?

Hormone	Source	Target organ
(a) Gastrin	Duodenum	Stomach
(b) Enterogastrone	Stomach	Duodenum
(c) Secretin	Stomach	Liver
(d) Cholecystokinin	Duodenum	Gall bladder

NCERT EXEMPLAR QUESTIONS

1. Select the right match of endocrine gland and their hormone among the options given below:

A. Pineal	(i) Epinephrine
B. Thyroid	(ii) Melatonin
C. Ovary	(iii) Estrogen
D. Adrenal medulla	(iv) Tetraiodothyronine

(a) A – (iv), B – (ii), C – (iii), D – (i)
(b) A – (ii), B – (iv), C – (i), D – (iii)
(c) A – (iv), B – (ii), C – (i), D – (iii)
(d) A – (ii), B – (iv), C – (iii), D – (i)

2. Which of the following hormones is not sereted by anterior pituitary:
(a) Growth hormone
(b) Follicle stimulating hormone
(c) Oxytocin
(d) Adrenocorticotrophic hormone

3. Mary is about to face an interview. But during the first five minutes before the interview, she experiences sweating, increased rate of heart beat, respiration, etc. Which of the following hormones are responsible for her restlessness?
(a) Estrogen and progesterone
(b) Oxytocin and vasopressin
(c) Adrenaline and noradrenaline
(d) Insulin and glucagon

4. The steroid responsible for balance of water and electrolytes in our body is:
(a) Insulin (b) Melatonin
(c) Testosterone (d) Aldosterone

5. Thymosin is responsible for:
(a) Raising the blood sugar level
(b) Raising the blood calcium level
(c) Differentiation of T lymphocytes
(d) Decrease in blood RBC

6. In the mechanism of action of a protein hormone, one of the second messengers is:
(a) Cyclic AMP (b) Insulin
(c) T_3 (d) Gastrin

7. Leydig cells produce a group of hormones called:
(a) Androgens (b) Estrogens
(c) Aldosterone (d) Gonadotropins

8. Corpus luteum secretes a hormone called:
(a) Prolactin (b) Progesterone
(c) Aldosterone (d) Testosterone

9. Cortisol is secreted by:
(a) Pancreas (b) Thyroid
(c) Adrenal (d) Thymus

10. A hormone responsible for normal sleep–wake cycle is:
(a) Epinephrine (b) Gastrin
(c) Melatonin (d) Insulin

11. Hormones are called chemical signals that stimulate specific target tissues. Their action depends on the presence of receptors on the respective target tissues. Which of the following is the correct location of the receptors in the case of protein hormones?
(a) Extra cellular matrix (b) Blood
(c) Plasma membrane (d) Nucleus

12. Choose the correct option among the following options:

Column I	Column II
A. Epinephrine	(i) Stimulates muscle growth
B. Testosterone	(ii) Decrease in blood pressure
C. Glucagon	(iii) Breakdown of liver glycogen content
D. Atrial natriuretic factor	(iv) Increases heart beat

(a) A – (ii), B – (i), C – (iii), D – (i)
(b) A – (iv), B (i), C – (iii), D (ii)
(c) A – (i), B – (ii), C – (iii), D – (iv)
(d) A (i), B (iv), C (ii), D – (iii)

13. Which of the following does not regulate calcium balance in the human body?
(a) Vitamin D
(b) Parathyroid hormone
(c) Thyrocalcitonin
(d) Thymosin

14. Which of the following organs in mammals does not consist of a central medullary region surrounded by a cortical region?
(a) Ovary (b) Adrenal
(c) Liver (d) Kidney

15. Which of the following conditions is not linked to deficiency of thyroid hormones?
(a) Cretinism (b) Goitre
(c) Myxedema (d) Exophthalmia

PRACTICE TEST PAPER

1. Which of the following hormones is not produced by the hypothalamus?
(a) Somatostatin
(b) Vasopressin
(c) Somatotrophic hormone
(d) Oxytocin

2. Which of the following is an incorrect statement?
(a) Pars nervosa is also called neurohypophysis
(b) Adenohypophysis consists of pars distalis and pars intermedia
(c) Pituitary gland is located in a bony cavity called sella turcica of the sphenoid bone
(d) Hormones secreted by neurohypophysis are transported from the hypothalamus through portal veins

3. Which of the following is incorrect with respect to the pineal gland?
(a) It is located at the dorsal side of the forebrain
(b) It regulates the diurnal rhythm
(c) It regulates defense capability and pigmentation
(d) It plays an osmoregulatory role by promoting water reabsorption from the DCT

4. Atrial natriuretic factor released from the atrial walls of the heart:
(a) Is a steroidal hormone
(b) Lowers blood pressure
(c) Checks the renin-angiotensin mechanism
(d) More than one option is correct

5. Which of the following is not a function of the thyroid gland?
(a) Regulates the metabolism of carbohydrates, proteins and lipids
(b) Promotes RBC synthesis
(c) Promotes body growth and mental development
(d) Increases blood calcium by promoting dissolution of calcium from bones

6. Zona glomerulosa of the adrenal cortex secretes:
(a) Glucocorticoids
(b) Mineralocorticoids
(c) Sex corticoids
(d) Cortisol

7. Which of the following conditions is not observed as a result of hypothyroidism during pregnancy?
(a) Mental retardation in growing baby
(b) Low intelligence quotient and abnormal skin of growing baby
(c) Obesity and deaf-mutism in growing baby
(d) Loss of weight and exophthalmic goitre in growing baby

8. Identify the incorrect match:
(a) Oxytocin – Milk ejection
(b) Melatonin – Circadian rhythm
(c) Cortisol – Hypercalcemic hormone
(d) Aldosterone – Salt and electrolyte balance

9. Patients with Grave's disease do not have:
(a) Hyperthyroidism
(b) Exophthalmos
(c) Weight gain
(d) Anxiety and tremors

10. Diabetes mellitus is caused by:
(a) Hypersecretion of insulin
(b) Hypersecretion of vasopressin
(c) Insulin deficiency
(d) Hyposecretion of thyroxine

11. Find the odd one out with respect to the peptide hormones:
(a) Insulin (b) Oxytocin
(c) Growth hormone (d) Cortisol

12. The α-cells of the pancreas secrete a hormone which:
(a) Stimulates glycogenolysis
(b) Stimulates glycogenesis
(c) Increases cellular glucose uptake and utilisation
(d) Causes lipid synthesis

13. __________ is a hormone that acts on the mammary glands and stimulates the formation of alveoli (sac-like structures which store milk) and milk secretion:
(a) Luteinizing hormone
(b) Testosterone
(c) Estrogen
(d) Progesterone

14. The circulating levels of calcium in blood are regulated by:
(a) Thyroxine and thyrocalcitonin
(b) Thyrocalcitonin and parathyroid hormone
(c) Cortisol and insulin
(d) Prolactin and thyroxine

15. Which of the following hormones is not an amino acid derivative?

(a) Adrenaline (b) Dopamine
(c) Thyroxine (d) Vasopressin

16. All of the following are related to the thymosin hormone, except:
(a) Promote production of antibodies
(b) Promote T-cell differentiation
(c) Levels increase in old age
(d) Is a peptide hormone

17. Which of the following sets of hormones acts via intracellular receptors?
(a) Iodothyronines and insulin
(b) Insulin and catecholamines
(c) Steroidal hormones and iodothyronines
(d) Adrenaline and glucagon

18. Which of the following is not a function of catecholamines?
(a) Increased sweat secretion
(b) Pupillary dilation
(c) Piloerection
(d) Glycogen synthesis

19. Which of the following is an incorrect match with respect to hormones and their function?
(a) Aldosterone – Reabsorption of Na^+ and excretion of K^+
(b) Prolactin – Milk synthesis
(c) Testosterone – Regulates male sexual behaviour
(d) Cholecystokinin – Increased gastric secretion and motility

20. Which of the following hormones is not produced from the pars distalis region of the pituitary?
(a) Prolactin
(b) Follicle stimulating hormone
(c) Adrenocorticotrophic hormone
(d) Gonadotrophin releasing hormone

21. ACTH stimulates ____**A**____ and promotes the secretion of ____**B**____.

Choose the option which correctly fills in the blanks A and B:

	A	B
(a)	Adrenal medulla	Adrenaline
(b)	Adrenal cortex	Cortisol
(c)	Adrenal cortex	Noradrenaline
(d)	Adenohypophysis	Aldosterone

22. Destruction of adrenal cortex causes disturbed carbohydrate metabolism resulting in weakness and fatigue and is called:
(a) Diabetes insipidus (b) Addison's disease
(c) Conn's syndrome (d) Cushing's syndrome

23. Identify the incorrect match:
(a) The two lobes of the thyroid gland are interconnected by a thin flap of connective tissue called isthmus
(b) Thyroid follicle secretes hormones called thymosins
(c) Thyroid lobes are composed of follicles and stromal tissues
(d) Deficiency of iodine results in enlargement of thyroid, called goitre

24. Luteinizing hormone (LH) stimulates:
(a) Synthesis and secretion of androgens from testis
(b) Ovulation of fully mature follicles in females
(c) Growth of corpus luteum
(d) All of the above

25. Which of the following hormones are called gonadotropins?
(a) Androgen and oestrogen
(b) Oestrogen and progesterone
(c) FSH and LH
(d) Both (a) and (b)

26. Which of the following hormones produces anti-inflammatory reactions and suppresses the immune response?
(a) Aldosterone (b) Androgenic steroids
(c) Cortisol (d) Insulin

27. Which hormone acts on the exocrine pancreas and stimulates secretion of water and bicarbonate ions?
(a) Gastrin
(b) Cholecystokinin
(c) Secretin
(d) Gastric inhibitory peptide (GIP)

28. Identify the incorrect statement:
(a) Oestrogen stimulates growth and development of female accessory sex organs and secondary sex characters
(b) Progesterone plays a major role in maintenance of pregnancy
(c) Androgens produce catabolic effects on protein and carbohydrate metabolism
(d) Leydig cells are present in intertubular spaces and produce androgens

29. Read the following statements and choose the correct option:
I: Insulin acts mainly on hepatocytes and adipocytes (cells of adipose tissue) and enhances cellular glucose uptake and utilisation
II: Insulin stimulates glycogenolysis, resulting in increased blood sugar (hyperglycemia)

(a) Both statements are correct
(b) Both statements are incorrect
(c) Only statement I is correct
(d) Only statement II is correct

30. The hormones secreted during emergency situations, or in response to stress of any kind and called hormones of fight or flight, are
(a) Thyrocalcitonin and thymosin
(b) Thyroxine and thyrocalcitonin
(c) Adrenaline and noradrenaline
(d) Parathyroid hormone and aldosterone

31. Erythropoietin:
(a) Is a peptide hormone
(b) Is secreted by juxtaglomerular cells of the kidneys
(c) Stimulates formation of RBCs
(d) All of the above are correct

32. Which of the following hormones produces second messengers and interacts with membrane-bound receptors?
(a) Thyroxine (b) Glucagon
(c) Progesterone (d) Estradiol

33. Which of the following is incorrect with respect to the functions of vasopressin?
(a) Vasoconstriction
(b) Prevents diuresis
(c) Dilates blood vessels
(d) Stimulates reabsorption of water

34. Which of the following is/are correct statement(s)?
(a) Somatostatin inhibits release of growth hormone
(b) The posterior pituitary is under direct neural regulation of the hypothalamus
(c) Hypothalamic hormones reach the anterior pituitary through portal vessels
(d) All are correct

35. Adrenal sex hormones are responsible for:
(a) Degradation of proteins to increase the level of amino acids
(b) Gluconeogenesis
(c) Growth of axial hair, facial hair and pubic hair
(d) Immunosuppressant action

36. Patients with hypothyroidism do not have:
(a) Weight gain due to slow metabolism
(b) Anaemia and slow heart rate
(c) Myxedema
(d) Rapid heart rate and excessive sweating

37. Mark the odd one out with respect to secondary messengers used to mediate the action of peptide hormones and catecholamines:
(a) cGMP and cAMP
(b) Calcium and diacylglycerol
(c) Inositol triphosphate and cAMP
(d) ATP and sodium

38. In males, the role of follicle stimulating hormone (FSH) is:
(a) Formation of androgens from Sertoli cells
(b) Maturation of Graafian follicle
(c) Regulation of spermatogenesis
(d) Resorption of water and electrolytes by the distal tubules

39. Which of the following is an incorrect statement?
(a) In humans, pars intermedia is almost merged with pars distalis
(b) Over-secretion of growth hormone stimulates abnormal growth of the body leading to gigantism
(c) Glucocorticoids stimulate gluconeogenesis, lipolysis and proteolysis
(d) Hyperparathyroidism during pregnancy causes defective development of growing baby resulting in cretinism

40. Identify the incorrect statement with respect to the adrenal glands:
(a) Our body has one pair of adrenal glands, one at the anterior part of each kidney
(b) The centrally located tissue is called the adrenal medulla and outside this lies the adrenal cortex
(c) The adrenal medulla secretes two hormones called adrenaline and noradrenaline
(d) The adrenal cortex secretes many hormones, commonly called as catecholamines

41. Identify the incorrect statement with respect to the testis:
(a) A pair of testes is present in the scrotal sac of males
(b) Testes perform dual function as a primary sex organ as well as an endocrine organ
(c) Testis is composed of seminiferous tubules and stromal or interstitial tissue
(d) The Leydig cells are present in intratubular spaces

42. Prolactin is also called:
(a) Hormone of maternity
(b) Mammotropin hormone
(c) Luteotropin hormone
(d) All of the above

43. Hyperparathyroidism does not cause:
(a) Hypercalcemia
(b) Tetany

(c) Softening and bending of bones
(d) Demineralisation of bones

44. Excessive secretion of androgens in females causes male-like characters such as low-pitched (hoarseness) voice, beard and moustaches. This disease is called:
(a) Gynecomastia
(b) Adrenal virilism
(c) Addison's disease
(d) Diabetes insipidus

45. Select the mismatched pair:

(a) Insulin	–	Hypoglycemia and glycogenesis
(b) Glucagon	–	Hyperglycemia and glycogenolysis
(c) Thyroxine	–	Regulates blood calcium levels
(d) Oxytocin	–	Uterine contraction during parturition

ANSWER KEYS

Exercises

1. (d) **2.** (d) **3.** (b) **4.** (c) **5.** (d) **6.** (c) **7.** (d) **8.** (b) **9.** (c) **10.** (d)
11. (c) **12.** (c) **13.** (d) **14.** (d) **15.** (d) **16.** (d) **17.** (c) **18.** (b) **19.** (b) **20.** (c)
21. (d) **22.** (b) **23.** (c) **24.** (c) **25.** (b) **26.** (d) **27.** (b) **28.** (b) **29.** (b) **30.** (c)
31. (b) **32.** (d) **33.** (b) **34.** (d) **35.** (d) **36.** (d) **37.** (b) **38.** (b) **39.** (a) **40.** (d)
41. (d) **42.** (b) **43.** (d) **44.** (d) **45.** (b) **46.** (b) **47.** (d) **48.** (c) **49.** (b) **50.** (d)
51. (b) **52.** (c) **53.** (c) **54.** (d) **55.** (b) **56.** (b) **57.** (b) **58.** (a) **59.** (c) **60.** (c)
61. (c) **62.** (d) **63.** (d) **64.** (b) **65.** (a) **66.** (c) **67.** (b) **68.** (a) **69.** (b) **70.** (b)
71. (b) **72.** (b) **73.** (c) **74.** (b) **75.** (a) **76.** (b) **77.** (c) **78.** (d) **79.** (a) **80.** (d)
81. (d) **82.** (c) **83.** (c) **84.** (d) **85.** (d) **86.** (d) **87.** (d) **88.** (b) **89.** (d) **90.** (c)
91. (b) **92.** (d) **93.** (b) **94.** (b) **95.** (d) **96.** (b) **97.** (b) **98.** (c) **99.** (d) **100.** (d)

NCERT Exemplar Questions

1. (d) **2.** (c) **3.** (c) **4.** (d) **5.** (c) **6.** (a) **7.** (a) **8.** (b) **9.** (c) **10.** (c)
11. (c) **12.** (b) **13.** (d) **14.** (c) **15.** (d)

Practice Test Paper

1. (c) **2.** (d) **3.** (d) **4.** (d) **5.** (d) **6.** (b) **7.** (d) **8.** (c) **9.** (c) **10.** (c)
11. (d) **12.** (a) **13.** (d) **14.** (b) **15.** (d) **16.** (c) **17.** (c) **18.** (d) **19.** (d) **20.** (d)
21. (b) **22.** (b) **23.** (b) **24.** (d) **25.** (c) **26.** (c) **27.** (c) **28.** (c) **29.** (c) **30.** (c)
31. (d) **32.** (b) **33.** (c) **34.** (d) **35.** (c) **36.** (d) **37.** (d) **38.** (c) **39.** (d) **40.** (d)
41. (d) **42.** (d) **43.** (b) **44.** (b) **45.** (c)

SOLUTIONS

Endocrine Glands and Hormones, Hormone Action

1. Hormones are non-nutrient chemicals which act as intercellular messengers and are produced in trace amounts.
2. Endocrine glands are ductless and secrete hormones.
3. The nerve cells do not innervate all cells of the body.
4. Invertebrates possess a very simple endocrine system with few hormones whereas a large number of chemicals act as hormones and provide coordination in vertebrates.
5. Hormones may be amino acid derivatives, proteins or steroidal in nature.
6. Hormones produce their effects on target tissues by binding to specific proteins called hormone receptors located in the target tissues only. Hormone receptors present on the cell membrane of the target cells are membrane-bound receptors and the receptors present inside the target cell are called intracellular receptors, mostly nuclear receptors.
7. Steroid hormones and iodothyronines interact with intracellular receptors and mostly regulate gene expression or chromosome function by the interaction of hormone-receptor complex with the genome.
8. Adrenal cortex hormones like cortisol and aldosterone are steroidal compounds.
9. Hormones which interact with membrane-bound receptors generate secondary messengers like cAMP, cGMP, calcium, etc.
10. Hormones that interact with intracellular receptors regulate gene expression and chromosomal function by the interaction of hormone-receptor complex with the genome.
11. Cortisol, testosterone and estradiol are steroidal hormones. Adrenaline is an amino acid derivative.
12. Erythropoietin is a peptide hormone.
13. Aldosterone is a steroidal hormone.
14. Hormones are intercellular messengers that are carried to target cells via blood.
15. Option (d) is correct.
16. Epinephrine binds to membrane-bound receptors and mediates their action by generating cAMP (second messenger).
17. Steroid hormones like testosterone and estradiol bind with intracellular receptors.
18. Both insulin and growth hormone of peptide hormones bind with membrane-bound receptors.
19. Option (b) is correct.
20. Option (c) is correct.

Hypothalamus and Pituitary Gland

21. Hypothalamus is a neuroendocrine structure. It is part of the diencephalon (forebrain) and has centres for hunger, thirst and thermoregulation. It also secretes neurohormones.
22. Hypothalamus forms the floor of the diencephalon; the forebrain regulates a wide spectrum of body functions.
23. The hypothalamic hormones reach the adenohypophysis through the portal circulatory system.
24. GnRH stimulates the anterior pituitary to release gonadotropins like LH and FSH.
25. Growth hormone inhibitory hormone (GHIH) is also called somatostatin.
26. Growth hormone (GH)/Somatotropin is not present in the portal vein as it is produced by the anterior pituitary.
27. Prolactin is responsible for milk production in the mammary glands. Oxytocin assists parturition and is called the birth hormone. Vasopressin is responsible for osmoregulation and water reabsorption from the distal tubules. TSH promotes secretion of thyroid hormones.

28. Oxytocin is secreted by pars nervosa though it is actually synthesised in the hypothalamus.

29. Excessive secretion of GH in middle age causes acromegaly. Oversecretion of GH before puberty causes gigantism.

30. In females, LH maintains the corpus luteum formed from remnants of the Graafian follicle after ovulation.

31. ACTH stimulates synthesis of steroid hormones like cortisol from the adrenal cortex.

32. Option (d) is correct.

33. Melanocyte stimulating hormone (MSH) is secreted by pars intermedia of the pituitary gland.

34. Adenohypophysis consists of two parts, pars distalis and pars intermedia.

35. FSH stimulates growth of the ovarian follicles while LH induces ovulation in mature follicles.

36. Oxytocin and vasopressin are synthesised in the hypothalamus and are transported to the posterior pituitary axonally.

37. Option (b) is correct.

38. Deficiency of vasopressin causes diabetes insipidus.

39. The posterior lobe of the pituitary gland is connected to the hypothalamus through neurosecretory cells.

40. Patients with diabetes insipidus have polyuria.

Thyroid Gland, Parathyroid Gland

41. Option (d) is correct.

42. The two lobes of the thyroid gland are attached to each other by a thin flap of connective tissue called isthmus.

43. The thyroid gland is composed of two lobes which are located on either side of the trachea.

44. Babies born to mothers who develop hypothyroidism during pregnancy, have cretinism. These babies are obese, dwarfish, with low IQ and deaf-mutism.

45. Thyroid glands secrete a protein hormone called thyrocalcitonin which regulates blood calcium levels.

46. Parathyroid hormone has hypercalcemic effect, that is, it increases blood calcium levels.

47. Option (d) is correct.

48. Grave's disease is an autoimmune disease of the thyroid gland in which antibodies mimic TSH.

49. Deficiency of iodine in diet causes goitre.

50. Parathyroid hormone and thyrocalcitonin are antagonistic hormones.

51. Tetra-iodothyronine is also called thyroxine.

52. Patients with hypoparathyroidism develop hypocalcemic tetany.

53. Parathyroid hormone is released in response to low calcium levels in blood.

54. Patients with hypothyroidism in adulthood develop myxedema. These patients have weight gain, slow heart rate and swollen puffy appearance of face and limbs.

55. Parathyroid hormone is a polypeptide hormone containing 84 amino acids.

Adrenal Gland, Pineal Gland, Thymus

56. Adrenal gland is also called suprarenal gland.

57. Hormones of the adrenal medulla are called catecholamines.

58. Cortisol is the main glucocorticoid in our body.

59. Corticoids which regulate the balance of water and electrolytes are called mineralocorticoids.

60. Glucocorticoids stimulate gluconeogenesis, lipolysis and proteolysis, and inhibit cellular uptake and utilisation of amino acids.

61. Aldosterone acts mainly at the renal tubules and stimulates the reabsorption of Na^+ and water and excretion of K^+ and phosphate ions. Thus, aldosterone helps in maintenance of electrolytes, body fluid volume, osmotic pressure and blood pressure.

62. Aldosterone is the main mineralocorticoid in our body.

63. The emergency hormones or hormones of fght or fight are adrenaline and noradrenaline.

64. Glucocorticoids are secreted by zona fasciculata of the adrenal cortex.

65. Thymus gland degenerates in old age, causing weakening of immune response.

66. Catecholamines increase the heart beat and also the strength of heart contraction.

67. Zona reticularis is the innermost layer of the adrenal cortex and secretes sex corticoids.

68. Cortisol produces anti-inflammatory reactions and suppresses the immune response. Cortisol stimulates RBC production.

69. Our body has one pair of adrenal glands, one at the anterior part of each kidney.

70. Addison's disease is caused due to destruction of the adrenal cortex.

71. Gynecomastia refers to enlargement of a man's breasts, usually due to hormonal imbalance or hormone therapy.

72. Adrenal cortex is derived from the mesoderm while adrenal medulla is derived from the ectoderm.

73. Cortisol inhibits cellular uptake and utilisation of amino acids.

74. Excessive secretion of androgens in females causes adrenal virilism.

75. Option (a) is correct.

76. Hassall's corpuscles are found in the thymus.

Pancreas

77. There are 1 to 2 million Islets of Langerhans in normal human pancreas, representing 1 to 2 per cent of pancreatic tissue.

78. Glucagon reduces cellular glucose uptake and utilisation.

79. Option (a) is correct.

80. Insulin is a hypoglycemic hormone that causes rapid movement of glucose from blood to hepatocytes and adipocytes, resulting in decreased blood glucose levels.

81. Patients with diabetes mellitus have raised blood glucose levels (hyperglycemia), loss of glucose in urine (glycosuria) and excessive urination (polyuria).

82. The α-cells of the pancreas secrete glucagon, which stimulates gluconeogenesis, causing hyperglycemia.

83. Somatostatin inhibits the release of both insulin and glucagon.

84. Option (d) is correct.

85. Polyuria is seen in patients with diabetes mellitus as well as in diabetes insipidus patients.

86. Glucagon is a peptide hormone.

Testis, Ovary and Hormones of the Heart, Kidneys and Gastrointestinal Tract

87. The atrial walls of the heart secrete atrial natriuretic factor (ANF), which decreases blood pressure.

88. Juxtaglomerular cells (JG cells) of the kidneys secrete erythropoietin.

89. LH is the hormone responsible for ovulation.

90. Progesterone is called pregnancy hormone.

91. Leydig cells in the testis secrete androgen under the influence of the luteinizing hormone (LH).

92. Option (d) is correct.

93. Decreased levels of estrogen in females after menopause is called osteoporosis.

94. Atrial natriuretic factor (ANF) lowers blood pressure by causing vasodilation.

95. Gastrin increases secretion of hydrochloric acid from the oxyntic cells of the stomach.

96. Gastrin and enterogastrone are antagonistic hormones.

97. Cholecystokinin causes contraction of the gall bladder.

98. Secretin acts on the exocrine pancreas and stimulates secretion of water and bicarbonate ions.

99. Angiotensinogen is secreted by the liver.

100. Option (d) is correct.